Organometallic Chemistry

PERIODIC TABLE OF THE ELEMENTS

GROUPS

PERIODS	1 IA	2 IIA	3 IIIB	4 IVB	5 VB	6 VIB	7 VIIB	8 VIIIB	9 VIIIB	10 VIIIB	11 IB	12 IIB	13 IIIA	14 IVA	15 VA	16 VIA	17 VIIA	18 VIIIA
1	1.008 H 1																	4.0023 He 2
2	6.941 Li 3	9.018 Be 4											10.811 B 5	12.011 C 6	14.007 N 7	15.999 O 8	18.998 F 9	20.180 Ne 10
3	22.9990 Na 11	24.305 Mg 12											26.982 Al 13	28.0855 Si 14	30.9738 P 15	32.06 S 16	35.453 Cl 17	39.948 Ar 18
4	39.0983 K 19	40.078 Ca 20	44.956 Sc 21	47.88 Ti 22	50.9415 V 23	51.996 Cr 24	54.938 Mn 25	55.847 Fe 26	58.933 Co 27	58.69 Ni 28	63.546 Cu 29	65.39 Zn 30	69.723 Ga 31	72.61 Ge 32	74.922 As 33	78.96 Se 34	79.904 Br 35	83.80 Kr 36
5	85.468 Rb 37	87.62 Sr 38	88.906 Y 39	91.224 Zr 40	92.9064 Nb 41	95.94 Mo 42	98.907 Tc 43	101.07 Ru 44	102.906 Rh 45	106.42 Pd 46	107.868 Ag 47	112.41 Cd 48	114.82 In 49	118.71 Sn 50	121.75 Sb 51	127.60 Te 52	126.904 I 53	131.29 Xe 54
6	132.905 Cs 55	137.33 Ba 56	138.906 *La 57	178.49 Hf 72	180.948 Ta 73	183.85 W 74	186.207 Re 75	190.2 Os 76	192.22 Ir 77	195.08 Pt 78	196.967 Au 79	200.59 Hg 80	204.383 Tl 81	207.2 Pb 82	208.980 Bi 83	(209) Po 84	(210) At 85	(222) Rn 86
7	(223) Fr 87	226.025 Ra 88	(227) **Ac 89	(261) Rf 104	(262) Ha 105	(263) Sg 106	(262) Ns 107	(264) Hs 108	(266) Mt 109									

TRANSITION ELEMENTS

140.115 Ce 58	140.908 Pr 59	144.24 Nd 60	(145) Pm 61	150.36 Sm 62	151.96 Eu 63	157.25 Gd 64	158.925 Tb 65	162.50 Dy 66	164.930 Ho 67	167.26 Er 68	168.934 Tm 69	173.04 Yb 70	174.967 Lu 71
232.038 Th 90	231.036 Pa 91	238.029 U 92	237.048 Np 93	(244) Pu 94	(243) Am 95	(247) Cm 96	(247) Bk 97	(251) Cf 98	(252) Es 99	(257) Fm 100	(258) Md 101	(259) No 102	(260) Lr 103

* Lanthanide series

** Actinide series

Numbers below the symbol of the element indicates the atomic numbers. Atomic masses, above the symbol of the element, are based on the assigned relative atomic mass of 12C = exactly 12; ( ) indicates the mass number of the isotope with the longest half-life.

Organometallic Chemistry

Gary O. Spessard
Gary L. Miessler

St. Olaf College
Northfield, Minnesota

Prentice Hall, Upper Saddle River, New Jersey 07458

Library of Congress Cataloging-in-Publication Data

Spessard, Gary O.
   Organometallic Chemistry / by Gary O. Spessard and Gary L.
Miessler.
     p.   cm.
   Includes bibliographical references and index.
   ISBN 0-13-640178-3
    1. Organometallic chemistry.   I. Miessler, Gary L.,
 II. Title.
 QD411.S65   1996
 547'.05--dc20

96-3164
CIP

Editorial Director: Tim Bozik
Editor-in-Chief: Paul F. Corey
Acquisitions Editor: John Challice
Director of Production/Manufacturing: David W. Riccardi
Special Projects Manager: Barbara Murray
Total Concept Coordinator: Kimberly P. Karpovich
Formatting Manager: John J. Jordan
Formatting/Art Production: Michael J. Bertrand, Cindy Dunn, Richard Foster,
                Jeff Henn, Karen Stephens, David Tay, Rebecca Wald
Art Director/Cover Design: Jayne Conte

©1997 by Prentice-Hall, Inc.
Upper Saddle River, New Jersey 07458

Reprinted with corrections October, 2000.

ISBN 0-13-640178-3

Printed in the United States of America
10 9 8 7 6 5 4 3

Prentice-Hall International (UK) Limited, *London*
Prentice-Hall of Australia Pty. Limited, *Sydney*
Prentice-Hall Canada Inc., *Toronto*
Prentice-Hall Hispanoamericana, S.A., *Mexico*
Prentice-Hall of India Private Limited, *New Delhi*
Prentice-Hall of Japan, Inc., *Tokyo*
Prentice-Hall Asia Pte. Ltd., *Singapore*
Editora Prentice-Hall do Brasil, Ltda., *Rio de Janeiro*

Contents

Preface

The field of organometallic chemistry is growing rapidly, particularly in the areas of organic synthesis and catalysis. No textbook could possibly cover all aspects of this important branch of chemistry in a completely up-to-date manner. We have, however, endeavored to produce a book that provides the student with a solid introduction into the field. We have narrowed the focus of our coverage to the chemistry of organotransition metal complexes. Although maingroup organometallic chemistry is also an important area, we believe that adequate coverage exists elsewhere in traditional organic and inorganic texts. Organotransition metal complexes are unique, useful, and aesthetically appealing. We hope our book provides some insight as to why this is so.

To the Student

The transition metals as well as the lanthanides and actinides comprise over half of the elements of the Periodic Table. At first glance, any systematic study of how these metals bond to carbon and then undergo reactions would seem to be overwhelming. This book introduces you early to the use of qualitative molecular orbital theory (Chapter 2). We can understand much about the chemistry of organotransition metal complexes by appreciating the power of molecular orbital theory to not only make sense of the structure of these complexes, but also to explain much about their reactivity. You will also discover that there are many trends and similarities among these metals that make the study of their chemistry much easier and more pleasing to the intellect. There are relatively few important ligand types that bind to metals, and these are introduced in Chapters 4 to 6. Once these ligands attach to the metal, there are again relatively few types of reactions that occur that involve the ligand and the metal. You will encounter these reaction types in Chapters 7 to 8. You will discover useful applications of organotransition metal complexes in Chapter 9 (catalysis) and Chapter 11 (organic synthesis). Chapter 10 considers metal carbene complexes, which now have unique and important roles in catalyzing the polymerization of alkenes and in the synthesis of carbon rings. Chapter 12 ties together much of the earlier material in the text as it discusses analogies between organic and organometallic compounds and how several metals bind together into cluster complexes. The last chapter introduces you to some of the "cutting edge" areas of organometallic chemistry, particularly how organotransition metal complexes play a role in biology and surface science.

Organometallic chemistry only comes alive when practiced in the laboratory. We have provided extensive footnotes that refer you to some of the interesting work that has occurred over the years. We urge you to check at least some of these references for more complete information. As you read through each chapter, take time to complete the exercises. They are designed to get you to think about what you have just read; complete answers are provided in Appendix B. The problems at the end of the chapter are provided to test your overall understanding of the material in that chapter. Several of these problems include references to original work in the literature.

To the Instructor

We wrote this book with the idea that it would appeal to students with a variety of backgrounds in chemistry. For several years at St. Olaf College we have offered organometallic chemistry to our students, many of whom are sophomores having had only one and a half years of college chemistry. These students have reacted well to our course and have been able to grasp relatively sophisticated theories related to bonding and reaction mechanisms. Instructors of advanced students may wish to skip Chapter 2, which discusses qualitative molecular orbital theory. We feel, however, that a solid background in MO theory is vital in understanding organometallic chemistry and we use aspects of MO theory extensively throughout the text.

Chapters 7 and 8 assume that students have had at least some acquaintance with organic chemistry and elementary thermodynamics. Topics relating to the study of reaction mechanisms such as kinetics and stereochemistry may require more extensive coverage than our text provides. Chapter 9 begins with a primer on catalysis designed to bring all students up to speed before detailed discussion of important catalytic cycles begins. Other catalytic processes are discussed in Chapter 10 (carbene complexes) and Chapter 11 (applications to synthesis). The isolobal analogy (Chapter 12) and the last section in Chapter 4 show that there are several analogies among organotransition metal chemistry and that of organic and main group compounds. Chapter 13 contains a great deal of interesting information on the interface between organometallic chemistry and other areas of science. Coverage of that chapter by the instructor, while desirable, is not essential for a good understanding of organotransition metal chemistry.

Several problems are placed at the end of Chapters 2 to 13. No answers to these problems are included, although some problems include references to the literature (a solutions manual is available to instructors). We feel that it is important to challenge students to work problems out without the benefit of readily available answers. The text does contain numerous exercises within chapters, and these answers are provided in Appendix B.

Acknowledgments

No textbook can be written without extensive help from a number of individuals. We offer our thanks to Mitsuru Kubota, the esteemed teacher of one of us (G.S.), who reviewed this book during its preparation and offered many helpful suggestions and much encouragement, and Charles Casey, who also reviewed the manuscript and provided both helpful comments and valuable references. Phil Hampton, an Ole grad and now professor at the University of New Mexico, also provided helpful insight through reviewing the manuscript. We appreciate the encouragement throughout the project of Robert Angelici and the many useful literature references provided by Lou Hegedus.

Several of the pictures of molecular orbitals in this book (including the front cover) were produced using CAChe Worksystem software, which is also used by our students in the organometallic chemistry course at St. Olaf. CAChe Scientific is a division of the Oxford Molecular Group, and we are indebted to that organization for their helpful suggestions. We especially appreciate the assistance of Evelyn Brosnan and Rick DeHoff.

At St. Olaf we are fortunate to have as students many who were able to assist us in critical parts of writing this book. We specifically acknowledge the work of Karla Miller and Robert Owen who skillfully composed many of the drawings throughout the text. Another student, Amy Roos, provided her expertise in generating some of the CAChe illustrations we have used. We owe a special debt of gratitude to Beth Truesdale, an English and Chemistry major, who provided critical and insightful comments on style and grammar. We have received countless useful suggestions from our students as they have struggled with draft versions of the manuscript. We appreciate the efforts of our editors at Prentice-Hall, Deirdre Cavanaugh and John Challice, and Barbara Murray in keeping the project moving along in its final stages. Finally, we wish to acknowledge the support and love of our families (Carol, Sarah, Becky, Naomi, and Rachel) who helped to see us through this long process.

Gary O. Spessard
Gary L. Miessler

1

Overview of
Organometallic Chemistry

Organometallic chemistry, the chemistry of compounds containing metal-carbon bonds, is one of the most interesting and certainly most rapidly growing areas of chemical research. It encompasses a wide variety of chemical compounds and their reactions: compounds containing both sigma (σ) and pi (π) bonds between metal atoms and carbon; many cluster compounds, containing one or more metal-metal bonds; molecules of structural types unusual or unknown in organic chemistry; and reactions that in some cases bear similarities to known organic reactions and in other cases are dramatically different. Aside from their intrinsically interesting nature, many organometallic compounds form useful catalysts and consequently are of significant industrial interest. Over the last several years, organometallic reagents have fostered key steps in the total synthesis of numerous molecules, many of which are biologically active.

1-1   Striking Difference

Several examples should illustrate how organometallic molecules are strikingly different from those encountered in "classical" inorganic and organic chemistry. Cyclic organic ligands,[1] containing delocalized π systems, can team up with metal atoms to form "sandwich" compounds, with a metal sandwiched between, for example, benzene or cyclopentadienyl rings. Sometimes atoms of other ele-

[1]A ligand is a molecule, ion, or molecular fragment bound to a central atom, usually a metal atom.

Figure 1-1
Examples of Sandwich
Compounds

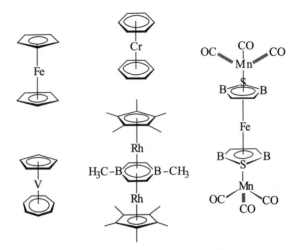

ments, such as phosphorus or sulfur, can be included as well. Examples of these double- and multiple-decker sandwich compounds are shown in Figure **1-1**.

A characteristic of metal atoms bonded to organic ligands, especially CO (the most common of all ligands in organometallic chemistry), is that they often exhibit the capacity to form covalent bonds to other metal atoms to form **cluster compounds** (some cluster compounds are also known that contain no organic ligands). These clusters may contain only two or three metal atoms or as many as several dozen; there is no limit to their size or variety. They may contain single, double, triple, or quadruple bonds between the metal atoms and may in some cases have ligands that bridge two or more of the metals. Examples of metal cluster compounds containing organic ligands are shown in Figure **1-2**; clusters will be discussed further in Chapter 12.

Carbon itself may play quite a different role than commonly encountered in organic chemistry. Certain metal clusters encapsulate carbon atoms; the resulting molecules, called **carbide clusters**, in some cases contain carbon bonded to five, six, or more surrounding metals. The traditional notion of carbon forming bonds to (at most) four additional atoms, must be reconsidered (a few examples of carbon bonded to more than four atoms are also known in organic chemistry). Two examples of carbide clusters are included in Figure **1-2**.

Strictly speaking, the only compounds classified as organometallic are those that contain metal–carbon bonds, but in practice, complexes containing several other ligands similar to CO in their bonding, such as NO and N_2, are frequently included. Other ligands, such as phosphines (PR_3) and dihydrogen (H_2), often occur in organometallic complexes, and their chemistry may be studied in association with the chemistry of organic ligands.

We will include examples of these and other nonorganic ligands as appropriate in our discussion of organometallic chemistry.

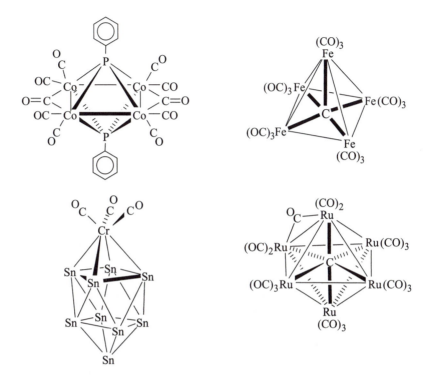

Figure 1-2
Examples of Cluster Compounds

1-2   Historical Background

The first organometallic compound to be reported was synthesized in 1827[2] by Zeise, who obtained yellow needle-like crystals after refluxing a mixture of $PtCl_4$ and $PtCl_2$ in ethanol, followed by addition of KCl solution.[3] Zeise correctly asserted that this yellow product (subsequently dubbed "Zeise's salt") contained an ethylene group. This assertion was questioned by other chemists, most notably Liebig, and was not verified conclusively until experiments performed by Birnbaum in 1868. The structure of the compound proved extremely elusive, however, and was not determined until more than 140 years

[2]To place this in historical perspective, John Quincy Adams was U. S. President at this time. The following year, Friedrich Wöhler reported that the organic compound urea could be made from the inorganic reagents HOCN and NH_3—the "birth" of organic chemistry.
[3]W.C. Zeise, *Annalen der Physik und Chemie*, **1831**, *21*, 497-541. A translation of excerpts from this paper can be found in G.B. Kauffman, Ed., *Classics in Coordination Chemistry*, part 2. Dover, New York, 1976, 21-37.

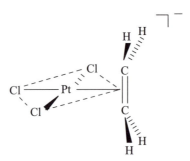

Figure 1-3
Anion of Zeise's Compound

after Zeise's discovery![4] Zeise's salt proved to be the first compound containing an organic molecule attached to a metal using the π electrons of the former. It is an ionic compound of formula $K[Pt(C_2H_4)Cl_3] \cdot H_2O$; the structure of the anion, shown in Figure **1-3**, is based on a square plane, with three chloro ligands occupying corners of the square and the ethylene occupying the fourth corner, but perpendicular to the plane.

The first compound containing carbon monoxide as a ligand was another platinum-chloro complex, reported in 1867. In 1890, Mond[5] reported the preparation of $Ni(CO)_4$, a compound that became commercially useful for the purification of nickel. Other metal-CO ("carbonyl") complexes were soon obtained.

Reactions between magnesium and alkyl halides performed by Barbier[6] in 1898 and 1899 and subsequently by Grignard,[7] led to the synthesis of alkyl magnesium complexes now known as Grignard reagents. These complexes contain magnesium-carbon σ bonds; they have been found to be extremely complex in their structure and function. In solution they participate in a variety of chemical equilibria, some of which are summarized in Figure **1-4**. Their synthetic utility was recognized early; by 1905, more than 200 research papers had appeared on the topic. Grignard reagents and other reagents containing metal-alkyl σ bonds (such as organolithium, organozinc, and organocadmium reagents) have proven of immense importance in the development of organic chemistry.

From the discovery of Zeise's salt in 1827 to approximately 1950, organometallic chemistry developed rather slowly. Some organometallic compounds, such as the Grignard reagents, found utility in organic synthesis, but there was little study of compounds containing metal-carbon bonds as a distinct

[4]R.A. Love, T.F. Koetzle, G.J.B. Williams, L.C. Andrews, and R. Bau, *Inorg. Chem.* **1975**, *14*, 2653.
[5]L. Mond, *J. Chem. Soc.*, **1890**, *57*, 749.
[6]P. Barbier, *Comp. Rend.*, **1899**, *128*, 110.
[7]V. Grignard, Comp. Rend. **1900**, *130*, 1322; *Ann. Chim.* **1901**, *24*, 433. An English translation of most of the latter paper is in *J. Chem. Ed.*, **1970**, *47*, 290.

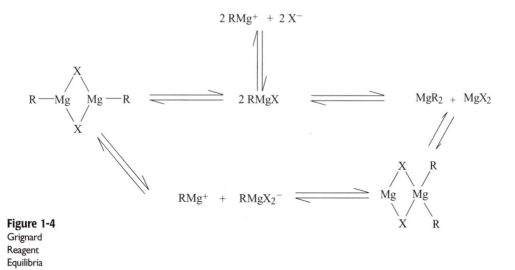

Figure 1-4
Grignard
Reagent
Equilibria

research area. In 1951, in an attempt to synthesize fulvalene (shown below) from cyclopentadienyl bromide, Kealy and Pauson reacted the Grignard reagent *cyclo*-C_5H_5MgBr with $FeCl_3$, using anhydrous diethyl ether as solvent.[8] This reaction did not yield the desired fulvalene but rather an orange solid of formula $(C_5H_5)_2Fe$, ferrocene, as shown below:

$$cyclo\text{-}C_5H_5MgBr + FeCl_3 \rightarrow (C_5H_5)_2Fe$$

Ferrocene   Fulvalene

The product was surprisingly stable; it could be sublimed in air without decomposition and was resistant to catalytic hydrogenation and Diels–Alder reactions. In 1956, X-ray diffraction showed the structure to consist of an iron atom sandwiched between two parallel C_5H_5 rings.[9] The details of the structure proved somewhat controversial, with the initial study indicating the rings to be in a staggered conformation. Electron diffraction studies of gas phase ferrocene, however, showed the rings to be eclipsed, or very nearly so. More recent X-ray diffraction studies of solid ferrocene have identified several crystalline phases, with an eclipsed conformation at 98 K and with conformations having the rings slightly twisted in higher temperature crystalline modifications (Figure **1-5**).[10]

[8]T.J. Kealy and P.L. Pauson, *Nature*, **1951**, *168*, 1039.
[9]J.D. Dunitz, L.E. Orgel, and R.A. Rich, *Acta Crystallogr.*, **1956**, *9*, 373.
[10]E.A.V. Ebsworth, D.W.H. Rankin, and S. Cradock, *Structural Methods in Inorganic Chemistry*, Blackwell Scientific Publications, Oxford, 1987.

Figure 1-5
Conformations of
Ferrocene (Adapted with
permission from G. L.
Miessler and D. A. Tarr,
Inorganic Chemistry,
Prentice-Hall, Englewood
Cliffs, NJ, 1991, 415.)

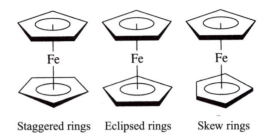

Staggered rings   Eclipsed rings   Skew rings

The discovery of the prototype sandwich compound ferrocene rapidly led to the discovery of other sandwich compounds, other compounds containing metal atoms bonded to cyclic organic ligands, and a vast array of additional compounds containing other organic ligands. It is often stated, with justification, that the discovery of ferrocene began the era of "modern" organometallic chemistry, an area that has grown with increasing rapidity in the succeeding decades.

In recent years, organotransition metal complexes containing chiral ligands have been found to catalyze the formation of specific enantiomers of chiral compounds. In some cases the enantioselectivity of these reactions has approached that of enzymatic systems. A striking example is the catalytic reduction of β-keto carboxylic esters to β-hydroxyesters using ruthenium(II) com-

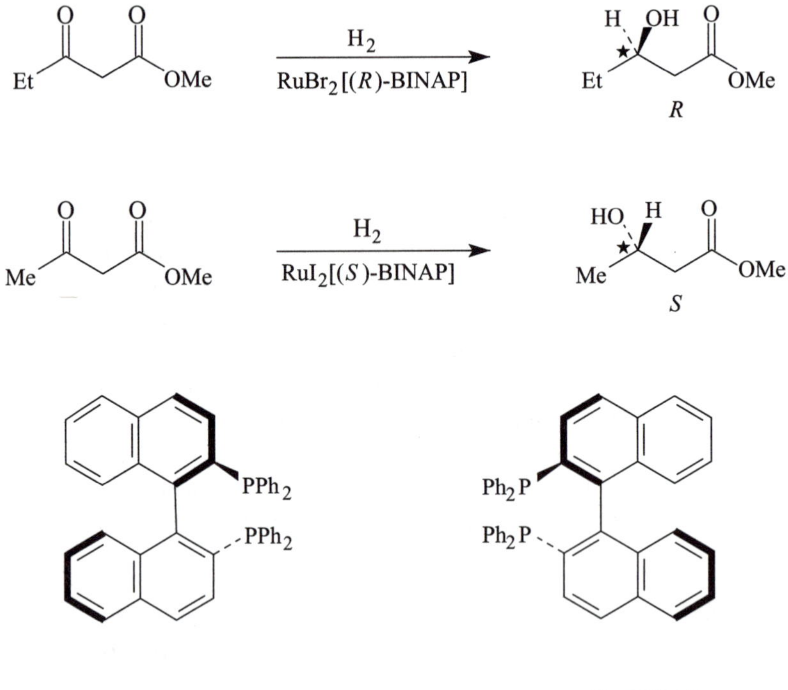

Figure 1-6
Enantioselective
Synthesis using
Chiral Ru-BINAP
Catalysts

(*R*)-BINAP

(*S*)-BINAP

plexes containing the chiral ligands (R)-BINAP and (S)-BINAP (see Figure **1-6**).[11] Selective synthesis of a single enantiomer in greater than 99% excess over the opposite enantiomer has been achieved. These results are comparable to the enantioselectivity achieved using enzymes from baker's yeast![12] Chiral syntheses using organotransition metal complexes are discussed in Chapters 9 and 11.

Finally, a discussion of the historical background of organometallic chemistry would be incomplete without mention of what surely qualifies as the oldest known organometallic compound, vitamin B_{12} coenzyme (Figure **1-7**). This naturally occurring cobalt complex contains a cobalt–carbon σ bond. It is a cofactor in a number of enzymes that catalyze 1,2 shifts in biochemical systems:

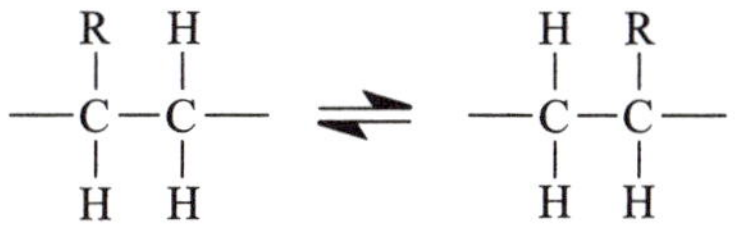

$$
\begin{array}{ccc}
\overset{\displaystyle R}{\underset{\displaystyle H}{|}}\ \overset{\displaystyle H}{\underset{\displaystyle H}{|}} & & \overset{\displaystyle H}{\underset{\displaystyle H}{|}}\ \overset{\displaystyle R}{\underset{\displaystyle H}{|}} \\
-\text{C}-\text{C}- & \rightleftharpoons & -\text{C}-\text{C}-
\end{array}
$$

The chemistry of vitamin B_{12} will be described briefly in Chapter 13.

Figure 1-7
Vitamin B_{12} Coenzyme

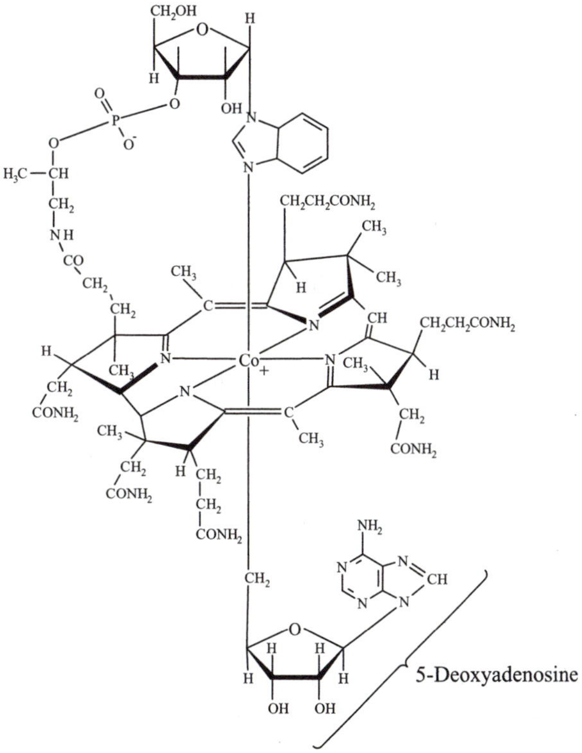

[11]R. Noyori, T. Ohkuma, M. Kitamura, H. Takaya, N. Sayo, H. Kumobayashi, and S. Akutagawal, *J. Am. Chem. Soc.*, **1987**, *109*, 5856.
[12]Stereoselective reductions of β-keto esters are of great importance in the biosynthesis of antibiotics.

2

Fundamentals of Structure and Bonding

An essential objective of organometallic chemistry is to understand how organic ligands can bond to metal atoms. An examination of interactions between metal orbitals and orbitals on these ligands can provide valuable insight into how organometallic molecules form and react; such an examination may also suggest future avenues for study and potential uses for these compounds.

Before considering how ligands bond to metals, we will find it useful to look at the types of orbitals involved in such bonds. This chapter includes a brief review of atomic orbitals, followed by a discussion of the ways in which atomic orbitals can interact to form molecular orbitals. In subsequent chapters we will consider how molecular orbitals of a variety of ligands can interact with transition metal orbitals. In these cases we will pay particular attention to how metal *d* orbitals can be involved.

2-1   Atomic Orbitals

The modern view of atoms holds that although atoms are composed of three types of subatomic particles—protons, neutrons, and electrons—the chemical behavior of atoms is governed by the behavior of the electrons.

Furthermore, the electrons do not behave according to the traditional concept of "particles" but, rather they exhibit the characteristics of waves. In a single atom, these waves can be described by the Schrödinger wave equation, which is shown as follows:

$$H\psi = E\psi$$

where:

H = Hamiltonian operator
E = Energy of the electron
ψ = Wave function

The wave function ψ is a mathematical expression describing the wave characteristics of an electron in an atom. The Hamiltonian operator H is a set of mathematical instructions to be performed on ψ in such a way as to give a result that is a numerical value (the energy of the electron) multiplied by ψ. The details of the Hamiltonian operator need not concern us here; elaborate discussions can be found in a variety of physical chemistry texts. The characteristics of the wave function ψ, however, are essential background for an understanding of the discussion of chemical bonding that will follow.

The wave function ψ has the following important characteristics:

1. It describes the *probable* location of electrons; it is not capable of predicting exactly where an electron is at a given place and time.
2. For any atom there are many solutions to the wave equation. Each solution describes one of the **orbitals** of the atom.
3. ψ is a mathematical description of a region of space (an orbital) that can be occupied by up to two electrons.
4. The square of the wave function, ψ^2, evaluated at a given set of coordinates (x, y, z), is proportional to the probability that an electron will be at that point.
5. The mathematical expression of ψ incorporates **quantum numbers**, which are related to the energy, size, and shape of atomic orbitals.

Solutions to the Schrödinger equation are described by quantum numbers, as summarized in Table **2-1**. Values for the first quantum numbers, n, l, m_l, are obtained by solving the Schrödinger equation and define an atomic orbital; the fourth, m_s, describes electron spin within an orbital.

Table 2-1 Quantum Numbers

Symbol	Name	Possible Values	Description
n	Principal	1, 2, 3, ...	Describes size and energy of orbitals. (In older terminology, defined *shell* of electrons.)
l	Azimuthal (Angular momentum)	0, 1, 2, ..., $(n-1)$ $l = 0$ s orbitals $l = 1$ p orbitals $l = 2$ d orbitals $l = 3$ f orbitals	Describes shape of orbitals; plays secondary role (after n) in determining energy. (In older terminology, defined *subshell* of electrons.)
m_l	Magnetic	$0, \pm1, \pm2, ..., \pm l$ (integer values from $-l$ to $+l$)	Describes orientation of orbitals. (The number of values of m_l is the number of orbitals in a subshell.)
m_s	Spin	$\frac{1}{2}, -\frac{1}{2}$	Describes spin of electron in orbital.

The quantum number l gives the classification of an orbital (s, p, d, etc.) and determines the orbital's shape. The number of values of the quantum number m_l is equal to the number of orbitals having that classification, as shown below:

		Classification	Number of orbitals
$l = 0$	$m_l = 0$ (one value)	s	1
$l = 1$	$m_l = -1, 0, 1$ (three values)	p	3
$l = 2$	$m_l = -2, -1, 0, 1, 2$ (five values)	d	5

The shapes and orientations of s, p, and d orbitals are extremely important in organometallic chemistry and are shown in Figure **2-1**.

Atomic orbitals on neighboring atoms may interact with each other—if certain conditions are satisfied:

- Atomic orbitals must have an appropriate orientation with respect to each other and be close enough to interact.
- Atomic orbitals should have similar energies, for such an interaction to be strong.

The **molecular orbital** concept of chemical bonding is based on this fundamental assumption. As the name implies, molecular orbitals (MOs) are similar in concept to atomic orbitals; in fact, molecular orbitals are derived from atomic orbitals.

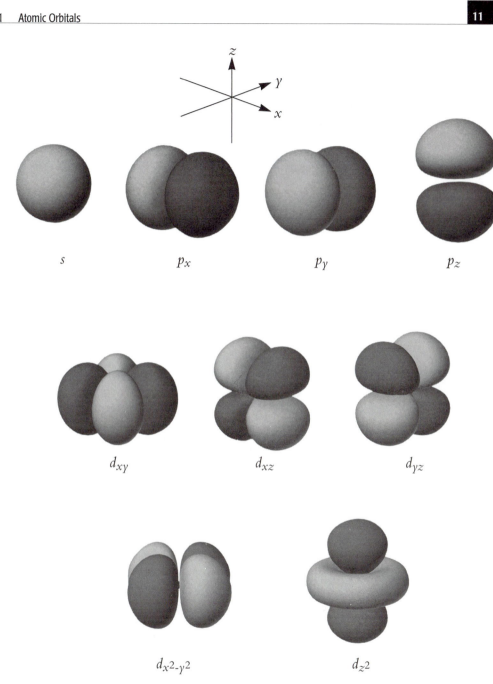

Figure 2-1
s, *p*, and *d* Orbitals

As stated, atomic orbitals on two atoms may interact if they are oriented appropriately with respect to each other.[1] In general, if atomic orbitals point toward each other and interact directly between the nuclei, they participate in a sigma (σ) interaction; if atomic orbitals are oriented parallel to each other such that they interact in two regions off to the side, they participate in a pi (π) interaction. The following are examples:

<u>Sigma Interactions:</u>   <u>Pi Interactions:</u>

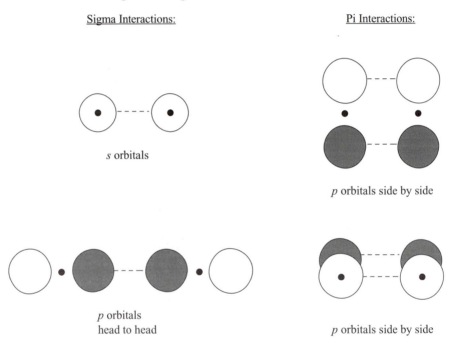

s orbitals

p orbitals side by side

p orbitals
head to head

p orbitals side by side

In addition, whenever orbitals on two atoms interact, they do so in two ways: a **bonding** fashion, in which the signs of the orbital waves match up, and an **antibonding** fashion, in which the signs of the orbital waves are opposite. In a bonding interaction, electrons are concentrated between the nuclei and tend to hold the nuclei together; in an antibonding interaction, electrons avoid the region of space between the nuclei and therefore expose the nuclei to each other's positive charges, thus tending to cause the nuclei to repel each other. Thus a bonding interaction stabilizes a molecule, while an antibonding interaction destabilizes it.

[1]In other words, the orbitals must have the same symmetry to interact. Background on chemical symmetry may be found in H. H. Jaffé and M. Orchin, *Symmetry in Chemistry*, Wiley, New York, 1965; applications of symmetry to orbital interactions may be found in T. H. Lowry and K. S. Richardson, *Mechanism and Theory in Organic Chemistry*, Harper & Row, New York, 1987, 55-83.

2-1-1 Sigma (σ) Interactions

The simplest case of a **σ** interaction is between *s* orbitals on neighboring atoms. Two types of interaction occur: bonding and antibonding. The bonding interaction leads to a molecular orbital (a bonding orbital, designated **σ**) lower in energy than the atomic orbitals from which it is formed; the antibonding interaction gives rise to a molecular orbital (an antibonding orbital, designated **σ***; an asterisk is commonly used to indicate antibonding) higher in energy than the atomic orbitals from which it is formed. These interactions are shown below.

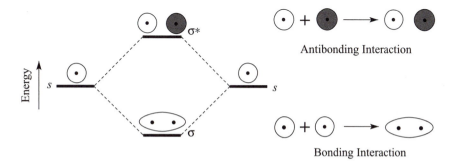

Similarly, *p* orbitals on adjacent atoms can interact. If the *p* orbitals are pointed directly toward each other, the interaction is classified as **σ**, and bonding and antibonding molecular orbitals are formed as follows:

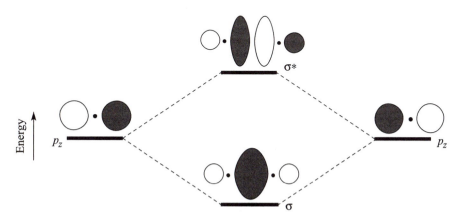

These interactions may be compared conceptually to the way in which overlapping waves interact (Figure **2-2**). If (electron) waves on neighboring atoms overlap in such a way that their signs are the same, the result is constructive interference of the waves; the resulting wave, representing a bonding interac-

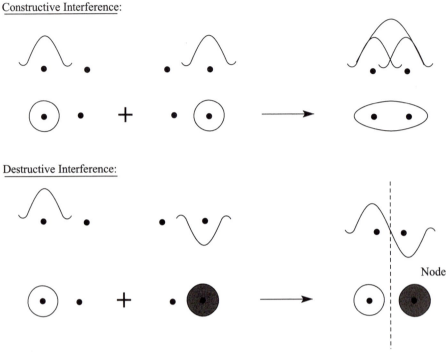

Figure 2-2
Interactions Between
Waves

tion, has a greater amplitude in the region between the nuclei. If the waves overlap such that their signs are opposite, the waves cancel each other out in the middle, creating a node of zero amplitude. This is an example of destructive interference and corresponds to an antibonding interaction.

2-1-2   Pi (π) Interactions

Similarly, parallel p orbitals on neighboring atoms can interact. In this case, the interaction occurs—not directly between the nuclei as in the σ case—but in two regions off to the side. When the signs on the neighboring orbital lobes match, the interaction is a bonding one, leading to the formation of a π molecular orbital lower in energy than the p orbitals; when the signs on the neighboring lobes are opposite, an antibonding ($\pi*$) orbital is formed of higher energy than the original atomic orbitals. These are shown for p orbitals in two orientations on the following page:

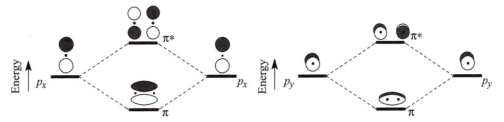

If the orientations of atomic orbitals are such that a bonding interaction would be canceled by an equal antibonding interaction, there is no net interaction, and the orbitals are designated **nonbonding**. For example, a p_x orbital on one atom is not oriented suitably to interact with a p_z orbital on a neighboring atom as shown below:[2]

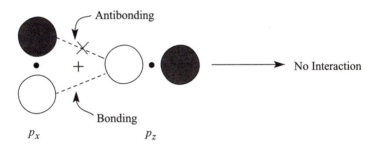

2-2   Molecular Orbitals

When all possible orbital interactions between neighboring atoms are considered, the result is a molecular orbital diagram. In this chapter we will consider first the simplest case: interactions between the orbitals of two atoms in diatomic molecules and ions. We will then extend this approach to consider the bonding in polyatomic organic systems.

2-2-1   Diatomic Molecules

Diatomic molecules are among the most important ligands occurring in organometallic compounds. We will therefore give special attention to the bonding in diatomics, considering first the **homonuclear** cases such as H_2 and O_2, then **heteronuclear** diatomics such as CO.

In general, the atomic orbitals of the interacting atoms will be shown on the far left and far right of these diagrams, and the molecular orbitals themselves in the middle. Relative energies of the orbitals will be indicated on the vertical scale of the diagrams.

[2]In general, we will choose the z axis to be the axis joining the atomic nuclei.

Figure 2-3
Molecular Orbitals of H_2

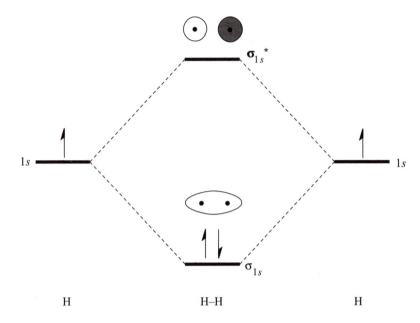

Homonuclear Diatomics

Molecular Orbitals of H_2. The simplest example of a diatomic molecule is H_2. For this molecule, the only atomic orbitals available are the $1s$ orbitals of the hydrogens. These orbitals can interact to give bonding σ_{1s} and antibonding σ_{1s}^* molecular orbitals; the molecular orbital energy level diagram is shown in Figure **2-3**. (Subscripts are often used to designate the atomic orbitals from which the molecular orbitals are derived.)

According to this diagram, H_2 has one pair of electrons in a bonding molecular orbital; this is the same as saying that it has a single bond. Since H_2 has only two electrons, it has no electrons left to occupy the higher energy antibonding orbital.

This picture of bonding can be compared with the Lewis dot structure, H : H, in which a single bond is designated by a pair of electrons shared between two atoms. In terms of molecular orbitals, a single bond is defined in H_2 as a pair of electrons occupying a bonding molecular orbital.

Bond Order

In the molecular orbital model, the number of bonds between two atoms is designated as the bond order and depends not only on the number of bonding electrons, but also on the number of antibonding electrons. In general, the bond order can be determined from the following equation:[3]

---

[3]In general, it is sufficient to consider only the valence electrons in the calculation of the bond order; inner electrons may be assumed to belong to individual atoms rather than to participate significantly in bonding.

Bond order = ½ (number of bonding electrons − number of antibonding electrons)

In the example of H_2, the bond order = ½ $(2 − 0)$ = 1 (a single bond). A bond order of 1 corresponds to a single bond, a bond order of 2 to a double bond, and so forth.

Suppose one wanted to consider whether He_2 is likely to be a stable molecule. This molecule, if it existed, would have a similar molecular orbital diagram to H_2, but it would have a total of four electrons: two in the bonding molecular orbital, two in the antibonding orbital. Its bond order would be = ½ $(2 − 2)$ = 0, or no bond. In other words, He_2 would have no bond at all; not surprisingly, molecules of He_2 have not been observed.

Second-Row Homonuclear Diatomic Molecules. Second-row elements in the periodic table have $2s$ and $2p$ valence orbitals. Interactions between these orbitals give the molecular orbital energy level diagrams shown in Figure **2-4**. The molecule O_2 is an example having this arrangement.

In general, σ interactions between $2p$ orbitals are stronger than π interactions; the σ interaction occurs directly between the atomic nuclei and has a somewhat stronger effect than the π interactions, which occur off to the side. Consequently, the σ_{2p} orbital is lower in energy than the π_{2p} orbitals, and the

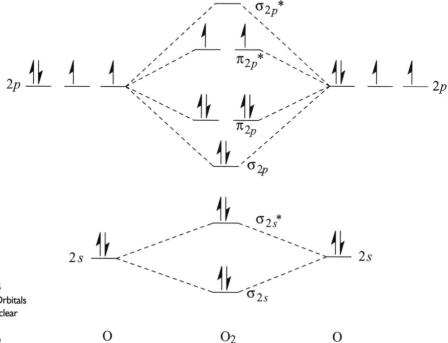

Figure 2-4
Molecular Orbitals
of Homonuclear
Diatomics.
Example: O_2

$\sigma_{2p}*$ is higher than the $\pi_{2p}*$ orbitals, as shown.[4] When the electrons are placed in the orbitals in order of increasing energy for O_2, the result is that two electrons occupy separate $\pi_{2p}*$ orbitals and have parallel spins. The existence of unpaired electrons in a molecule causes the molecule to behave as a tiny magnet that can be attracted by an external magnetic field; the molecule is **paramagnetic**. Experiments have shown that O_2 is, indeed, attracted by magnetic fields, lending support to the molecular orbital picture. The Lewis dot picture of O_2, shown below, does not provide a way of predicting these unpaired electrons.

The bond order in $O_2 = \frac{1}{2}(8 - 4) = 2$ (a double bond) $\overset{..}{\text{O}}{=}\overset{..}{\text{O}}{:}$

The oxygen–oxygen distance of 120.7 pm in O_2 is comparable to the distance expected for a double bond.

Diatomic Ions. Homonuclear diatomic ions can be treated similarly to the neutral molecules. Although the relative energies of molecular orbitals in such ions may be somewhat different than in the neutral molecules, in general the molecular orbital diagram for the molecules may be used for the ions, simply by adjusting the electron count.

Heteronuclear Diatomics

There are many diatomic molecules and ions containing two different elements; several of these play crucial roles in organometallic chemistry. See Table **2-2**.

Table 2-2 Examples of Heteronuclear Diatomics of the Second-Row Elements

Formula	Name	Name as Ligand
CN^-	Cyanide	Cyano
CO	Carbon monoxide	Carbonyl
NO^+	Nitrosyl	Nitrosyl[a]
NO	Nitric oxide	Nitrosyl

[a]The charge on the ligand when NO is bonded to a metal can be a matter of uncertainty. The name "nitrosyl" is used in general for NO ligands, without implications for the charge on the ligand or the oxidation state of the metal.

---

[4]The molecular orbital diagram shown in Figure **2-4** considers only interactions between atomic orbitals of identical energy: $2s$ with $2s$ and $2p$ with $2p$. Interactions also occur between s and p_z orbitals, however, as shown at right. When these interactions are taken into account, the net effect is to raise the energy of the σ_{2p} orbital relative to the π_{2p}. In some cases this interaction may be strong enough to push the σ_{2p} higher than the π_{2p}.

$s \qquad p_z$

Figure 2-5
Molecular Orbitals of CO

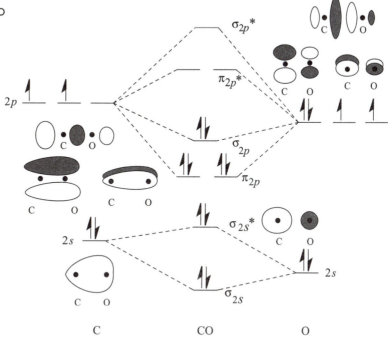

Molecular orbital diagrams for these second-row heteronuclear diatomics can be drawn rather easily by modifying the homonuclear pattern (as described for O_2, Figure **2-4**) slightly. The relative energies of the atomic orbitals should be shown indicating that the more electronegative element has lower energy orbitals.

For example, in CO (Figure **2-5**) the atomic orbitals of the more electronegative oxygen are lower, a reflection of the greater effective nuclear charge (oxygen has two more protons than carbon) pulling its orbitals to lower energy. The relative energies of the molecular orbitals of CO are similar to those of O_2.[5]

The shapes of the molecular orbitals should also be noted, especially since they have much chemical significance. In carbon monoxide the bonding π orbitals are skewed toward the more electronegative oxygen; the antibonding (and empty) π^* orbitals are skewed toward the carbon. The large, empty lobes on carbon have immense importance in the numerous compounds containing CO bonded to metals (the order of atoms in these cases is almost always M–C–O), as will be discussed in Chapter 4. In addition, the highest energy lone pair of electrons is concentrated on the carbon; it, too, plays an important role in bonding to metals.

[5]The one exception: the σ_{2p} orbital is higher in energy than the π_{2p} orbitals in CO; the opposite is true in O_2—see footnote 4.

Table 2-3 Valence Orbital Potential Energies

Atomic Number	Element	Orbital Potential Energy (eV)		
		1s	2s	2p
1	H	-13.6		
2	He	-24.5		
3	Li		-5.5	
4	Be		-9.3	
5	B		-14.0	-8.3
6	C		-19.5	-10.7
7	N		-25.5	-13.1
8	O		-32.4	-15.9
9	F		-46.4	-18.7
10	Ne		-48.5	-21.6

A useful reference is provided by the valence orbital potential energy; this is a measure of average potential energy of atomic orbitals. Valence orbital potential energies are negative, representing attractions between valence electrons and the nuclei; the more negative the value, the stronger the attraction, and the lower the energy of the orbital. Valence orbital potential energies for the first ten elements are given in Table **2-3**. In general, the more electronegative the element, the lower (more negative) the potential energies of the corresponding valence orbitals. For example, in CO the 2p orbitals of oxygen (–15.9 eV) are lower in potential energy than the 2p orbitals of carbon (–10.7 eV); this is shown in the molecular orbital diagram of CO (Figure **2–5**) by indicating that the energies of the 2p atomic orbitals are lower in energy for oxygen than for carbon.

Comparison of Atomic Orbitals with Molecular Orbitals

In many ways molecular orbitals are like atomic orbitals:

1. Both types of orbitals have a definite energy and shape.
2. Both types of orbitals can be occupied by up to two electrons; if two electrons occupy the same orbital, they must have opposite spin (the Pauli exclusion principle applies).
3. In filling orbitals of the same energy (degenerate levels), electrons in both types of orbitals tend to occupy separate orbitals and to have parallel spins (Hund's rule applies).
4. Both types of orbitals describe probable locations of electrons, rather than exact locations (an orbital does not designate an orbit).

5. Both types of orbitals describe the wave nature of electrons; in some cases, a single orbital may have positive and negative portions representing positive and negative values of the corresponding wave functions (like peaks and valleys of waves on an ocean). This is the case for π molecular orbitals and p atomic orbitals.

Molecular orbitals differ from atomic orbitals only in that the former arise from interactions between the latter, and that molecular orbitals therefore describe the behavior of electrons in molecules rather than in single atoms.

Suggested Procedure for Writing Molecular Orbital Diagrams

The molecular orbital concept is extremely important. The exercises that follow provide useful practice in drawing MO energy level diagrams. The following steps outline a recommended procedure:

1. Indicate the relative energies of the atomic orbitals from which the molecular orbitals are derived. These are ordinarily shown on the far left and on the far right of the MO diagram. Generally, it is sufficient to use only the valence orbitals, especially in homonuclear diatomic molecules; the core electrons are located much closer to the nuclei than the valence electrons, and consequently, these inner electrons have only minor effects on the bonding (they are also filled with electrons and generate an equal number of bonding and antibonding electrons, giving no net effect on the bond order).

2. Show the bonding and antibonding molecular orbitals that result from interactions of the atomic orbitals. These are shown in the center of the diagram. (For convenience, the axis joining the nuclei is generally chosen as the z axis.)

3. Identify the molecular orbitals with appropriate labels:

    σ for sigma interactions π for pi interactions

    Antibonding MOs are indicated by an asterisk superscript. For example, an antibonding σ orbital resulting from overlap of $2s$ atomic orbitals is designated $\sigma_{2s}{}^*$.

4. Place the appropriate number of electrons in the molecular orbitals.

    Simply determine the total number of valence electrons in the molecule or ion and place these in the MOs, starting with the lowest energy MO. Hund's rule and the Pauli exclusion principle should be followed.

5. Check: Be sure that the total number of molecular orbitals is equal to the sum of the number of atomic orbitals used, and that the total number of valence electrons in the molecular orbitals is correct.

Exercise 2-1

Prepare a molecular orbital energy level diagram of N_2. Include labels for the molecular orbitals and all valence electrons. Predict the bond order for N_2.

Example 2-1

Prepare a molecular orbital energy level diagram of the peroxide ion, O_2^{2-}, and predict the bond order of this ion.

Solution: The molecular orbital energy level diagram is similar to that for O_2, shown in Figure **2–4**. The peroxide ion has two more antibonding electrons than neutral O_2, giving a bond order of 1 as shown below:

Bond order = $\frac{1}{2}$ (8 – 6) = 1 (a single bond)

Exercise 2-2

Prepare a molecular orbital energy level diagram for the acetylide ion, C_2^{2-}, and predict the bond order of this ion.

2-2-2 Polyatomic Molecules: The Group Orbital Approach

Most molecules, of course, have far more than two atoms, and the method used for diatomic species is not sufficient to devise an appropriate molecular orbital picture. However, the assumptions fundamental to diatomics still apply: orbitals interact if their lobes are oriented appropriately with respect to each other, and interactions are stronger if the orbitals interacting are closer in energy.

One way to view interactions between atomic orbitals in polyatomic molecules and ions is to consider separately the orbitals on a central atom and the orbitals on surrounding atoms. In this approach the orbitals on surrounding atoms, considered as a group, are labeled **group orbitals.**[6]

Examples

Molecular Orbitals of H_3^+. As a simple example, we will consider the orbitals of H_3^+, a known (although unstable) ion. One possible structure of H_3^+ would be linear: H–H–H. In this ion we have only the $1s$ orbitals to consider. There are two outer hydrogens (to contribute the group orbitals) plus the central hydrogen. The $1s$ orbitals of the outer hydrogens may have the same sign of their wave functions (group orbital 1) or opposite signs (group orbital 2):

Group Orbitals in H_3^+:

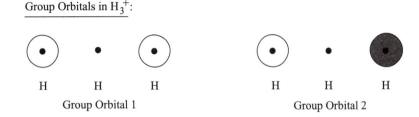

H H H	H H H
Group Orbital 1	Group Orbital 2

In linear H_3^+, the only orbital available on the central hydrogen is its $1s$. This orbital is capable of interacting, in both a bonding and an antibonding fashion, with Group Orbital 1, as shown below:

H H H	H H H
Bonding Interaction	Antibonding Interaction

The $1s$ orbital on the central hydrogen, however, cannot interact with Group Orbital 2; a bonding interaction on one side would be canceled by an antibonding interaction on the opposite side.

When sketching the molecular orbitals of polyatomic species, we begin by placing the valence orbitals of the central atom on the far left and the group

---

[6]The designation "group orbital" does not imply direct bonding between the orbitals involved in the group; group orbitals should rather be viewed as collections of similar orbitals.

orbitals of the surrounding atoms on the far right. We then show the resulting molecular orbital diagram in the middle. For H_3^+ the molecular orbital diagram[7] is shown in Figure **2-6**:

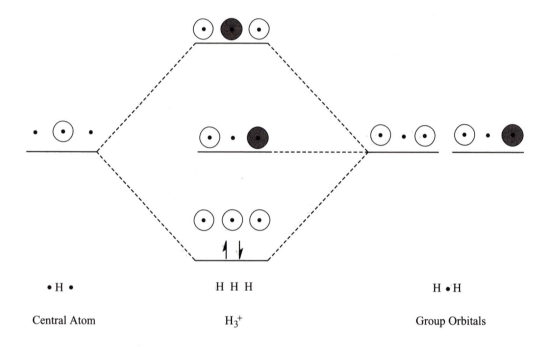

Figure 2-6
Molecular Orbitals of Linear H_3^+

In this ion the single pair of electrons occupies the lowest-energy molecular orbital, a bonding orbital distributed over all three atoms. This implies a bond order of approximately one-half (half an electron pair per bond), and would be consistent with the observation that H_3^+ is known, but not stable.[8]

[7]Note that the MO at the middle energy level in Figure **2-6** is Group Orbital 2. This MO is designated as *nonbonding*. The nonbonding MO in this case is energy equivalent to a situation in which two $1s$ orbitals are positioned an infinite distance apart and incapable of interacting. Nonbonding energy levels typically result when an odd number of orbitals interact to form MOs. The concept of nonbonding MOs is introduced at the end of Section **2-1**.

[8]H_3^+ is actually believed to be cyclic rather than linear, as will be discussed later in this chapter.

Molecular Orbitals of CO₂. In most cases of polyatomic molecules, p orbitals cause the orbital analysis to be much more complex than in the case of H_3^+. Carbon dioxide is such a situation, involving valence p orbitals on both the central and surrounding atoms. The group orbitals are on the oxygens and are of three types as shown below:

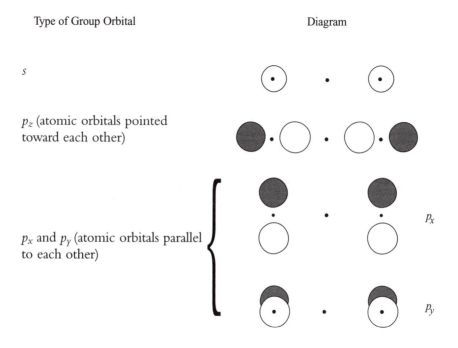

Type of Group Orbital	Diagram

s

p_z (atomic orbitals pointed toward each other)

p_x and p_y (atomic orbitals parallel to each other)

To determine the types of interactions possible between the central carbon and the group orbitals, we need to consider each of the types of group orbitals in turn.

s **Group Orbitals**

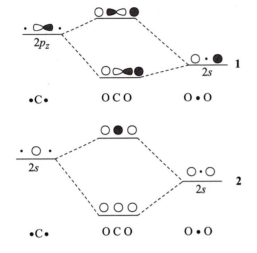

Group orbital 1 can interact with the $2p_z$ orbital of carbon. This interaction is likely to be weak, since the $2s$ orbitals of oxygen are much lower in energy than the $2p$ orbitals of carbon.

Group orbital 2 can interact with the $2s$ orbital of carbon.

p_z Group Orbitals

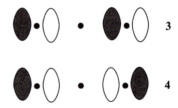

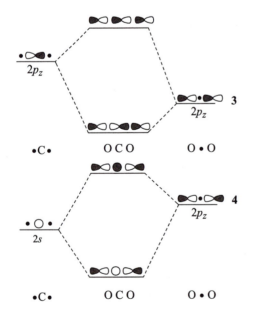

Interactions of the p_z group orbitals are similar to the case of the s group orbitals. Group orbital 3 can interact with the $2p_z$ orbital of carbon, and group orbital 4 can interact with the $2s$ orbital of carbon.

p_x and p_y Group Orbitals

The p_x and p_y interactions differ only in their orientation. The p_x group orbitals are:

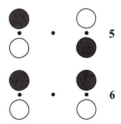

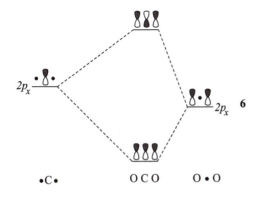

No orbital on the central carbon can interact with group orbital 5; this group orbital is non-bonding.

The $2p_x$ orbital of carbon, however, can interact with group orbital 6, as shown at left.

The p_y interactions are similar to the p_x. The group orbitals are:

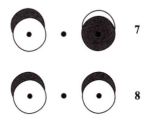

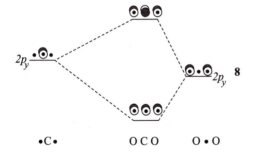

Since no atomic orbital on carbon is capable of interacting with group orbital 7, this group orbital is nonbonding.

Group orbital 8 can, however, interact with the $2p_y$ orbital of carbon, in an interaction similar (except for orientation) to that of group orbital 6.

Figure 2-7
The Molecular Orbitals of CO_2

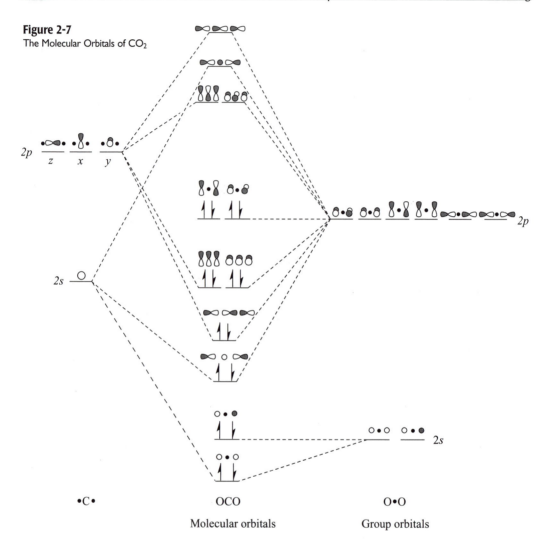

The overall molecular orbital diagram of CO_2 is shown in Figure **2-7**. The molecular orbital picture of other linear triatomic species, such as N_3^-, and CS_2, can be determined similarly. A similar approach can also be used in "linear" π systems, as described later in this chapter.

Exercise 2-3

For the linear ion FHF^-, sketch the group orbitals on the fluorines, and determine which of these orbitals can interact with the 1s orbital of hydrogen. Sketch a molecular orbital diagram.

Exercise 2-4

For the linear azide ion, N_3^-, sketch the group orbitals, and determine which of these orbitals can interact with the valence orbitals of the central nitrogen.

2-2-3 Ligands Having Extended Pi Systems

While it is a relatively simple matter to describe pictorially how σ and π bonds occur between pairs of atoms, it is a somewhat more involved process to explain bonding between metals and organic ligands having extended π systems. How, for example, are the C_5H_5 rings attached to Fe in ferrocene (Figure **1-5**), and how can 1,3-butadiene bond to metals? In order to understand the bonding between metals and π systems, it is necessary to first consider the π bonding within the ligands themselves. Fortunately, the group orbital approach can be adapted to simplify these situations. In the following discussion we first describe linear, then cyclic π systems; in later chapters we consider the question of how molecules containing such systems can bond to metals.

Examples

Linear Pi Systems. The simplest case of an organic molecule having a linear π system is ethylene, which has a single π bond resulting from the interactions of two $2p$ orbitals on its carbon atoms. Interactions of these p orbitals result in one bonding and one antibonding π orbital, as shown:

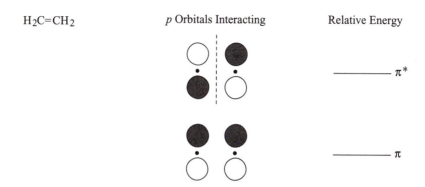

The antibonding interaction has a nodal plane perpendicular to the internuclear axis, while the bonding interaction has no such nodal plane.

Next is the three–atom π system, π–allyl, C_3H_5. In this case, there are three $2p$ orbitals to be considered, one from each of the carbon atoms participating in the π system. This situation can be viewed using the group orbital approach. The group orbitals here are derived from the parallel $2p$ orbitals on the terminal carbon atoms:

Group Orbital 1   Group Orbital 2

Group orbital 1 can interact, in both a bonding and an antibonding fashion, with the corresponding $2p$ orbital on the central carbon. Group orbital 2, on the other hand, is not suitable for interacting with any of the central carbon orbitals; this group orbital is nonbonding. The resulting molecular orbital diagram for this three–atom π system is shown in Figure **2–8**.

Figure 2-8
Pi Molecular Orbitals for
π-C_3H_5

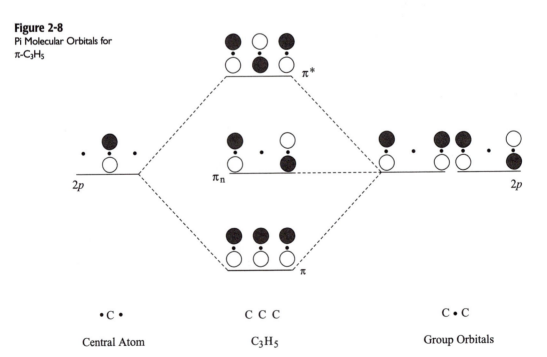

To extend this approach, we will find it useful to view these π interactions in a slightly different way. For the π-allyl system the possible interactions are as follows:

$H_2C=CH-CH_2$ *p* Orbitals Interacting   Relative Energy

—————— π^*

—————— π_n

—————— π

The lowest energy π molecular orbital for this system has all three *p* orbitals interacting constructively, to give a bonding molecular orbital. Higher in energy is the nonbonding situation (π_n), in which a nodal plane bisects the molecule, cutting through the central carbon. In this case the *p* orbital on the central carbon does not participate in the molecular orbital (in general, nodal planes passing through the center of *p* orbitals and perpendicular to internuclear axes will "cancel" these orbitals from participation in the π molecular orbitals). Highest in energy is the antibonding π^* orbital, in which there is an antibonding interaction between each neighboring pair of carbon *p* orbitals. This can be compared with the π^* orbital of ethylene, in which an antibonding interaction also occurs.

In these π systems, there is an increase in the number of nodes in going from lower energy to higher energy orbitals—for example, in the π-allyl system, the number of nodes increases from zero to one to two, from the lowest energy to the highest energy orbital.[9] This is a trend that will also occur in more extended π systems.

---

[9]This does not include the nodal plane that is coplanar with the carbon chain, bisecting each *p* orbital participating in the π system.

One additional example should suffice to illustrate this procedure. 1,3-Butadiene may exist in *cisoid* or *transoid* forms (with respect to the central C-C bond). For our purposes, it is sufficient to treat both as "linear" systems; the nodal behavior of the molecular orbitals will be the same in each case as in a linear π system of four atoms. As for ethylene and π-allyl, the $2p$ orbitals of the carbon atoms in the chain may interact in a variety of ways, with the lowest energy π molecular orbital having all constructive interactions between neighboring p orbitals, and the energy of the other π orbitals increasing with the number of nodes between the atoms.

$H_2C = CH - CH = CH_2$ *p* Orbitals Interacting   Relative Energy

This picture can be derived by a group orbital approach, beginning with the π and π^* orbitals of ethylene, as shown in Figure **2-9**.[10] The π orbitals of two interacting ethylenes form the two lowest energy π orbitals of butadiene, and similarly, the π^* orbitals of the ethylenes can form the two highest energy π orbitals of butadiene.

[10]T.H. Lowry and K.S. Richardson, *Mechanism and Theory in Organic Chemistry*, Harper & Row, New York, 1987, 79.

Figure 2-9
Construction of π
Orbitals of Butadiene
from Group Orbitals of
Ethylene

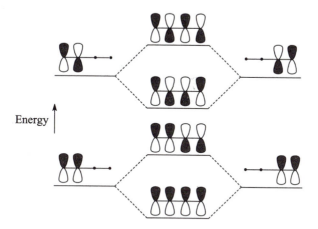

Energy

Similar patterns can be obtained for longer π systems.

Exercise 2-5

Sketch the π molecular orbitals for linear C_5H_7, and predict the relative energies of these orbitals.

Cyclic Pi Systems. The procedure for obtaining a pictorial representation of the orbitals of cyclic π systems of hydrocarbons is similar to the procedure for the linear systems described in the preceding section. Before discussing cyclic π systems, however, let's consider again the simple case of H_3^+, now examining a possible cyclic structure for this ion. One possible interaction for *cyclo*-H_3^+ would have a bonding interaction between each participating 1s orbital:

Three atomic orbitals are involved; three molecular orbitals must also be formed. What are the other two? In the linear case, the middle MO had a single node, the highest energy MO, two nodes (Figure **2–6**). Suppose the linear arrangement is wrapped into an equilateral triangle, with the nodes maintained as follows:

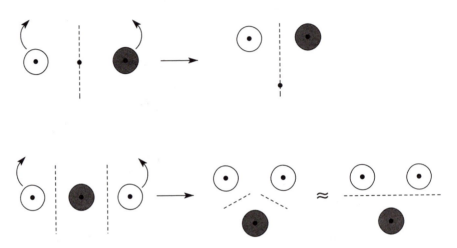

The result is a pair of molecular orbitals in cyclic H_3^+, each having a single node. Since they have the same number of nodes, they have the same energy (are degenerate), as shown in Figure **2–10**. One pair of electrons occupies a bonding orbital, giving rise to the equivalent of one-third bond between each pair of hydrogen atoms. This is believed to be the correct geometry of H_3^+.

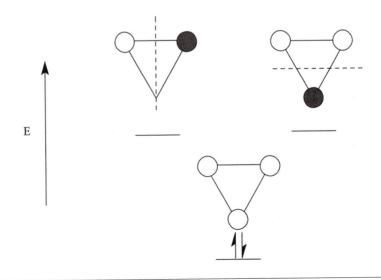

Figure 2-10
Molecular Orbitals of
Cyclic H_3^+

Cyclic Hydrocarbons. The smallest cyclic hydrocarbon having an extended π system is *cyclo*-C_3H_3. The lowest energy π molecular orbital for this system is the one resulting from constructive interaction between each of the $2p$ orbitals in the ring (top view) as shown below:

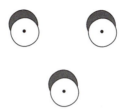

Two additional π molecular orbitals are needed (since the number of molecular orbitals must equal the number of atomic orbitals used). Each of these has a single nodal plane that is perpendicular to the plane of the molecule and bisects the molecule; as in *cyclo*-H_3^+, the nodes for these two molecular orbitals are perpendicular to each other:

These molecular orbitals have the same energy; in general, π molecular orbitals having the same number of nodes in cyclic π systems of hydrocarbons are degenerate. The total π molecular orbital diagram for *cyclo*-C_3H_3 can therefore be summarized as follows:

cyclo-C_3H_3 p Orbitals Interacting   Relative Energy

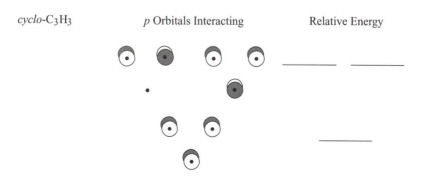

A simple way to determine the p orbital interactions and the relative energies of the cyclic π systems having regular polyhedral geometry is to inscribe the polyhedron inside a circle with one vertex pointed down; each vertex tangent to the circle then corresponds to the relative energy of a molecular orbital.[11] Furthermore, the number of nodal planes bisecting the molecule (and perpendicular to the plane of the molecule) increases with higher energy, with the bottom orbital having zero nodes, the next pair of orbitals a single node, and so on. For example, the next cyclic π system, *cyclo*-C_4H_4 (cyclobutadiene), would be predicted by this scheme to have molecular orbitals as follows[12]:

Relative Energy

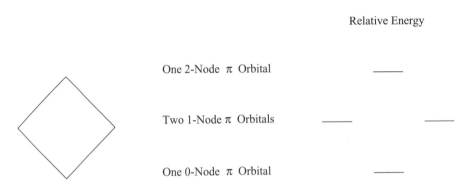

One 2-Node π Orbital   —————

Two 1-Node π Orbitals   —————   —————

One 0-Node π Orbital   —————

Similar results are obtained for other cyclic π systems as shown in Figure **2–11**. In these diagrams nodal planes are disposed symmetrically. For example, in *cyclo*-C_4H_4 the single node molecular orbitals bisect the molecule through opposite sides; the nodal planes are oriented at 90° angles to each other. The two-node orbital for this molecule also has the nodal planes at 90° angles.

This method may seem oversimplified, but the nodal behavior and relative energies are the same as obtained from molecular orbital calculations. Throughout this discussion, we have shown in some cases, not the actual shapes of the π molecular orbitals, but rather the p orbitals used. The nodal behavior of both sets (the π orbitals and the p orbitals are used) is identical and therefore sufficient to discuss how these ligands can bond to metals. Additional diagrams of numerous molecular orbitals for linear and cyclic π systems can be found in reference 13.

[11]This operation is called the Frost-Hückel mnemonic or simply the *circle trick*.

[12]This approach would predict a diradical for cyclobutadiene (one electron in each 1-node orbital). While cyclobutadiene itself is very reactive (P. Reeves, T. Devon, and R. Pettit, *J. Am. Chem. Soc.*, **1969**, *91*, 5890), complexes containing derivatives of cyclobutadiene are known. At 8 K cyclobutadiene has been isolated in a solid argon matrix (O.L. Chapman, C.L. McIntosh, and J. Pacansky, *J. Am. Chem. Soc.*, **1973**, *95*, 614; A. Krantz, C.Y. Lin, and M.D. Newton, *ibid.*, **1973**, *95*, 2746).

[13]W.L. Jorgenson and L. Salem, *The Organic Chemist's Book of Orbitals*, Academic Press, New York, 1973.

p Orbitals Interacting   Relative Energy

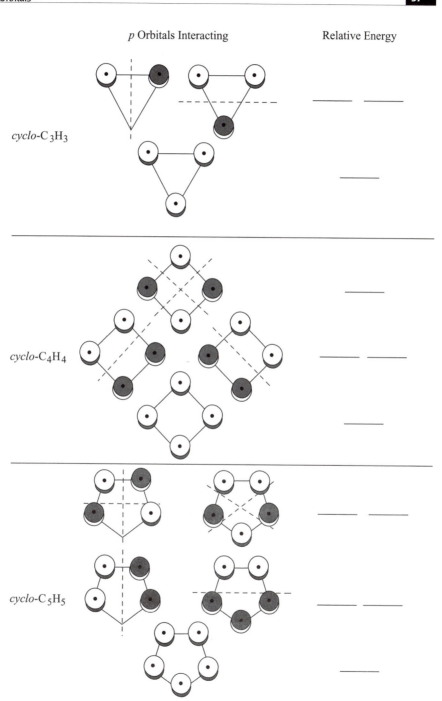

cyclo-C$_3$H$_3$

cyclo-C$_4$H$_4$

cyclo-C$_5$H$_5$

Figure 2-11
Molecular Orbitals for Cyclic π Systems

Benzene. In the molecular orbital approach for benzene, each carbon in the ring is considered to use sp^2 hybrid orbitals. These orbitals are involved in carbon-carbon σ bonding (from overlap of sp^2 hybrids on adjacent carbons) and carbon-hydrogen σ bonding (from overlap of an sp^2 hybrid on each carbon with the $1s$ orbital of hydrogen). This leaves, on each carbon, a p orbital *not* participating in the hybrids and available to participate in a cyclic π system. When these six p orbitals interact, six π molecular orbitals are formed, as shown in Figure **2-12**:

Relative Energy

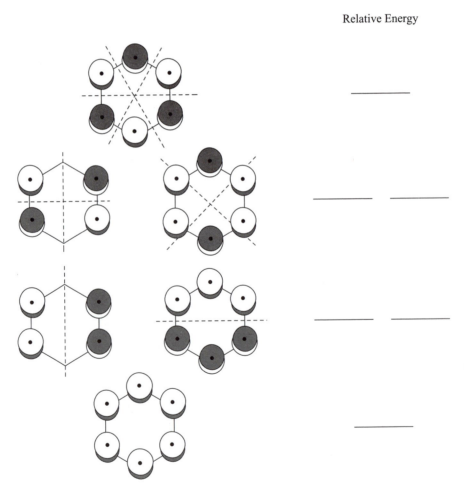

Figure 2-12
Pi Molecular Orbitals of Benzene

Benzene has six π electrons (equivalent to the six electrons used in double bonding in the Lewis structure); these occupy the three lowest-energy π molecular orbitals. The lowest-energy orbital is bonding with respect to each carbon–carbon bond (and corresponds to one bonding pair distributed over six bonds). The next two molecular orbitals have the same energy and are principally bonding, but each has a node bisecting the molecule and adding a degree of antibonding character. Overall, there are six electrons, or three pairs, occupying bonding orbitals spread over six bonds. The net effect is an approximate π bond order of one-half (actually, slightly less because of the nodes in the second and third orbitals), similar to the prediction of the Lewis model.

Many other cyclic molecules like benzene are known. Some of these are involved in another "classic" family of molecules: the sandwich compounds, in which atoms, most commonly of metals, are sandwiched between rings. These remarkable compounds have proven extraordinarily interesting since the prototype, ferrocene, was synthesized in the 1950s. Examples of sandwich compounds are shown in Chapter 1 (Figure **1-1**); these are discussed further in Chapter 5.

Problems

2-1   Prepare a molecular orbital diagram for nitric oxide (NO) and predict the bond order of this molecule. On the basis of the molecular orbital diagram, what do you predict for the bond orders of the ions NO^+ and NO^-? Which of these diatomic species would you expect to have the shortest bond length? Why?

2-2   The shapes of the π and $\pi*$ orbitals of diatomic ligands are important in organometallic chemistry.

       a.   Sketch the expected shapes of the π and $\pi*$ orbitals of the following:

$$N_2 \quad CN^- \quad CO \quad BO^-$$

       b.   What trends do you predict in the shapes of these orbitals? Why do you expect such trends? (Hint: consider relative electronegativities.)

2-3   Is H_4 a plausible molecule? By the approach described in this chapter, sketch the molecular orbitals (show interacting atomic orbitals, nodes of molecular orbitals, and relative energies of molecular orbitals) for the following geometries:

       a.   Linear

       b.   Square

     Which, if either, would you expect to be a more likely structure?

2-4   Sketch the π orbitals of 1,3,5-hexatriene, C_6H_8, and predict the relative energies of the orbitals. Indicate clearly the locations of the nodes.

2-5   Sketch the π orbitals of *cyclo*-C_7H_7, and predict the relative energies of these orbitals. Indicate clearly the location of the nodes.

3

18-Electron Rule

In main group chemistry, electron counts in molecules are often (but not always) described by the octet rule, in which the electronic structures can be rationalized on the basis of a valence shell requirement of eight electrons (two valence s electrons plus six valence p electrons). Similarly, in organometallic chemistry the electronic structures of many compounds are based on a total valence electron count of 18 on the central metal atom (ten valence d electrons in addition to the s and p electrons of the "octet"). As in the case of the octet rule, there are many exceptions to the **18-electron rule**,[1] but the rule nevertheless provides some useful guidelines to the chemistry of many organometallic complexes. In this chaper we will first examine how electrons are counted according to this rule. Then we will consider the basis for its usefulness (and some of the reasons why it is not always valid).

3-1   Counting Electrons

Several schemes exist for counting electrons in organometallic compounds. We will describe two of these using several examples. The first two examples will be of classic 18-electron species.

[1]Often called the effective atomic number (EAN) rule.

3-1-1   Method A: Donor Pair Method

This method considers ligands to donate electron pairs to the metal. To determine the total electron count, one must take into account the charge on each ligand and determine the formal oxidation state of the metal. We demonstrate this method using two examples of classic 18–electron species: $Cr(CO)_6$ and $(\eta^5\text{-}C_5H_5)Fe(CO)_2Cl$.

$Cr(CO)_6$

A chromium atom has six electrons outside its noble gas core. Each CO is considered to act as a donor of two electrons (from an electron dot standpoint, :C≡O:, these correspond to the lone pair on carbon). Thus the total electron count is

Cr	6 electrons
6(CO)	6×2 electrons = 12 electrons
	Total = 18 electrons

$Cr(CO)_6$ is therefore considered to be an 18–electron complex. It is thermally stable; for example, it can be sublimed without decomposition. $Cr(CO)_5$, a 16-electron species, and $Cr(CO)_7$, a 20-electron species, are, on the other hand, much less stable and known only as transient species. Likewise, the 17-electron $[Cr(CO)_6]^+$ and 19-electron $[Cr(CO)_6]^-$ are much less stable than the neutral, 18-electron $Cr(CO)_6$. The bonding in $Cr(CO)_6$, which provides a rationale for the special stability of many 18–electron systems, is discussed in Section **3-2**.

$(\eta^5\text{-}C_5H_5)Fe(CO)_2Cl$

As usual, CO is counted as a two-electron donor. Chloride is considered Cl^-, also a donor of two electrons. Pentahapto-C_5H_5 (see diagram on p. 43) is considered by this method to be $C_5H_5^-$, a donor of three electron pairs; it is a six-electron donor. Therefore, since this complex[2] is considered to contain the two negative ligands, Cl^- and $C_5H_5^-$, the oxidation state of iron in (η^5-$C_5H_5)Fe(CO)_2Cl$ is 2+. Iron(II) has six electrons beyond its noble gas core:

iron(0) has the electron configuration $[Ar] \, 4s^2 3d^6$

iron(II) has the electron configuration $[Ar] \, 3d^6$

---

[2]The η^5 notation (read "pentahapto" and signifying that the ligand has a *hapticity* of 5) designates that all five carbon atoms in the C_5H_5 ring are bonded to the iron (in general the superscript in this notation indicates the number of atoms in a ligand bonded to a metal; this type of notation will be discussed further in Chapter 5).

The electron count in the molecule $(\eta^5\text{-}C_5H_5)Fe(CO)_2Cl$ therefore is

Fe(II)	6 electrons
$\eta^5\text{-}C_5H_5^-$	6 electrons
2 (CO)	4 electrons
Cl^-	2 electrons
Total =	18 electrons

3-1-2   Method B: Neutral Ligand Method

This method uses the number of electrons that would be donated by ligands *if they were neutral*. For simple, inorganic ligands this usually means that ligands are considered to donate the number of electrons equal to their negative charge as free ions (Cl is a one-electron donor if singly bonded to a metal, O is a two-electron donor if doubly bonded, N is a three-electron donor if triply bonded, etc.). To determine the total electron count by this method, one does not need to determine the oxidation state of the metal.

$(\eta^5\text{-}C_5H_5)Fe(CO)_2Cl$

In this method, $\eta^5\text{-}C_5H_5$ is now considered as if it were a neutral ligand (or radical), in which case it would contribute five electrons. CO is a two-electron donor and Cl (counted as if it were a neutral species) is a one-electron donor. An iron atom has eight electrons beyond its noble gas core. The electron count is as follows:

Fe atom	8 electrons
$\eta^5\text{-}C_5H_5$	5 electrons
2 (CO)	4 electrons
Cl	1 electron
Total =	18 electrons

Method B gives the same result as Method A; $(\eta^5\text{-}C_5H_5)Fe(CO)_2Cl$ is an 18-electron species.

3-1-3   Other Considerations

Many organometallic complexes are charged, and this charge must be included when determining the total electron count. We can verify, by either method of electron counting, that $[Mn(CO)_6]^+$ and $[(\eta^5\text{-}C_5H_5)Fe(CO)_2]^-$ are both 18-electron ions.

In addition, metal–metal bonds must be counted. For example, in the dimeric complex $(CO)_5Mn\text{-}Mn(CO)_5$ the electron count per manganese atom is (by both methods):

Mn	7 electrons
5 (CO)	10 electrons
Mn–Mn bond	1 electron[3]
	Total = 18 electrons

For future reference, the electron counts for common ligands according to both schemes are given in Table **3–1**.

Table 3-1 Electron Counting Schemes for Common Ligands

Ligand	Method A	Method B
H	2 (:H⁻)	1
F, Cl, Br, I	2 (:Ẍ:⁻)	1
OH	2 (:Ö:H⁻)	1
CN	2 (:C≡N:⁻)	1
CH₃	2 (:CH₃⁻)	1
NO (bent M - N - O)	2 (:N̈ = Ö:⁻)	1
CO, PR₃	2	2
NH₃, H₂O	2	2
= CRR′ (carbene)	2	2
H₂C = CH₂	2	2
= O, = S	4 (:Ö:²⁻, :S̈:²⁻)	2
NO (linear M - N - O)	2 (:N ≡ O:⁺)	3
η³-C₃H₅	4 (C₃H₅⁻)	3
≡ CR (carbyne)	3	3
≡ N	6 (N³⁻)	3
butadiene	4	4
η⁵-C₅H₅	6 (C₅H₅⁻)	5
η⁶-C₆H₆	6	6
η⁷-C₇H₇	6 (C₇H₇⁺)	7

[3]For a Mn=Mn bond, each metal atom contributes two electrons; for a Mn≡Mn bond, each metal contributes three electrons.

Example 3-1

Both methods of electron counting are illustrated for the following three complexes:

Complex	Method A		Method B	
$ClMn(CO)_5$	Mn(I)	6 e$^-$	Mn	7 e$^-$
	Cl$^-$	2 e$^-$	Cl	1 e$^-$
	5 CO	10 e$^-$	5 CO	10 e$^-$
		18 e$^-$		18 e$^-$
$(\eta^5\text{-}C_5H_5)_2Fe$	Fe(II)	6 e$^-$	Fe	8 e$^-$
(Ferrocene)	2 η^5-C$_5$H$_5^-$	12 e$^-$	2 η^5-C$_5$H$_5$	10 e$^-$
		18 e$^-$		18 e$^-$
$[Re(CO)_5(PF_3)]^+$	Re(I)	6 e$^-$	Re	7 e$^-$
	5 CO	10 e$^-$	5 CO	10 e$^-$
	PF$_3$	2 e$^-$	PF$_3$	2 e$^-$
	+ charge	a	+ charge	– 1 e$^-$
		18 e$^-$		18 e$^-$

[a]Charge on ion is accounted for in assignment of oxidation state to Re.

The electron counting method used is a matter of individual preference. Method A has the advantage of including the formal oxidation state of the metal but may tend to overemphasize the ionic nature of some metal-ligand bonds. Counting electrons for some otherwise simple ligands (such as O^{2-} and N^{3-}) may seem cumbersome and unrealistic. Method B is often quicker, especially for ligands having extended π systems; for example, η^5 ligands have an electron count of five, η^3 ligands an electron count of three, and so on (see footnote 2). Also, Method B has the advantage of not requiring that the oxidation state of the metal be assigned. Other electron counting schemes have also been developed. It is generally best to select one method and to use it consistently.

Electron counting (by any method) does *not* imply anything about the degree of covalent or ionic bonding; it is strictly a bookkeeping procedure, as are the oxidation numbers that may be used in the counting. Physical measurements are necessary to provide evidence about the actual electron distribution in molecules. Linear and cyclic organic π systems interact with metals in more complicated ways and will be discussed in Chapter 5.

Exercise 3-1

Determine the valence electron counts for the metals in the following complexes:

a. $[Fe(CO)_4]^{2-}$

b. $[(\eta^5-C_5H_5)_2Co]^+$

c. $(\eta^3-C_5H_5)(\eta^5-C_5H_5)Fe(CO)$

Exercise 3-2

Identify the first-row transition metal for the following 18-electron species:

a. $[M(CO)_3(PPh_3)]^-$

b. $HM(CO)_5$

c. $(\eta^4-C_8H_8)M(CO)_3$

d. $[(\eta^5-C_5H_5)M(CO)_3]_2$ (assume single M-M bond)

L-X Notation

It is useful to introduce some symbolism that will appear often in later chapters in this book. Most ligands may be classified as "L-type" or "X-type."[4] L-Type ligands are neutral, two-electron donors such as CO or PR_3. Ligands such as Cl or CH_3 are designated as X-type. X-Type ligands typically carry a negative charge and would be two-electron donors according to the Donor Pair Method (Method A) and one-electron donors according to the Neutral Ligand Method (Method B). Some ligands, such as $\eta^5-C_5H_5$, contain both types of classifications. If we consider the structure of $\eta^5-C_5H_5$ to have the following structural representation, it would be symbolized in L-X notation as L_2X.

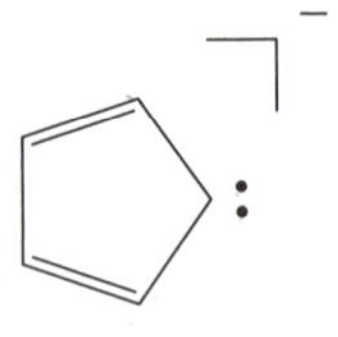

---

[4]For a more complete discussion of this notation, see R.H. Crabtree, *The Organometallic Chemistry of the Transition Metals*, 2nd. ed. John Wiley & Sons, New York, 1994, 24–38.

In general, any organometallic complex containing L- and X-type ligands may be represented by the general formula:

$$[MX_aL_b]^c$$

where   a = the number of X-type ligands,
               b = the number of L-type ligands,
and   c = the charge.

Several useful relationships result from the following notation:

1. Electron count (EAN):
   $$EAN = N + a + 2b - c$$
   where N = the group number of the metal in the Periodic Table

Example:

$HFe(CO)_4^- \equiv [MXL_4]^-$
$EAN = 8 + 1 + 8 + 1 = 18$ electrons

2. Coordination number (CN):
   $$CN = a + b$$

Example:

$[ReH(PPh_3)_3(CO)_3]^+ \equiv [MXL_6]^+$
$CN = 1 + 6 = 7$

3. Oxidation state of the metal (OS):
   $$OS = a + c$$

Example:

$Rh(H)(H)(PPh_3)_2(CO)Cl \equiv [MX_3L_3]^0$
$OS = 3 + 0 = 3$

4. Number of d electrons (d^n):
   $$d^n = N - OS = N - (a + c)$$

Example:

$(\eta^5-C_5H_5)_2Zr(H)(Cl) \equiv [MX_4L_4]^0$
$d^n = 4 - (4 + 0) = 0$

Exercise 3-3

Represent the complex $[Ir(CO)(PPh_3)_2(Cl)(NO)]^+$ in L-X notation. Calculate the electron count, coordination number, oxidation state of the metal, and the number of d electrons for the metal.

3-2   Why 18 Electrons?

An oversimplified rationale for the special significance of 18 electrons can be made by analogy with the octet rule in main group chemistry. If the octet represents a complete valence electron shell configuration (s^2p^6), then the number 18 can be considered to correspond to a filled valence shell for a transition metal $(s^2p^6d^{10})$. This analogy, while perhaps useful to relate the electron configurations to the idea of valence shells of electrons for atoms, does not explain why so many complexes violate this "rule." In particular, the valence shell rationale does not distinguish among types of interactions that may occur between metal and ligand orbitals; this distinction is an important consideration in determining which complexes obey the rule and which ones violate it.

3-2-1   Types of Metal–Ligand Interactions

Metal and ligand orbitals can interact in several ways. The type of interaction depends on the orientation of the orbitals with respect to each other. Most such interactions can be classified into three types: sigma donor ligands, pi donor ligands, and pi acceptor ligands. These are discussed in the following sections.

Sigma (σ) Donor Ligands

These ligands have an electron pair capable of being donated directly toward an empty (or partly empty) metal orbital. The following is an example of the interaction of a donor orbital of NH_3, occupied by a lone pair of electrons, and an empty d orbital of suitable orientation on a metal:

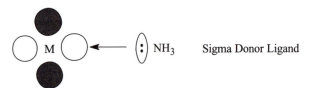

In a σ donor interaction, the electron pair on the ligand is stabilized by the formation of a bonding molecular orbital, and the empty metal orbital (a d orbital in the example above) is destabilized, as shown below:

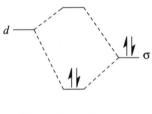

Pi (π) Donor Ligands

In some cases ligands may donate electrons in a π fashion, for example using a filled *p* orbital on a ligand as shown below. Halide ions may participate in this type of interaction, with an electron pair donated in a π fashion to an empty metal *d* orbital (this *d* orbital must have a different orientation than the *d* orbital used in σ interactions).

Pi Donor Ligand

The π donor interaction is in one respect similar to the σ donor case; the electron pair on the ligand is stabilized by the formation of a bonding molecular orbital, and the empty metal orbital is destabilized:

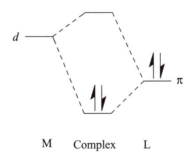

M   Complex   L

Pi (π) Acceptor Ligands

Although ligands are often thought of as donating electron pairs to metals, such is not always the case. Many types of ligands are known in which σ donation is complemented by the ability of a ligand to accept electron density from a metal using suitable acceptor orbitals. The classic example of such a ligand is CO, the carbonyl ligand. The CO ligand can function as a σ donor using an electron pair in its highest energy molecular orbital (HOMO) as shown below:

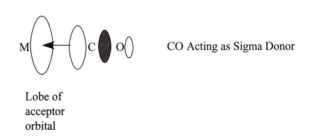

CO Acting as Sigma Donor

Lobe of
acceptor
orbital

At the same time, CO has empty π^* orbitals of suitable orientation to accept electron density from the metal:

CO Acting as Pi Acceptor

The effect of the π acceptor interaction is the opposite of the σ interactions; an electron pair on the metal now is stabilized when a molecular orbital is formed, while the energy of the (empty) ligand orbital is *destabilized* as it becomes a higher energy (antibonding) molecular orbital:

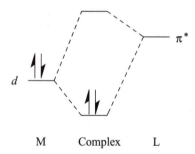

M   Complex   L

These two ligand functions, σ donor and π acceptor, are synergistic. The more effective the σ donation, the greater the electron density on the metal. The resulting electron-rich metal in turn is capable of donating electrons back to the ligand,[5] which then acts as a π acceptor.

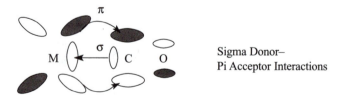

Sigma Donor–
Pi Acceptor Interactions

[5]This phenomenon is sometimes called "back bonding" or "back donation."

Table 3-2 Examples of Donor and Acceptor Ligands

Sigma Donor	Pi Donor[a]	Pi Acceptor[a]
NH_3	OH^-	CO
H_2O	Cl^-	CN^-
H^-	RCO_2^-	PR_3

[a]These ligands also act as σ donors.

Examples of common ligands of all three types are given in Table **3-2**. In many instances, ligands behave in more than one way. Hydroxide, for example, can be considered both a σ donor and π donor, while a phosphine (PR_3) is both a σ donor and π acceptor.

3-2-2   Molecular Orbitals and the 18-Electron Rule

Octahedral Complexes

As we pointed out earlier in this chapter, $Cr(CO)_6$ is a good example of a complex that obeys the 18-electron rule. The molecular orbitals of interest in this molecule are those that result primarily from interactions between the d orbitals of Cr and the σ donor (HOMO) and π acceptor orbitals (LUMO) of the six CO ligands. The molecular orbitals corresponding to these interactions are shown in Figure **3-1** (p. 52) .[6] (See also Chapter 2, Figure **2-5**.)

Chromium (0) has six electrons outside its noble gas core. Each of the six CO ligands contributes a pair of electrons to give a total electron count of 18. In the molecular orbital diagram, these 18 electrons appear as 12 σ electrons (the σ electrons of the CO ligands, stabilized by their interaction with the metal orbitals) and six metal-ligand bonding electrons (these electrons occupy orbitals having the symmetry label t_{2g}).

Adding one or more electrons to $Cr(CO)_6$ would populate metal-ligand antibonding orbitals (these orbitals have the symmetry label e_g*); the consequence would be destabilization of the molecule. Removing electrons from $Cr(CO)_6$ would depopulate the t_{2g} orbitals. These are bonding in nature as a consequence of the strong π acceptor ability of the CO ligands; a decrease in electron density in these orbitals would also tend to destabilize the complex. The result is that the 18-electron configuration for this molecule is the most stable.

[6]Molecular orbitals are often designated by symmetry labels, such as the t_{2g} and e_g* labels describing orbitals with significant d character in Figure 3-1. The method of assigning these labels is beyond the scope of this text. For further information on symmetry labels, see F.A. Cotton, *Chemical Applications of Group Theory*, 3rd ed. Wiley, New York, 1990.

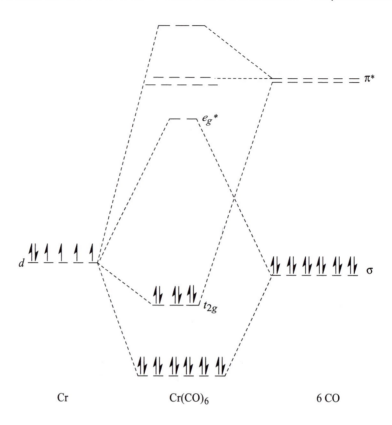

Figure 3-1
Molecular Orbitals of Cr(CO)$_6$
(*Only interactions between ligand
orbitals and metal d orbitals are shown.)

The shapes of the t_{2g} and e_g* orbitals support this description of bonding in Cr(CO)$_6$. One of the t_{2g} orbitals is shown in Figure **3-2**. An electron pair in this orbital will spend the majority of its time in one of the four three-lobed regions near the chromium (these regions result from interaction of the d_{xy} orbital of Cr with four π* orbitals of CO). As the figure shows, electrons in each of these regions would be expected to be attracted strongly by three nuclei, the chromium and two carbons; the resulting effect would be to hold these nuclei together, in effect keeping the CO ligands attached to the metal. The two other t_{2g} orbitals have the same shape, but different orientations, than the one shown in Figure **3-2**; these orbitals involve the d_{xz} and d_{yz} orbitals of the metal. Collectively, the electron pairs in the three t_{2g} orbitals are crucial in bonding the six carbonyls to the chromium; in the absence of these electrons, the carbonyls would readily dissociate.

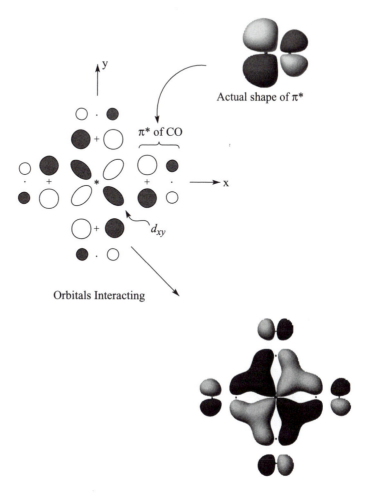

Figure 3-2
A t_{2g} Orbital of $Cr(CO)_6$ t_{2g}

One of the e_g* orbitals of $Cr(CO)_6$ is shown in Figure **3-3**. The principal interaction in this case is antibonding between the σ orbitals of the carbonyls and the d orbital lobes of the chromium. Electrons in this orbital, or in the other e_g* orbital, would therefore destabilize the molecule, also leading to dissociation of CO.

Continuing for the moment to consider six-coordinate molecules of octahedral geometry, we can gain some insight as to when the 18-electron rule can be expected to be most valid. $Cr(CO)_6$ obeys the rule because of two factors: the strong σ donor ability of CO raises the e_g* orbitals in energy, making them considerably antibonding; and the strong π acceptor ability of

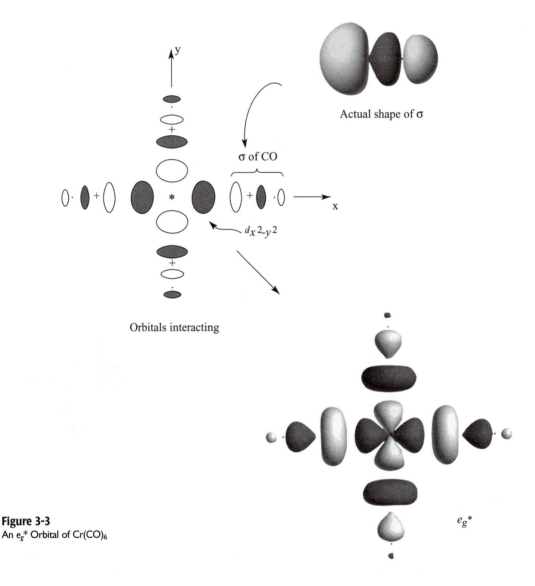

Actual shape of σ

σ of CO

$d_{x^2-y^2}$

Orbitals interacting

e_g*

Figure 3-3
An e_g* Orbital of Cr(CO)$_6$

CO lowers the t_{2g} orbitals in energy, making them bonding. Ligands that are both strong σ donors and π acceptors therefore should be the most effective at forcing adherence to the 18-electron rule. Other ligands, including some organic ligands, do not have these features and consequently their compounds may or may not adhere to the rule.

Figure 3-4
Exceptions to the 18-Electron Rule

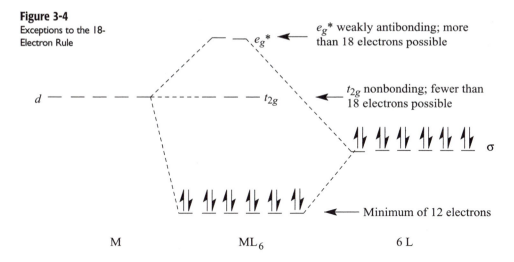

e_g^* weakly antibonding; more than 18 electrons possible

t_{2g} nonbonding; fewer than 18 electrons possible

σ

Minimum of 12 electrons

M   ML$_6$ 6 L

Examples of exceptions may be noted. $Zn(en)_3^{2+}$ is a 22-electron species; it has both the t_{2g} and e_g^* orbitals filled. While en (ethylenediamine = $NH_2CH_2CH_2NH_2$) is a good σ donor, it is not as strong a donor as CO. As a result, the e_g^* orbital is not sufficiently antibonding to cause significant destabilization of the complex, and the 22-electron species, with 4 electrons in e_g^* orbitals, is stable. An example of a 12-electron species is TiF_6^{2-}. In this case the fluoride ligand is a π donor as well as a σ donor. The π donor ability of F⁻ destabilizes the t_{2g} orbitals of the complex, making them slightly antibonding. The species TiF_6^{2-} has 12 electrons in the bonding σ orbitals and no electrons in the antibonding t_{2g} or e_g^* orbitals.

These examples of exceptions to the 18-electron rule are shown schematically in Figure **3-4**.[7]

Other Geometries

The same type of argument can be made for complexes of geometry other than octahedral; in most, but not all cases there is an 18-electron configuration of special stability for complexes of strongly π accepting ligands. Examples include trigonal bipyramidal geometry [for example, $Fe(CO)_5$] and tetrahedral geometry [for example, $Ni(CO)_4$]. The most common exception is square planar geometry, in which a 16-electron configuration may be the most stable, especially for complexes of d^8 metals.

[7] P.R. Mitchell and R.V. Parish, *J. Chem. Ed.* **1969**, *46*, 311

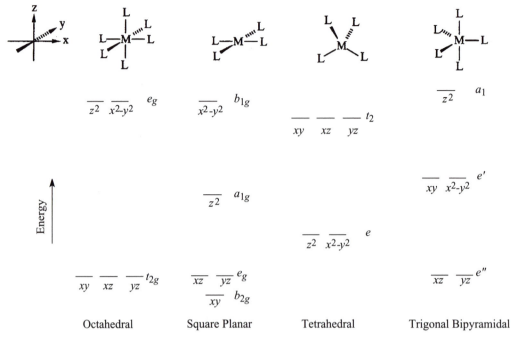

Figure 3-5
Relative Energies of Metal *d* Orbitals for
Complexes of Common Geometries.
(*Only σ donor and π acceptor interactions
are shown.)

The relative energies of *d* orbitals of transition metal complexes can be predicted approximately by the *crystal field approach*, a model of bonding that considers how the orbitals on a metal would interact with negative charges located at the sites of the ligands. By this model, the metal *d* orbitals pointing most directly toward the ligands will interact most strongly; electron pairs on the ligands will be stabilized, and a metal *d* orbital destabilized by such interactions. The result: the more directly the *d* orbital points toward the ligands, the higher the energy of the resulting antibonding orbital.

For example, in octahedral geometry, the two *d* orbitals pointing most directly toward the ligands are the $d_{x^2-y^2}$ and d_{z^2}. These metal orbitals interact more strongly with the ligands than the other *d* orbitals and give rise to the highest energy (antibonding) molecular orbitals (labeled e_g^* ; see Figure **3–1**). Similar reasoning can predict the relative energies of molecular orbitals derived from metal *d* orbitals for other geometries; examples of common geometries are shown in Figure **3–5**. Symbols such as a_1, a_{1g}, and b_{2g} are additional symmetry labels useful in describing molecular orbitals with significant *d* character.[8]

[8]See footnote 6, this chapter.

3-3   Square Planar Complexes

Chapter 9 discusses square planar complexes, which are extremely important in the field of catalysis. Examples include the d^8, 16-electron complexes shown in Figure **3-6.**

To understand why 16-electron square planar complexes might be especially stable, it is necessary to examine the molecular orbitals of such a complex. An example of a molecular orbital diagram for a square planar molecule of formula ML_4 (L= ligand that can function as both σ donor and π acceptor) is shown in Figure **3-7**.

Four molecular orbitals of ML_4 are derived primarily from the σ donor orbitals of the ligands; electrons occupying such orbitals are bonding in nature. Three additional orbitals are slightly bonding (derived primarily from d_{xz}, d_{yz}, and d_{xy} orbitals of the metal) and one is essentially nonbonding (derived primarily from the d_{z^2} orbital of the metal). These bonding and nonbonding orbitals can be filled by 16 electrons. Additional electrons would occupy an antibonding orbital derived from the antibonding interaction of a metal $d_{x^2-y^2}$ orbital with the σ donor orbitals of the ligands. Consequently, for square planar complexes of ligands having both σ donor and π acceptor characteristics, a 16-electron configuration may be significantly more stable than an 18-electron configuration. Sixteen-electron square planar complexes may also be capable of accepting one or two ligands at the vacant coordination sites (along the z axis), thereby achieving an 18-electron configuration. As will be shown in Chapter 7, this is a common reaction of 16-electron square planar complexes.

Figure 3-6
Examples of Square Planar
d^8 Molecules

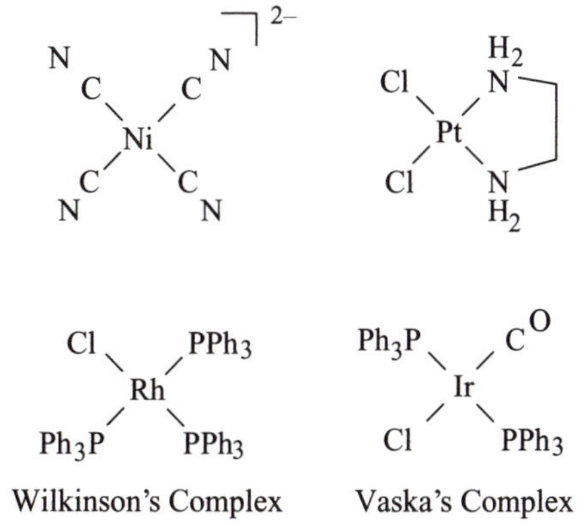

Wilkinson's Complex

Vaska's Complex

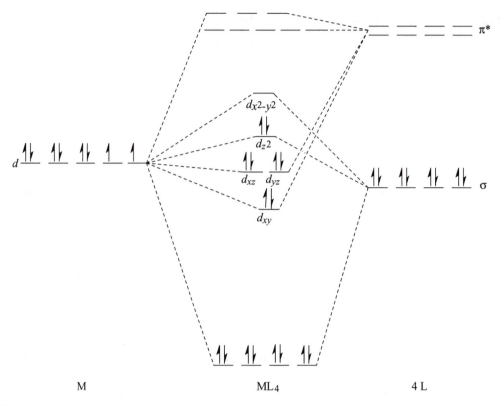

Figure 3-7
Molecular Orbitals of Square Planar
Complexes (*Only σ donor and π
acceptor interactions are shown.)

Exercise 3-4

Verify that the complexes in Figure **3-6** are 16–electron species.

Sixteen–electron square planar complexes are most commonly found for d^8 metals, in particular those metals having formal oxidation states of 2+ (Ni^{2+}, Pd^{2+}, and Pt^{2+}) and 1+ (Rh^+, Ir^+). Some of these complexes have important catalytic behavior, as will be discussed in Chapter 9.

Problems

3-1   Octahedral transition metal complexes containing strong π acceptor ligands obey the 18-electron rule much more often than complexes containing π donor ligands. Why?

3-2   Assuming that the 18-electron rule applies, identify the first-row transition metal:
   a. $M(CO)(CS)(PPh_3)_2Br$
   b. $[M(CO)_7]^+$
   c. $[(\eta^3-C_3Ph_3)(\eta^4-C_4H_4)M(NH_3)_2]^+$
   d. $[(\eta^5-C_5H_5)(\eta^4-C_5H_6)M]^+$
   e. $[(\eta^3-C_3H_5)M(CN)_4]^{2-}$

3-3   Assuming that the 18-electron rule applies, identify the second-row transition metal:
   a. $[(\eta^5-C_5H_5)M(CO)_3]_2$ (assume single M-M bond)
   b. $[(\eta^5-C_5H_5)M(CO)_2]_2$ (assume double M=M bond)
   c. $[M(CO)_3(NO)]^-$ (assume linear M–N–O)
   d. $(\eta^4-C_8H_8)M(CO)_3$
   e. $[M(CO)_3(PMe_3)]^-$

3-4   What charge, z, would be necessary for the following to obey the 18-electron rule?
   a. $[Ru(CO)_4(SiMe_3)]^z$
   b. $[(\eta^3-C_3H_5)V(CNCH_3)_5]^z$
   c. $[(\eta^5-C_6H_7)Fe(CO)_3]^z$
   d. $[(\eta^6-C_6H_6)_2Ru]^z$

3-5   Determine the specified quantity:
   a.   The metal-metal bond order in $[(\eta^5-C_5Me_5)Rh(CO)]_2$
   b.   The expected charge on $[W(CO)_5(SnPh_3)]^z$
   c.   The identity of the second-row transition metal in $(\eta^5-C_5H_5)(\eta^1-C_3H_5)(\eta^3-C_3H_5)_2M$, a 16-electron molecule

3-6   For a square planar complex ML_4, where L is a σ donor and π acceptor, sketch
               the following interactions (assume that the z axis is perpendicular to the plane
               of the molecule):

   a.   Interaction of d_{xy} orbital on M with π* orbitals on ligands
   b.   Interaction of d_{xz} orbital on M with π* orbitals on ligands
   c.   Interaction of $d_{x^2-y^2}$ orbital on M with σ orbitals on ligands
   d.   Interaction of d_{z^2} orbital on M with σ orbitals on ligands

4

Carbonyl Ligand

Carbon monoxide is the most common ligand in organometallic chemistry. It may serve as the only ligand in **binary carbonyls** such as $Ni(CO)_4$, $W(CO)_6$, and $Fe_2(CO)_9$ or, more commonly, in combination with other ligands, both organic and inorganic. In this chapter, we consider the bonding between metals and CO, the synthesis and some reactions of CO complexes, and examples of the various types of CO complexes formed.

4-1 Bonding

It is useful to review the bonding in CO. CO may bond to a single metal or it may serve as a bridge between two or more metals. First, we consider carbonyl bonds to a single metal.

4-1-1 CO as a Terminal Ligand

As described in Chapter 2 (Figure **2-5**), the molecular orbital picture of CO is similar to that of N_2 (Exercise **2-1**); the molecular orbitals derived primarily from the $2p$ atomic orbitals of these molecules are shown in Figure **4-1** (p.62). Two

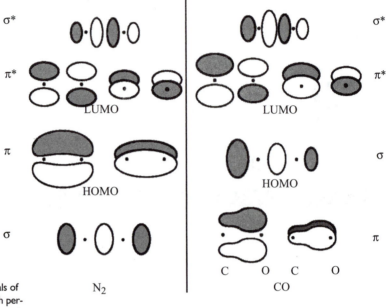

Figure 4-1
Selected Molecular Orbitals of
CO and N_2 (Adapted with per-
mission from G.L. Miessler and
D.A. Tarr, *Inorganic Chemistry*,
Prentice-Hall, Englewood Cliffs,
NJ, 1991, 426.)

features of these molecular orbitals of CO deserve attention. First, the highest
energy occupied orbital (the HOMO) has its largest lobe on carbon. It is
through this orbital, occupied by an electron pair, that CO exerts its σ donor
function, donating electron density directly toward an appropriate metal orbital
(such as an unfilled *p*, *d*, or hybrid orbital). At the same time, CO has two
empty π* orbitals (the lowest unoccupied molecular orbitals, or LUMO); these
have larger lobes on carbon than on oxygen. As a consequence of the localiza-
tion of π* orbitals on carbon, the carbon acts as the principal site of the π
acceptor function of the ligand: a metal atom having electrons in a *d* orbital of
suitable symmetry can donate electron density to these π* orbitals. These σ
donor and π acceptor interactions are illustrated in Figure **4-2**.

The overall effect is synergistic: CO can donate electron density via a σ
orbital to a metal atom; the greater the electron density on the metal, the more
effectively it is able to return electron density to the π* orbitals of CO. The net
effect can be rather strong bonding between the metal and CO; however, as
will be described later, the strength of this bonding is dependent on several factors, including the charge on the complex and the ligand environment of the
metal.

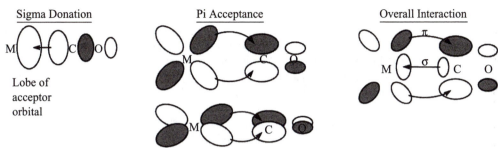

Figure 4-2
Sigma and Pi Interactions between CO and a Metal Atom (Adapted with permission from G. L. Miessler and D. A. Tarr, *Inorganic Chemistry*, Prentice-Hall, Englewood Cliffs, NJ, 1991, 426.)

Exercise 4-1

N_2 has molecular orbitals slightly different than those of CO, as shown in Figure **4-1**. Would you expect N_2 to be a stronger or weaker π acceptor than CO?

If this picture of bonding between CO and metal atoms is valid, it should be supported by experimental evidence. Two sources of such evidence are infrared spectroscopy and X-ray crystallography. First, any change in the bonding between carbon and oxygen should be reflected in the C–O (carbonyl) stretching vibration as observed by IR. As in organic compounds, the C–O stretch in organometallic compounds is often very intense (stretching the C–O bond results in a substantial change in dipole moment), and its energy often provides valuable information about the molecular structure. Free carbon monoxide has a C–O stretch at 2143 cm^{-1}; Cr(CO)$_6$, on the other hand, has its carbonyl stretch at 2000 cm^{-1}. The lower energy for the stretching mode means that the C–O bond is weaker in Cr(CO)$_6$.

The energy necessary to stretch a bond is proportional to $\sqrt{\dfrac{k}{\mu}}$, where:

k = force constant

μ = reduced mass; for atoms of mass m_1 and m_2 the reduced mass is given by:

$$\mu = \frac{m_1 m_2}{m_1 + m_2}$$

The stronger the bond between two atoms, the larger the force constant; consequently the greater the energy necessary to stretch the bond and the higher the energy of the corresponding band (the higher the wave number, cm^{-1}) in the infrared spectrum. Similarly, the more massive the atoms involved in the bond, as reflected in a higher reduced mass, the less energy necessary to stretch the bond and the lower the energy of the absorption in the infrared spectrum.

Both σ donation (which donates electron density from a bonding orbital on CO) and π acceptance (which places electron density in C-O *anti*bonding orbitals) would be expected to weaken the C-O bond and to decrease the energy necessary to stretch that bond.

Additional evidence is provided by X-ray crystallography, which gives information on relative positions of atoms in molecules. In carbon monoxide the C-O distance has been measured at 112.8 pm. Weakening of the C-O bond by the factors described above would be expected to cause this distance to increase. Such an increase in bond length is found in complexes containing CO, with C-O distances approximately 115 pm for many carbonyls. While such measurements provide definitive measures of bond distances, in practice it is far more convenient to use infrared spectra to obtain data on the strength of C-O bonds.

The charge on a carbonyl complex is also reflected in its infrared spectrum. Three isoelectronic hexacarbonyls have the following C-O stretching bands:[1]

Complex	$\nu(CO)$, cm^{-1}
$[V(CO)_6]^-$	1858
$[Cr(CO)_6]$	2000
$[Mn(CO)_5]^+$	2095
Free carbonyl	2143

Of these three, $[V(CO)_6]^-$ has the metal with smallest nuclear charge, meaning that vanadium has the weakest ability to attract electrons and the greatest tendency to "back" donate electron density to CO. The consequence is strong population of the π^* orbitals of CO and reduction of the strength of the C-O bond. In general, the more negative the charge on organometallic species, the greater the tendency of the metal to donate electrons to the π^* orbitals of CO and the lower the energy of the C-O stretching vibrations.

---

[1]K. Nakamoto, *Infrared and Raman Spectra of Inorganic and Coordination Compounds*, 4th ed. Wiley, New York, 1986, 292-93.

4-1-2   Bridging Modes of CO

Although CO is most commonly found as a "terminal" ligand attached to a single metal atom, many cases are known in which CO forms bridges between two or more metals. Many such bridging modes are known; the most common are shown in Table **4-1**.[2]

Table 4-1 Bonding Modes of CO

Type of CO	Approximate Range for $\nu(CO)$ in Neutral Complexes (cm^{-1})
Free CO	2143
Terminal M - CO	1850 - 2120
Bridging:	
Symmetrical μ_2-CO	1700 - 1860

$$\begin{array}{c} O \\ \| \\ C \\ M \qquad M \end{array}$$

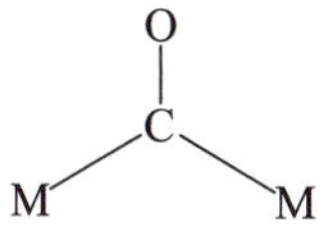	
Symmetrical μ_3-CO 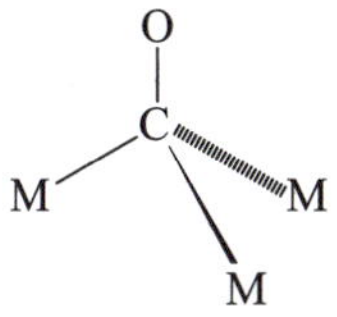	1600 - 1700

$$\begin{array}{c} O \\ \| \\ C \\ M \qquad M \\ M \end{array}$$

In addition to the symmetrical bridging modes of CO shown in Table **4-1**, CO sometimes bridges metals asymmetrically; in these cases CO is considered *semibridging* (shown below).

Semibridging Modes of CO:

$$\begin{array}{cc} O & O \\ C & C \\ M \quad M & M \quad M \\ & M \end{array}$$

[2]Bridging ligands are designated using the Greek letter μ (mu), followed by a subscript indicating the number of atoms bridged. For example, μ_2 indicates that a ligand bridges two atoms.

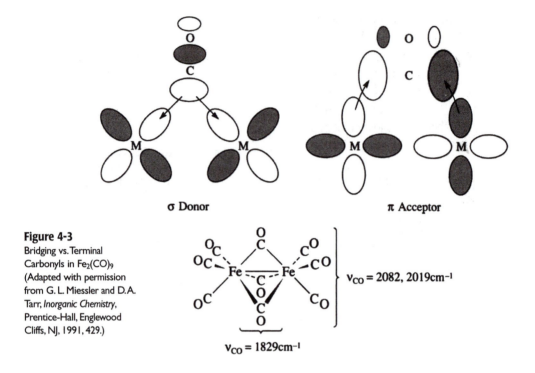

σ Donor   π Acceptor

Figure 4-3
Bridging vs. Terminal
Carbonyls in Fe₂(CO)₉
(Adapted with permission
from G. L. Miessler and D. A.
Tarr, *Inorganic Chemistry,*
Prentice-Hall, Englewood
Cliffs, NJ, 1991, 429.)

v_{CO} = 2082, 2019cm⁻¹

v_{CO} = 1829cm⁻¹

A variety of factors can result in an asymmetric carbonyl bridge, including inherent asymmetry in molecules (such as CO bridging two or three different metals) and crowding (where moving of a CO away from a symmetrically bridging position would help reduce crowding in part of a molecule).[3]

The bridging mode is strongly correlated with the position of the C—O stretching band. In cases where CO bridges two metal atoms, both metals can contribute electron density into π* orbitals of CO to weaken the C—O bond and lower the energy of the stretch. Consequently, the C—O stretch for doubly bridging CO is at much lower energy than for terminal carbonyls. An example is shown in Figure **4-3**.

Interaction of three metal atoms with a triply bridging CO further weakens the C—O bond; the infrared band for the C—O stretch is still lower than in the doubly bridging case. Ordinarily, terminal and bridging carbonyls can be considered two-electron donors, with the two donated electrons shared by the metals in the bridging cases.[4]

---

[3] For a discussion of various factors involved in asymmetrically bridging CO ligands, see F. A. Cotton and G. Wilkinson, *Advanced Inorganic Chemistry,* 5th ed. Wiley-Interscience, New York, 1988, 1028-1032.
[4] Triply bridging carbonyls also function as two-electron donors, formally two thirds of an electron per metal. In such cases it is often best to consider the two electrons donated to the complex as a whole rather than to specific metal atoms.

A particularly interesting situation is that of the nearly linear bridging carbonyls such as in $[(\eta^5\text{-}C_5H_5)Mo(CO)_2]_2$. When a sample of $[(\eta^5\text{-}C_5H_5)Mo(CO)_3]_2$ is heated, some carbon monoxide is driven off; the product, $[(\eta^5\text{-}C_5H_5)Mo(CO)_2]_2$, reacts readily with CO to reverse this reaction:[5]

$$[(\eta^5\text{-}C_5H_5)Mo(CO)_3]_2 \overset{\Delta}{\rightleftharpoons} [(\eta^5\text{-}C_5H_5)Mo(CO)_2]_2 + 2CO$$
$$1960, 1915 \text{ cm}^{-1} \qquad\qquad 1889, 1859 \text{ cm}^{-1}$$

This reaction is accompanied by a shift of the infrared bands in the carbonyl region to lower energies, as shown. The Mo–Mo bonds also shorten by approximately 80 pm, consistent with an increase in the metal-metal bond order from one to three. Although it was originally proposed that the "linear" CO ligands may donate some electron density to the neighboring metal from π orbitals, subsequent calculations have indicated that a more important interaction is the donation from a metal d orbital to the π^* orbital of CO, as shown in Figure **4-4**.[6] Such donation weakens the carbon-oxygen bond in the ligand and results in the observed shift of the C–O stretching bands to lower energy.

Additional examples of the utility of infrared spectra in characterizing carbonyl complexes are included in Section **4-5**.

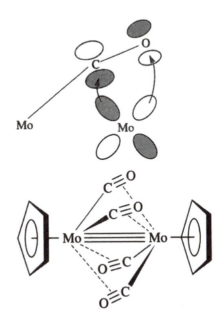

Figure 4-4
Bridging CO in $[(\eta^5\text{-}C_5H_5)Mo(CO)_2]_2$ (Adapted with permission from G. L. Miessler and D. A. Tarr, *Inorganic Chemistry*, Prentice-Hall, Englewood Cliffs, NJ, 1991, 429.)

[5]D.S. Ginley and M.S. Wrighton, *J. Am. Chem. Soc.*, **1975**, *97*, 3533; R.J. Klingler, W. Butler, and M.D. Curtis, *J. Am. Chem. Soc.*, **1975**, *97*, 3535.
[6]A.L. Sargent and M.B. Hall, *J. Am. Chem. Soc.*, **1989**, *111*, 1563 and references therein.

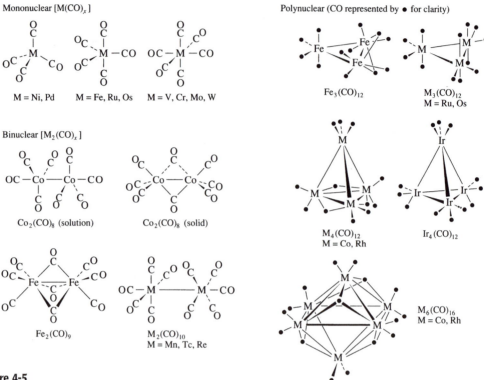

Mononuclear [M(CO)$_x$]

M = Ni, Pd   M = Fe, Ru, Os   M = V, Cr, Mo, W

Binuclear [M$_2$(CO)$_x$]

Co$_2$(CO)$_8$ (solution)   Co$_2$(CO)$_8$ (solid)

Fe$_2$(CO)$_9$ M$_2$(CO)$_{10}$
M = Mn, Tc, Re

Polynuclear (CO represented by ● for clarity)

Fe$_3$(CO)$_{12}$ M$_3$(CO)$_{12}$
M = Ru, Os

M$_4$(CO)$_{12}$
M = Co, Rh   Ir$_4$(CO)$_{12}$

M$_6$(CO)$_{16}$
M = Co, Rh

Figure 4-5
Binary Carbonyl Complexes (Adapted with permission
from G. L. Miessler and D. A. Tarr, *Inorganic Chemistry*,
Prentice-Hall, Englewood Cliffs, NJ, 1991, 430.)

4-2   Binary Carbonyl Complexes

Binary carbonyls, containing only metal atoms and CO, are fairly numerous.
Structures of some binary carbonyl complexes are shown in Figure **4-5**. Most
of these complexes obey the 18-electron rule. The cluster compounds
Co$_6$(CO)$_{16}$ and Rh$_6$(CO)$_{16}$ do not obey the rule, however; more detailed
analysis of the bonding in cluster compounds is necessary to satisfactorily
account for the electron counting in these and other cluster compounds. This
question is considered in Chapter 12.

One other binary carbonyl does not obey the rule: the 17-electron
V(CO)$_6$. This complex is one of a few cases in which strong π acceptor ligands
do not succeed in requiring an 18-electron configuration. In V(CO)$_6$ the vanadium is apparently too small to permit a seventh coordination site; hence, no
metal–metal bonded dimer (which would give an 18-electron configuration) is
possible. However, V(CO)$_6$ is easily reduced to [V(CO)$_6$]$^-$, a well-studied 18-
electron complex.

Exercise 4-2

Verify the 18-electron rule for five of the binary carbonyls [other than $Co_6(CO)_{16}$ and $Rh_6(CO)_{16}$] shown in Figure **4-5**.

An interesting feature of the structures of binary carbonyl complexes is that the tendency of CO to bridge transition metals tends to decrease moving down the Periodic Table. For example, in $Fe_2(CO)_9$ there are three bridging carbonyls, but in $Ru_2(CO)_9$ and $Os_2(CO)_9$ there is a single bridging CO.

4-2-1   Synthesis

Binary carbonyl complexes can be synthesized in many ways. Several of the most common methods are described below.

1.   Direct reaction of a transition metal with CO. The most facile of these reactions involves nickel, which reacts with CO at ambient temperature and 1 atm:

     $Ni + 4\ CO \rightarrow Ni(CO)_4$

     $Ni(CO)_4$ is a volatile, extremely toxic liquid that must be handled with great caution. It was first observed by Mond, who found that CO reacted with nickel valves. The reverse reaction, involving thermal decomposition of $Ni(CO)_4$, can be used to prepare nickel of very high purity. Coupling of the forward and reverse reactions has been used commercially in the Mond process for obtaining purified nickel from ores. Other binary carbonyls can be obtained from direct reaction of metal powders with CO, but elevated temperatures and pressures are necessary.

2.   Reductive carbonylation. This synthesis involves reduction of a metal compound in the presence of CO and an appropriate reducing agent. Examples:

     $CrCl_3 + 6\ CO + Al \rightarrow Cr(CO)_6 + AlCl_3$ (catalyzed by $AlCl_3$)

     $Re_2O_7 + 17\ CO \rightarrow Re_2(CO)_{10} + 7\ CO_2$ (CO acts as reducing agent; high temperature and pressure required)

3.   Thermal or photochemical reaction of other binary carbonyls. Examples:

     $Fe(CO)_5 \xrightarrow{h\nu} Fe_2(CO)_9$ (involves dissociation of CO)

     $Fe(CO)_5 \xrightarrow{\Delta} Fe_3(CO)_{12}$ (involves dissociation of CO)

4-2-2   Reactions

The most common reaction of carbonyl complexes is CO dissociation. This reaction, which may be initiated thermally or by absorption of ultraviolet light, characteristically involves loss of CO from an 18-electron complex to give a 16-electron intermediate, which may react in a variety of ways, depending on the nature of the complex and its environment. A common reaction is replacement of the lost CO by another ligand to form a new 18-electron species as product. The following are examples:[7]

$$Cr(CO)_6 + PPh_3 \xrightarrow{\Delta \text{ or } h\nu} Cr(CO)_5(PPh_3) + CO$$

$$Re(CO)_5Br + en \xrightarrow{\Delta} fac\text{-}Re(CO)_3(en)Br + 2\ CO$$

fac-Re(CO)$_3$(en)Br mer-Re(CO)$_3$(en)Br

This type of reaction, therefore, provides a pathway in which CO complexes can be used as precursors for a variety of complexes of other ligands. Additional aspects of CO dissociation reactions will be discussed in Chapter 7.

4-3   Oxygen-Bonded Carbonyls

This chapter would be incomplete without the mention of one additional aspect of CO as a ligand: it can sometimes bond through oxygen as well as carbon. This phenomenon was first noted in the ability of the oxygen of a metal carbonyl complex to act as a donor toward Lewis acids such as $AlCl_3$, with the overall function of CO serving as a bridge between the two metals. Numerous examples are now known in which CO bonds through its oxygen to transition metal atoms, with the C-O-metal arrangement generally bent. Attachment of a Lewis acid to the oxygen results in significant weakening and lengthening of the C-O bond. The result is a shift of the C-O stretching vibration to lower energy in the infrared. The magnitude of this shift is typically

---

[7]The prefix *fac* designates a *facial* stereoisomer—in this case an octahedron with three CO groups at the corners of the same triangular face. *Meridional* isomers (abbreviated *mer*) correspond to stereoisomers in which three identical groups (e.g., COs) lie in the same plane.

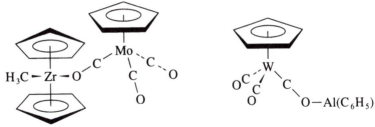

Figure 4-6
Oxygen-Bonded Carbonyls
(Adapted with permission from
G. L. Miessler and D. A. Tarr,
Inorganic Chemistry, Prentice-Hall,
Englewood Cliffs, NJ, 1991, 432.)

between 100 and 200 cm^{-1}. Examples of O-bonded carbonyls (sometimes called isocarbonyls) are shown in Figure **4-6**. The physical and chemical properties of oxygen–bonded carbonyls have been reviewed.[8]

4-4   Ligands Similar to CO

Several diatomic ligands similar to CO are worth brief mention. Two of these, CS (thiocarbonyl) and CSe (selenocarbonyl), are of interest in part for purposes of comparison with CO. Several other common ligands are isoelectronic with CO and, not surprisingly, exhibit structural and chemical parallels with CO. Two examples are CN$^-$ (cyanide) and N$_2$ (dinitrogen). Although not an organic ligand, the NO (nitrosyl) ligand deserves discussion because of its similarities to CO. In addition NS (thionitrosyl) is discussed briefly.

4-4-1   CS and CSe Complexes

In most cases synthesis of CS and CSe complexes is somewhat more difficult than for analogous CO complexes, since CS and CSe do not exist as stable, free molecules and do not, therefore, provide a ready ligand source.[9] Consequently, the comparatively small number of such complexes should not be viewed as an indication of their stability; the chemistry of CS and CSe complexes may eventually rival that of CO complexes in breadth and utility. Thiocarbonyl complexes are also of interest as possible intermediates in certain sulfur-transfer reactions in the removal of sulfur from natural fuels. In recent years the chemistry of complexes containing these ligands has developed more rapidly, as avenues for their synthesis have been devised.

[8] C.P. Horwitz and D.F. Shriver, *Adv. Organomet. Chem.*, **1984**, *23*, 219.
[9] E. K. Moltzen, K. J. Klabunde, and A. Senning, *Chem. Rev.*, **1988**, *88*, 391.

CS and CSe are similar to CO in their bonding modes; they behave as both σ donors and π acceptors and can bond to metals in terminal or bridging modes. Of these two ligands, CS has been studied more closely. It usually functions as a stronger σ donor and π acceptor than CO.[10]

4-4-2   CN⁻ and N₂ Complexes

Cyanide is a stronger σ donor and a somewhat weaker π acceptor than CO; overall it is similar to CO in its ability to interact with metal orbitals. Unlike most organic ligands, which bond to metals in low formal oxidation states, cyanide bonds readily to metals having higher oxidation states. As a good σ donor, CN⁻ interacts strongly with positively charged metal ions; as a weaker π acceptor than CO (largely a consequence of the negative charge of CN⁻), cyanide is not as able to stabilize metals in low oxidation states. Therefore its compounds are often studied in the context of "classical" coordination chemistry rather than organometallic chemistry. Dinitrogen is a weaker donor and acceptor. However, N₂ complexes are of great interest, especially as possible intermediates in reactions that may simulate natural processes of nitrogen fixation.[11]

4-4-3   NO Complexes

Like CO, the NO (nitrosyl ligand) is both a σ donor and π acceptor and can serve as a terminal or bridging ligand; useful information can be obtained about its compounds by analysis of its infrared spectra. Unlike CO, however, terminal NO has two common coordination modes: linear (like CO), and bent. Useful information about the linear and bent bonding modes of NO is summarized in Figure **4-7**.

Linear   Bent

[10]P. V. Broadhurst, *Polyhedron*, **1985**, *4*, 1801.

[11]R. A. Henderson, G. J. Leigh, and C. J. Pickett, *Adv. Inorg. Chem. Radiochem.*, **1983**, 27, 197.

Figure 4-7
Linear and Bent NO (Adapted
with permission from G. L.
Miessler and D. A. Tarr, *Inorganic
Chemistry*, Prentice-Hall,
Englewood Cliffs, NJ, 1991, 434.)

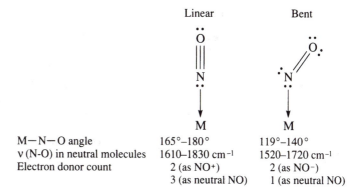

	Linear	Bent
M—N—O angle	165°–180°	119°–140°
v (N-O) in neutral molecules	1610–1830 cm^{-1}	1520–1720 cm^{-1}
Electron donor count	2 (as NO+)	2 (as NO-)
	3 (as neutral NO)	1 (as neutral NO)

A formal analogy is often drawn between the linear bonding modes of both ligands: NO$^+$ is isoelectronic with CO; therefore, in its bonding to metals, linear NO is considered by electron counting method A (Chapter 3) as NO$^+$, a two-electron donor. By the neutral ligand method (method B, Chapter 3), linear NO is counted as a three-electron donor (it has one more electron than the two-electron donor CO).

The bent coordination mode of NO is often considered to arise formally from NO$^-$, with the bent geometry suggesting sp^2 hybridization at the nitrogen. By electron counting method A, therefore, bent NO is considered the two-electron donor NO$^-$; by the neutral ligand model it is considered a one-electron donor.

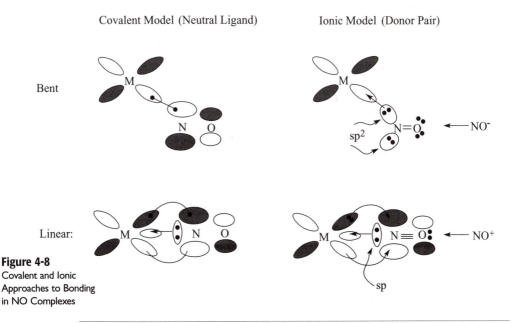

Covalent Model (Neutral Ligand)   Ionic Model (Donor Pair)

Figure 4-8
Covalent and Ionic
Approaches to Bonding
in NO Complexes

These views of bonding can be illustrated as in Figure **4-8**. In the neutral ligand model, which may be described as emphasizing the covalent nature of metal–ligand bonds, the electron configuration of NO is the starting point. NO has similar molecular orbitals to CO, and one more electron, occupying a $\pi *$ orbital. When NO bonds in a bent fashion to a metal, the $\pi*$ orbital interacts with a metal d orbital; if each orbital has a single electron, a covalent bond between the ligand and metal is formed. In the linear coordination mode, the HOMO of NO, occupied by a pair of electrons, can donate this pair to a suitable empty orbital on the metal. In addition, the singly occupied $\pi*$ orbital of NO can participate in covalent bonding with another d orbital on the metal. The result is the same type of synergistic σ donor – π acceptor bonding so common in CO complexes.

In the ionic approach (electron counting method A), NO in the bent coordination mode is considered NO^-. The nitrogen in NO^- is considered to be sp^2 hybridized and can donate an electron pair to a metal d orbital as shown in the figure. In the linear mode, the ligand is considered NO^+, with sp hybridization on nitrogen. An electron pair is donated through this hybrid. The metal (which is formally considered to have one more electron than in the covalent, neutral ligand case) can then donate an electron pair to an empty $\pi*$ orbital on the ligand.

Numerous complexes containing each mode are known, and examples are also known in which both linear and bent NO occur in the same complex. While linear coordination usually gives rise to N-O stretching vibrations at higher energy than in the bent mode, there is enough overlap in the ranges of these bands that infrared spectra alone may not be sufficient to distinguish between the two. Furthermore, the manner of packing in crystals may give rise to considerable bending of the metal-N-O bond from 180° in the "linear" coordination mode.

One compound containing only a metal and NO ligands is known: $Cr(NO)_4$, a tetrahedral molecule isoelectronic with $Ni(CO)_4$. Complexes containing bridging nitrosyl ligands are also known, with the bridging ligand generally considered formally a three-electron donor.

Examples in which NO plays different roles in the same molecule are shown below.

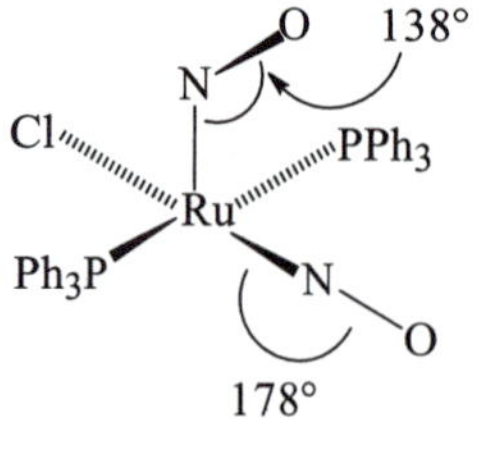

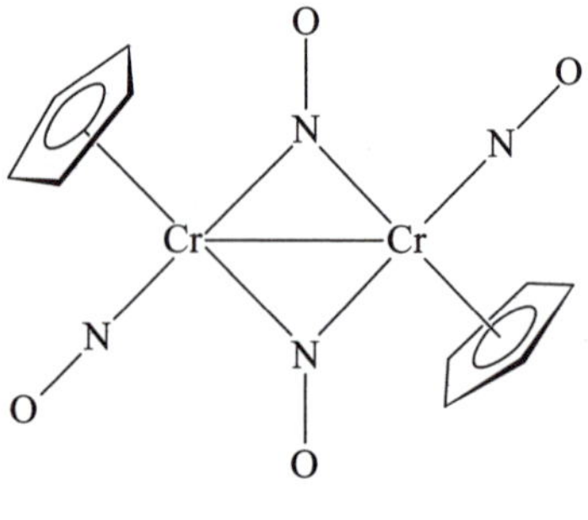

4-4-4 NS Complexes

In recent years several dozen compounds containing the isoelectronic NS (thionitrosyl) ligand have been synthesized. Infrared data have indicated that, like NO, NS can function in linear, bent, and bridging modes. In general, NS is similar to NO in its ability to act as a π acceptor ligand; the relative abilities of NO and NS to accept π electrons depend on the electronic environment of the compounds being compared.[12]

4-5 Infrared Spectra

Infrared (IR) spectra can be useful in two respects. The number of infrared bands depends on molecular symmetry; consequently, by determining the number of such bands for a particular ligand (such as CO), one may be able to decide among several alternative geometries for a compound—or at least reduce the number of possibilities. In addition, the position of the IR band can indicate the function of a ligand (for example, terminal vs. bridging modes) and, in the case of π acceptor ligands, can describe the electron environment of the metal.

4-5-1 Number of IR Bands

Vibrational modes, to be infrared-active, must result in a change in the dipole moment of the molecule. Not all vibrations give rise to a change in dipole moment; these consequently do not appear in the infrared spectrum.

Carbonyl complexes provide convenient examples of vibrations that are visible and invisible in the infrared spectrum. Identical reasoning applies to other linear monodentate ligands (such as CN^- and NO), and to more complex ligands as well. We will begin by considering several simple cases.

Monocarbonyl Complexes

These complexes have a single possible C–O stretching mode. Consequently, they show a single band in the IR.

Dicarbonyl Complexes

Two geometries, linear and bent, must be considered.

$$O\!-\!C\!-\!M\!-\!C\!-\!O$$

[12]H. W. Roesky and K. K. Pandey, *Adv. Inorg. Chem. Radiochem.* **1983**, *26*, 337.

In the case of two CO ligands arranged linearly, only an antisymmetric vibration of the ligands is IR-active; a symmetric vibrational mode results in no change in dipole moment and hence is inactive. If two COs are oriented in a nonlinear fashion, however, both symmetric and antisymmetric vibrations result in changes in dipole moment, and both are IR-active. These modes are shown below:

Symmetric Stretch:

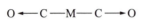

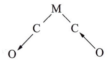

No change in dipole moment:
IR-inactive

Change in dipole moment;
IR-active

Antisymmetric Stretch:

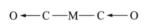

Change in dipole moment;
IR-active

Change in dipole moment;
IR-active

Therefore, an infrared spectrum can be a convenient tool for determining structure for molecules known to have exactly two CO ligands: a single band indicates linear orientation of the carbonyls; two bands, nonlinear orientation.

Complexes Containing Three or More Carbonyls

The predictions in these cases are not quite so simple. The exact number of carbonyl bands can be determined by the methods of group theory.[13] For convenient reference, the numbers of bands expected for a variety of CO complexes are given in Table **4-2**.

Several additional points relating to the number of infrared bands are worth noting. First, while one can predict the number of infrared-active bands by using Table **4-2**, fewer bands may sometimes be observed. In some cases bands may overlap to such a degree that they become indistinguishable; alternatively, one or more bands may have very low intensity and be difficult to observe. In some cases, isomers may be present in the same sample, and it may

[13]G. L. Miessler and D. A. Tarr, *Inorganic Chemistry*, Prentice Hall, New York, **1991**, 110-116.

Table 4-2 Carbonyl Stretching Bands[a]

Number of Carbonyls	Coordination Number		
	4	5	6
3			
IR bands:	2	1	2
IR bands:		3	3
IR bands:		3	
4			
IR bands:	1	4	1
IR bands:		3	4
5			
IR bands:		2	3
6			
IR bands:			1

[a]Adapted with permission from G. L. Miessler and D. A. Tarr, Inorganic Chemistry, Prentice-Hall, Englewood Cliffs, NJ, 1991, 457.

be difficult to sort out which IR absorptions belong to which compound. In carbonyl complexes the number of C–O stretching bands cannot exceed the number of CO ligands. The alternative is possible in some cases (more CO groups than IR bands), when vibrational modes are not IR active (do not cause a change in dipole moment). Examples can be found in Table **4–2**. The highly symmetric tetrahedral [M(CO)$_4$] and octahedral [M(CO)$_6$] complexes have a single carbonyl band in the infrared spectrum.

4-5-2   Positions of Infrared Bands

In this chapter we have already encountered two examples in which the position of the carbonyl stretching band provides useful information. In the case of the isoelectronic species —[V(CO)$_6$]$^-$, Cr(CO)$_6$, and [Mn(CO)$_6$]$^+$— an increase in negative charge on the complex causes a significant reduction in the energy of the C–O band as a consequence of additional π back bonding from the metal to the ligands. The bonding mode is also reflected in the infrared spectrum, with energy decreasing in the order as follows:

terminal CO > doubly bridging CO > triply bridging CO

The positions of infrared bands are also a function of other ligands present, as shown in the following examples:

Complex	v(CO), cm^{-1}
fac-Mo(CO)$_3$(PF$_3$)$_3$	2074, 2026
fac-Mo(CO)$_3$(PCl$_3$)$_3$	2041, 1989
fac-Mo(CO)$_3$(PPh$_3$)$_3$	1937, 1841

Moving down this series, the σ donor ability of the phosphine ligands increases, and the π acceptor ability decreases. PF$_3$ is the weakest of the donors (as a consequence of the highly electronegative fluorines) and the strongest of the acceptors. As a result, the molybdenum in Mo(CO)$_3$(PPh$_3$)$_3$ carries the greatest electron density; it is the most able to donate electron density to the $\pi*$ orbitals of the CO ligands. Consequently, the CO ligands in Mo(CO)$_3$(PPh$_3$)$_3$ have the weakest C–O bonds and the lowest energy stretching bands. Many comparable series are known.

The important point is that the position of the carbonyl bands can provide important clues to the electronic environment of the metal. The greater the electron density on the metal (and the greater the negative charge), the greater the back bonding to CO, and the lower the energy of the carbonyl stretching vibrations. Similar correlations between metal environment and infrared spectra can be drawn for a variety of other ligands, both organic and inorganic. NO, for example, has an infrared spectrum that is strongly correlated with environment in a manner similar to that of CO.[14] In combination with information on the number of infrared bands, the positions of such bands for CO and other ligands can therefore be extremely useful in characterizing organometallic compounds.

4-6   Main Group Parallels with Binary Carbonyl Complexes

Chemical similarities occur between main group and transition metal species that are "electronically equivalent," i.e., species that require the same number of electrons to achieve a filled valence configuration.[15] For example, a halogen atom, one electron short of a valence shell octet, may be considered electronically equivalent to $Mn(CO)_5$, a 17-electron species one electron short of an 18-electron configuration. In this section we will briefly discuss some parallels between main group atoms and ions and electronically equivalent binary carbonyl complexes.

Much chemistry of main group and metal carbonyl species can be rationalized from the way in which these species can achieve closed shell (octet or 18-electron) configurations. These methods of achieving more stable configurations will be illustrated for the following electronically equivalent species:

Electrons Short of Filled Shell	Examples of Electronically Equivalent Species	
	Main Group	**Metal Carbonyl**
1	Cl, Br, I	$Mn(CO)_5$, $Co(CO)_4$
2	S	$Fe(CO)_4$, $Os(CO)_4$
3	P	$Co(CO)_3$, $Ir(CO)_3$

Halogen atoms, one electron short of a valence shell octet, exhibit chemical similarities with 17-electron organometallic species; some of the most striking are the parallels between halogen atoms and $Co(CO)_4$, as summarized in Table **4-3**. Both can reach filled shell electron configurations by acquiring an electron or by dimerization to form either a Co-Co or Cl-Cl bond. The neu-

---

[14]However, interpretation of infrared spectra of NO complexes is made more complicated by the possibility of both bent and linear coordination modes of NO.
[15]J.E. Ellis, *J. Chem. Ed.*, **1976**, *53*, 2.

Table 4-3 Parallels Between Cl and Co(CO)$_4$[a]

Characteristic	Examples	Examples
Ion of 1$^-$ charge	Cl$^-$	[Co(CO)$_4$]$^-$
Neutral dimeric species	Cl$_2$	[Co(CO)$_4$]$_2$
Hydrohalic acid	HCl (strong acid in aqueous solution)	HCo(CO)$_4$ (strong acid in aqueous solution)[b]
Formation of inter-halogen compounds	Br$_2$ + Cl$_2$ $\rightleftharpoons$ 2 BrCl	I$_2$ + [Co(CO)$_4$]$_2$ $\longrightarrow$ 2 ICo(CO)$_4$ (unstable)
Formation of heavy metal salts of low solubility in water	AgCl	AgCo(CO)$_4$
Addition to unsaturated species	Cl$_2$ + H$_2$C=CH$_2$ $\longrightarrow$ H$-$C$-$C$-$H (Cl, Cl, H, H)	[Co(CO)$_4$]$_2$ + F$_2$C=CF$_2$ $\longrightarrow$ (CO)$_4$Co$-$C$-$C$-$Co(CO)$_4$ (F, F, F, F)
Disproportionation by Lewis bases	Cl$_2$ + N(CH$_3$)$_3$ $\longrightarrow$ [ClN(CH$_3$)$_3$]Cl	[Co(CO)$_4$]$_2$ + C$_5$H$_{10}$NH $\longrightarrow$ [(CO)$_4$Co(C$_5$H$_{10}$NH)][Co(CO)$_4$] (piperidine)

[a]Adapted with permission from G. L. Miessler and D. A. Tarr, Inorganic Chemistry, Prentice-Hall, Englewood Cliffs, NJ, 1991, 502.
[b]Note, however, HCo(CO)$_4$ is only slightly soluble in H$_2$O.

tral dimers are capable of adding across multiple carbon–carbon bonds and can undergo disproportionation by Lewis bases. Anions of both electronically equivalent species have a 1- charge and can combine with H$^+$ to form acids: both HX (X = Cl, Br, or I) and HCo(CO)$_4$ are strong acids in aqueous solution. Both types of anions form precipitates with heavy metal ions such as Ag$^+$ in aqueous solution. The parallels between seven-electron halogen atoms and 17-electron binary carbonyl species are sufficiently strong to justify applying the label *pseudohalogen* to these carbonyls.[16]

Similarly, six-electron main group species show chemical similarities with 16-electron organometallic species. As for the halogens and 17-electron organometallic complexes, many of these similarities can be accounted for on the basis of ways in which the species can acquire or share electrons to achieve filled shell configurations. Some similarities between sulfur and the electronically equivalent Fe(CO)$_4$ are listed in Table **4-4**.

The concept of electronically equivalent groups can also be extended to five-electron main group elements [Group 15 (VA)] and 15-electron organometallic species. For example, phosphorus and Ir(CO)$_3$ both form tetrahedral tetramers, as shown in Figure **4-9**. The 15-electron Co(CO)$_3$ [which is isoelectronic with Ir(CO)$_3$] can replace one or more phosphorus atoms in the P$_4$ tetrahedron, as also shown in this figure.

[16]The label *pseudohalogen* is also applied to main group species having electronic similarities to halogen atoms. Examples include CN and SCN.

Table 4-4 Parallels Between Sulfur and Fe(CO)$_4$[a]

Characteristic	Examples	Examples
Ion of 2$^-$ charge	S^{2-}	$[Fe(CO)_4]^{2-}$
Neutral compound	S_8	$Fe_2(CO)_9$, $Fe_3(CO)_{12}$
Hydride	H_2S: $pK_1 = 7.24$[b] $pK_2 = 14.92$	$H_2Fe(CO)_4$: $pK_1 = 4.44$[b] $pK_2 = 14$
Phosphine adduct	Ph_3PS	$Ph_3PFe(CO)_4$
Polymeric mercury compound	(structure)	(structure)
Compound with ethylene	(ethylene sulfide structure)	(π complex structure)

[a](Adapted with permission from G. L. Miessler and D. A. Tarr, Inorganic Chemistry, Prentice-Hall, Englewood Cliffs, NJ, 1991, 503.)
[b]pK values in aqueous solution at 25°C.

The parallels between electronically equivalent main group and organometallic species are interesting, and summarize a considerable amount of their chemistry. The limitations of these parallels should, however, also be recognized. For example, main group compounds having expanded octets may not have organometallic analogues; organometallic analogues of such compounds as IF$_7$ and XeF$_4$ are not known. Organometallic complexes of ligands significantly weaker than CO in the spectrochemical series may not follow the 18-electron rule and may consequently behave quite differently than electronically equivalent main group species. In addition, the reaction chemistry of organometallic compounds may be very different than main group chemistry. For example, loss of ligands such as CO is far more common in organometallic chemistry than in main group chemistry. Therefore, as in any scheme based on as simple a framework as electron counting, the concept of electronically equivalent groups (while useful) has its limitations. It serves as valuable background, however, for a potentially more versatile way to seek parallels between main group and organometallic chemistry—the concept of isolobal groups—to be discussed in Chapter 12.

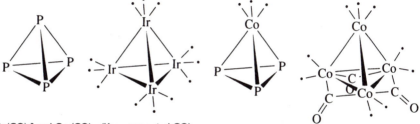

Figure 4-9
P$_4$, [Ir(CO)$_3$]$_4$, P$_3$[Co(CO)$_3$], and Co$_4$(CO)$_{12}$ (Key: · = terminal CO)
(Adapted with permission from G. L. Miessler and D. A. Tarr, *Inorganic Chemistry*, Prentice-Hall, Englewood Cliffs, NJ, 1991, 503.)

Problems

4-1 $V(CO)_6$ and $[V(CO)_6]^-$ are both octahedral. Which has the shorter carbon-oxygen distance? The shorter vanadium–carbon distance?

4-2   Predict the metal-containing product of the following reactions:
   a. $Mo(CO)_6$ + ethylenediamine $\rightarrow$
        [In this reaction two moles of gas are liberated for every mole of $Mo(CO)_6$ reacting.]
   b. $V(CO)_6$ + NO $\rightarrow$

4-3   Select the best choice in each of the following, and briefly justify the reason for your selection.
   a.   Shortest C-O bond: $Ni(CO)_4$ $[Co(CO)_4]^-$ $[Fe(CO)_4]^{2-}$
   b.   Highest C-O stretching frequency: $Ni(CO)_3(PH_3)$ $Ni(CO)_3(PF_3)$ $Ni(CO)_3(PCl_3)$ $Ni(CO)_3(PMe_3)$
   c.   Best π acceptor ligand: NO^- O_2 O_2^- O_2^{2-}

4-4   When heated at low pressure, the compound $(\eta^5\text{-}C_5Me_5)Rh(CO)_2$ reacts to give a gas and another product having a single peak in the ^{1}H-NMR and a single band near 1850 cm^{-1} in the infrared. Suggest a structure for this product.

4-5   The following questions concern the complex shown at right. 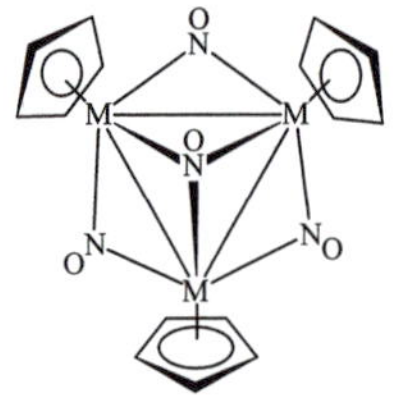
   a.   Identify the first-row transition metal.
   b.   This compound shows infrared absorptions at 1320 and 1495 cm^{-1}. Account for these bands.
   c.   Suppose each C_5H_5 ligand were replaced by C_5Me_5. Would you expect the IR bands described in part b to shift to higher or lower energies? Why?

4-6   The ion $[Ru(Cl)(NO)_2(PPh_3)_2]^+$ has N-O stretching bands at 1687 and 1845 cm^{-1}. The C-O stretching bands of dicarbonyl complexes are typically much closer than this in energy. Why are the N-O band frequencies so much farther apart?

4-7   Account for the observation that only a single carbonyl stretching band is observed for the ion $[Co(CO)_3(PPh_3)_2]^+$.

4-8   Remarkably, not until after more than a century of carbonyl chemistry were stable salts finally isolated[17] containing the cations $[M(CO)_2]^+$ (M = Ag, Au) and $[M(CO)_4]^{2+}$ (M = Pd, Pt). Where would you predict the carbonyl stretching bands in these ions to occur?

4-9   An unusual bonding mode of the carbonyl ligand has recently been reported[18] in which the carbon bridges two metals, while oxygen is bonded to a third. Would you predict that the carbonyl stretching frequency in this bridging mode would be higher or lower than for ordinary doubly bridging CO ligands? Explain briefly.

[17]L. Weber, *Angew. Chem. Int. Ed. Engl.*, **1994**, *33*, 1077.
[18]R. D. Adams, Z. Li, J-C-Lii, and W. Wu, *J. Am. Chem. Soc.*, **1992**, *114*, 4918.

5

Pi Ligands

Among the most distinctive aspects of organometallic chemistry is the ability of a wide variety of ligands containing π electron systems to form bonds to metals. The most common of these π ligands are hydrocarbons, both linear (such as ethylene and butadiene) and cyclic (such as benzene and derivatives of the cyclopropenyl group, C_3H_3). Many π ligands containing "heteroatoms" (such as sulfur, boron, and nitrogen) are also known. In some cases these π bonded ligands are far more stable when attached to metals than when free (some ligands are essentially unknown in the free state). In general, the ability of π ligands to undergo reactions is quite different than when these ligands are unattached to metals. Finally, the structures of many of these metal-π ligand complexes are striking, lending an almost artistic interest to this realm within organometallic chemistry.

In this chapter, we will consider interactions between ligand π systems and metals, beginning with the simplest of the linear systems—ethylene—and then proceeding with more complex linear and cyclic systems. During this discussion, we will pay special attention to the classic example of the compound ferrocene.

5-1 Linear Pi Systems

5-1-1 Pi-Ethylene Complexes

Many complexes involve ethylene (C_2H_4) as a ligand—including the anion of Zeise's salt $[Pt(\eta^2\text{-}C_2H_4)Cl_3]^-$ (Figure **1–3**)—one of the earliest organometallic compounds to be synthesized. In such complexes, ethylene most commonly acts as a *sidebound* ligand with the following geometry with respect to the metal:

Ethylene donates electron density to the metal in a σ fashion, using its π bonding electron pair, as shown in Figure **5–1**. At the same time, electron density can be donated back to the ligand in a π fashion from a metal d orbital to the empty π^* orbital of the ligand. This is another example of the synergistic effect of σ donation and π acceptance encountered earlier with the CO ligand.

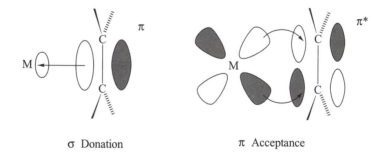

Figure 5-1
Bonding in Ethylene Complexes (Adapted with permission from G.L. Miessler and D.A. Tarr, *Inorganic Chemistry*, Prentice-Hall, Englewood Cliffs, NJ, 1991, 439.)

σ Donation π Acceptance

This picture of bonding is in agreement with measured C–C distances. Free ethylene has a C–C distance of 133.7 pm, while the corresponding distance in the anion of Zeise's salt is 137.5 pm. The lengthening of this bond can be explained by a combination of the two factors involved in the synergistic σ donor–π acceptor nature of the ligand: (a) donation of electron density to the metal in a σ fashion reduces the electron density in a filled π bonding orbital within the ligand, weakening the C–C bond; (b) the back donation of electron density from the metal to the $\pi*$ orbital of the ligand also reduces the C-C bond strength by populating this antibonding orbital. The net effect weakens and hence lengthens the C-C bond in the C_2H_4 ligand.

5-1-2 Pi-Allyl Complexes

The allyl group can function as a trihapto ligand: using delocalized π orbitals as described previously; as a monohapto ligand, primarily σ bonded to a metal; or as a bridging ligand. Examples of these types of coordination are shown in Figure 5-2.

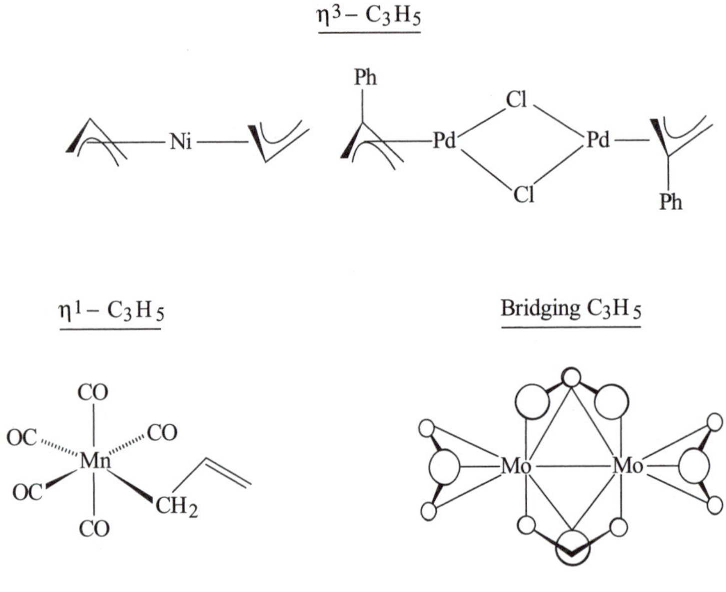

$\eta^3 - C_3H_5$

$\eta^1 - C_3H_5$

Bridging C_3H_5

Figure 5-2
Examples of Allyl Complexes

Bonding between η^3-C_3H_5 and a metal atom is shown schematically in Figure 5-3. The lowest energy π orbital (Figure 2-8) can donate electron density to a suitable orbital on the metal (the bottom interaction shown in Figure 5-3). The next orbital, nonbonding in free allyl, can act as a donor or acceptor, depending on the electron distribution between the metal and the ligand; frequently its primary function is as a filled donor orbital. The empty, highest energy π orbital acts as an acceptor; thus there can be synergistic σ and π interactions between allyl and the metal. In π-allyl complexes the carbon-metal distances reflect the overall environment of the metal. While in the majority of cases, the central carbon is closer to the metal than to the end carbons, in some cases the reverse is true. The C–C–C angle within the ligand is generally near 120°, consistent with sp^2 hybridization on the central carbon.

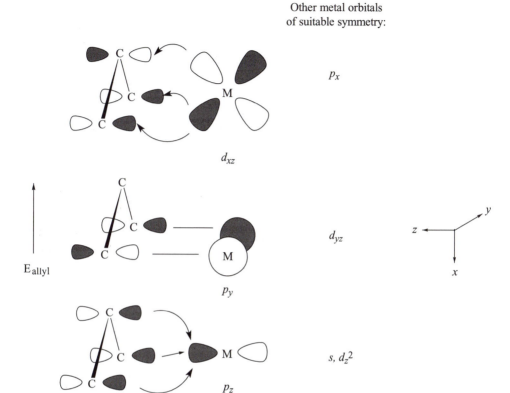

Other metal orbitals
of suitable symmetry:

p_x

d_{xz}

E_{allyl}

d_{yz}

p_y

s, d_{z^2}

p_z

Figure 5-3
Bonding in η^3-Allyl Complexes (Adapted with permission
from G.L. Miessler and D.A. Tarr, *Inorganic Chemistry*,
Prentice-Hall, Englewood Cliffs, NJ, 1991, 441.)

Allyl complexes (or complexes of substituted allyls) are intermediates in
many reactions—some of which take advantage of the capacity of this ligand
to function in both a η^3 and η^1 fashion. Loss of CO from carbonyl complexes
containing η^1-allyl ligands often results in conversion of η^1- to η^3-allyl. For
example:

$$[Mn(CO)_5]^- + C_3H_5Cl \rightarrow (\eta^1\text{-}C_3H_5)Mn(CO)_5 \xrightarrow{\Delta \text{ or } h\nu} (\eta^3\text{-}C_3H_5)Mn(CO)_4$$
$$+ Cl^- \qquad\qquad\qquad + CO$$

(Note that all manganese-containing species in this sequence of reactions are
18-electron species. The mechanism of this reaction is discussed in Chapter 7,
Section **7-2-2**.)

5-1-3   Other Linear Pi Systems

Many other such systems are known. Several examples of organic ligands having longer π systems are shown in Figure **5-4**. Butadiene and longer conjugated π systems have the possibility of isomeric ligand forms (*cis* and *trans* for

cis-Butadiene complex   *trans*-Butadiene complex

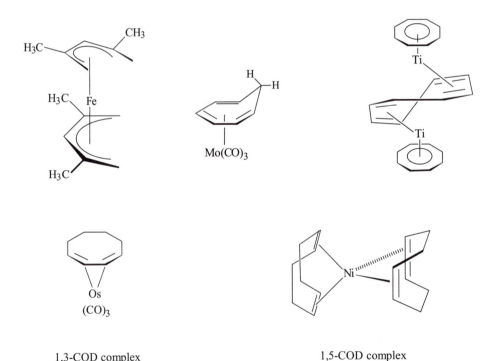

1,3-COD complex   1,5-COD complex

Figure 5-4
Examples of Molecules
Containing Linear π Systems

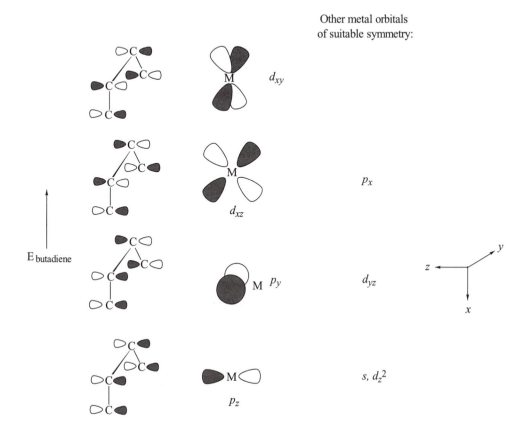

Other metal orbitals
of suitable symmetry:

Figure 5-5
Bonding in *cis*-Butadiene Complexes

butadiene). Larger cyclic ligands may have a π system extending through part of the ring. An example is cyclooctadiene ("COD"): the 1,3–isomer has a four-atom π system comparable to butadiene; 1,5–cyclooctadiene has two isolated double bonds, one or both of which may interact with a metal in a manner similar to ethylene. A schematic diagram outlining the metal-ligand interactions for a *cis*-butadiene complex is shown in Figure **5-5**.

5-2   Cyclic Pi Systems

5-2-1   Cyclopentadienyl (Cp) Complexes

The cyclopentadienyl group, C_5H_5, may bond to metals in a variety of ways, with examples known of the η^1-, η^3-modes, as well as the most common η^5-bonding mode. As described in Chapter 1, the discovery of the first cyclopentadienyl complex, ferrocene, was a landmark in the development of organometallic chemistry and stimulated the search for other compounds containing π–bonded organic ligands. Numerous substituted cyclopentadienyl ligands are also known, such as $C_5(CH_3)_5$ (often abbreviated Cp*) and $C_5(benzyl)_5$.

Ferrocene

Ferrocene is the prototype of a series of sandwich compounds called the **metallocenes**, with the formula $(C_5H_5)_2M$. The bonding in ferrocene can be viewed in a variety of ways. One possibility is to consider it an iron(II) complex with two cyclopentadienide ($C_5H_5^-$) ions, while another is to view it as iron(0) coordinated by two neutral C_5H_5 ligands. The actual bonding situation in ferrocene is much more complicated and requires an analysis of the various metal-ligand interactions in this molecule. As usual, orbitals on the central Fe and on the two C_5H_5 rings interact if they are of appropriate symmetry; furthermore, we expect interactions to be strongest if they are between orbitals of similar energy.

Ferrocene, $(\eta^5–C_5H_5)_2Fe$

Figure **2-11** in Chapter 2 shows sketches of the molecular orbitals of a C_5H_5 ring; two of these rings are arranged in a parallel fashion in ferrocene to "sandwich in" the metal atom. The following discussion is based on the eclipsed conformation of ferrocene, the conformation consistent with gas-phase and low-temperature data on this molecule.[1,2] Descriptions of the bonding in ferrocene based on the staggered geometry are common in the chemical literature, since this was once believed to be the molecule's most stable conformation.[3]

[1]A. Haaland and J.E. Nilsson, *Acta Chem. Scand.*, **1968**, *22*, 2653; see also A. Haaland, *Acc. Chem. Res.*, **1979**, *12*, 415.
[2]P. Seiler and J. Dunitz, *Acta Crystallogr.*, *Sect. B*, **1982**, *38*, 1741.
[3]The $C_5(CH_3)_5$ and $C_5(benzyl)_5$ analogs of ferrocene do have staggered geometries, as do several other metallocenes; see M.D. Rausch, W-M. Tsai, J.W. Chambers, R.D. Rogers, and H.G. Alt, *Organometallics* **1989**, *8*, 816 for some examples.

The group orbitals are derived from the π orbitals of the two C_5H_5 rings by pairing C_5H_5 orbitals of the same energy and same number of nodes, for example, pairing the zero-node orbital of one ring with the zero-node orbital of the other.[4] The molecular orbitals must be paired in such a way that *the nodal planes are coincident*. Furthermore, in each pairing, there are two possible orientations of the ring molecular orbitals: one in which lobes of like sign are pointed toward each other, and one in which lobes of opposite sign are pointed toward each other. For example, the zero-node orbitals of the C_5H_5 rings may be paired in the following ways to generate two of the group orbitals:

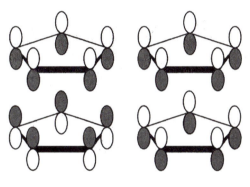

Orbital lobes of like   Orbital lobes of opposite
sign pointed toward   sign pointed toward each
each other   other

The ten group orbitals arising from the C_5H_5 ligands are shown in Figure **5-6**. Note that the two π systems for each group orbital do not interact directly with each other.

The process of developing the molecular orbital picture of ferrocene now becomes one of matching the group orbitals with the s, p, and d orbitals of appropriate symmetry on Fe.

Exercise 5-1

Determine which orbitals on Fe are appropriate for interaction with each of the group orbitals in Figure **5-6**.

---

[4]Not counting the nodal planes that are coplanar with the C_5H_5 rings.

Figure 5-6
Group Orbitals for the
C₅H₅ Ligands of
Ferrocene (Adapted with
permission from G. L.
Miessler and D. A. Tarr,
Inorganic Chemistry,
Prentice-Hall, Englewood
Cliffs, NJ, 1991, 443.)

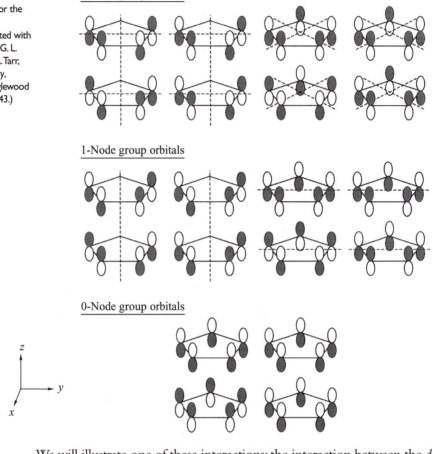

2-Node group orbitals

1-Node group orbitals

0-Node group orbitals

We will illustrate one of these interactions: the interaction between the d_{yz} orbital of Fe and its appropriate group orbital (one of the one-node group orbitals shown in Figure **5-6**). This interaction can occur in a bonding and an antibonding fashion as shown below:

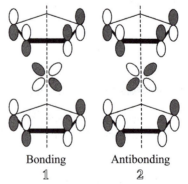

Bonding   Antibonding
1   2

Figure 5-7
Bonding Molecular
Orbital Formed
from d_{yz} Orbital of
Iron in Ferrocene

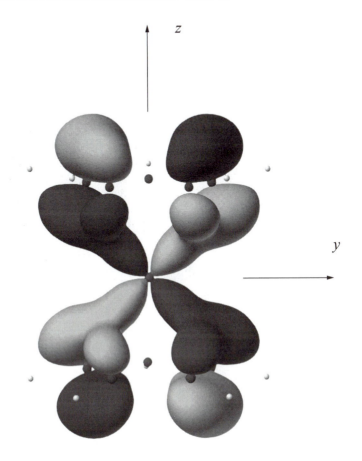

In the bonding orbital the lobes of the d_{yz} orbital of iron merge with the lobes of the group orbital to which they point, giving the striking appearance shown in Figure **5-7**.

The nodal behavior of the interacting metal and group orbitals must be preserved in the molecular orbitals that result. Thus, the molecular orbital in Figure **5-7** must have two nodal planes, the same as those of the original d_{yz} orbital: the xy plane (shown horizontally in the figure) and the xz plane (cutting vertically through the molecule in Figure **5-7**). The latter is shown more clearly in a top view of the molecular orbital in Figure **5-8.**

The complete energy-level diagram for the molecular orbitals of ferrocene is shown in Figure **5-9**. The molecular orbital resulting from the d_{yz} bonding interaction, labeled 1 in the MO diagram, contains a pair of electrons. Its antibonding counterpart, 2, is empty. It is a useful exercise to match the

Figure 5-8
Top View of Ferrocene
Molecular Orbital (d_{yz})

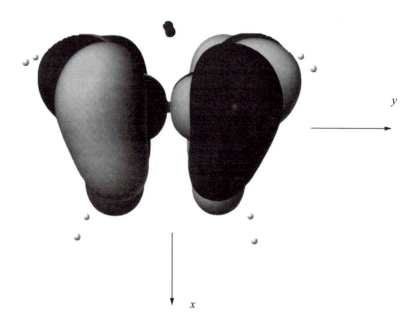

other group orbitals from Figure **5-6** with the molecular orbitals in Figure **5-9** to verify the types of metal–ligand interactions that occur.

The most interesting orbitals of ferrocene are those having the greatest d orbital character; these are highlighted in the box in Figure **5-9**. Two of these orbitals, having largely d_{xy} and $d_{x^2-y^2}$ character, are weakly bonding and are occupied by electron pairs; one, having largely d_{z^2} character, is essentially non-bonding and occupied by one electron pair; and two, having primarily d_{xz} and d_{yz} character, are empty. The relative energies of these orbitals and their d orbital–group orbital interactions are shown in Figure **5-10**.[5,6]

---

[5]The relative energies of the lowest three orbitals shown in Figure **5-10** have been a matter of controversy. UV photoelectron spectroscopy indicates the order as shown, with the orbital having largely d_{z^2} character slightly higher in energy than the degenerate pair having substantial d_{xy} and $d_{x^2-y^2}$ character. This order may be reversed for some metallocenes. (See, A. Haaland, *Accts. Chem. Res.*, **1979**, *12*, 415.)

[6]J. C. Giordan, J. H. Moore, and J. A. Tossell, *Acc. Chem. Res.*, **1986**, *19*, 281; E. Rühl and A. P. Hitchcock, *J. Am. Chem. Soc.*, **1989**, *111*, 5069.

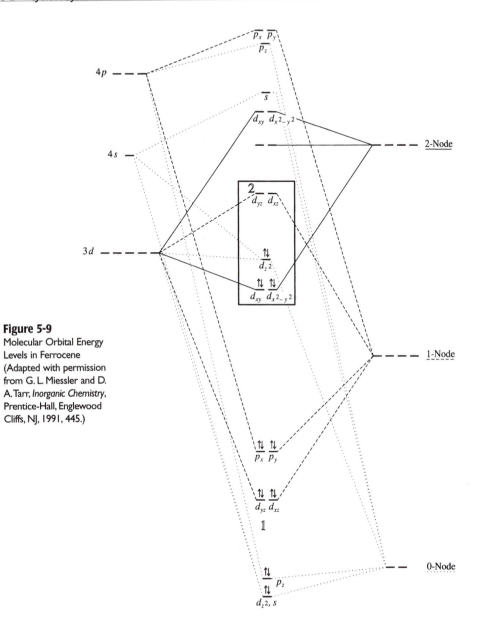

Figure 5-9
Molecular Orbital Energy
Levels in Ferrocene
(Adapted with permission
from G. L. Miessler and D.
A. Tarr, *Inorganic Chemistry*,
Prentice-Hall, Englewood
Cliffs, NJ, 1991, 445.)

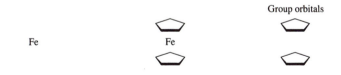

Figure 5-10
Molecular Orbitals of
Ferrocene Greatest *d*
Character (Adapted
with permission from
G. L. Miessler and D. A.
Tarr, *Inorganic Chemistry*,
Prentice-Hall,
Englewood Cliffs, NJ,
1991, 446.)

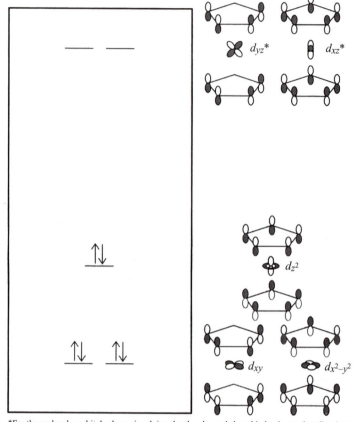

*For the molecular orbitals shown involving the d_{yz}, d_{xz}, and d_{z^2} orbitals, the *anti*bonding interactions are shown; the bonding interactions result in molecular orbitals of lower energy.

The overall bonding in ferrocene can now be summarized. The occupied orbitals of the cyclopentadienyl ligands—the zero-node and one-node group orbitals—are stabilized by their interactions with iron. In addition, six electrons occupy molecular orbitals that are largely derived from iron *d* orbitals (as one would expect for d^6 iron(II)), but these occupied orbitals, shown in Figure **5-10**, have significant ligand character too. The molecular orbital picture in this case is consistent with the 18-electron rule; however, as will be evident in other metallocenes, the cyclopentadienyl ligand is not as effective as other ligands, notably CO, at favoring the 18-electron arrangement.

Other Metallocenes

Other metallocenes have similar structures but do not necessarily obey the 18-electron rule. For example, cobaltocene, $(\eta^5\text{-}C_5H_5)_2Co$, and nickelocene, $(\eta^5\text{-}C_5H_5)_2Ni$, are structurally similar 19- and 20-electron species. The "extra" electrons have important chemical and physical consequences, as seen in comparative data:[7]

Complex	Color	Electron Count	M–C Distance (pm)	ΔH for M^{2+}–$C_5H_5^-$ Dissociation (kcal/mol)
$(\eta^5\text{-}C_5H_5)_2Fe$	orange	18	206.4	351
$(\eta^5\text{-}C_5H_5)_2Co$	purple	19	211.9	335
$(\eta^5\text{-}C_5H_5)_2Ni$	green	20	219.6	315

The nineteenth and twentieth electrons of the metallocenes occupy the slightly antibonding orbitals. As a consequence, the metal–ligand distance increases (the ligands in cobaltocene and nickelocene are less tightly held to the metal), and ΔH for metal–ligand dissociation decreases.

Ferrocene itself shows much more chemical stability than cobaltocene and nickelocene; many of the chemical reactions of the latter are characterized by a tendency to yield 18-electron products. For example, ferrocene is unreactive toward iodine and rarely participates in reactions in which other ligands substitute for the cyclopentadienyl ligand. However, cobaltocene and nickelocene undergo the following reactions to give 18-electron products:

$$2\ (\eta^5\text{-}C_5H_5)_2Co + I_2 \rightarrow 2\ [(\eta^5\text{-}C_5H_5)_2\ Co]^+ + 2\ I^-$$

$$\underset{19\ e^-}{} \qquad \underset{\substack{18\ e^- \\ \text{cobalticinium} \\ \text{ion}}}{}$$

$$(\eta^5\text{-}C_5H_5)_2Ni + 4\ PF_3 \rightarrow Ni(PF_3)_4 + \text{organic products}$$

$$\underset{20\ e^-}{} \qquad \underset{18\ e^-}{}$$

Interestingly, cobalticinium reacts with hydride to give a neutral, 18-electron sandwich compound in which one cyclopentadienyl ligand has been modified into $\eta^4\text{-}C_5H_6$, as shown in Figure **5–11**.

Ferrocene, however, is by no means chemically inert. It undergoes a variety of reactions, including many on the cyclopentadienyl rings. A good example is that of electrophilic acyl substitution (Figure **5–12**), a reaction paralleling Friedel-Crafts acylation of benzene and its derivatives. In general, electrophilic aromatic substitution reactions are much more rapid for ferrocene than for benzene, an indication of greater concentration of electron density in the rings of the sandwich compound than in benzene.

[7]A. Haaland, *op. cit., Acc. Chem. Res.*

Figure 5-11
Reaction of
Cobalticinium with
Hydride (Adapted
with permission from
G.L. Miessler and D.A.
Tarr, *Inorganic
Chemistry*, Prentice-
Hall, Englewood Cliffs,
NJ, 1991, 447).

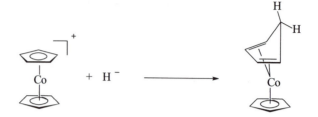

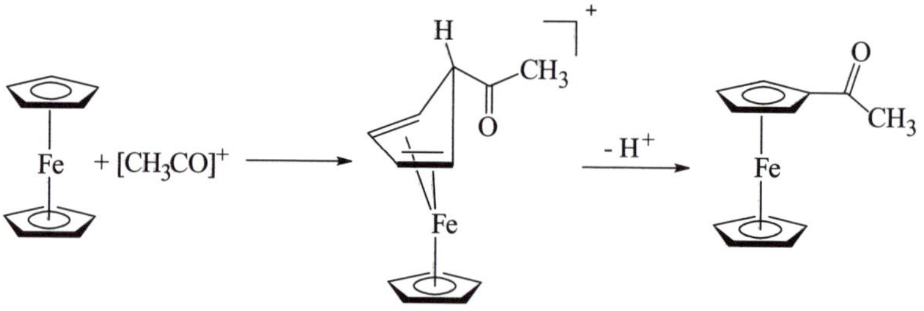

Figure 5-12
Electrophilic Substitution
with Acylium Ion on
Ferrocene (Adapted
with permission from
G.L. Miessler and D.A.
Tarr, *Inorganic Chemistry*,
Prentice-Hall, Englewood
Cliffs, NJ, 1991, 447.)

5-2-2   Complexes Containing Cyclopentadienyl and CO Ligands

Not surprisingly, many complexes are known that contain both Cp and CO ligands. These include "half–sandwich" compounds such as (η^5-C_5H_5)Mn(CO)$_3$ and dimeric and larger cluster molecules. Examples are shown in Figure **5-13**. As for the binary CO complexes, those of the second- and third-row transition metals show a decreasing tendency of CO to act as a bridging ligand.

5-2-3   Other Cyclic Pi Ligands

Many other cyclic π ligands are known. The most common cyclic hydrocarbon ligands are listed in Table **5-1**. Depending on the ligand and the electron requirements of the metal (or metals), these ligands may be capable of bonding in a monohapto or polyhapto fashion, and they may bridge two or more metals. The different sized rings have unique features and are worth a brief survey.

Cyclo-C_3R_3

The number of cyclopropenyl complexes has remained relatively small; principally these are of the phenyl derivative C_3Ph_3.[8] The difficulty in synthesizing these complexes has paralleled the difficulty in preparing the parent aromatic

[8]A. Efraty, *Chem. Rev.*, **1977**, 77, 691.

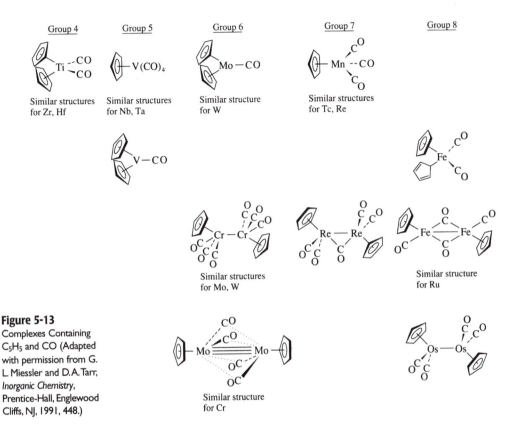

Figure 5-13
Complexes Containing
C_5H_5 and CO (Adapted
with permission from G.
L. Miessler and D. A. Tarr,
Inorganic Chemistry,
Prentice-Hall, Englewood
Cliffs, NJ, 1991, 448.)

ions, since they are small, highly strained ring systems. The first complex containing a η^3-C_3Ph_3 ligand was the dimeric nickel complex $[(\eta^3$-$C_3Ph_3)$ $Ni(CO)Br]_2$, prepared by the following reaction:[9]

$$2\ C_3Ph_3Br\ +\ 2\ Ni(CO)_4\ \longrightarrow\ \text{[complex]}\ +\ 6\ CO$$

Sandwich compounds involving η^3-C_3Ph_3 and larger rings are also known; one example is shown at right.

[9] S. F. A. Kettle, *Inorg. Chem.*, **1964**, *3*, 604.

Table 5-1 Cyclic Pi Ligands

Formula	Structure	Name	Electron Count	
			Method A[a]	Method B
C_3H_3		Cyclopropenyl R=alkyl, phenyl	$2\,(C_3R_3^{+})$	3
C_4H_4		Cyclobutadiene	$6\,(C_4H_4^{2-})$	4
C_5H_5		Cyclopentadienyl (Cp)	$6\,(C_5H_5^{-})$	5
C_6H_6		Benzene	6	6
C_7H_7		Tropylium	$6\,(C_7H_7^{+})$	7
C_8H_8		Cyclooctatetraene (COT)	$10\,(C_8H_8^{2-})$	8

[a] Method A counts these cyclic pi systems by assigning them the number of pi electrons (2, 6, or 10) they would have as aromatic rings.

Because of the small size of the π system in *cyclo*-C_3R_3, overlap with metal d orbitals is unfavorable for second- and third-row transition metals; consequently, only the first-row transition metals are known to form η^3-C_3R_3 complexes.[10]

Cyclo-C₄H₄

Cyclobutadiene transition metal complexes were actually synthesized long before cyclobutadiene itself.[11] Interestingly, one of these complexes was also a halogen-bridged dimeric nickel complex as shown below:

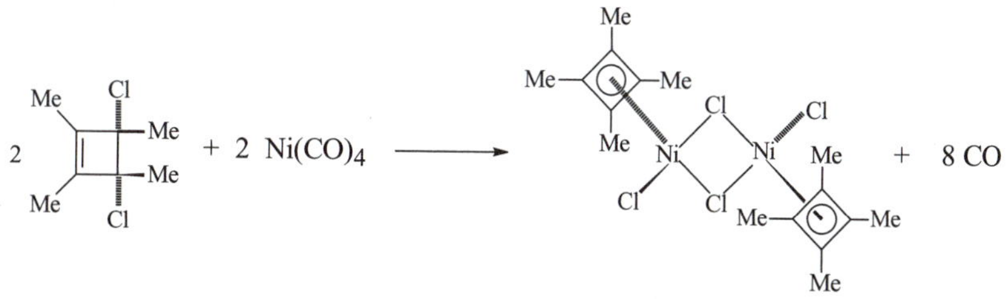

The η^4-C_4H_4 ligand is ordinarily square;[12] its bonding to metals may be viewed in a manner similar to that observed in ferrocene, with the principal interaction between the π orbitals of the ring and d orbitals of the metal.

Exercise 5-2

Sketch the group orbitals on the ligands[13] for the sandwich compound $(\eta^4$-$C_4Ph_4)_2Ni$. Identify the metal s, p, and d orbitals suitable for interaction with each of the group orbitals. (Assume that the rings are parallel and eclipsed.)

One of the more intriguing ways to synthesize η^4-C_4H_4 complexes is by coupling two acetylenes via a "template" reaction such as shown below.[14]

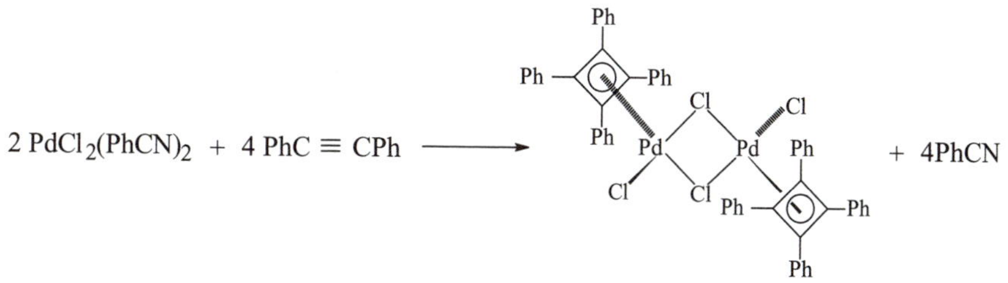

---

[10]However, second- and third-row transition metals can form unsymmetrical complexes with C_3R_3 in which the metal-carbon distances differ significantly.

[11]R. Criegee, *Angew. Chem.*, **1959**, *71*, 70.

[12]This is in contrast to the rectangular shape of free cyclobutadiene.

[13]Reference to Figure **2-11** in Chapter 2 may be helpful.

[14] P. M. Maitlis, *Acc. Chem. Res.*, **1976**, *9*, 93.

Like ferrocene, η^4-C_4H_4 complexes readily undergo electrophilic substitution at the ring; this is undoubtedly a consequence of significant electron donation from the metal to the ring. A particularly useful feature of η^4-C_4H_4 complexes is that they undergo decomposition reactions that can be used as a source of free cyclobutadiene for organic synthesis as shown below:[15]

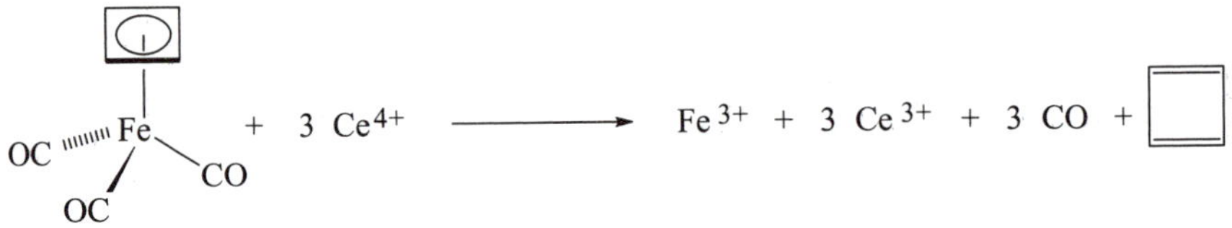

$$\text{OC} \cdots \text{Fe}(\text{CO})_2 \cdots \text{OC} \quad + \quad 3\ Ce^{4+} \quad \longrightarrow \quad Fe^{3+} \ + \ 3\ Ce^{3+} \ + \ 3\ CO \ + \ \square$$

Cyclo-C_5H_5

Cyclopentadienyl complexes, as previously discussed, are exceptionally numerous and have been the subject of extensive study. Several routes are available for introducing this ligand into a metal complex. One approach is to react a metal compound with the cyclopentadienylide ion, $C_5H_5^-$. This ion can be purchased as the sodium salt in solution; it can also be prepared by the two-step process shown below:

1. Cracking of dicyclopentadiene (Diels–Alder dimer of cyclopentadiene):

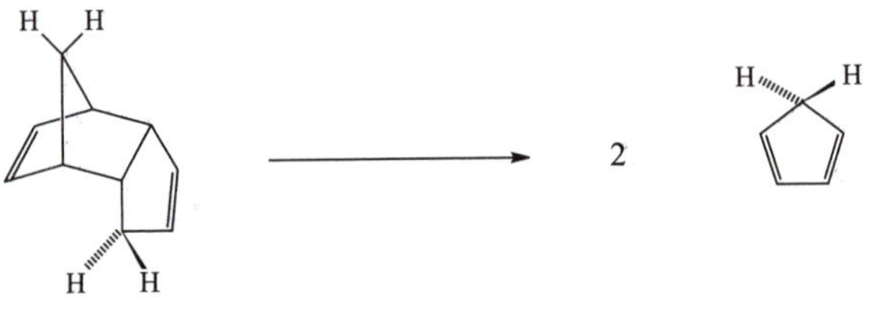

$C_{10}H_{12}$

2. Reduction of cyclopentadiene by sodium:

$$2\ C_5H_6 + 2\ Na \rightarrow 2\ NaC_5H_5 + H_2$$

The sodium salt can then react with the appropriate metal complex or ion:

$$Fe^{2+} + 2\ C_5H_5^- \rightarrow Fe(C_5H_5)_2$$

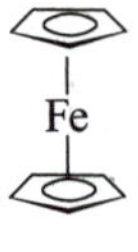

[15] J.C. Barborak, L. Watts, and R. Pettit, *J. Am. Chem. Soc.*, **1966**, *88*, 1328.

A second method is by direct reaction of a metal or metal complex with the cyclopentadiene monomer. The following is an example:

$$Mo(CO)_6 + 2\ C_5H_6 \rightarrow [(C_5H_5)Mo(CO)_3]_2 + H_2$$

Many examples of cyclopentadienyl complexes will be found later in this text.

Cyclo-C_6H_6 (Benzene)

Benzene and its derivatives are among the better known of the many η^6-arene complexes. The best known of these is diben-zenechromium, $(C_6H_6)_2Cr$. Like other compounds of formula $(C_6H_6)_2M$, dibenzenechromium can be prepared by the Fischer-Hafner synthesis, using a transition metal halide, aluminum as a reducing agent, and the Lewis acid $AlCl_3$:

$$3\ CrCl_3 + 2\ Al + AlCl_3 + 6\ C_6H_6 \rightarrow 3\ [(C_6H_6)_2Cr]^+ + 3\ AlCl_4^-$$

The cation is then reduced to the neutral $(C_6H_6)_2Cr$ using a reducing agent, most commonly dithionite, $S_2O_4^{2-}$.

Dibenzenechromium, like ferrocene, exhibits eclipsed rings in its most stable conformation. The metal-ligand bonding in this compound may be interpreted using the group orbital approach applied to ferrocene earlier in this chapter.

Exercise 5-3

Show how the d_{xy} orbital of chromium can interact with a two-node group orbital derived from the benzene rings (see Figure **2-12**) to form bonding and antibonding molecular orbitals. Sketch the shape of the bonding molecular orbital that would result from this interaction.

Dibenzenechromium is less thermally stable than ferrocene. Furthermore, unlike ferrocene, dibenzenechromium is not subject to electrophilic aromatic substitution; the electrophile oxidizes the chromium(0) to chromium(I) instead of attacking the rings.

In some cases cyclization reactions occur in which three 2-membered π systems fuse into a η^6 six-membered ring. An example of this very interesting phenomenon is shown below;[16] this type of mechanism will be discussed in Chapter 11.

[16]K. H. Dötz, *Angew. Chem. Int. Ed. Engl.*, **1984**, *23*, 587.

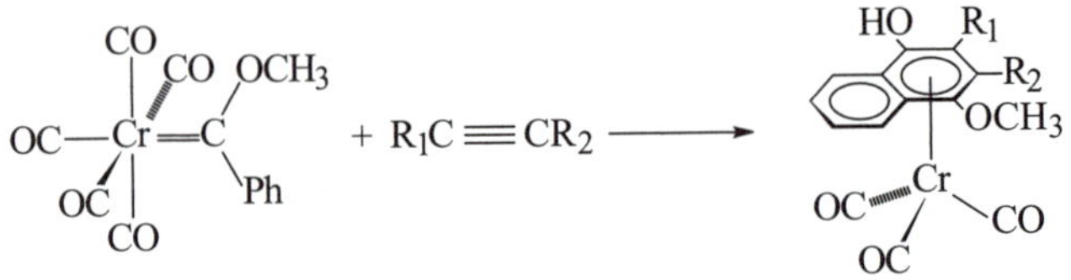

Chromium and other metals also form 18-electron complexes containing linked and fused hexahapto six-membered rings with a variety of geometries; examples are shown in Figure **5-14**. Many sandwich complexes containing $\eta^6\text{-}C_6H_6$ and rings of other hapticity [for example, $(\eta^6\text{-}C_6H_6)(\eta^5\text{-}C_5H_5)Mn]$ and half–sandwich complexes containing $\eta^6\text{-}C_6H_6$ and carbonyl ligands are also known.

Figure 5-14
Examples of
Chromium
Complexes Containing
Linked and Fused
Six-Membered Rings

Exercise 5-4

Predict the charges of the following 18-electron complexes containing $\eta^6\text{-}C_6H_6$:

a. $[(C_6H_6)_2Ru]^z$ b. $[(C_6H_6)V(CO)_4]^z$ c. $[(C_6H_6)(\eta^5\text{-}C_5H_5)Co]^z$

Cyclo-C_7H_7 (Tropylium[17])
Hexahapto cycloheptatriene complexes $(\eta^6\text{-}C_7H_8)$ can be treated with $Ph_3C^+BF_4^-$ to abstract a hydrogen and yield the $\eta^7\text{-}C_7H_7$ ligand. A well-studied example is the synthesis of $[(\eta^7\text{-}C_7H_7)Mo(CO)_3]^+$ as shown below:

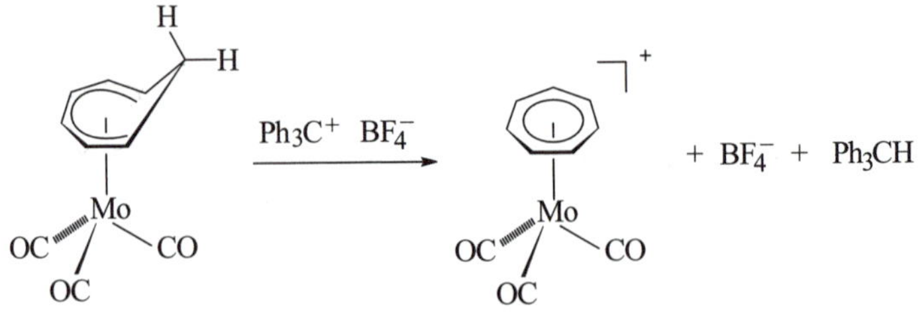

A cation is produced in this reaction, making the ring susceptible to nucleophilic addition by methoxide as shown on the next page:

[17]"Tropylium" is the name of the aromatic cation, $C_7H_7^+$.

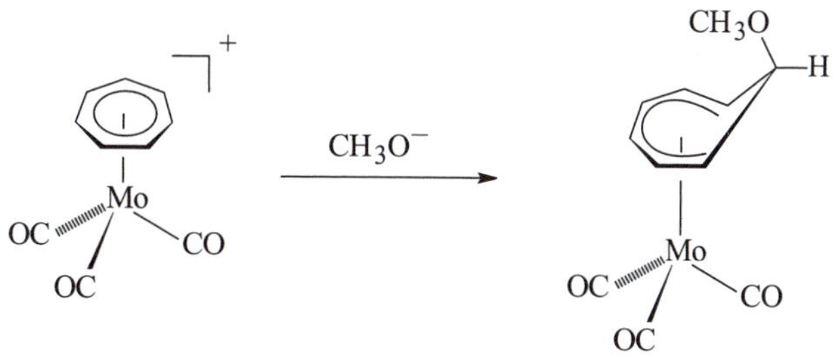

In some cases the seven–membered ring is subject to contraction, as in the example[18] below:

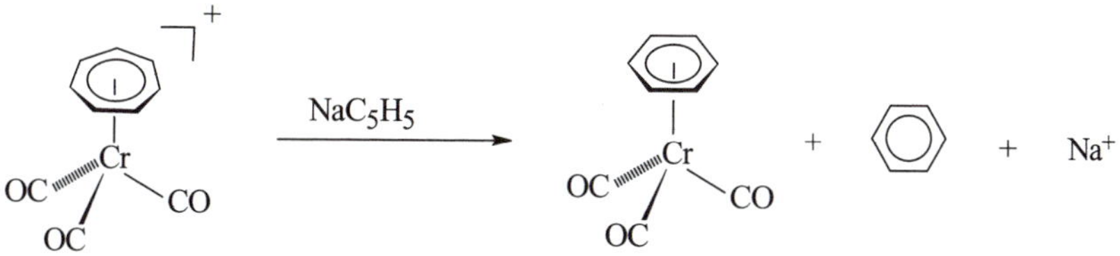

Few complexes containing a metal sandwiched between two η^7-C_7H_7 rings are known. One example is $[(\eta^7$-$C_7H_7)_2V]^{2+}$. Several sandwich complexes have, however, been synthesized containing η^7-C_7H_7 in combination with rings of other size.

Exercise 5-5

Identify the first-row transition metal in the following 18–electron sandwich complexes:

a. $(\eta^7$-$C_7H_7)(\eta^5$-$C_5H_5)M$ b. $[(\eta^7$-$C_7H_6CH_3)(\eta^5$-$C_5H_5)M]^+$

Cyclo-C_8H_8 (Cyclooctatetraene, COT)

Cyclooctatetraene has the most diverse array of bonding modes of any of the cyclic hydrocarbons of formula C_nH_n. In addition to the η^8- mode of principal interest in this chapter (formally involving $C_8H_8^{2-}$), this ligand can also function in η^2-, η^4-, η^6- modes and sometimes modes of odd hapticity; in addition, it can form a variety of bridges between metals. Examples of bonding modes of cyclo-C_8H_8 are given in Figure **5–15**.

---

[18] J. D. Munro and P.L. Pauson, J. *Chem. Soc., Dalton Trans.,* **1961**, 3479.

Figure 5-15
Examples of
Bonding Modes of
Cyclooctatetraene

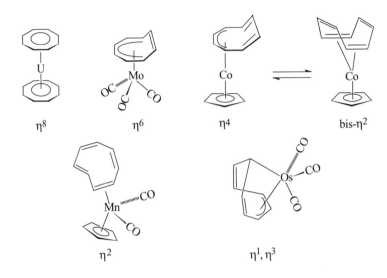

η^8 η^6 η^4 bis-η^2

η^2 η^1, η^3

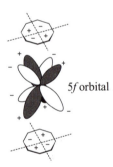

5*f* orbital

Although examples of molecules containing planar C_8H_8 ligands are known for some *d* block transition metals, the most
interesting complexes of this ligand are with the actinide elements. These elements have 5*f* orbitals suitable for interaction
with some π orbitals on the ligand rings. Among the most
important interactions in these complexes are the interactions
between *f* orbitals and two-node group orbitals, as shown at left;
in addition, weaker interactions may occur between other *f*
orbitals and three-node group orbitals.

(Adapted from A. Streitwieser, Jr., U. Muller-Westerhoff,
G. Sonnichsen, F. Mares, D.G. Morrell, K.O. Hodgson, and
C.A. Harmon, *J. Am. Chem. Soc.,* 1973, *95,* 8644. Copyright
1973 American Chemical Society.)

Exercise 5-6

Sketch the π orbitals for cyclooctatetraenide, $C_8H_8^{2-}$, and indicate the
positions of the nodal planes cutting through the ring.

The most famous of the sandwich compounds of cyclic C_8H_8 is uranocene,
(η^8-C_8H_8)$_2$U. This air-sensitive compound is actually a 22-electron species—
clearly the 18-electron rule does not apply in the realm of the actinides. A
schematic diagram of selected orbital interactions in uranocene is given in
Figure **5-16**.

Figure 5-16
Bonding in Uranocene

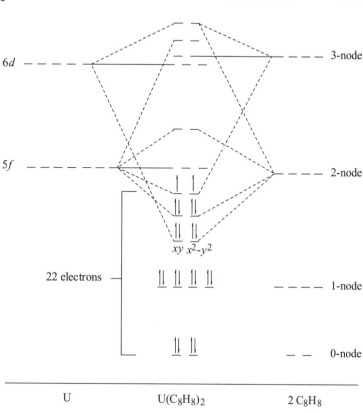

Similar metallocenes are also known for other actinide elements, although their study is inhibited somewhat by the radioactivity of these elements.

Particularly interesting are the cases in which cyclic ligands can bridge metals resulting in "triple-decker" and higher order sandwich compounds (Figure **5-17**).

Figure 5-17
Multiple-Decker Sandwich Compounds (Adapted with permission from G. L. Miessler and D. A. Tarr, *Inorganic Chemistry*, Prentice-Hall, Englewood Cliffs, NJ, 1991, 4, 412.)

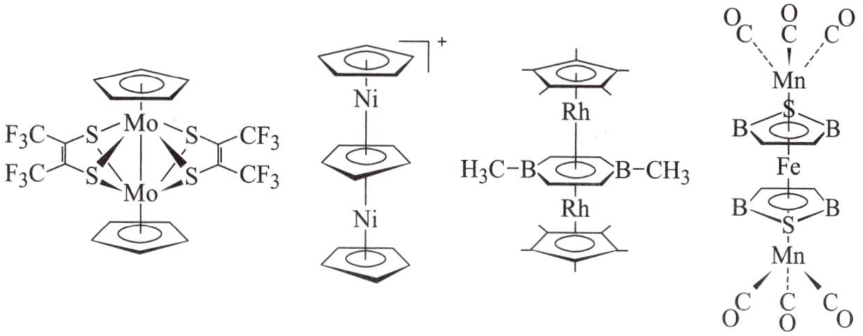

5-3   NMR Spectra of Organometallic Compounds

Nuclear magnetic resonance (NMR) is an extremely valuable tool in characterizing organometallic complexes. The advent of high-field NMR instruments using superconducting magnets has, in many ways, revolutionized the study of these compounds. Convenient NMR spectra can now be taken using several metal nuclei as well as the more traditional nuclei such as ^{1}H, ^{13}C, ^{19}F, and ^{31}P; the combined spectral data of several nuclei make it possible to identify many compounds by their NMR spectra alone.

As in organic chemistry, chemical shifts, splitting patterns, and coupling constants are useful in characterizing the environments of individual atoms in organometallic compounds. The reader may find it useful to review the basic theory of NMR as presented in an organic chemistry text. More advanced discussions of NMR, especially those relating to ^{13}C, have been surveyed elsewhere.[19]

5-3-1 ^{13}C NMR

Carbon-13 NMR has become increasingly useful with the advent of modern instrumentation. Although the isotope ^{13}C has a low natural abundance (approximately 1.1%) and low sensitivity for the NMR experiment (about 1.6% the sensitivity of ^{1}H), Fourier transform techniques now make it possible to obtain useful ^{13}C spectra for a wide range of organometallic complexes. Nevertheless, the time necessary to obtain a ^{13}C spectrum may still be an experimental difficulty for compounds present in very small amounts or those having low solubility. Rapid reactions may also be inaccessible by this technique. Some useful features of ^{13}C spectra include the following:

- An opportunity to observe organic ligands that do not contain hydrogen (such as CO and CF_3).
- Direct observation of the carbon skeleton of organic ligands. This possibility is enhanced when a spectrum is acquired with complete proton decoupling, since decoupled spectra show singlets for atoms in each environment.
- ^{13}C chemical shifts are much more widely dispersed than ^{1}H shifts. Thus proton decoupled ^{13}C spectra may give well-separated singlets and allow themselves more straightforward analysis than proton spectra. Proton spectra are much more likely to have overlapping peaks. The consequence is that more complex structures may be elucidated by ^{13}C than by ^{1}H spectra. In addition, the wide dispersion of chemical shifts often

---

[19]B.E. Mann, *Advances in Organometallic Chemistry*, **1974**, *12*, 135; P.W. Jolly and R. Mynott, *Advances in Organometallic Chemistry*, **1981**, *19*, 257; E. Breitmaier and W. Voelter, *Carbon 13 NMR Spectroscopy*, VCH, New York, 1987.

Table 5-2 [13]C Chemical Shifts for Organometallic Compounds

Ligand	[13]C Chemical Shift Range[a]	
M–CH_3	-28.9 to +23.5	
M=CR_2	190 to 400	
M≡CR	235 to 401	
M–CO	177 to 275	
in neutral binary CO complexes	183 to 223	
M–(η^5-C_5H_5)	-790 to +1430	
Fe(η^5-C_5H_5)$_2$	69.2	
M–(η^3-C_3H_5)	C_2: 91 to 129	C_1 and C_3: 46 to 79
M–C_6H_5 M–C: 130 to 193 *ortho*: 132 to 141		
meta: 127 to 130 *para*: 121 to 131		

[a]Parts per million relative to $Si(CH_3)_4$

makes it easy to distinguish different ligands in compounds containing several different types of organic ligands.

- ^{13}C NMR is also a valuable tool for observing rapid intramolecular rearrangement processes.[20] Because ^{13}C peaks are typically more widely dispersed than ^{1}H peaks, faster exchange processes can be observed by ^{13}C NMR.

Approximate ranges of chemical shifts for ^{13}C spectra of some categories of organometallic complexes are listed in Table **5-2**.[21] Several features of these data deserve comment. The wide ranges of chemical shifts should be noted, a reflection of the dramatic effect that the molecular environment can have. In addition, the following aspects of ^{13}C spectra should be noted:

- Terminal carbonyl peaks are frequently in the range δ 195 to 225 ppm, a range sufficiently distinctive that the CO resonances are usually easy to distinguish from those of other ligands.
- One factor correlated with the ^{13}C chemical shift is the strength of the C–O bond; in general the stronger the bond, the lower the chemical shift.[22]
- Bridging carbonyls have slightly greater chemical shifts than terminal carbonyls and consequently may lend themselves to easy identification (however, infrared spectra are usually better than NMR for distinguishing between bridging and terminal carbonyls).
- Cyclopentadienyl ligands have a wide range of chemical shifts, with the value for ferrocene nearer the low end for such values. Other organic ligands may also have wide ranges in ^{13}C chemical shifts.

[20]E. Breitmaier and W. Voelter, *Carbon 13 NMR Spectroscopy*, VCH, New York, 1987.
[21]Mann, *op. cit.*, has extensive tables of chemical shifts for organic ligands.
[22]P.C. Lauterbur and R.B. King, *J. Am. Chem. Soc.*, **1965**, *87*, 3266.

Table 5-3 Examples of ¹H Chemical Shifts for Organometallic Compounds

Complex	¹H Chemical Shift[a]
$Mn(CO)_5$**H**	-7.5
$W(CH_3)_6$	1.80
$Ni(\eta^2\text{-}C_2H_4)_3$	3.06
$(\eta^5\text{-}C_5H_5)_2Fe$	4.04
$(\eta^6\text{-}C_6H_6)_2Cr$	4.12
$(\eta^5\text{-}C_5H_5)_2Ta(CH_3)(=CH_2)$	10.22

[a]Parts per million relative to $Si(CH_3)_4$.

5-3-2   Proton NMR

The ¹H spectra of organometallic compounds containing hydrogens can also provide useful structural information.[23] For example, protons bonded directly to metals (hydride complexes, to be discussed in Chapter 6) are very strongly shielded, with chemical shifts commonly in the approximate range -5 to -20 ppm relative to $Si(CH_3)_4$. Such protons are easy to detect, since few other protons typically appear in this region.

Protons in methyl complexes (M–CH_3, also discussed in Chapter 6) typically have chemical shifts between 1 and 4 ppm, similar to their positions in organic molecules. Cyclic π ligands, such as η^5-C_5H_5 and η^6-C_6H_6, most commonly have ¹H chemical shifts between 4 and 7 ppm and, because of the relatively large number of protons involved, may lend themselves to easy identification. Protons in other types of organic ligands also have characteristic chemical shifts; examples are given in Table **5–3**.

As in organic chemistry, integration of the various peaks in organometallic compounds can provide the ratio of atoms in different environments; it is usually accurate, for example, to assume that the area of a ¹H peak (or set of peaks) is proportional to the number of nuclei giving rise to that peak. However, for ¹³C this approach is somewhat less reliable. For instance, relaxation times of different carbon atoms in organometallic complexes vary widely, and this may lead to inaccuracy in the correlation of peak area with number of atoms (the correlation between area and number of atoms is dependent on rapid relaxation).[24] Addition of paramagnetic reagents may speed up relaxation

[23]The ranges mentioned are for diamagnetic complexes. Paramagnetic complexes may have much larger chemical shifts, sometimes several hundred parts per million relative to tetramethylsilane.
[24] For more information on the problems associated with integration in ¹³C-NMR, see J. K. M. Sanders and B. K. Hunter, *Modern NMR Spectroscopy*, Saunders, New York, 1992.

and thereby improve the validity of integration data; one compound often used is $Cr(acac)_3$ [acac = acetylacetonate = $H_3CC(O)CH_2C(O)CH_3^-$].

5-3-3   Molecular Rearrangement Processes

Under certain conditions, NMR spectra may change significantly as the temperature is changed. As an example of this phenomenon, consider a molecule having two hydrogens in nonequivalent environments, as shown at right. In this example, the carbon to which the hydrogens are attached is bonded to another atom by a bond that has hindered rotation as a result of π bonding.

At low temperature, rotation about this bond is very slow, and the NMR shows two signals: one for each hydrogen. Since the magnetic environments of these hydrogens are different, their corresponding chemical shifts are different.

As the temperature is increased, the rate of rotation about the bond increases. Instead of showing the hydrogens in their original positions, the NMR now shows the peaks beginning to merge, or "coalesce."

At still higher temperatures, the rate of rotation about the bond becomes so rapid that the NMR can no longer distinguish the individual environments of the hydrogens; instead, it now shows a single signal corresponding to the average of the two original signals.

One of the most interesting complexes in Figure **5-13** is $(C_5H_5)_2Fe(CO)_2$. This compound contains both η^1- and η^5- C_5H_5 ligands (and consequently obeys the 18-electron rule). The ^1H NMR spectrum at 30°C shows two singlets of equal area. A singlet would be expected for the five equivalent protons of the η^5- C_5H_5 ring but is surprising for the η^1- C_5H_5 ring, since the protons are not all equivalent.

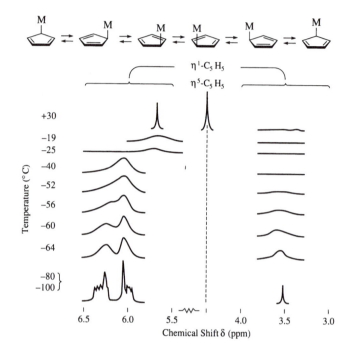

Figure 5-18

Ring Whizzer Mechanism and Variable Temperature NMR Spectra of $(\eta^1\text{-}C_5H_5)(\eta^5\text{-}C_5H_5)Fe(CO)_2$ (Adapted with permission from G. L. Miessler and D. A. Tarr, *Inorganic Chemistry*, Prentice-Hall, Englewood Cliffs, NJ, 1991, 449.)

A "ring whizzer" mechanism,[25] Figure **5-18**, has been proposed by which the five ring positions of the monohapto ring interchange via 1,2-metal shifts extremely rapidly, so rapidly that the NMR can see only the average signal for this ring.[26] At lower temperatures this process is slower, and the different resonances for the protons of $\eta^1\text{-}C_5H_5$ become apparent, as also shown in the figure. A more detailed discussion of NMR spectra in organometallic complexes, including nuclei not mentioned here, is available.[27]

[25]C. H. Campbell and M. L. H. Green, *J. Chem. Soc. (A)*, **1970**, 1318.

[26] M.J. Bennett, Jr., F.A. Cotton, A. Davison, J.W. Faller, S.J. Lippard, and S.M. Morehouse, *J. Am. Chem. Soc.*, **1966**, *88*, 4371.

[27]C. Elschenbroich and A. Salzer, *Organometallics*, 2nd ed. VCH Publishers, New York, 1992.

Problems

5-1   Homoleptic complexes of the ethylene ligand are not common. The simplest of these involving a transition metal observed to date is $(\eta^2\text{-}C_2H_4)_2Ni$, which can be trapped in cold matrices when $(\eta^2\text{-}C_2H_4)_3Ni$ is photolyzed.

a.   Assuming that the ethylene ligands are parallel, sketch the group orbitals of these ligands. For each group orbital, list the atomic orbitals of Ni of appropriate shape and orientation for interaction (consider s, p, and d orbitals of Ni).

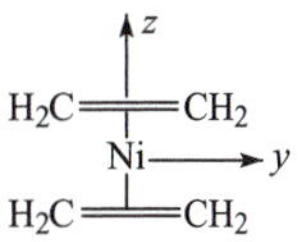

b.   Sketch the shape of a bonding molecular orbital involving one of the d orbitals of Ni.

c.   Is the C–C distance in $(\eta^2\text{-}C_2H_4)_2Ni$ likely to be longer or shorter than in free ethylene? Why?

5-2   Suppose a metal is bonded to a $\eta^3\text{-}C_3H_3$ ligand. For each of the π molecular orbitals of this ligand, determine which s, p, and d orbitals of a metal atom would be suitable for interaction. (For convenience, assign the z axis to join the metal to the center of the triangle of the ligand.)

5-3   When the positions of hydrogen atoms in cyclic ligands $\eta^n\text{-}C_nH_n$ are analyzed, in some cases they are above the plane of the carbon atoms (away from the metal), in some cases below. As the size of the ring increases, is the "up" or "down" orientation of the hydrogens likely to be favored? Why?

5-4   Predict the products of the following reactions:

a. $Co_2(CO)_8 + H_2 \rightarrow \mathbf{A}$
A is a strong acid, has a molecular weight less than 200, and has a single ^1H NMR resonance.

b. $(\eta^4\text{-}C_4H_6)Fe(CO)_3 + HBF_4 + CO \rightarrow \mathbf{B}$
B has four infrared bands near 2100 cm^{-1}. The hapticity of the hydrocarbon ligand in **B** is different than that of the butadiene ligand in the reactant. BF_4^- is a byproduct of this reaction.

c. $Cr(CO)_3(NCCH_3)_3 + \text{cycloheptatriene} \rightarrow \mathbf{C}$
The number of infrared bands near 2000 cm^{-1} in the product is the same as in the reactant.

d.   Cobaltocene $+ O_2 \rightarrow \mathbf{D}$
The molecular weight of **D** is more than double that of cobaltocene. **D** has protons in four magnetic environments with relative NMR intensities of 5:2:2:1. One half of **D** is a mirror image of the other half. **D** has an infrared band in the C–O single bond region.

5-5   Ferrocene reacts with the acylium ion, $H_3C–C{\equiv}O^+$, to give complex **X**, a cation; base extracts H^+ from **X** to give **Y**, which has the formula $C_{12}H_{12}FeO$. **Y** reacts further with the acylium ion, followed by base, to give **Z**. Suggest structures for **X**, **Y**, and **Z**.

5-6   Heating $[(\eta^5\text{-}C_5H_5)Fe(CO)_3]^+$ with NaH gives **A**, having formula $FeC_7H_6O_2$, plus colorless gas **B**. Molecule **A** reacts rapidly at room temperature to eliminate colorless gas **C**, forming solid **D**, which has empirical formula $FeC_7H_5O_2$. Compound **D** has two strong IR bands, one near 1850 cm^{-1}, the other near 2000 cm^{-1}. Treatment of **D** with iodine generates solid **E** of empirical formula $FeC_7H_5O_2I$. Reaction of NaC_5H_5 with **E** gives solid **F** of formula $FeC_{12}H_{10}O_2$. On heating **F** gives off **B**, leaving a sublimable, orange solid **G** of formula $FeC_{10}H_{10}$. Propose structures for **A** to **G**.

5-7   Photolysis of $(\eta^5\text{-}C_5H_5)(\eta^2\text{-}C_2H_4)_2Rh$ in the presence of benzene in an unreacting solvent such as pentane gives a remarkable product having the following characteristics:

- It has the empirical formula RhC_8H_8.
- It has a metal–metal bond and follows the 18–electron rule.
- Although the ^1H NMR spectrum has not been reported, it is likely to have the following characteristics:
  - Three types of resonances, with relative intensities of 5:2:1.
  - The largest peak should have a chemical shift similar to that of ring protons in the starting material.

Those who like the sea might be particularly fond of this molecule!

Propose a structure for this product.[28]

5-8   The ring whizzer mechanism of rearrangement of $(\eta^4\text{-}C_8H_8)Ru(CO)_3$ is substantially slower than for $(\eta^4\text{-}C_8H_8)Fe(CO)_3$ and can be observed by ^1H NMR as well as ^{13}C NMR. Suggest a reason for the slower rearrangment in the ruthenium compound.[29]

5-9   Three structures have been proposed having the formula $(C_nH_n)_3M_2$. One of these is the well-known triple-decker sandwich complex of nickel shown in Figure **5–17**. What are two other feasible structures having this formula?[30]

[28]J. Müller, P. E. Gaede, and K. Qiao, *Angew. Chem. Int. Ed. Engl.*, **1993**, *32*, 1697.

[29]F. A. Cotton and D. L. Hunter, *J. Am. Chem. Soc.*, **1976**, *98*, 1413 and references therein.

[30]J. Bieri, T. Egolf, W. von Philipsborn, U. Piantini, R. Prewo, U. Ruppli, and A. Salzer, *Organometallics*, **1986**, *5*, 2413 and U. Bertling, U. Englert, and A. Salzer, *Angew. Chem. Int. Ed. Engl.*, **1994**, *33*, 1003.

5-10   Both the ^1H and ^{13}C variable temperature NMR spectra of $(\eta^4-C_8H_8)Fe(CO)_3$ have been reported. What information could be determined using ^{13}C-NMR that could not be determined using ^1H-NMR?[31]

---

[31]F. A. Cotton, *Acct. Chem. Res.*, **1968**, *1*, 257 and F. A. Cotton and D. L. Hunter, *J. Am. Chem. Soc.*, **1976**, *98*, 1413)

6

Other Important Ligands

We have by no means exhausted the types of ligands encountered in organometallic chemistry. On the contrary, several additional classes of ligands are of immense importance. Three types of ligands containing metal-carbon σ bonds deserve particular attention: alkyl ligands; carbenes (containing metal-carbon double bonds); and carbynes (containing metal-carbon triple bonds). The latter two contain metal-carbon π interactions as well as σ interactions. In addition, several nonorganic ligands play important roles in organometallic chemistry. Examples include hydrogen atoms, dihydrogen (H_2), phosphines (PR_3), and the related arsenic and antimony compounds. In some cases these ligands exhibit behavior paralleling that of organic ligands; in other instances, complexes containing these ligands serve important functions in organometallic reactions, including catalytic processes.

6-1   Complexes Containing M–C, M=C, and M≡C Bonds

Complexes containing direct metal–carbon single, double, and triple bonds have been studied extensively; the most important ligands having these types of bonds are summarized in Table **6-1**.

Table 6-1 Complexes Containing M–C, M=C, and M≡C Bonds

Ligand	Formula	Example
Alkyl	$-CR_3$	$W(CH_3)_6$
Carbene (alkylidene)	$=CRR'$	$(CO)_5Cr=C\begin{smallmatrix}\nearrow OCH_3 \\ \searrow Ph\end{smallmatrix}$
Carbyne (alkylidyne)	$\equiv CR$	$\begin{smallmatrix} OC & CO \\ Br-Cr\equiv C-Ph \\ OC & CO \end{smallmatrix}$

6-1-1   Alkyl and Related Complexes

Some of the earliest known organometallic complexes were those having σ bonds between main group metal atoms and alkyl groups. Examples include Grignard reagents, which have magnesium-alkyl bonds, and alkyl complexes with alkali metals, such as methyllithium. The syntheses and reactions of these main group compounds are typically discussed in detail in organic chemistry texts.

The first stable transition metal alkyls were synthesized in the first decade of the twentieth century; many such complexes are now known. The metal-ligand bonding in these compounds may be viewed as primarily involving covalent sharing of electrons between the metal and the carbon in a σ fashion, as shown below:

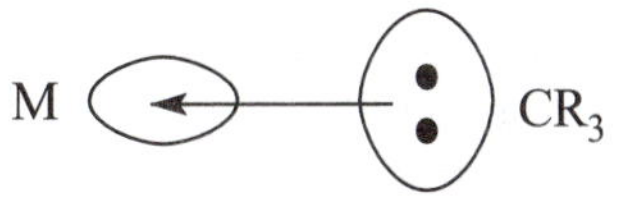

sp^3 orbital

In terms of electron counting, the alkyl ligand may be considered the two-electron donor: CR_3^- (method A) or the one-electron donor $\cdot CR_3$ (method B). Significant ionic contribution to the bonding may occur in complexes of highly electropositive elements such as the alkali metals and alkaline earths.

Exercise 6-1

Identify the transition metal in the following 18-electron complexes:

a. $CH_3M(CO)_5$ M = 1st row transition metal

b. $(\eta^5\text{-}C_5H_5)CH_3M(CO)_2$ M = 1st row transition metal

c. $CH_3M(CO)_2(PMe_3)_2Br$ M = 3rd row transition metal

Many synthetic routes to transition metal–alkyl complexes have been developed. Three of the most important of these methods are:

1. Reaction of transition metal halide with organolithium, organomagnesium, or organoaluminum reagent
    Example: $ZrCl_4 + 4\ PhCH_2MgCl \rightarrow Zr(CH_2Ph)_4 + 4\ MgCl_2$

2. Reaction of metal carbonyl anion with alkyl halide
    Example: $Na[Mn(CO)_5] + CH_3I \rightarrow CH_3Mn(CO)_5 + NaI$

3. Reaction of metal carbonyl anion with acyl halide
    Example:

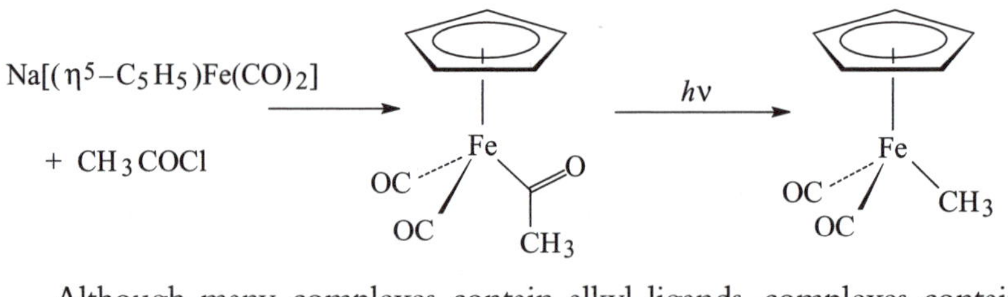

$Na[(\eta^5\text{-}C_5H_5)Fe(CO)_2]$

$+ CH_3COCl$

Although many complexes contain alkyl ligands, complexes containing alkyl groups as the only ligands[1] are relatively few and have a tendency to be kinetically unstable and difficult to isolate;[2] their stability is enhanced by structural crowding, which protects the coordination sites of the metal by blocking pathways to decomposition. For example, the six–coordinate $W(CH_3)_6$ can be melted at 30°C without decomposition, whereas the four-coordinate $Ti(CH_3)_4$ is subject to decomposition at approximately –40°C.[3]

In addition, alkyl complexes having β hydrogens tend to be *much* less stable (kinetically) than complexes lacking β hydrogens. Complexes with β hydrogens are subject to a method of decomposition called β-**elimination** as described in the following diagram (opposite):

---

[1]The term *homoleptic* has been coined to describe complexes in which all ligands are identical.
[2]An interesting historical perspective on alkyl complexes is in G. Wilkinson, *Science*, **1974**, *185*, 109.
[3]A. J. Shortland and G. Wilkinson, *J. Chem. Soc., Dalton Trans.*, **1973**, 872.

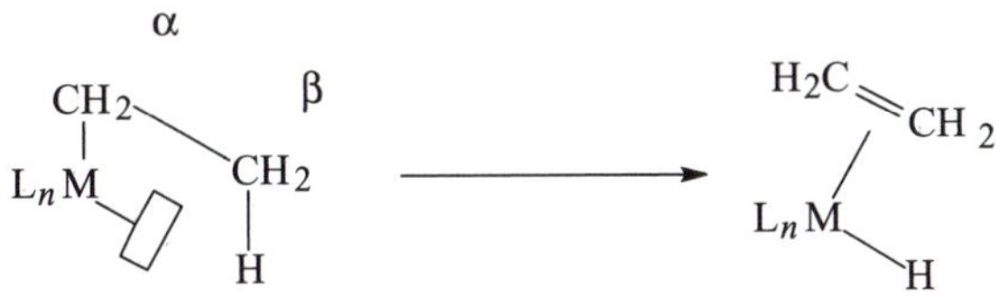

To provide stability, the β positions of alkyl ligands can be blocked in a variety of ways. The following are examples:

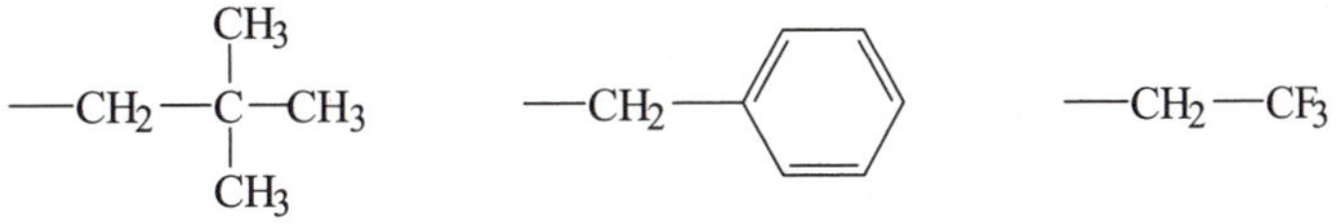

β-elimination reactions will be discussed in more detail in Chapter 8.

Sometimes the kinetic instability of complexes containing β hydrogens can be put to good use. As we will see in Chapter 9, the instability of alkyl complexes having β hydrogens can be a very useful feature in catalytic processes, where the rapid reactivity of an intermediate can be crucial to the effectiveness of the overall process. Many alkyl complexes therefore are important in catalytic cycles. One alkyl with an interesting use is diethyl zinc, which has been selected by the Library of Congress for long-term preservation of books since it neutralizes the acid in paper.

Several other ligands involve direct metal–carbon σ bonds. Examples are given in Table **6-2**:

Table 6-2 Other Organic Ligands Forming Sigma Bonds to Metal

Ligand	Formula	Example
Aryl	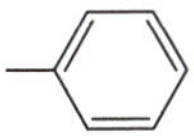	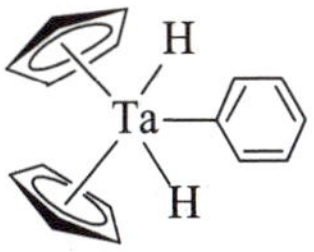
Alkenyl (vinyl)	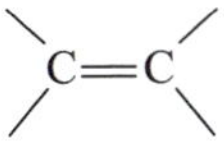	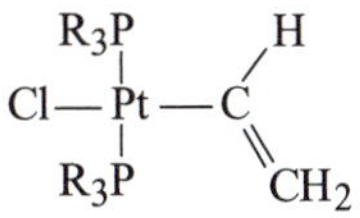
Alkynyl	$-C\equiv C-$	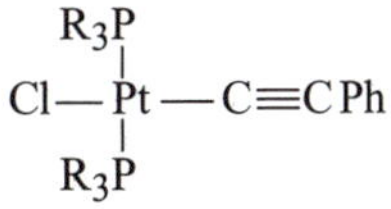

In addition, there are many known examples of **metallacycles**, complexes containing metals incorporated into rings. The following equation is an example:

$$PtCl_2(PR_3)_2 \; + \; Li \diagdown \diagup \diagdown \diagup Li \; \longrightarrow \; \underset{R_3P}{\overset{R_3P}{\diagdown}} Pt \diagup \diagdown \; + \; 2\ LiCl$$

Metallacyclopentane

In addition to being interesting in their own right, metallacycles are proposed as intermediates in a variety of catalytic processes. Additional examples of metallacycles will be encountered in later chapters.

6-1-2   Carbene (Alkylidene) Complexes

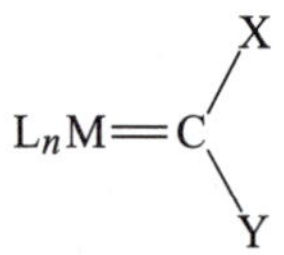

$$L_nM{=}C\overset{X}{\underset{Y}{\diagup}}$$

Carbene complexes contain metal–carbon double bonds;[4] they have the general structure shown at left (X,Y = alkyl, aryl, H, or highly electronegative atoms such as O, N, S, or halogens). First synthesized in 1964 by Fischer,[5] carbene complexes are now known for the majority of transition metals and for a wide range of ligands, including the prototype carbene :CH$_2$. The majority of such complexes, including those first synthesized by Fischer, contain one or two highly electronegative *heteroatoms* such as O, N, or S directly attached to the carbene carbon. These are commonly designated as *Fischer-type* carbene complexes and have been studied extensively over the past three decades. Other carbene complexes contain only carbon and/or hydrogen attached to the carbene carbon. First synthesized several years after the initial Fischer carbenes, these have been

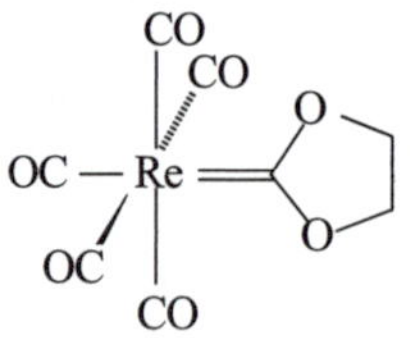

Fischer Carbene Complex 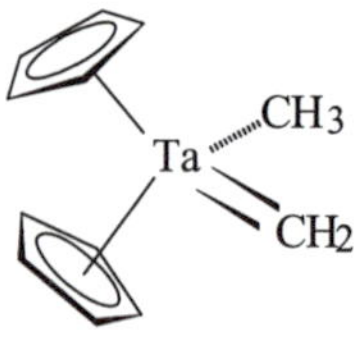   Schrock Carbene (Alkylidene) Complex

---

[4]IUPAC has recommended that the term "alkylidene" be used to describe all complexes containing metal–carbon double bonds and that "carbene" be restricted to free :CR$_2$. For a detailed description of the distinction between these two terms (and between "carbyne" and "alkylidyne," described below) see W. A. Nugent and J. M. Mayer, *Metal-Ligand Multiple Bonds*, Wiley-Interscience, New York, 1988, 11-16.

[5]E. O. Fischer and A. Maasbol, *Angew. Chem. Int. Ed. Engl.*, **1964**, *3*, 580.

Figure 6-1
Pi Bonding in Carbene
Complexes and in
Alkenes (Adapted from
G.L. Miessler and D.A.
Tarr, *Inorganic Chemistry*,
Prentice-Hall,
Englewood Cliffs, NJ,
1991, 450.)

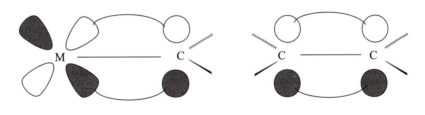

studied extensively by Schrock's research group and several others; they are sometimes designated *Schrock-type* carbene complexes, commonly referred to as alkylidenes.[6]

In this chapter, we confine the discussion to a very qualitative view of the bonding in carbenes; more detail on the structure, bonding, and reaction chemistry of carbene complexes will be discussed in Chapter 10.

The formal double bond in carbene complexes may be compared with the double bond in alkenes; in the case of the carbene, the metal must use a *d* orbital (rather than a *p* orbital) in forming the π bond with carbon, as illustrated in Figure **6-1**.

Another important aspect of bonding in carbene complexes is that complexes having a highly electronegative atom such as O, N, or S attached to the carbene carbon tend to be more stable than complexes lacking such an atom.

$$(CO)_5Cr = C \overset{\displaystyle OCH_3}{\underset{\displaystyle Ph}{\Big\backslash}} \qquad\qquad (CO)_5Cr = C \overset{\displaystyle H}{\underset{\displaystyle Ph}{\Big\backslash}}$$

For example, $Cr(CO)_5[C(OCH_3)C_6H_5]$, with an oxygen on the carbene carbon, is much more stable than $Cr(CO)_5[C(H)C_6H_5]$. The stability of the complex is enhanced if the highly electronegative atom can participate in the π bonding, with the result a delocalized, three-atom π system involving a *d* orbital on the metal and *p* orbitals on carbon and on the heteroatom.[7] An example of such a π system is shown in Figure **6-2**. Such a delocalized three-atom system provides more stability than would a simple metal-carbon π bond. As will be discussed in Chapter 10, the presence of a highly electronegative atom can stabilize the complex by lowering the energy of the M-C-heteroatom bonding orbital.

The methoxycarbene complex $Cr(CO)_5[C(OCH_3)C_6H_5]$, shown above, illustrates some important characteristics of bonding in transition metal

[6]R. R. Schrock, *J. Am. Chem. Soc.*, **1974**, *96*, 6796; *Acc. Chem. Res.*, **1979**, *12*, 98.
[7]In some cases, *d* orbitals on the heteroatom can also participate.

Figure 6-2
Delocalized π Bonding in
Carbene Complexes. E
Designates a Highly
Electronegative
Heteroatom Such as O,
N, or S (Adapted from
G.L. Miessler and D.A.
Tarr, *Inorganic Chemistry*,
Prentice-Hall, Englewood
Cliffs, NJ, 1991, 451.)

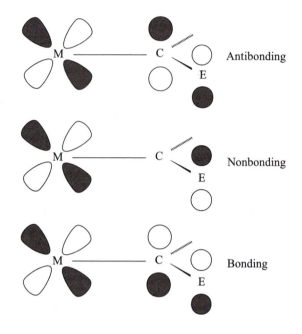

carbene complexes.[8] Evidence for double bonding between chromium and carbon is provided by X-ray crystallography, which measures this distance at 204 pm, compared with a typical Cr–C single bond distance of approximately 220 pm.

One very interesting aspect of this complex is its ability to exhibit a proton NMR spectrum that is temperature dependent. At room temperature a single resonance is found for the methyl protons; however, as the temperature is lowered this peak first broadens, then splits into two peaks. How can this behavior be explained?

A single methyl proton resonance, corresponding to one magnetic environment, is expected for the carbene complex as illustrated, with a double bond between chromium and carbon, and a single bond (permitting rapid rotation about the bond) between carbon and oxygen. The room temperature NMR is therefore as expected, with a single methyl signal. The splitting of this peak at lower temperature into two peaks, however, suggests two different proton environments.[9] Two environments are possible if there is hindered rotation about the C–O bond. A resonance structure for the complex can be drawn showing some double bonding between C and O; such double bonding is

[8] E. O. Fischer, *Adv. Organometallic Chem.*, **1976**, *14*, 1.
[9] C. G. Kreiter and E. O. Fischer, *Angew. Chem. Int. Ed. Engl.*, **1969**, *8*, 761.

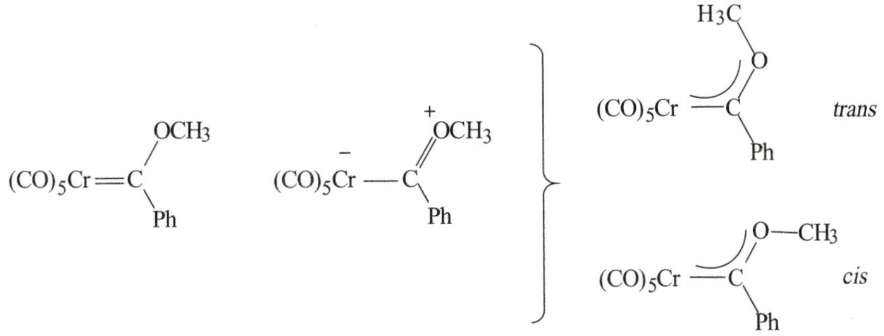

Figure 6-3
Resonance Structures and
cis and *trans* Isomers for
Cr(CO)$_5$[C(OCH$_3$)Ph]

significant enough for *cis* and *trans* isomers, as shown in Figure **6-3**, to be observable at low temperatures.

Evidence for double bond character in the C-O bond is also provided by crystal structure data, which show a C-O bond distance of 133 pm, compared with a typical C-O single bond distance of 143 pm.[10] The double bonding between C and O, although weak (typical C=O bonds are much shorter, approximately 116 pm), is sufficient to slow down rotation about the bond so that at low temperatures proton NMR detects the *cis* and *trans* methyl protons separately. At higher temperatures, there is sufficient energy to cause rapid rotation about the C-O bond so that the NMR sees only an average signal, which is observed as a single peak.

X-ray crystallographic data, as mentioned, show double bond character in both the Cr-C and C-O bonds. This supports the statement made early in the discussion of carbene complexes that π bonding in complexes of this type (containing a highly electronegative atom—in this case oxygen) may be considered delocalized over three atoms. While not absolutely essential for all carbene complexes, the delocalization of π electron density over three or more atoms provides an additional measure of stability to many of these compounds.[11]

6-1-3   Carbyne Complexes

Nine years after Fischer's report of the first stable transition metal-carbene complex, the first report was made of the synthesis of complexes containing a

[10] O. S. Mills and A. D. Redhouse, *J. Chem. Soc. (A)*, **1968**, 642.
[11] *Transition Metal Carbene Complexes*, Verlag Chemie, Weinheim, West Germany, 1983, 120-122.

transition metal–carbon triple bond—fittingly enough, this report was also made by the Fischer group.[12] Carbyne complexes have metal-carbon triple bonds; they are formally analogous to alkynes.[13] Many carbyne complexes are now known; examples of carbyne ligands include the following:

$$M \equiv C\text{-}R \quad R = \text{aryl (first discovered), alkyl, H, SiMe}_3, \text{NEt}_2, \text{PMe}_3, \text{SPh, and Cl}$$

In Fischer's original synthesis, carbyne complexes were obtained fortuitously as products of the reactions of carbene complexes with Lewis acids. For example, the methoxycarbene complex $Cr(CO)_5[C(OCH_3)C_6H_5]$ was found to react with the Lewis acids BX_3 (X=Cl, Br, or I):

First the Lewis acid attacks the oxygen, the basic site on the carbene as in the following equation:

$$Cr(CO)_5 = C \begin{smallmatrix} OCH_3 \\ \\ Ph \end{smallmatrix} + BX_3 \longrightarrow [(CO)_5 Cr \equiv C\text{-}Ph]^+ X^- + X_2BOCH_3$$

Subsequently the intermediate loses CO, with the halogen coordinating in a position *trans* to the carbyne as follows:

$$[(CO)_5 Cr \equiv C\text{-}Ph]^+ X^- \longrightarrow \begin{smallmatrix} OC \quad CO \\ \diagdown \quad \\ X\text{—}Cr \equiv C\text{-}Ph \\ \diagup \quad \\ OC \quad CO \end{smallmatrix}$$

The best evidence for the carbyne nature of the complex is provided by X-ray crystallography, which gives a Cr–C bond distance of 168 pm (for X = Cl), considerably shorter than the comparable distance for the parent carbene complex. The $Cr \equiv C\text{-}C$ angle is, as expected, 180° for this complex; however, slight deviations from linearity are observed for many complexes in crystalline form—in part a consequence of the manner of packing in the crystal.

Bonding in carbyne complexes may be viewed as a combination of a σ bond plus two π bonds, as illustrated in Figure **6-4**.

---

[12]E. O. Fischer, G. Kreis, C.G. Müller, G. Huttner, and H. Lorentz., *Angew. Chem. Int. Ed. Engl.*, **1973**, *12*, 64.

[13]Complexes containing metal-carbon triple bonds in which the metal is in a relatively high formal oxidation state are frequently designated *alkylidynes*. Carbyne complexes are also sometimes classified as *Fischer-type* or *Schrock-type*, as in the case of carbene complexes. The distinction between these will be discussed in Chapter 10.

Figure 6-4
Bonding in Carbyne
Complexes (Adapted
from G. L. Miessler and D.
A. Tarr, *Inorganic Chemistry,*
Prentice-Hall, Englewood
Cliffs, NJ, 1991, 453.)

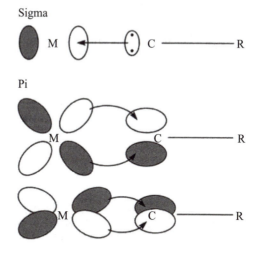

The carbyne ligand has a lone pair of electrons in an *sp* hybrid on carbon; this lone pair can donate to a suitable hybrid orbital on Cr to form a σ bond. In addition, the carbon has two *p* orbitals that can accept electron density from *d* orbitals on Cr to form π bonds. Thus the overall function of the carbyne ligand is as both a σ donor and π acceptor. For electron counting purposes a :CR⁺ ligand can be considered a two-electron donor (method A); it is usually more convenient to count neutral CR as a three–electron donor (method B).

Example 6-1

Verify that $Cr(CO)_4Br(\equiv CPh)$ satisfies the 18–electron rule.

Method A		**Method B**	
Cr	6 e	Cr	6 e
4 CO	8 e	4 CO	8 e
:Br⁻	2 e	Br	1 e
:CC₆H₅	⁺2 e	≡CC₆H₅	3 e
	18 e		18 e

Exercise 6-2

Identify the third-row transition metal in the following carbyne complexes:
a. $(\eta^5\text{-}C_5Me_5)M(CCMe_3)(H)(PR_3)_2$
b. $M(\text{C-}o\text{-tolyl})(CO)(PPh_3)_2Cl$

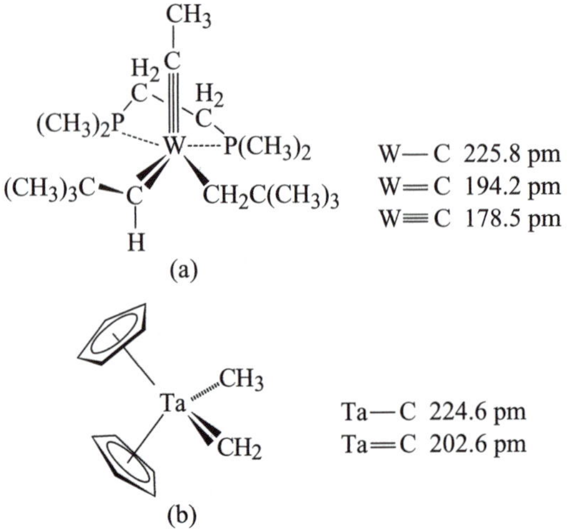
CH3
H2 C
C
H2
C
(CH3)2P
W
P(CH3)2
(CH3)3C
C
CH2C(CH3)3
H
(a)
W—C 225.8 pm
W═C 194.2 pm
W≡C 178.5 pm
Ta
CH3
CH2
(b)
Ta—C 224.6 pm
Ta═C 202.6 pm

Figure 6-5
Complexes Containing Alkyl, Carbene, and Carbyne Ligands (Adapted from G. L. Miessler and D. A. Tarr, *Inorganic Chemistry*, Prentice-Hall, Englewood Cliffs, NJ, 1991, 454.)

Carbyne complexes can be synthesized in a variety of ways in addition to Lewis acid attack on carbenes, as described above. Synthetic routes for carbynes and the reactions of these compounds have been reviewed in the literature.[14] Reactions of carbynes will be discussed in more detail in Chapter 10.

In some cases molecules have been synthesized containing two or three of the types of ligands discussed in this section: alkyl, carbene, and carbyne. Such molecules provide an opportunity to make direct comparisons of lengths of metal-carbon single, double, and triple bonds, as shown in Figure **6-5**.

6-1-4 Other Organic Ligands

The list of ligands capable of bonding to metals through carbon is still not complete. Examples of several other important ligands not yet considered are given in Table **6-3**.

Acyl complexes can be formed from migration of alkyl groups to CO ligands as in the following equation:

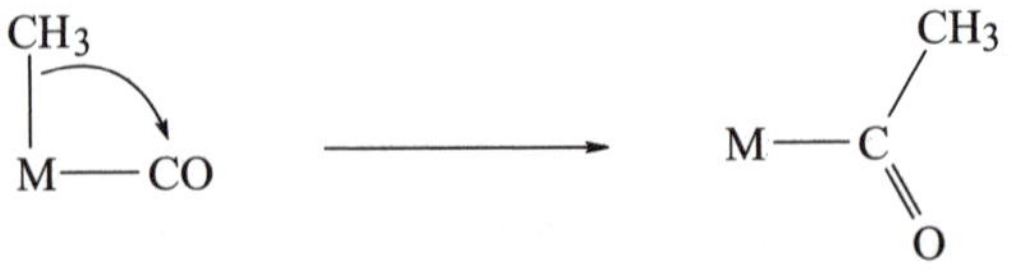
CH3
M—CO
CH3
M—C
O

[14]H.P. Kim and R.J. Angelici, "Transition Metal Complexes with Terminal Carbyne Ligands," *Advances in Organometallic Chemistry,* **1987**, *27*, 51; H. Fischer, P. Hoffmann, F.R. Kreissl, R.R. Schrock, U. Schubert, and K. Weiss, *Carbyne Complexes,* VCH Publishers, Weinheim, West Germany, 1988.

Table 6-3 Other Organic Ligands

Ligand	Formula	Example
Acyl	$-\overset{\overset{\displaystyle O}{\|\|}}{C}-R$ (R = alkyl, aryl)	
Perfluoroalkyl, perfluoroaryl	$-CF_3, -C_6F_5$	
Vinylidene	$= C=CR_2$	

The formation of acyl complexes is an important step in "carbonyl insertion" reactions,[15] as will be described in Chapter 8 (Section **8-1-1**).

Trifluoromethyl, $-CF_3$, and other perfluoroalkyl and perfluoroaryl complexes are, in general, much more thermally stable than their alkyl and aryl counterparts. For example, $CF_3Co(CO)_4$ can be distilled without decomposing at its boiling point, 91°C, while $CH_3Co(CO)_4$ decomposes even at temperatures as low as –30°C. The reasons for this enhanced stability in the perfluoro complexes are complex, but a major factor is clearly the greater bond dissociation energy of the M–ligand bond in the fluorine-containing ligands. In the perfluoroalkyl complexes, the highly electronegative fluorines cause the ligands to pull electrons away from the metal, inducing significant polarity into the M–C bond, a polarity that increases the bond strength. In the perfluoroaryl complexes, low-lying π* orbitals on the ligands are available to participate in back bonding to the metal, likewise strengthening the M–C bond.

Vinylidene complexes are examples of carbene complexes having cumulated (consecutive) double bonds. Examples of their formation will also be encountered in later chapters.

[15] As we will see, this name for this reaction is misleading!

6-2   Hydride and Dihydrogen Complexes

The simplest of all possible ligands is the hydrogen atom; similarly, the simplest possible diatomic ligand is H_2. It is perhaps not surprising that these ligands have gained attention by virtue of their apparent simplicity, as models for bonding schemes in coordination compounds. Moreover, both ligands have played immense roles in the development of applications of organometallic chemistry to organic synthesis, and especially catalytic processes. While the hydrogen atom (ordinarily designated the *hydride* ligand) has been recognized as an important ligand for many years, the significance of the *dihydrogen* ligand has become recognized relatively recently, and its chemistry is now developing rapidly.[16,17]

6-2-1   Hydride Complexes

Although hydrogen atoms form bonds with nearly every element in the Periodic Table, we will specifically consider coordination compounds containing H bonded to transition metals. Because the hydrogen atom has only a $1s$ orbital of suitable energy for bonding, the bond between H and a transition metal must, by necessity, be a σ interaction, involving metal s, p, and/or d orbitals (or a hybrid orbital). As a ligand, H may be considered a two-electron donor as hydride (:H^-, method A) or a one-electron neutral donor (H atom, method B).

Although some homoleptic transition metal complexes of the hydride ligand are known—an example of some structural interest is the nine-coordinate $[ReH_9]^{2-}$ ion[18]—we are principally concerned about complexes containing H in combination with other ligands. Such complexes may be made in a variety of ways, as will be seen in succeeding chapters. Probably the most common synthesis is by reaction of a transition metal complex with H_2. The following are examples:

$$Co_2(CO)_8 + H_2 \rightarrow 2 \; HCo(CO)_4$$

$$trans\text{-}Ir(CO)Cl(PEt_3)_2 + H_2 \rightarrow Ir(CO)Cl(H)_2(PEt_3)_2$$

The hydride ligand can often be readily recognized by ^1H NMR because the H experiences strong shielding. Typical chemical shifts of hydride are between

[16]G. J. Kubas, *Comments Inorg. Chem.*, **1988**, *7*, 17; R. H. Crabtree, *Acc. Chem. Res.*, **1990**, *23*, 95; G. J. Kubas, *Acc. Chem. Res.*, **1988**, *21*, 120.
[17]J. K. Burdett, O. Eisenstein, and S. A. Jackson, "Transition Metal Dihydrogen Complexes: Theoretical Studies," in A. Dedieu, Ed., *Transition Metal Hydrides*, VCH Publishers, New York, 1992.
[18]S. C. Abrahams et al., *Inorg. Chem.*, **1964**, *3*, 558.

ca. -2 and -12 parts per million for terminal hydride, with bridging hydride absorbing at still higher field.

One of the most interesting aspects of transition metal–hydride chemistry is the relationship between this ligand and the rapidly developing chemistry of the dihydrogen ligand, H_2.

6-2-2   Dihydrogen Complexes

Although complexes containing H_2 molecules coordinated to transition metals had been proposed for many years, and many complexes containing hydride ligands had been prepared, the first structural characterization of a dihydrogen complex did not occur until 1984, when Kubas and co-workers synthesized the complexes $M(CO)_3(PR_3)_2(H_2)$ (M=Mo, W; R=cyclohexyl, isopropyl).[19] Subsequently many H_2 complexes have been identified, and the chemistry of this ligand has developed rapidly.

The bonding between dihydrogen and a transition metal can be described as shown in Figure **6-6**. The σ electrons in H_2 can be donated to a suitable empty orbital on the metal (such as a *d* orbital or a hybrid orbital), while the empty σ* orbital of the ligand can accept electron density from occupied *d* orbitals of the metal. The result is an overall weakening and lengthening of the H-H bond in comparison with free H_2. Typical H-H distances in complexes containing coordinated dihydrogen are in the range 82-90 pm, in comparison with 74.14 pm in free H_2.

This bonding scheme leads to interesting ramifications quite distinctive from other donor-acceptor ligands such as CO and cyclic π systems that are held together by multiple bonds. If the metal is electron rich and donates strongly to the σ* of H_2, the H-H bond in the ligand can rupture, giving separate H atoms. Consequently, the search for stable H_2 complexes has focused on metals likely to be relatively poor donors, such as those in high oxidation states or those surrounded by ligands that function as strong electron acceptors.

Figure 6-6
Bonding in Dihydrogen Complexes (Adapted from G.L. Miessler and D.A. Tarr, *Inorganic Chemistry*, Prentice-Hall, Englewood Cliffs, NJ, 1991, 435.)

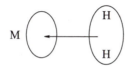

Sigma Donation

Pi Acceptance

[19]G. J. Kubas, R. R. Ryan, B. I. Swanson, P. J. Vergamini, and H. J. Wasserman, *J. Am. Chem. Soc.*, **1984**, *106*, 451.

In particular, good π acceptors such CO and NO can be effective at stabilizing the dihydrogen ligand.

Exercise 6-3

Explain why $Mo(PMe_3)_5H_2$ is a dihydride (contains two separate H ligands), while $Mo(CO)_3(PR_3)_2(H_2)$ contains the dihydrogen ligand. (R=isopropyl)

The dividing line between whether an $M(H)_2$ or $M(H_2)$ structure will be favored is a very narrow one, and subtle differences in environment can be important. For example, in $[Rh(P(CH_2CH_2PPh_2)_3)H_2]^+$, two isomers are found as shown below. If hydrogen occupies an apex of the trigonal bipyramidal isomer, it is present as H_2; if hydrogen occupies *cis* positions of the octahedral isomer, it is present as separate H atoms.[20] Calculations have shown that there is less donation from the metal to the ligand at the apical site of the trigonal bipyramid than in the *cis* positions of the octahedral isomer.[21]

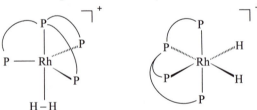

Dihydrogen complexes have frequently been suggested as possible intermediates in a variety of reactions of hydrogen at metal centers. Some of these reactions are steps in catalytic processes of significant commercial interest. As this ligand becomes more completely understood, the applications of its chemistry are likely to become extremely important.

6-2-3   Agostic Hydrogens

In some cases hydrogen atoms may form "bent" linkages between carbon atoms on ligands and metal atoms; such interactions have come to be described as *agostic*. An example is shown in Figure **6-7**.[22]

[20]C. Bianchini, C. Mealli, M. Peruzzini, and F. Zanobini, *J. Am. Chem. Soc.*, **1987**, *109*, 5548.

[21]F. Maseras, M. Duran, A. Lledos, and J. Bertran, *Inorg. Chem.*, **1989**, *28*, 2984.

[22]G. D. Stucky and S. D. Ittel, *J. Am. Chem. Soc.*, **1980**, *102*, 981.

Figure 6-7

Example of an Agostic Interaction (Adapted with permission from C. Elschenbroich and A. Salzer, *Organometallics*, 2nd ed., VCH Publishers, Weinheim, Germany, 1992, 268.)

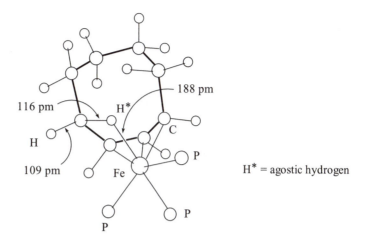

H* = agostic hydrogen

Some such interactions have been substantiated by neutron diffraction; many others have been proposed, especially in reaction intermediates. Because the hydrogen involved in an agostic interaction with a metal is, in effect, acting as a donor to form a weak bond with the metal, the C-H bond becomes elongated and the bond is weakened. The result is a tendency to "activate" the C-H bond as a step toward further reaction. Several examples of these types of interactions will be discussed in later chapters.

6-3   Phosphines and Related Ligands

Among the most important of all ligands are the phosphines, PR_3. Together with other phosphorus-containing ligands and related compounds of arsenic and antimony, phosphines parallel the CO ligand in many ways. Like CO, phosphines are σ donors (via a hybrid orbital containing a lone pair on phosphorus) and π acceptors. For many years it was thought that empty $3d$ orbitals of phosphorus functioned as the acceptor orbitals, as shown in Figure **6-8**. By this view, as the R groups attached to phosphorus become more electronegative, they withdraw electrons from phosphorus, making the phosphorus more positive and better able to accept electrons from the metal via a d orbital. The nature of the R groups, therefore, determines the relative donor/acceptor ability of the ligand. $P(CH_3)_3$, for example, is a strong σ donor by virtue of the electron-releasing nature of the methyl groups; at the same time, it is a relatively weak π acceptor. PF_3, on the other hand, is a strong π acceptor (and weak σ donor) and rivals CO in the overall strength of its interaction with metal d orbitals. Not surprisingly, complexes containing PF_3 tend to obey the 18-electron rule. By changing the R groups, one can therefore "fine tune" the

Figure 6-8
Bonding in Phosphines
(traditional view)

Sigma Donation   Pi Acceptance

phosphine to be a donor/acceptor of a desired strength.

In 1985, a revised view of the bonding in phosphines was proposed.[23] By this proposal, the important acceptor orbital of the phosphine is not a pure $3d$ orbital, but rather a combination of a $3d$ orbital with a $\sigma*$ orbital involved in the P-R bonding, as shown in Figure **6-9**. This orbital has two acceptor lobes, similar to those of a $3d$ orbital, but is antibonding with respect to the P-R bond. To test this scheme, crystal structures of a variety of phosphine complexes in different oxidation states were compared. In most of these cases, as the charge on the metal became more negative, the P-R distance increased—as would be expected if additional electron density were pushed into an orbital having P-R antibonding character.

Figure 6-9
Pi Acceptor Orbitals of
Phosphines (revised view)

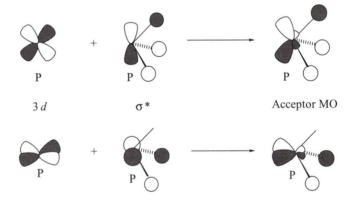

Exercise 6-4

The M-P distance in $(\eta^5\text{-}C_5H_5)Co(PEt_3)_2$ is 221.8 pm and the P–C distance is 184.6 pm. The corresponding distances in $[(\eta^5\text{-}C_5H_5)Co(PEt_3)_2]^+$ are 223.0 pm and 182.9 pm.[24] Account for the changes in these distances as the former complex is oxidized.

[23]A. G. Orpen and N. G. Connelly, *J. Chem. Soc., Chem. Commun.*, **1985**, 1310.
[24]R. L. Harlow, R. J. McKinney, and J. F. Whitney, *Organometallics*, **1983**, *2*, 1839.

Table 6-4 Selected Ligands of Phosphorus, Arsenic, and Antimony

Formula	Name
PR_3	phosphine
$P(OR)_3$	phosphite
$Ph_2P\diagup\diagdown PPh_2$	dppe[a] (diphenylphosphinoethane)
AsR_3	arsine
As(CH₃)₂ structure	diars
SbR_3	stibine

[a] There are many similiar ligands with 4-letter acronyms to summarize their structures.
For Example: **dmpe** = **d**i**m**ethyl**p**hosphino**e**thane **depp** = **d**i**e**thylphosphino**p**ropane

Similar interactions occur in other phosphorus–containing ligands and related ligands containing arsenic and antimony. Names and formulas of some of these ligands are given in Table **6-4.**

An additional factor very important in phosphine chemistry is the amount of space occupied by the R group. This factor is important in a variety of contexts; for example, the rate at which phosphine dissociates from a metal is related to the amount of space occupied by the phosphine and the resultant crowding around the metal. To describe the steric effects of phosphines and other ligands, Tolman has defined the **cone angle** as the apex angle θ of a

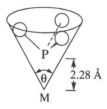

Figure 6-10
Ligand Cone Angle (Adapted from G. L. Miessler and D. A. Tarr, *Inorganic Chemistry*, Prentice-Hall, Englewood Cliffs, NJ, 1991, 472.)

cone that encompasses the van der Waals radii of the outermost atoms of a ligand, as shown in Figure **6–10**.[25] Values of cone angles of selected ligands are given in Table **6–5**.

Table 6-5 Ligand Cone Angles

Ligand	θ	Ligand	θ
PH_3	87°	$P(CH_3)(C_6H_5)_2$	136°
PF_3	104°	$P(CF_3)_3$	137°
$P(OCH_3)_3$	107°	$P(O\text{-}o\text{-}C_6H_4CH_3)_3$	141°
$P(OC_2H_5)_3$	109°	$P(C_6H_5)_3$	145°
$P(CH_3)_3$	118°	$P(cyclo\text{-}C_6H_{11})_3$	170°
PCl_3	124°	$P(t\text{-}C_4H_9)_3$	182°
$P(CH_3)_2(C_6H_5)$	127°	$P(C_6F_5)_3$	184°
PBr_3	131°	$P(o\text{-}C_6H_4CH_3)_3$	194°
$P(C_2H_5)_3$	132°	$P(mesityl)_3$	212°

As might be expected, the presence of bulky ligands can lead to more rapid ligand dissociation as a consequence of crowding around the metal. For example, the rate of the reaction

$$cis\text{-}Mo(CO)_4L_2 + CO \rightarrow Mo(CO)_5L + L \text{ (L=phosphine or phosphite)},$$

which is first order in $cis\text{-}Mo(CO)_4L_2$, increases with increasing ligand bulk, as shown by the cone angles.[26]

Table 6-6 Rates of Reaction of $cis\text{-}Mo(CO)_4L_2$

Ligand	Rate Constant (x $10^{-5}s^{-1}$)	Cone Angle
$P(OPh)_3$	<1.0	128°
$PMePh_2$	1.3	136°
$P(O\text{-}o\text{-tolyl})_3$	16	141°
PPh_3	320	145°
$PPh(cyclohexyl)_2$	6400	162°

[25]C.A. Tolman, *J. Am. Chem. Soc.,* **1970**, *92*, 2953; *Chem. Rev.,* **1977**, *77*, 313.
[26]D.J. Darensbourg and A.H. Graves, *Inorg. Chem.,* **1979**, *18*, 1257.

Numerous other examples of the effect of ligand bulk on dissociation of ligands have been reported in the chemical literature.[27] We will see additional examples of this effect as we examine reactions of organometallic complexes in later chapters.

---

[27]For example, M. J. Wovkulich and J.D. Atwood, *Organometallics*, **1982**, *1*, 1316; J.D. Atwood, M.J. Wovkulich, and D.C. Sonnenberger, *Acc. Chem. Res.*, **1983**, *16*, 350.

Problems

6-1   Identify the first-row transition metal in each of the following:
- a. $H_3CM(CO)_5$
- b. $M(CO)(CS)(PF_3)(PPh_3)Br$
- c. $(CO)_5M=C(OCH_3)C_6H_5$
- d. $(\eta^5\text{-}C_5H_5)(CO)_2M=C=C=C(CMe_3)_2$
- e. $M(CO)_5(COCH_3)$

6-2   Predict the products of the following reactions:
- a. $[(\eta^5\text{-}C_5H_5)W(CO)_3]^- + C_2H_5I \rightarrow$
- b. $Mn_2(CO)_{10} + H_2 \rightarrow$
- c. $Mo(CO)_6 + PPh_3 \rightarrow$

6-3   Which complex would you predict to have the longer Fe-P bonds, $(\eta^4\text{-}C_4Ph_4)Fe(CO)(PR_3)_2$ or $[(\eta^4\text{-}C_4Ph_4)Fe(CO)(PR_3)_2]^+$, where $R = OCH_3$? Explain.

6-4   Give likely structures for **X** and **Y**. (see equation 10.5, p. 313)
$$Fe(CO)_5 + LiCH_3 \rightarrow \mathbf{X}$$
$$\mathbf{X} + [(CH_3)_3O][BF_4] \rightarrow \mathbf{Y}$$

6-5   The complex $Ir(CO)(Cl)(PEt_3)_2$ reacts with H_2 to give a product having two Ir-H stretching bands in the infrared and a single ^{31}P-NMR resonance. Suggest a structure for this product.

6-6 $NaMn(CO)_5$ reacts with $H_2C=CHCH_2Cl$ to give **A** + **B**. Compound **A** obeys the 18-electron rule and shows protons in three distinct magnetic environments. Water-soluble compound **B** reacts with aqueous $AgNO_3$ to give a white precipitate that turns gray on exposure to light. When heated, **A** gives off gas **C** and converts to **D**, which has protons in two distinct magnetic environments. Identify compounds **A** to **D**.

6-7   The complex $[(\eta^5\text{-}C_5H_5)Mo(CO)_3]_2$ reacts with I_2 to give a product **A** having three infrared bands near 2000 cm^{-1}. This product reacts with triphenylphosphine to give **B**, which has two bands near 2000 cm^{-1}. Identify **A** and **B**.

6-8   Reaction of $[\eta^5\text{-}C_5(CH_3)_5]Ru(PR_3)Br$ with H_2 in toluene at 25°C gives a square pyramidal product with the $\eta^5\text{-}C_5(CH_3)_5$ ligand at the top of the pyramid. ^1H-NMR shows two peaks near $\delta = -6ppm$ but no evidence of H-H coupling or other indication of a bond between hydrogens.

$[PR_3 = P(isopropyl)_2(C_6H_5)]$
a. Propose a reasonable structure for the product.
b. Suppose you wanted to prepare a dihydrogen complex by using a similar reaction. What type of phosphine would you use in your reactant, and why?

6-9 For complexes having the formula *trans*-$Rh(CO)ClL_2$ (L = phosphine), arrange the following phosphines in order of decreasing energy of the carbonyl stretching band in the IR for their corresponding complexes: PPh_3, $P(t$-$C_4H_9)_3$, $P(p$-$C_6H_4F)_3$, $P(p$-$C_6H_4Me)_3$, $P(C_6F_5)_3$

6-10 Equilibrium constants for the reaction $Co(CO)Br_2L_2 \rightleftharpoons CoBr_2L_2 + CO$ are given below:

L	K
PEt_3	1
PEt_2Ph	2.5
$PEtPh_2$	24.2

Account for this trend in equilibrium constants.

7

Organometallic Reactions I

Reactions That Occur at the Metal

The diversity of transition metals and organic ligands suggests that the chemistry of organotransition metal complexes is also diverse. So much so, that trying to understand the reactions of organotransition metal compounds would seem to be a daunting task. Over the last 30 years, much progress has been made in the study of organometallic reaction mechanisms. The good news from these investigations is that organometallic compounds undergo relatively few kinds of reactions. Moreover, many of these reaction types have direct parallels with organic chemistry. We will try to point out these similarities during our discussion of basic kinds of reactions in this chapter and the next. Comprehension of these reaction types at this point in the text should greatly assist readers in grasping later topics, such as catalysis and organic synthesis.

In Chapter 7, we will consider reactions where the center of action occurs primarily at the metal and not at the ligands. Reactions that occur mainly at the ligands will be discussed in Chapter 8.

Table **7-1**[1] briefly shows several important kinds of organometallic reactions. Note that some of these reactions actually are the same, though the

[1]A note on symbolism used throughout the remainder of the text is in order. Recall from Chapter 3 that the symbol "L" represents a ligand that donates two electrons to a complex, e.g., CO or phospine. "X" (or sometimes "Y" or "Z") represents a one-electron donor such as Cl or alkyl. To indicate a collection of unspecified L-type ligands attached to a metal, the symbol "L_n" is employed.

Table 7-1 Common Reactions of Organotransition Metal Complexes

Reaction Type	Schematic Example	ΔOxidation State of M	ΔCoordination Number of M	Δe^- Count	Text Section
Ligand substitution	$L_2 + L_n ML_1 \rightarrow L_n ML_2 + L_1$	0	0	0	7-1
Oxidative addition	$L_n M + X\text{-}Y \rightarrow L_n M(X)(Y)$	+2	+2	+2	7-2
Reductive elimination	$L_n M(X)(Y) \rightarrow L_n M + X\text{-}Y$	-2	-2	-2	7-3
1,1-Insertion[a]	$L_n M\text{-}X{=}Z \rightarrow L_n M\text{-}X\text{-}Y$	0	-1	-2	8-1-1
1,2-Insertion[a]	$L_n M\text{-} \rightarrow L_n M\text{-}X\text{-}Z\text{-}Y$	0	-1	-2	8-1-2
Nucleophilic addition[b]	$L_n M\text{-} + Nuc:^- \rightarrow [L_n M\text{-}X\text{-}Z\text{-}Nuc]^-$	0[c]	0	0	8-2
Nucleophilic abstraction[b]	$L_n MCR + Nuc\text{-}H \rightarrow L_n M\text{-}H + Nuc\text{-}CR$	0	0	0	8-3
Electrophilic addition[b]	$L_n M\text{-}X{=}Z + E^+ \rightarrow [L_n M{=}X\text{-}Z\text{-}E]^+$	0	0	0	8-4
Electrophilic abstraction[b]	$L_n M\text{-}X\text{-}Z\text{-}Y + E^+ \rightarrow L_n M^+ \text{-} + E\text{-}Y$	0	0	0	8-4

[a] The reverse reaction, called *deinsertion* or *elimination*, is also possible.
[b] There is no one general reaction; several variations are possible.
[c] Depends upon change in hapticity; e.g., X=Z going from $\eta^3 \rightarrow \eta^2$ results in a 2 e$^-$ reduction.

forward reaction and the reverse reactions have different names (i.e., *oxidative addition* and *reductive elimination*). Also note the change in electron count and oxidation state during the course of these reactions.[2]

7-1 Ligand Substitution

Often the first (and last) step in a catalytic cycle (to be discussed in Chapter 9), ligand substitution is a very common reaction type. This reaction is comparable in many ways to substitutions that occur at the carbon atom in organic chemistry. Equation **7.1** shows the general reaction.

$$L_n ML_1 + L_2 \rightarrow L_n ML_2 + L_1 \qquad\qquad \textbf{7.1}$$

[2]For a good discussion of a number of common types of organometallic reactions, see C.A. Tolman, *Chem. Soc. Rev.,* **1972,** *1,* 337.

Much of the work on ligand substitution of organometallic complexes has involved metal carbonyls with a trialkyl or a triarylphosphine serving as the replacement ligand. There are two main mechanistic pathways by which substitution may occur—associative (A) and dissociative (D).

Associative (A):

$$L_nML_1 \;+\; L_2 \xrightarrow[\text{slow}]{} L_nML_1L_2 \xrightarrow[\text{fast}]{} L_nML_2 \;+\; L_1$$

Dissociative (D):

$$L_nML_1 \underset{\text{slow}}{\overset{}{\rightleftharpoons}} \begin{array}{c} L_nM \\ + \\ L_1 \end{array} \xrightarrow[\text{fast}]{+L_2} L_nML_2$$

The A pathway is characterized by a bimolecular rate-determining step involving the substrate and incoming ligand. Sixteen-electron complexes usually undergo ligand substitution by an A mechanism. D mechanisms, typically involving 18-electron complexes, show a rate-limiting step in which one ligand on the substrate must depart before another can bind to the metal. Although there are clean-cut examples of dissociative and associative reactions, several ligand substitutions do not fit neatly into either type. The pathways of these reactions seem to lie between the mechanistic extremes of A and D. These mechanisms are called interchange (I) pathways, specifically labeled I_a (associative interchange) and I_d (dissociative interchange), where, for example, I_a stands for *interchange with an increase in coordination number in the rate-determining step*. This situation is similar to nucleophilic substitution at carbon where a spectrum of pathways exists between the mechanistic extremes of S_N2 and S_N1.

7-1-1 *Trans* Effect and *Trans* Influence

Before we consider the mechanistic types of substitution reactions in more detail, it is worthwhile to consider an effect that can direct the course of ligand substitution known as the *trans effect*. It has long been known[3] that, particularly in square planar transition metal complexes, certain ligands seem to direct substitution preferentially to a position *trans* to themselves. Both thermodynamics and kinetics provide the basis for this tendency. When the reaction is controlled primarily by thermodynamic factors, the *trans* effect is more correctly termed the *trans influence*. The true *trans* effect is associated with reactions where factors

---

[3]I.I. Chernyaev, *Ann. Inst. Platine*, **1926**, *4*, 261.

affecting the rate of reaction control the product outcome. Over the years, the two phenomena have been lumped together as the *trans* effect. In this section we will discuss the two factors that govern the product distribution, dividing them into the *trans* influence and the *trans* effect.

Ligands that form strong σ bonds, such as hydride and alkyl, or π acceptor ligands, such as CN⁻, CO, or PR₃ that also bond strongly to the metal, tend to weaken the metal-ligand bond *trans* to the first ligand. In the ground state, this is a thermodynamic property called the *trans* influence. That a ligand does weaken the metal-ligand bond of the group situated *trans* to it has been demonstrated in studies of square planar platinum complexes.[4] These investigations have shown that a ligand with a strong *trans* influence causes, in the infrared region of the electromagnetic spectrum, a decrease in M-L stretching frequency of the ligand *trans* to it (thus indicating a weakening of the M-L bond).

Moreover, NMR studies have measured the coupling constant J_{Pt-P} of a number of trialkylphosphine platinum complexes. Trialkyl phosphines, as shown in Figures **7-1a** and **b**, have a greater *trans* influence than Cl, and this is reflected in the higher coupling constant (J_{Pt-P}) and *lower* IR Pt-Cl stretching frequency (v_{Cl-Pt}) present when Cl is *trans* to the phosphine rather than when PEt₃ is *trans* to itself. The higher coupling constant in the *cis* isomer reflects the higher s character of the ligand-metal bond. The methyl group, on the other hand, according to Figure **7-1c** has a higher *trans* influence than Cl as reflected in the small J_{Pt-P} of 1719 Hz, when methyl is *trans* to the phosphine, and 4179 Hz, when phosphine is *trans* to Cl.

Since the methyl group is a strong σ bonding ligand, it causes the σ bonding character of the Pt-P bond to decrease, thus decreasing the coupling constant. Phosphines in general have a high *trans* influence because they exert both strong σ donation and π acceptance from the metal, thereby weakening the bond *trans* to the phosphine.

Ligands that have a strong *trans* influence based on σ electron donating ability are listed in approximate descending order of their effect according to the following example:

$$H^- > PR_3 > SCN^- > I^- > CH_3^- > CO > CN^- > Br^- > Cl^- > NH_3 > OH^-$$

The tendency of certain ligands to direct incoming groups to the *trans* position also occurs with reactions under kinetic control. This case is known as the true *trans* effect whereby the influence of the ligand *trans* to the incoming one is felt due to the difference in energy between the ground state and the transition state in the rate-determining step. Not only are σ donation effects important, but also the effect of π donation from the metal to the ligand are

[4] A. Yamamoto, *Organotransition Metal Chemistry*, Wiley-Interscience, New York, 1986, 179-182.

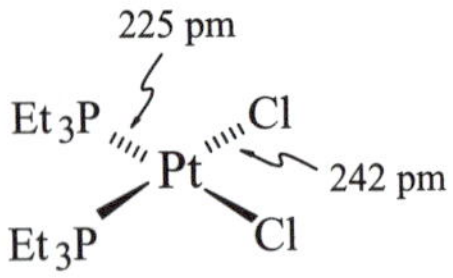

$$Et_3P \overset{225\ pm}{\underset{Et_3P}{\diagup}} \overset{Cl}{\underset{Cl}{\overset{Pt}{\diagdown}}} 242\ pm$$

a

$$Et_3P \overset{231\ pm}{\diagup} \overset{Cl}{\underset{Cl}{\overset{Pt}{\diagdown}}} \overset{232\ pm}{\underset{PEt_3}{}}$$

b

$\nu\,(Pt\text{-}P) = 442$ and $427\ cm^{-1}$

$\nu\,(Pt\text{-}Cl) = 303$ and $281\ cm^{-1}$

$J\,(Pt\text{-}P) = 3508\ Hz$

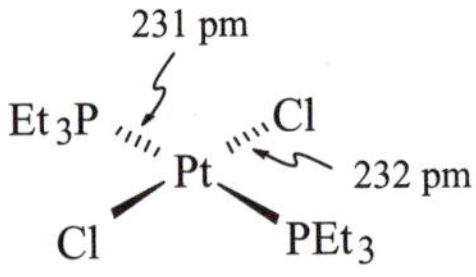

$\nu\,(Pt\text{-}P) = 419\ cm^{-1}$

$\nu\,(Pt\text{-}Cl) = 339\ cm^{-1}$

$J\,(Pt\text{-}P) = 2380\ Hz$

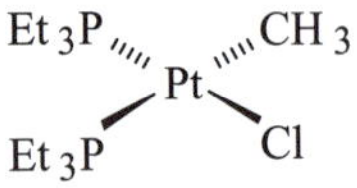

Figure 7-1
Spectral and Structural
Characteristics of Pt(II)
Complexes

$$Et_3P \overset{}{\underset{Et_3P}{\diagup}} \overset{CH_3}{\underset{Cl}{\overset{Pt}{\diagdown}}}$$

c

$J\,(Pt\text{-}P, PEt_3\ \textit{trans}\ to\ CH_3) = 1719\ Hz$

$J\,(Pt\text{-}P, PEt_3\ \textit{trans}\ to\ Cl) = 4179\ Hz$

important. When a ligand forms strong π acceptor bonds with platinum, for instance, charge is removed from the metal. The effect on the energy of the ground state is relatively small, but it is significant on the transition state energy. This is because, during ligand substitution of a square planar Pt complex, we usually expect an increase in coordination to trigonal bipyramidal geometry in the transition state (see Section **7-1-2**). A π bonding ligand originally present in the substrate (the complex undergoing substitution) can contribute to stabilizing the transition state, **1**, via the d_{xz} orbital from the metal.

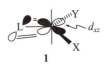

1

Overall, this effect increases the rate of reaction *trans* to the strong π acceptor ligand. The order of π bonding effects for various ligands is as follows:

$$C_2H_4, CO > CN^- > NO_2^- > SCN^- > I^- > Br^- > Cl^- > NH_3 > OH^-$$

Table 7-2 Effect of *trans* Ligand on Rate for *trans*-Pt(PEt₃)₂L(Cl) at 25°C

L	k (M⁻¹s⁻¹)
H⁻	4.2
PEt₃	3.8
CH₃⁻	6.7 × 10⁻²
Ph⁻	1.6 × 10⁻²
Cl⁻	4 × 10⁻⁴

$$\begin{array}{ccc} L & & PEt_3 \\ & \diagdown\!\!Pt\!\!\diagup & \\ Et_3P & & Cl \end{array}$$

When σ donation and π acceptance effects are combined, the overall *trans* effect list is as follows:

$$CO, CN^-, C_2H_4, > PR_3, H^- > CH_3^-, S=C(NH_2)_2 > C_6H_5^- > NO_2^-, SCN^-, I^- > Br^- > Cl^- > Py, NH_3, OH^-, H_2O$$

Ligands highest in the series are strong π acceptors, followed by strong σ donors. Ligands at the low end of the series possess neither strong σ nor π bonding abilities. The *trans* effect can be very large—rates in platinum complexes may differ as much as 10^4 between the complexes with strong *trans* effect ligands and those with weak ones. Table **7-2** shows the relative rate of pyridine (Py) substitution on *trans*-Pt(PEt₃)₂(Cl)L when L is varied.[5]

To assess the nature of the *trans* effect, one must investigate the influence of a ligand in destabilizing the ground state (really the *trans* influence) and in stabilizing the transition state. Groups such as alkyl and hydride do not have π acceptor orbitals and thus are unable to form π bonds to the metal that could effectively stabilize the trigonal bipyramidal transition state in an associative reaction involving a coordinatively unsaturated[6] square planar complex. Presumably these ligands are effective solely by lowering the bond energy of the M-X bond and destabilizing the reactant. Ligands such as CO and the phosphines can stabilize the trigonal bipyramidal transition state through π bonding and can also decrease the energy of the M-X bond in the ground state through strong σ donation. Which effect is dominant is sometimes difficult to determine, but Figure **7-2** summarizes the possible situations that could occur. When all is said and done, however, the phenomenon is usually called the *trans* effect regardless of whether the reason for the effect of the directing ligand is kinetic or thermodynamic in origin.

[5] F. Basolo, J. Chatt, H.B. Gray, R.G. Pearson, and B.L. Shaw, *J. Chem. Soc.*, **1961**, 2207.

[6] A coordinatively unsaturated complex possesses less than 18 electrons and has sites available for ligand bonding, e.g., a 16-electron square planar Pt complex. An octahedral, 18-electron complex, such as Cr(CO)₆, is deemed coordinatively saturated.

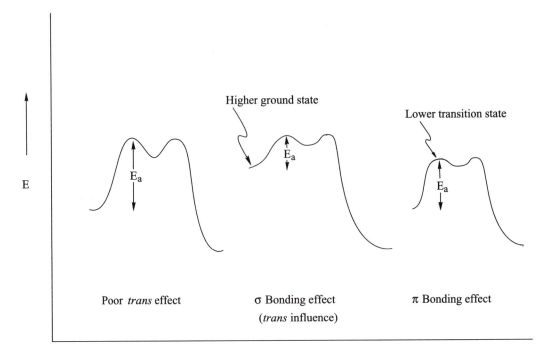

Figure 7-2
Activation Energy and the *trans* Effect
The depth of the energy curve for the intermediate
and the relative heights of the two maxima will vary
with the specific reactants. (Adapted from G.L.
Miessler and D.A. Tarr, *Inorganic Chemistry*,
Prentice-Hall, Englewood Cliffs, NJ, 1991, 397.)

The *trans* effect is somewhat different in octahedral complexes due to the smaller *s* character of each bond and possibly due to steric effects caused by the relatively greater crowding found in octahedral versus square planar environments. In general, however, it appears that *trans* effects are related to the ability of the ligand to stabilize the transition state during rate-limiting dissociation. Studies on octahedral Cr complexes[7] indicate that the order of the *trans* effect is rather similar and follows the general trend as shown above.

Equations **7.2** and **7.3** provide two examples showing how one could perform a synthesis taking advantage of the *trans* effect.

[7]J.D. Atwood, M.J. Wovkulich, and D.C. Sonnenberger, *Acc. Chem. Res.*, **1983**, *16*, 350.

$$\begin{bmatrix} & \underset{|}{\text{NH}_3} & \\ \text{H}_3\text{N} - & \text{Pt} & -\text{NH}_3 \\ & \underset{|}{\text{NH}_3} & \end{bmatrix}^{2+} \xrightarrow{\text{Cl}^-} \begin{bmatrix} & \underset{|}{\text{NH}_3} & \\ \text{Cl} - & \text{Pt} & -\text{NH}_3 \\ & \underset{|}{\text{NH}_3} & \end{bmatrix}^{+} \xrightarrow{\text{Cl}^-} \begin{matrix} & \underset{|}{\text{NH}_3} & \\ \text{Cl} - & \text{Pt} & -\text{Cl} \\ & \underset{|}{\text{NH}_3} & \end{matrix} \quad \textbf{7.2}$$

$$\begin{bmatrix} & \underset{|}{\text{NH}_3} & \\ \text{Py} - & \text{Pt} & -\text{Py} \\ & \underset{|}{\text{NH}_3} & \end{bmatrix}^{2+} \xrightarrow{\text{Cl}^-} \begin{bmatrix} & \underset{|}{\text{NH}_3} & \\ \text{Py} - & \text{Pt} & -\text{Cl} \\ & \underset{|}{\text{NH}_3} & \end{bmatrix}^{+} \xrightarrow{\text{Cl}^-} \begin{matrix} & \underset{|}{\text{NH}_3} & \\ \text{Cl} - & \text{Pt} & -\text{Cl} \\ & \underset{|}{\text{NH}_3} & \end{matrix} \quad \textbf{7.3}$$

Exercise 7-1

Propose synthetic routes for the following:

a.

$$[\text{PtCl}_3(\text{NH}_3)]^- \rightarrow \quad \begin{matrix} & \underset{|}{\text{Br}} & \\ \text{Cl} - & \text{Pt} & -\text{NH}_3 \\ & \underset{|}{\text{Py}} & \end{matrix}$$

b. $\text{PtCl}_4^{2-} \rightarrow trans\text{-}[\text{PtCl}_2(\text{NO}_2)(\text{NH}_3)]^-$

7-1-2   Associative Substitution

Coordinatively unsaturated 16-electron complexes typically undergo associative substitution. Here the mechanism involves a slow bimolecular step where the incoming ligand and 16-electron complex combine to form a coordinatively saturated 18-electron intermediate. The intermediate rapidly expels the leaving group to give the new substituted 16-electron product. This process is outlined in equation **7.4**.

$$\text{ML}_4 + \text{Y} \xrightarrow{k_1} \text{ML}_4\text{Y} \xrightarrow{k_2 \text{ (fast)}} \text{ML}_3\text{Y} + \text{L} \qquad \textbf{7.4}$$

A two-term rate law lends support to the mechanism outlined above and often takes the following general form:

$$\text{Rate} = k_S[\text{ML}_4] + k_1[\text{ML}_4][\text{Y}]$$

The k_s term arises because the solvent may also act as a nucleophile. Since a large excess of solvent is usually present in a reaction, the first term is called *pseudo* first order. The rate law expression is identical to that for an S_N2 reaction in organic chemistry. These reactions typically have highly negative entropies of activation,[8] which provide additional support for the associative nature of the reaction.[9]

Associative substitution is usually found with square planar d^8 metal complexes such as those of Ni(II), Pd(II), Pt(II), Ir(I), and Au(III). Substitution reactions of these complexes have been thoroughly investigated.[10] As in the case of S_N2 reactions, the rate of substitution depends upon several factors including nucleophilicity (*soft* nucleophiles such as SCN⁻ and PR_3 are particularly effective when the metal center is *soft*, e.g., Pt(II)), leaving group ability (a function of Brønsted-Lowry base strength—the weaker the base, the better the leaving group), and *trans* effect.

Hard and Soft Acids and Bases

The concept of hard and soft acids and bases needs some explanation here. We are really talking about the hardness or softness of *Lewis* acids or bases. Hard acids tend to be small in size, have a large positive charge, and possess an electron cloud that is not capable of significant distortion (non-polarizable). Examples of hard acids would be H^+, Al^{3+}, or BF_3. Soft acids are just the opposite in that they tend to be of large size with respect to their charge and possess polarizable electron clouds. Species such as Hg^{2+}, Tl^+ and low-valent transition metals typify soft acids.

Hard and soft bases are similar to their acid counterparts except that they are an electron-rich species instead of electron poor. Hard bases would include F⁻, OH⁻, or NH_3. Species such as I⁻, R_2S, CO, or R_3P constitute

---

[8] The enthalpy of activation ($\Delta H^{\ddagger}$) and the entropy of activation ($\Delta S^{\ddagger}$), both derived from rate studies on reactions, are useful in providing information about the mechanism of a reaction. Both arise from absolute rate theory (also known as transition state theory), which treats the activated complex at the transition state as a discrete entity in equilibrium with reactants. The value for $\Delta H^{\ddagger}$ corresponds to E_a from the Arrhenius equation according to $E_a = \Delta H^{\ddagger} + RT$. Its magnitude is a measure of bond energy changes in going from reactants to transition state. The entropy of activation indicates the change in order as the reaction proceeds to the transition state. A highly negative value for $\Delta S^{\ddagger}$ implies that order increases on the way to the transition state, as exemplified by a bimolecular transformation such as the Diels-Alder reaction. A reaction that gives a positive value for $\Delta S^{\ddagger}$ indicates that the path to the transition state leads to greater disorder. Dissociative processes such as the S_N1 reaction show large positive values of $\Delta S^{\ddagger}$. For more information on absolute rate theory, see F. A. Carey and R. J. Sundberg, *Advanced Organic Chemistry, Part A,* 3rd ed., Plenum Press, New York, 1990, 191-196.

[9] J. D. Atwood, *Inorganic and Organometallic Reaction Mechanisms*, Brooks/Cole, Monterey, Calif., 1985, 65-66.

[10] J. P. Collman, L. S. Hegedus, J. R. Norton, and R. G. Finke, *Principles and Applications of Organotransition Metal Chemistry*, 2nd ed., University Science Books, Mill Valley, Calif., 1987, 241-244 and references therein.

examples of soft bases. A good rule of thumb to apply when assessing the softness of an acid or base is to assume that softness increases as one proceeds down
a column or group in the Periodic Table.

Reactions between hard acids and hard bases readily occur because the
two reacting partners have a strong coulombic attraction for each other. Soft
acids and soft bases tend to react rapidly for a more subtle reason. Figure **7-3**
suggests that soft acids and bases interact well because the HOMO (highest
occupied molecular orbital) of the base is close in energy to the LUMO (lowest unoccupied molecular orbital) of the acid. Since the HOMO is filled with
two electrons and the LUMO lacks electrons, the interaction between the two
orbitals gives a doubly occupied bonding orbital and an empty antibonding
orbital. Overall, the interaction of the two orbitals results in a substantial energy
lowering. This would not be the case if the acid were hard (the LUMO would
be quite high in energy) and the base also hard (the HOMO is relatively low in
energy). If so, the HOMO and LUMO would be so far apart in energy that
effective interaction between them would be negligible, resulting in very little
energy lowering.

It is difficult to quantify the "hardness" or "softness" of a Lewis acid or
base. We cannot measure a pK_a value as in the case of assessing the strength of
Brønsted-Lowry acids. R.G. Pearson has reported a scale of absolute hardness
for acids and bases defining hardness to be one-half the difference between the
ionization potential and the electron affinity of an atom or species.[11] He has
obtained good agreement in many cases between the calculated values for hardness and the experimental behavior of the acid or base.

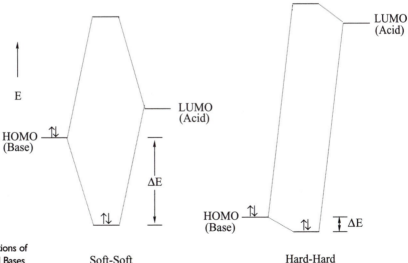

Figure 7-3
HOMO-LUMO Interactions of
Hard and Soft Acids and Bases   Soft-Soft   Hard-Hard

[11]R.G. Pearson, *Inorg. Chem.,* **1988,** *27,* 734. See also G.L. Miessler and D.A. Tarr, *Inorganic
Chemistry,* Prentice-Hall, Englewood Cliffs, NJ, 1991, 197-202, 213-217.

Figure 7-4
Ligand Substitution of a
Square Planar Complex

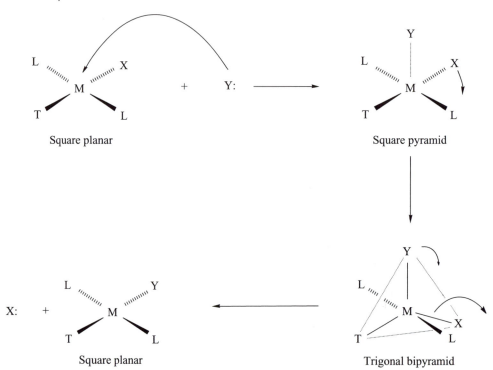

Unlike S_N2 reactions in organic chemistry, associative substitution of square planar complexes usually results in *retention* of configuration and not inversion. Figure **7-4** shows that the reaction is thought to proceed by attack of the incoming nucleophile **Y** (or solvent) from the top or up from the bottom to give a square-pyramidal intermediate that rearranges to a trigonal bipyramid. Note that the *trans* ligand, **T**, and the incoming and leaving ligands, **Y** and **X**, are all equatorial substituents of the trigonal bipyramid. Finally the leaving group leaves and the final product is a square planar complex with the incoming ligand *trans* to **T**.

Associative Substitution with 18-Electron Systems
The 18-electron rule makes it unlikely that coordinatively saturated, 18-electron complexes would undergo substitution by an associative mechanism.

An exception to this conventional wisdom was formulated by Basolo[12] who postulated that substitution in 18-electron complexes may occur associatively if the metal complex can delocalize a pair of electrons onto one of its ligands.

For example, equation **7.5** shows substitution of an 18-electron molybdenum complex by phosphorus ligands,[13] which proceeds with a first-order rate law in starting complex and first-order in incoming ligand—exactly as would be demonstrated in an associative substitution of a 16-electron square planar complex as discussed earlier in Section **7-1-2**. Note that the rate-determining step of this reaction involves a change in hapticity of the arene ligand from η^6 to η^4 coincident with attachment of the incoming phosphine ligand.

$$L = P(n\text{-butyl})_3 \qquad\qquad \text{rate} = k\,[(PhH)Mo(CO)_3][L] \qquad\qquad \textbf{7.5}$$

Such a change preserves the electron count in the coordination sphere when the extra incoming ligand is attached. Studies on similar molybdenum-triene complexes have resulted in the isolation of 18-electron η^4-(triene) $Mo(CO)_3(\eta^4\text{-}C_6H_6)[P(n\text{-Bu})_3]$; moreover, an η^4-arene complex has been isolated and characterized in the solid state.[14]

Equations **7.6** to **7.9** provide examples of associative substitution reactions on 18-electron complexes. In all cases modification of the electron contribution of one of the ligands occurs. In reaction **7.6** the nitrosyl group rearranges from its three-electron, linear (an 18-e$^-$ complex) form to a one-electron, bent (a 16-e$^-$ complex) form. This allows the phosphine to attack to give the 18-e$^-$ complex $Co(CO)_3(PR_3)(NO)$. Finally CO departs to give the product. The overall rate of the reaction is directly related to the nucleophilicity of the phosphine and follows the order: $PEt_3 > PEt_2Ph > PPh_3 > P(OMe)_3$ (see Chapter 6).

$$Co(CO)_3NO + PR_3 \rightarrow Co(CO)_2(PR_3)(NO) + CO \qquad\qquad \textbf{7.6}$$

[12]F. Basolo, *Inorg. Chim. Acta,* **1981**, *50*, 65.

[13]F. Zingales, A. Chiesa, and F. Basolo, *J. Am. Chem. Soc.,* **1966**, *88*, 2707.

[14]D.J. Darensbourg, "Ligand Substitution in Metal Carbonyls," in F.G.A. Stone and R. West, Eds., *Advances in Organometallic Chemistry,* vol. 21. Academic Press, New York, 123.

Reaction **7.7** involves conversion from η^5 to η^1-Cp, and reaction **7.8** shows "slippage" of η^5 to η^3-indenyl. One of the most interesting examples of this type involves reorganization of electrons in a tetrazole ligand attached to iron in reaction **7.9**. The tetrazole changes from an L_2 ligand to an X_2. The $\Delta H^{\ddagger}$ of 6.9 kcal/mole and $\Delta S^{\ddagger}$ of –31.4 eu clearly support a rapid associative mechanism for the reaction.

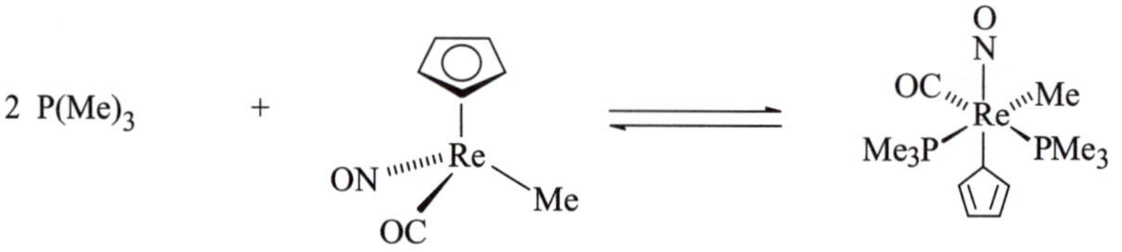

7.7

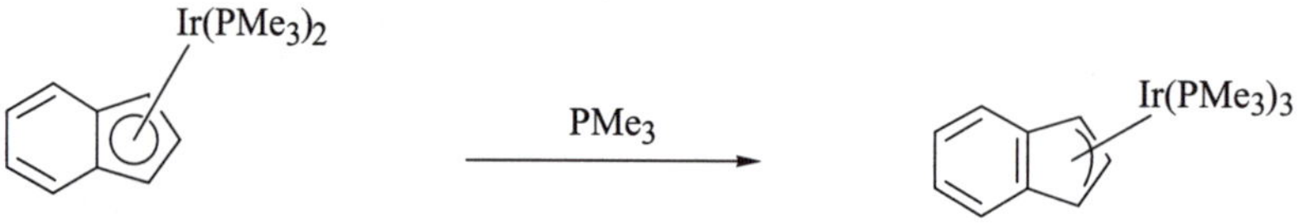

7.8

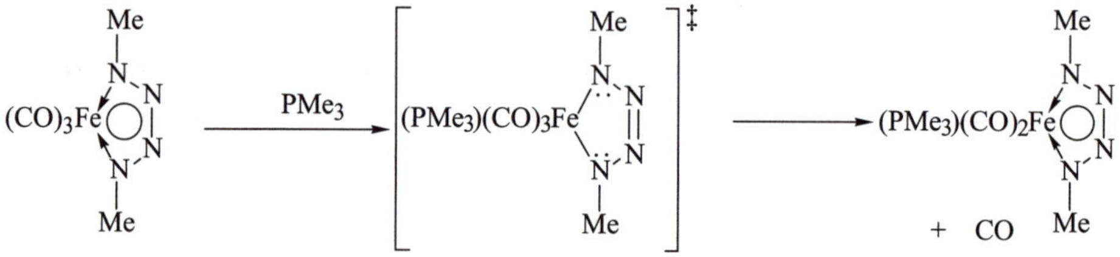

7.9

Exercise 7-2

a. Propose a structure for the starting material in equation **7.6**.

b. Calculate the electron count for each complex in equation **7.9**.

7-1-3   Dissociative Displacements

The D mechanism requires the formation of an intermediate with reduced coordination number, and is the most common pathway for 18-electron complexes as outlined in equations **7.10** and **7.11**.

$$ML_6 \underset{k_{-1}}{\overset{k_1}{\rightleftharpoons}} ML_5 + L \qquad\qquad\qquad 7.10$$

$$ML_5 + Y \xrightarrow{k_2} ML_5Y \qquad\qquad\qquad 7.11$$

Application of the steady state approximation[15] provides a rate law for this mechanism as:

$$Rate = \frac{k_1 k_2 [ML_6][Y]}{k_{-1}[L] + k_2[Y]}$$

If the concentration of leaving group (L) is relatively small compared to that of the incoming nucleophile or if k_{-1} is low, then the rate law reduces to the following:

$$Rate = k[ML_6]$$

The above equation is much like the rate law for a limiting S_N1 reaction in organic chemistry. Studies have shown, however, that the resulting 16–electron intermediates, formed upon ligand dissociation, are highly reactive and show

---

[15]The steady-state approximation assumes that, in a multistep reaction, the concentration of intermediates is small and constant because the rate of formation equals the rate of destruction. Application of the steady-state approximation for equations **7.10** and **7.11** proceeds as follows:

We first write the rate law as

$$Rate = k_2[ML_5][Y]$$

Since we do not know the concentration of the fleeting intermediate, ML_5, we apply the steady-state approximation to find its concentration in terms of variables that are measurable. We assume that the change in concentration of ML_5 with time is equal to zero (that is, the rate of formation of intermediate is equal to the rate of disappearance—this *is* the approximation). Thus we have the following equation:

$$d[ML_5]/dt = 0 = k_1[ML_6] - k_{-1}[ML_5][L] - k_2[ML_5][Y]$$

Rearranging terms in order to solve for $[ML_5]$, we have the following equations:

$$[ML_5]\{k_{-1}[L] + k_2[Y]\} = k_1[ML_6]$$

and

$$[ML_5] = k_1[ML_6]/\{k_{-1}[L] + k_2[Y]\}$$

Plugging the value for $[ML_5]$ back into the original rate law expression, we have a new equation:

$$Rate = k_1 k_2[ML_6][Y]/\{k_{-1}[L] + k_2[Y]\}$$

For more discussion of the steady-state approximation, see I.N. Levine, *Physical Chemistry*, 3rd ed. McGraw-Hill, New York, 1988, 532-533 and F.A. Carey and R.J. Sundberg, *Advanced Organic Chemistry, Part A*, 3rd ed., Plenum Press, New York, 1990, 186-188.

little selectivity toward achieving coordinative saturation. Thus, k_{-1} is often comparable in magnitude to k_2,[16] and the more complicated rate law holds unless the incoming ligand is in large concentration. Studies on the kinetics of ligand substitution reactions of metal carbonyl complexes have yielded significant information on the mechanism of displacement of other neutral ligands, (e.g., phosphines, alkenes, and amines) from low-valent metal centers. A typical example of a substitution that occurs by a dissociative mechanism is shown in equation **7.12**.

$$Cr(CO)_6 + PPh_3 \xrightarrow{130°} Cr(CO)_5PPh_3 + CO \qquad \textbf{7.12}$$

Relative Rates

Rates of substitution for various ML_6 complexes (the carbonyl complexes have been the most studied) differ considerably depending on the metal and the ligands present. The 18-electron species $[V(CO)_6]^-$ is resistant to substitution by even molten PPh_3.[17] Group 6 hexacoordinate complexes typically show k_1 values ranging from 10^{-12} to 10^{-3} sec^{-1}.[18]

Second-row complexes tend to react faster than first- or third-row complexes in simple ligand substitutions by a dissociative mechanism.[19] This trend has been well studied for Group 6 complexes and is often observed with d^9 and d^{10} metals. The reasons for this trend are unclear, however. For the series of metal carbonyls $M(CO)_6$ where M = Cr, Mo, and W, the observed order of reactivity does not mirror the order of M–C bond energies, which are W > Mo > Cr. The order does seem to correspond to the calculated values for the force constants of M–C bonds in Group 6 metal carbonyls.[20]

Steric Effects

Steric effects play a role in the rate of ligand displacement. The rate of reaction (equation **7.13**) has been measured for a number of phosphine ligands (PR_3), and the results indicate that the rate of substitution is accelerated by bulky phosphines. A good correlation between cone angle (see section **6-3**) of the phosphine ligand and rate of displacement by CO was obtained, as shown in Table **7-3**.

$$cis\text{-}Mo(CO)_4(PR_3)_2 + CO \rightarrow Mo(CO)_5(PR_3) + PR_3 \qquad \textbf{7.13}$$

[16]D.J. Darensbourg, *op. cit.*, 115.
[17]J.A.S. Howell and P.M. Burkinshaw, *Chem. Rev.*, **1983**, *83*, 559.
[18]D.J. Darensbourg, *op. cit.*, 117.
[19]D.J. Darensbourg, *op. cit.*, 116.
[20]J.A.S. Howell and P.M. Burkinshaw, *op. cit.*

Table 7-3 Rates of Ligand (L) Replacement by CO at 70°C[a]

L	Cone angle (°)	Rate constant (sec^{-1})	$\Delta H^{\ddagger}$ (kcal/mole)	$\Delta S^{\ddagger}$ (eu)[b]
Phosphines:				
PMe$_2$Ph	122	$<1.0 \times 10^{-6}$		
PMePh$_2$	136	1.33×10^{-5}		
PPh$_3$	145	3.16×10^{-3}	29.7	14.4
PPhCy$_2$[c]	162	6.40×10^{-2}	30.2	21.7
Phosphites:				
P(OPh)$_3$	128	$<1.0 \times 10^{-5}$		
P(O-o-tol)$_3$	141	1.60×10^{-4}	31.9	14.4

[a]D.J. Darensbourg, *op. cit.*, 119.
[b]eu = cal/mole-K
[c]Cy = cyclohexyl

Cis Effect

Octahedral complexes of the general formula M(CO)$_5$X often show a tendency to lose a CO substituted *cis* to the X ligand (Group 6 and 7 metal complexes have been well investigated[21]). This phenomenon of *cis* labilization is termed the *cis effect*. Loss of a carbonyl group from an octahedral complex provides two distinct square pyramidal intermediates: **a** and **b** (Figure 7–5).

Calculations show[22] that a ligand such as a halogen (a poor σ donor and a ligand that is not a π acceptor) stabilizes the geometry shown in **a** more than **b**, and thus as a consequence of the Hammond postulate (see the following page), the transition state leading to **a** is more stable than that leading to **b**. When *cis* ligands are present that do not share the electronic properties of the halogens

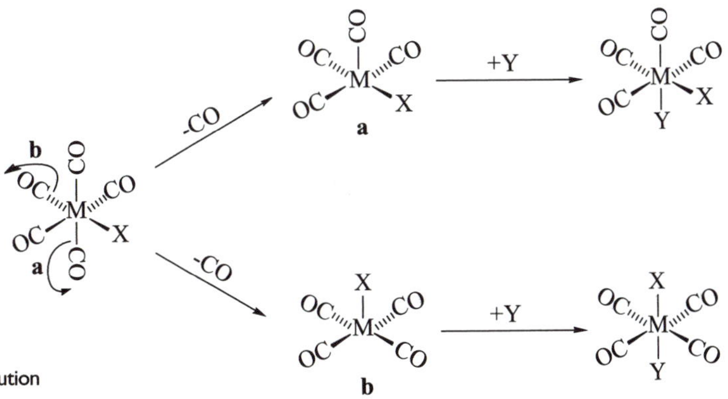

Figure 7-5
Intermediates in the Substitution
of M(CO)$_5$X Complexes

[21]J.D. Atwood and T.L. Brown, *J. Am. Chem. Soc.*, **1976**, *98*, 3160.
[22]M. Elian and R. Hoffmann, *Inorg. Chem.*, **1975**, *14*, 1058 and J.K. Burdett, *J. Chem. Soc., Faraday Trans. 2*, **1974**, *70*, 1599.

(i. e., they are better σ donors and/or have the capacity for π acceptance), the preference for *cis* labilization is reduced. Ligands that are stronger π acceptors than CO (such as $F_2C=CF_2$) actually labilize preferential loss of the *trans* CO.

The Hammond Postulate

A mechanism is a description of events—primarily bond breaking and making—that occur during the transformation of reactant to product. It is important to have knowledge about the structure of the transition state, especially the one involved in the rate-determining step. Such knowledge assists the chemist in predicting what effects different substituents will have on the rate of the reaction and the product distribution (if more than one product is possible), and in understanding the role of stereochemistry on the transformation of reactant to product.

The Hammond postulate allows us to ascertain the structure of the transition state under certain conditions if we know something about the structure and energy of the immediate prior or later species, which could be an unstable intermediate, reactant, or product. Proposed originally by George Hammond,[23] the postulate declares, "if two states, as for example, a transition state and an unstable intermediate, occur consecutively during a reaction process and have nearly the same energy content, their interconversion will involve only a small reorganization of molecular structure."

We can apply this postulate by looking at three cases described in Figure **7-6**. Figure **7-6a** shows a highly exothermic overall reaction or step in a reaction coordinate-energy diagram. The energy of the reactant is very similar to that of the transition state. In this case, we can say that the transition state comes early in the reaction coordinate and the structure of the starting material closely resembles that of the transition state. Figure **7-6b** shows the opposite situation where a highly endothermic step occurs. The transition state comes late in the reaction pathway and thus resembles that of the product. A good example here would be the ionization step that occurs in an S_N1 reaction where a substrate undergoes dissociation to give a leaving group and a carbocation intermediate (equation **7.14**).

$$(CH_3)_3C\text{-}Br \rightleftharpoons (CH_3)_3C^+ + Br^- \qquad\qquad \textbf{7.14}$$

The last case, Figure **7-6c**, indicates a transition state that is much higher in energy than either reactant or product. In this instance we can say little about the nature of the transition state as a function of the structure of reactant or product.

---

[23]G.S. Hammond, *J. Am. Chem. Soc.*, **1955**, 77, 334. The Hammond postulate is also discussed in organic chemistry textbooks, e.g., J. McMurry, *Organic Chemistry*, 4th ed. Brooks-Cole, Pacific Grove, CA, 1996, 206-210 and F.A. Carey and R.J. Sundberg, *Advanced Organic Chemistry, Part A*, 3rd ed., Plenum Press, New York, 1990, 211-215.

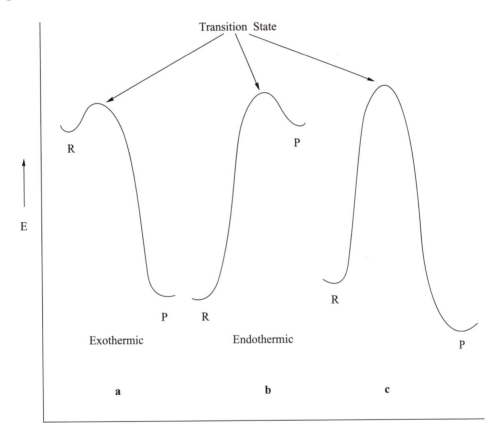

Figure 7-6
The Relationship of Transition State
to Reaction Coordinate and Energy—
The Hammond Postulate (Adapted
from F.A. Carey and R.J. Sundberg,
Advanced Organic Chemistry, part A,
3rd ed., Plenum Press, NY, 1990, 212.)

Other Ligands

The rate of substitution of a particular ligand is a function of ligand type. Carbon donor L-type ligands, such as CO or arene, dissociate rather easily because they are neutral in the free state. L_nX groups (such as Cp), on the other hand, dissociate more reluctantly as radicals or ions that are less stable.

A picture emerges from several lines of investigation on the displacement of ligands from coordinatively saturated species. That is, *the reaction appears to proceed by rate-determining loss of ligand to give a highly reactive 16-electron intermediate.* This intermediate probably closely resembles the transition state occurring during the rate-determining step, according to the Hammond postulate. The intermediate then captures a new nucleophile to complete the process of substitution.

ML$_5$ and ML$_4$ Complexes

ML$_5$ d^8 systems involving Group 8 and 9 metals are typically trigonal bipyramidal in structure, and because of this, low-energy barriers to exchange and interconversion between square-pyramidal and trigonal bipyramidal forms occur. The facility of these isomerizations makes mechanistic conclusions on the basis of product stereochemistries rather meaningless. Nevertheless, many ML$_5$ complexes undergo substitution reactions, and kinetic parameters often show a D mechanism. Equations **7.15** to **7.17** serve as examples for ML5 substitutions.

$$Ru(CO)_4PPh_3 + PPh_3 \rightarrow trans\text{-}Ru(CO)_3(PPh_3)_2 + CO \qquad \textbf{7.15}$$
$$\Delta S^{\ddagger} = +14 \text{ eu}$$

$$(CO)_4Fe(CH_2\text{=}CHPh) + CO \rightarrow Fe(CO)_5 + CH_2\text{=}CHPh \qquad \textbf{7.16}$$
$$\Delta S^{\ddagger} = +13 \text{ eu}$$

$$\overset{\displaystyle O}{\overset{\|}{MeCCo(CO)_4}} + PPh_3 \rightarrow \overset{\displaystyle O}{\overset{\|}{MeCCo(CO)_3PPh_3}} + CO \qquad \textbf{7.17}$$
$$\Delta S^{\ddagger} = +3.2 \text{ eu}$$

The dissociative substitution of d^{10} ML$_4$ complexes has been well studied, particularly with regard to M = Ni (see equation **7.18** below).

$$Ni(CO)_4 + {}^{14}CO \rightarrow Ni(CO)_3({}^{14}CO) + CO \qquad \textbf{7.18}$$
$$-10°$$

Here the geometry of the starting complex is invariably tetrahedral. Ni(CO)$_4$ also reacts readily with phosphines to give mono- and disubstituted complexes. The resulting mono- and disubstituted complexes react incrementally more slowly than Ni(CO)$_4$ toward further ligand exchange due to the stabilizing influence of phosphine ligands.

The rate of ligand dissociation of (L= PR$_3$) complexes to give ML$_3$ correlates strongly with the cone angle. As we have seen previously, the dissociation is slower when phosphite ligands, with the same cone angle as the corresponding phosphine, are compared. This is due to the greater π acceptance ability of phosphites compared to phosphines and results in ground state metal stabilization. For the Group 10 d^{10} ML$_4$ series, the rate of substitution typically varies in the order Pd > Pt ≈ Ni.

7-1-4   Interchange Pathway

We would not expect coordinatively saturated 18-electron complexes to increase coordination number or electron count during ligand substitution processes. Earlier in this chapter the interchange (I) pathway was mentioned as a mechanistic possibility for ligand substitution. There is evidence that indicates that something like an I pathway can happen in which there is an apparent increase in coordination number on the way from coordinatively saturated reactant to product. This is shown in outline form in equation **7.19**.

$$ML_6 + L' \overset{K_{diff}}{\rightleftharpoons} ML_6 \cdot L' \underset{k_{-3}}{\overset{k_3}{\rightleftharpoons}} ML_6L' \rightarrow ML_5L' + L \qquad \text{7.19}$$

The best-studied cases where interchange pathways may occur are those in which there is competition between I (actually an I_d, dissociative interchange) and D mechanisms (Scheme **7.1**)

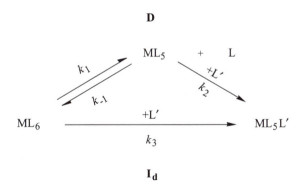

Scheme 7.1
Competing D and I_d Pathways

In the I_d mechanism (as shown in equation **7.19**), the substrate and **L'** combine under diffusion-control[24] to give a "cage complex" ($ML_6 \cdot L'$) in which **ML_6** and **L'** are loosely bound together and confined inside a "cage" of solvent molecules. **$ML_6 \cdot L'$** then undergoes rate-determining formation of **ML_6L'**, followed by rapid loss of leaving group, **L**. According to this scheme and using the steady-state approximation, the rate law would be as follows:

$$\text{Rate} = K_{diff}k_3[ML_6][L']$$

[24]Diffusion-controlled rate processes are very rapid, and are influenced solely by the rate at which molecules can diffuse toward one another and collide.

The I_d mechanism may be further described as a process that proceeds through a transition state, as in **2** below:

$$[L' \cdots\cdots ML_5 \cdots\cdots L]$$

2

Here the rate-determining step occurs when both the incoming and departing ligands are loosely held in the coordination sphere. The I_d mechanism distinguishes itself from the I_a in that there is considerable rupture of the M–L bond of the departing ligand and very little bond making between the metal and the incoming **L'** group in the transition state. It is difficult to find hard evidence for such a mechanism because isolation of the intermediates **$ML_6 \cdot L'$** or **$ML_6 L'$** would be problematic indeed.

More indirect kinetic evidence suggests that competing D and I_d pathways (Scheme **7.1**) can occur under some conditions. When L' = amines, phosphines, or nitriles, results indicate that a two–term rate law applies:

$$Rate = k_1[ML_6] + k_3[ML_6][L']$$

In this expression k_1 refers to the rate constant for rate-determining dissociation (equation **7.10**) and k_3 to that for the rate-determining step in an I_d process (equation **7:19**). Table **7-4** gives a few examples of Group 6 metal complexes that seem to undergo simultaneous D and I_d substitutions. Note that the enthalpy of activation, $\Delta H_3^\ddagger$, for the I_d pathway is less than that for the corresponding purely dissociative process, $\Delta H_1^\ddagger$. The $\Delta S^\ddagger$ terms for the two

Table 7-4[a] Activation Parameters for Ligand Substitution of $M(CO)_6$ with PBu_3

M	$\Delta H_1^{\ddagger b}$	$\Delta S_1^{\ddagger c}$	$\Delta H_3^\ddagger$	$\Delta S_3^\ddagger$
Cr	40.2	22	25.5	−15
Mo	31.6	6.7	21.7	−15
W	39.9	14	29.2	−6.9

[a]Taken from data reported by J.A.S. Howell and P.M. Burkinshaw, op. cit.,560.
[b]kcal/mole
[c]entropy units (eu), cal/mole-K

pathways, moreover, are vastly different—the purely dissociative process shows a highly positive entropy of activation, while $\Delta S^{\ddagger}$ for I_d processes is clearly negative, as would be expected for a transition state that increases in order on going from reactants to products.

The data in Table **7-4** also show that the activation enthalpy for both pathways is lowest when M = Mo, consistent with the general trend in rate of ligand substitution of first row, third row < second row. One reason for the increasing tendency for occurrence of the I_d pathway in going from Cr to Mo might be that Mo has a larger atomic radius than Cr and thus could accommodate a larger coordination number than its smaller homolog.

Scheme **7.2**, which describes another case where an I_d process may occur, involves substitution on complexes containing a polyhapto ligand such as a diene or triene.

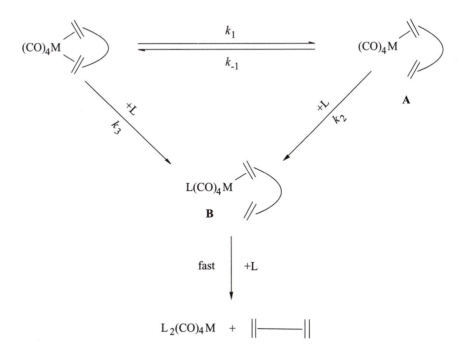

Scheme 7.2
Substitution Involving a
Substrate with a Polyhapto Ligand

Application of the steady-state approximation with respect to the two intermediates, **A** and **B**, gives a rate law of the following form:

Rate = $k_1 k_2[M(CO)_4 diene][L]/(k_{-1} + k_2[L]) + k_3[M(CO)_4 diene][L]$

If $k_2[L]$ is large (i.e., there is a large concentration of incoming ligand), then the rate law reduces to the following:

Rate = $k_1[M(CO)_4 diene] + k_3[M(CO)_4 diene][L]$

The above is an expression that corresponds to simultaneous D and I_d pathways. Another example of such a reaction is shown in equation **7.20**:

$$(cod)Mo(CO)_4 + 2\,AsPh_3 \rightarrow (AsPh_3)_2 Mo(CO)_4 + cod \qquad \textbf{7.20}$$

$$\Delta H_1^{\ddagger} = 25.0 \text{ kcal/mole} \quad \Delta S_1^{\ddagger} = 2.2 \text{ eu}$$
$$\Delta H_3^{\ddagger} = 17.0 \text{ kcal/mole} \quad \Delta S_3^{\ddagger} = -17.0 \text{ eu}$$

Again, the $\Delta S_1^{\ddagger}$ represents the entropy of activation for the D pathway, while $\Delta S_3^{\ddagger}$ corresponds to the I_d process.

7-1-5   17-Electron Complexes

In contrast to the 18-electron $V(CO)_6^-$, which is extremely sluggish in its reaction with even molten PPh_3, $V(CO)_6$, a 17-e$^-$ complex, reacts with PPh_3 readily at -70°C![25] Corresponding 17-e$^-$ complexes generally seem to react much faster toward ligand substitution than their 18-e$^-$ analogs. While dissociative or radical pathways are possible, more recent evidence points to an associative mechanism taking place in a number of cases involving Group 5 and 7 metals.

For example, in the substitution of $V(CO)_6$ with phosphines, the rate of reaction does change with respect to change in phosphine nucleophile according to the order below:[26]

$PMe_3 > PBu_3 > P(OMe)_3 > PPh_3$

If the reaction mechanism involved a dissociative pathway, a change in nucleophile would be expected to have little effect on the overall reaction rate, analogous to the S_N1 reaction in organic chemistry. The trend in reaction rate with respect to phosphine probably reflects a combination of steric and electronic factors. Trialkylphospines are more electron rich—and thus better σ bases (σ donors)—than triarylphosphines or phosphites. Steric effects seem to play a

[25]J.E. Ellis, R.A. Faltynek, G.L. Rochfort, R.E. Stevens, and G.A. Zank, *Inorg. Chem.*, **1980**, *19*, 1082.
[26]Q-Z Shi, T.G. Richmond, W.C. Trogler, F. Basolo, *J. Am. Chem. Soc.*, **1984**, *106*, 71.

role, and we would expect this to be the case since an associative reaction would involve a sterically congested ML_7 (19 e$^-$) intermediate or transition state. That is apparently why PMe_3 reacts faster than PBu_3 when both have about the same σ basicity. The reaction also shows negative values for the $\Delta S^{\ddagger}$ regardless of the nucleophile, consistent with an associative mechanism. Finally, there are examples of 19-electron complexes having been isolated and characterized.[27]

Equation **7.21**[28] provides another example of an associative ligand substitution involving a 17-e$^-$ substrate.

$$(CO)_5Mn-Mn(CO)_5 \xrightarrow{h\nu} 2\ Mn(CO)_5 \xrightarrow{PPh_3} Mn(CO)_4(PPh_3) + CO$$
$$\qquad\qquad\qquad\qquad (17e^-)\qquad\qquad\qquad (17e^-)$$

7.21

$$2\ Mn(CO)_4(PPh_3) \longrightarrow [Mn(CO)_4(PPh_3)]_2$$

7-2   Oxidative Addition

Oxidative addition (OA) and its reverse-reaction counterpart, reductive elimination (RE), play important roles as key steps in catalytic cycles (see Chapter 9) and in synthetic transformations (see Chapter 11). In the broadest sense OA involves the attachment of two groups (X-Y) to a metal complex of relatively low oxidation state. This produces a new complex with an oxidation state two units higher than before, an increase in coordination number of two, and an electron count two higher than present in starting material. Equation **7.22** outlines the essential changes present in an oxidative addition (and, of course, in the reverse direction—reductive elimination).

$$L_nM\ +\ X-Y \underset{RE}{\overset{OA}{\rightleftharpoons}} L_nM\diagdown^X_Y \qquad\qquad \textbf{7.22}$$

(Ox. State = 0)   (Ox. State = +2)

Readers familiar with organic chemistry will recognize that the formation of a Grignard reagent is really an OA as shown in equation **7.23**. Here phenyl magnesium bromide forms when the phenyl group and bromine add to elemental magnesium. Note that the oxidation state of the metal changes

[27]J.P. Collman, L.S. Hegedus, J.R. Norton, and R.G. Finke, *op. cit.*,244 and references therein.
[28]T.R. Herrinton and T.L. Brown, *J. Am. Chem. Soc.*, **1985**, *107*, 5700.

from zero to +2. Group 1 and 2 metals are capable of undergoing OA in the metallic state and differ from the transition metals, which require the prior association of ligands to the metal atom before reaction occurs.

$$Ph–Br + Mg(0) \xrightarrow[\text{Ether}]{} Ph–Mg(II)–Br \qquad\qquad \textbf{7.23}$$

Oxidative additions typically occur on transition metal complexes with electron counts of 16 or less. Addition is possible to 18-electron complexes; however, loss of a ligand (dissociation) must occur first (equation **7.24**):[29]

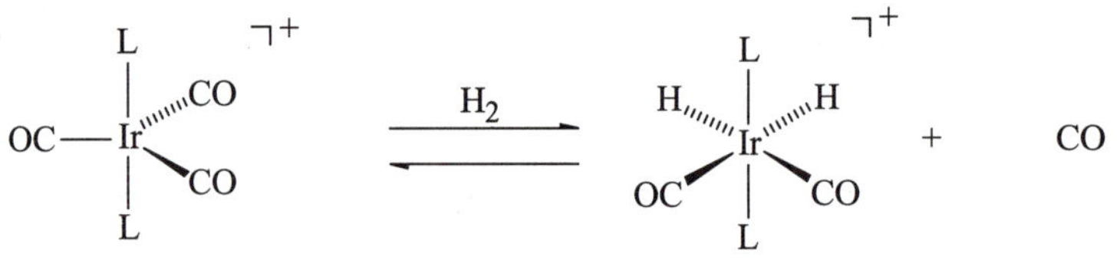

L = PMePh$_2$

A binuclear variant of OA may occur where each metal increases in oxidation state by one unit. Equation **7.25** shows one such example:[30]

$$(CO)_5Mn–Mn(CO)_5 + Br_2 \rightarrow 2\ BrMn(CO)_5 \qquad\qquad \textbf{7.25}$$

Finally, OA may occur intramolecularly as equation **7.26** demonstrates. Intramolecular additions of this type are called *cyclometallations*:

Exercise 7-3

Verify that equations **7.24** and **7.26** are oxidative additions.

---

[29]M.J. Church, M.J. Mays, R.N.F. Simpson, and F.P. Stephanini, *J. Chem. Soc., A*, **1970**, 2909.
[30]It should be pointed out that in equation **7.25** the metal undergoes oxidation, but the coordination number of the metal does not change.

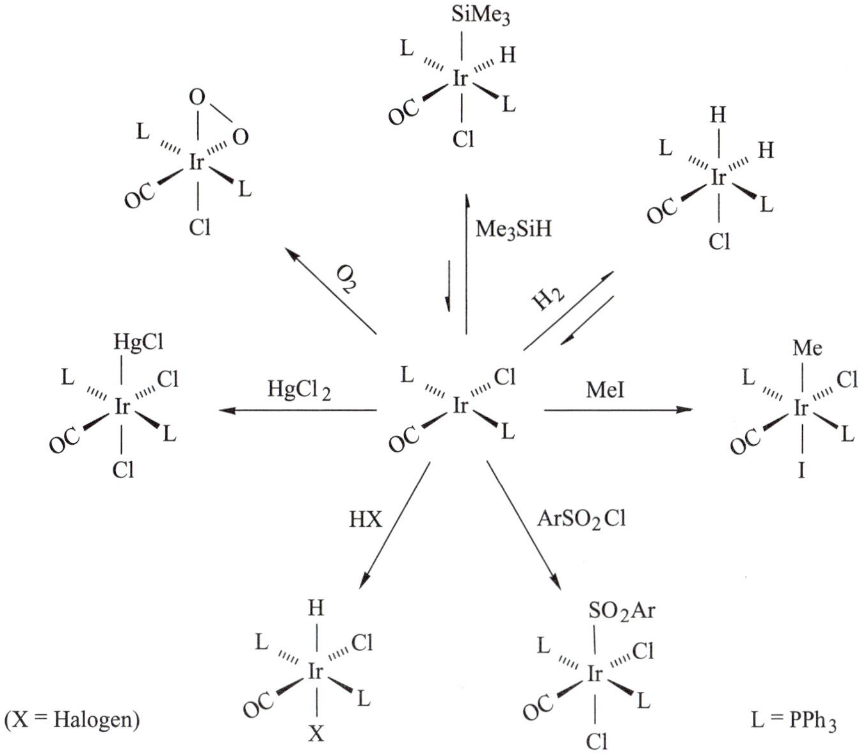

Scheme 7.3
Oxidative Addition Reactions
of Vaska's Compound

One of the first detailed studies of OA was performed by Vaska[31] on 16-electron square planar iridium complexes, the most notable of which is referred to as Vaska's compound, the central compound in Scheme **7.3**. The reactions portrayed in Scheme **7.3** show that several different kinds of molecules react with the iridium complex. Mechanistic pathways leading to the products shown and to products resulting from other metal complexes are many and varied. We will consider a few of these mechanistic types, trying to point out as we go along similarities to other reactions from the realm of organic chemistry.

[31] L. Vaska and J. W. Diluzio, *J. Am. Chem. Soc.*, **1962**, *84*, 679.

7-2-1   Three-Center Concerted Addition

Addition of H_2

The addition of dihydrogen to Vaska's compound (Scheme **7.3**) is an excellent example of concerted, three-center addition in which the transition state is a three-membered ring consisting of the metal and the two hydrogen atoms. Addition of H_2 to Wilkinson's catalyst, an early step in the homogeneous catalytic hydrogenation of alkenes (to be discussed in Chapter 9), serves as another well-known example (equation **7.27**):

$$L = PPh_3 \qquad\qquad\qquad\qquad\qquad\qquad\qquad\qquad \textbf{7.27}$$

Addition of dihydrogen to Vaska's compound is feasible thermodynamically when one compares the dissociation energy of dihydrogen (104 kcal/mole) with the energy released when two Ir-H bonds form (120 kcal/mole total). Since the reaction is of the associative type, the entropy change ought to be negative; experiment finds $\Delta S = -30$ eu. Thus the free energy change turns out to be slightly negative ($\Delta G = -7$ kcal/mole at 25°C), implying that the reaction could be reversible under suitable conditions.

Rate studies of H_2 addition to square-planar complexes have been conducted.[32] Measurements at room temperature in several aromatic solvents yield a bimolecular rate law as follows:

$$\text{Rate} = k_{obs}[(PPh_3)_2(CO)(Cl)Ir][H_2]$$

The $\Delta H^{\ddagger}$ is low (10–12 kcal/mole) and entropies of activation are negative ($\Delta S^{\ddagger} = -20$ to -23 eu)–both of which suggest an early transition state (according to the Hammond postulate) and an OA reaction in the rate-determining step. Figure **7-7** shows a reaction coordinate-energy profile for concerted addition of dihydrogen to a metal complex.

Several theoretical investigations of dihydrogen addition have employed MO theory to show the interaction of H_2 with the orbitals on the metal complex.[33] Calculations involving square-planar molecules show that two interactions are most connected with the breaking of the H-H bond and the formation of two M-H bonds as dihydrogen approaches the metal in a "parallel" manner. Detailed analysis indicates that when the square-planar

[32]P. Zhou, A.A. Vitale, J. San Filippo, Jr., and W.H. Saunders, Jr., *J. Am. Chem. Soc.*, **1985**, *107*, 8049.
[33]J.J. Low and W.A. Goddard, III, *J. Am. Chem. Soc.*, **1984**, *106*, 6928.

geometry distorts to an angular geometry **3**,[34] two important interactions may occur that lower the energy of the pathway to the transition state as the dihydrogen molecule approaches the metal complex.

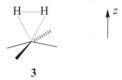

3

1.   The lowest unoccupied orbital (LUMO) of the metal (d_{z^2}), possessing a symmetry compatible with the σ bonding orbital of H_2 (HOMO), interacts in a bonding manner.

2.   The highest occupied molecular orbital of the metal (HOMO) becomes much the same as a d_{xz} orbital, allowing bonding interaction with the $\sigma*$ orbital of H_2, which possesses the same symmetry.

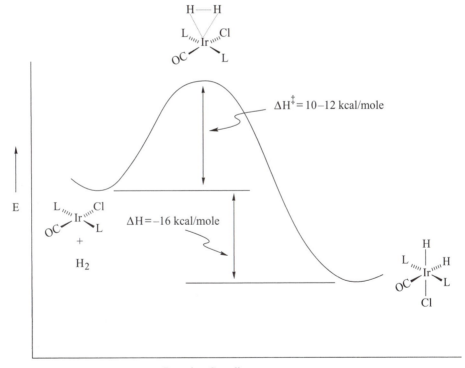

Figure 7-7
Reaction Coordinate-Energy Diagram for the Addition of H_2 to Vaska's Compound

[34]A. Didieu and A. Strich, *Inorg. Chem.*, **1979**, *18*, 2940.

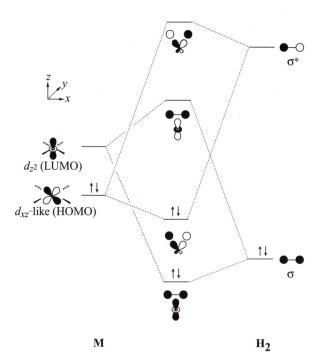

Figure 7-8
Interactions of
Frontier Metal and H_2
Molecular Orbitals

The interaction between the σ orbital of H_2 and the LUMO of the metal in Figure **7-8** represents σ electron donation from H_2 to the metal, and is bonding.[35] The second interaction is that between the HOMO of the metal and the σ^* orbital of H_2. This corresponds to a back bonding π donation from the metal to H_2, thus weakening the H–H bond. Attack of dihydrogen "perpendicularly" to the metal complex (Figure **7-9**) seems to require a much higher energy transition state based on MO theory. By now, such donor-acceptance ligand bonding interactions should be quite familiar (recall that in Chapter 6 it was established that the dividing line between a true dihydride and a dihydrogen-metal complex is a subtle one). Equations **7.28** and **7.29** provide other examples of OA of dihydrogen.

$$Os(CO)_5 + H_2 \rightarrow (CO)_4OsH_2 + CO \hspace{3cm} \textbf{7.28}$$

$$(dppe)Ir(CO)Br + H_2 \;\rightleftharpoons\; \underset{\mathbf{5}}{\overset{}{\text{H}\underset{\text{Br}}{\overset{\text{H}}{\underset{|}{\overset{|}{\text{Ir}}}}}\overset{\text{P}}{\underset{\text{P}}{\diagup}}}\; \longrightarrow \; \underset{\mathbf{6}}{\text{OC}\underset{\text{H}}{\overset{\text{H}}{\underset{\text{Br}}{\overset{|}{\text{Ir}}}}}\overset{\text{P}}{\underset{\text{P}}{\diagup}} \hspace{1.5cm} \textbf{7.29}$$

$$\mathbf{4}$$

[35]Net repulsive interactions of the H_2 σ orbital with filled metal orbitals are also possible and these would lower the overall bonding effectiveness of the bonding; see J-Y. Saillard and R. Hoffmann, *J. Am. Chem. Soc.*, **1984**, *106*, 2006.

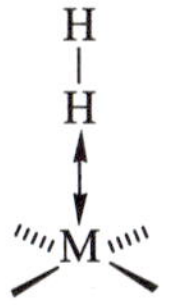

Figure 7-9
Perpendicular Approach
of H_2

The addition of H_2 to complex **4** to give **5** is interesting. Originally it was thought that **5** rearranged to its diastereomer, **6**, via reversible loss of H_2 followed by OA again. The second OA of H_2 was assumed to occur along a different trajectory than that for **4** → **5**. Closer examination of the reaction revealed that the rearrangement of **5** to **6** actually involved a bimolecular reaction between **4** and **5** (see problem **7-8**).[36]

Addition of C-H

While OA of a C-H bond to a metal is not, strictly speaking, the addition of a symmetrical addendum,[37] it is a reaction where the oxidizing agent possesses electronic characteristics similar to that of the H-H bond. It is potentially a reaction of great economic significance since OA of C-H allows the binding of a metal to relatively abundant and normally unreactive organic molecules (e.g., methane, ethane, and benzene). Subsequent transformations could lead to more complex hydrocarbons similar to those found in gasoline and diesel fuel. Addition of C-H takes place intra- or intermolecularly; the former, known as cyclometallation, being more common.

Examples of intramolecular C-H addition reactions (often a competing and more favorable reaction than corresponding intermolecular C-H addition) are abundant, probably due to the favorable entropy effect of having both the metal and the C-H group within the same molecule. The reaction described in equation **7.30** demonstrates cyclometallation followed by reductive elimination (see Section 7-3) of neopentane.[38]

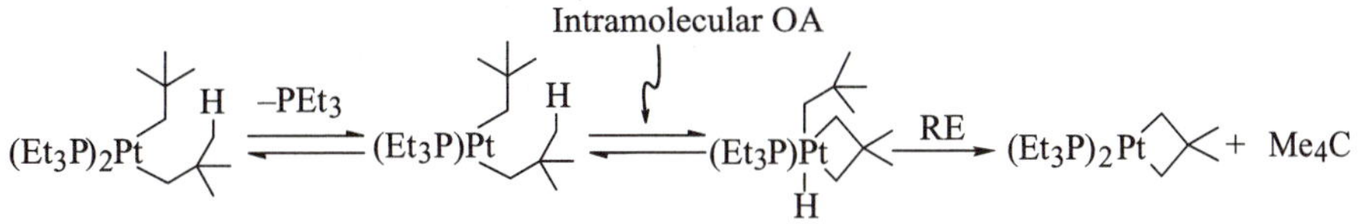

Ox. State:	+2	+2	+4	+2
e⁻ Count:	16	14	16	16

7.30

---

[36]A.J. Kunin, C.E. Johnson, J.A. Maquire, W.D. Jones, and R. Eisenberg, *J. Am. Chem. Soc.*, **1987**, *109*, 2963.

[37]The group X–Y, adding to the metal, is called the *addendum* or the oxidizing agent.

[38]J.A. Ibers, R. DiCosimo, and G.M. Whitesides, *Organometallics*, **1982**, *1*, 13.

Over the past decade, researchers have focused a great deal of effort toward understanding intermolecular OA—also known as a type of C–H activation—because such reactions are still relatively rare, theoretically interesting, and would be practically important if they could be performed on the industrial scale. In 1982 the first reports of intermolecular C–H addition appeared[39] and are described by equation **7.31**. The 16-electron Cp^*IrL acts much like carbene, $:CH_2$,[40] in its ability to insert into a C–H bond. Although the cyclohexyl iridium hydride has been characterized by X-ray crystallography,[41] the process of isolating and characterizing C–H addition products is not always straightforward. Equation **7.32** shows the transformation of an alkyl iridium hydride to the corresponding alkyl halide, a compound much more amenable to characterization.

$$(Cp^*)(PMe_3)Ir(H)_2 \xrightarrow[h\nu]{-H_2} [(Cp^*)Ir(PMe_3)] \xrightarrow{R-H} (Cp^*)(PMe_3)Ir(H)(R)$$

$$R = Ph, Me, Me_3CCH_2, Cyclohexyl \qquad \textbf{7.31}$$

$$(Cp^*)(PMe_3)Ir\overset{}{\underset{H}{\diagup}} + CHBr_3 \longrightarrow (Cp^*)^*(PMe_3)Ir\overset{}{\underset{Br}{\diagup}} + CH_2Br_2$$

$$\textbf{7.32}$$

Even though more than one mechanism for C–H addition has been discovered, evidence indicates that a concerted, three-centered process frequently occurs, analogous to the addition of dihydrogen. Experimental[42] and theoretical[43] studies, however, have shown that the trajectory of the C–H bond as it approaches the metal is not the same as dihydrogen. Figure **7–10** shows that, instead of a parallel approach, the hydrogen of the C–H bond approaches first, and then the carbon atom follows, resulting in retention of configuration at the carbon center.

Because metal-arene bonds are stronger than metal alkyl types, C–H addition is more favorable thermodynamically for aryl C–H bonds than for corresponding addition of alkanes. As we proceed from first- to the third-row

---

[39]A.H. Janowicz and R.G. Bergman, *J. Am. Chem. Soc.*, **1982**, *104*, 352 and J.K. Hoyano and W.A.G. Graham, *J. Am. Chem. Soc.*, **1982**, *104*, 3723.

[40]As discussed in Chapter 12, Cp^*IrL and $:CH_2$ are considered to be "isolobal." Both species are two electrons short of a complete valence shell complement of 18 and 8, respectively, and both have two orbitals available for additional bonding. For a good discussion of carbene chemistry, see T.H. Lowry and K.S. Richardson, *Mechanism and Theory in Organic Chemistry*, 3rd ed. Harper and Row, New York, **1987**, 546–562.

[41]J.M. Buchanan, J.M. Stryker, and R.G. Bergman, *J. Am. Chem. Soc.*, **1986**, *108*, 1537.

[42]R.H. Crabtree, E.M. Holt, M. Lavin, and S.M. Morehouse, *Inorg. Chem.*, **1985**, *24*, 1986.

[43]J.J. Low and W.A. Goddard, III, *J. Am. Chem. Soc.*, **1984**, *106*, 6928 and J-Y. Saillard and R. Hoffmann, *J. Am. Chem. Soc.*, **1984**, *106*, 2006.

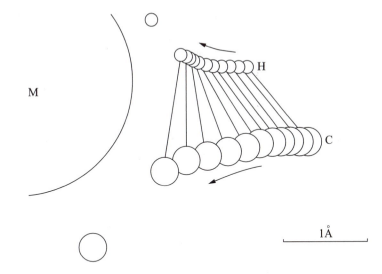

Figure 7-10
Trajectory of
Approach of the C-H
Bond to the Metal
(Figure reproduced
with permission from
the American
Chemical Society.)

transition metals, OA of all C-H bonds becomes more favorable, owing to the increasing strength of metal-hydrogen and metal-carbon bonds. Most examples to date of intermolecular C-H activation have involved third row metals. In spite of thermodynamic constraints, it appears that kinetic barriers to OA of C-H to a metal are relatively low. In general, C-H OA is favorable if: (1) the metal complex is coordinatively unsaturated; (2) the metal complex is sterically uncongested; (3) the metal is a second, or preferably, a third-row element; (4) the R groups on the phosphine are resistant to cyclometallation (e.g., R = CH_3); and (5) the metal has a filled orbital capable of interacting with the σ^* antibonding orbital of the C-H bond.[44]

There are other complications regarding the addition of alkyl C-H bonds to a metal. The resulting metal alkyl could either undergo reductive elimination back to starting material (Section **7-3**) or, if a β carbon-hydrogen bond is present, rapid 1,2-elimination (to be discussed in Chapter 8). This problem is demonstrated for the general case in equation **7.33**.

$$L_mM + CH_3\text{-}\overset{\overset{\displaystyle H}{|}}{C}H\text{-}CH_3 \;\rightleftharpoons\; L_mM\overset{\diagup H}{\underset{\diagdown \alpha\ \beta}{}} \xrightarrow{\text{1,2-Elim.}} L_mM\overset{\diagup H}{\underset{\diagdown H}{}} + \;=\!\diagup$$

$$\textbf{7.33}$$

The tendency for these side reactions to occur limits, for the time being, the use of OA as a means of activating hydrocarbons, and presents the chemist with a stiff challenge to make C-H activation synthetically useful.

[44]R.H. Crabtree, *Chem. Rev.*, **1985**, *85*, 245.

Other Additions of Symmetrical Species, Y–Y

Oxidative addition of a C–C bond to a metal, like C–H addition, is potentially significant on an industrial scale. If a metal complex were available that could react via OA with an internal C–C bond present in a long-chain hydrocarbon, the ultimate result could be equivalent to a process known as "cracking." Cracking reactions are used in oil refining to break down long chain alkanes into smaller molecules more suitable for combustion in automobile engines. Appropriate metal complexes could serve as homogeneous catalysts for this process under mild conditions. Unfortunately, unstrained C–C bonds—the type found in saturated hydrocarbons—typically do not react with metal complexes due to thermodynamic considerations. The energy cost of breaking the C–C bond is not regained by the formation of two new M–C bonds. Most examples of C–C OA have involved the reaction of cyclic hydrocarbons with at least one highly strained, relatively weak C–C bond. Here, relief of ring strain is a driving force in the reaction.

Reaction of the platinum complex known as Zeise's dimer with bicyclo[1.1.0]butane gives the corresponding platinabicyclo[1.1.1]pentane, **7**, (equation **7.34**) as a stable crystalline solid.[45] Equation **7.35** describes a transformation whereby C–H addition occurs first to the Rh complex, followed by rearrangement to the metallacycle; the last step involves the addition of the C–C bond.[46] Metallacyclobutanes are thought to be intermediates in some alkene polymerization and metathesis reactions; you will encounter these in Chapter 10.

$$\triangleleft\!\!\triangleright \;+\; [PtCl_2(C_2H_4)]_2 \longrightarrow [PtCl_2(C_2H_4)]_n \xrightarrow{+2\,Py} \qquad \mathbf{7} \qquad \mathbf{7.34}$$

Zeise's dimer

$$(Cp^*)(L)Rh(H) \;+\; \triangle \xrightarrow[-H_2]{h\nu} (Cp^*)(L)Rh\!\!\begin{array}{c}H\;\triangle\\ \diagdown H\end{array} \longrightarrow (Cp^*)(L)Rh\!\!\diamondsuit$$

$$\mathbf{7.35}$$

The OA of halogens, X_2 (X = Cl, Br, or I), to square-planar complexes, involving such d^8 metals as Rh(I) or Ir(I), typically gives *trans* addition, as shown in equation **7.36**.[47]

$$Ir(PPh_3)_2(CO)(Cl) + X_2 \rightarrow trans\text{-}Ir(PPh_3)_2(CO)(Cl)(X)_2 \qquad \mathbf{7.36}$$

[45] A. Miyashita, M. Takahashi, and J. Takaya, *J. Am. Chem. Soc.*, **1981**, *103*, 6257.
[46] R. A. Periana and R. G. Bergman, *J. Am. Chem. Soc.*, **1986**, *108*, 7346.
[47] J. P. Collman and C. T. Sears, Jr., *Inorg. Chem.*, **1968**, *7*, 27.

The mechanism probably involves a two-step process in which the relatively electron-rich metal abstracts X^+ from X_2 (much like addition of halogen to a double bond in organic chemistry) followed by capture of the positively charged metal complex by the remaining X^-. Interestingly, the stereochemistry resulting from halogen attack on d^8 trigonal bipyramid complexes (M = Fe, Ru, or Os) is *cis* (equation **7.37**). This apparently results from a two-step mechanism in which the complex attacks the halogen to form an octahedral cation complex, which is followed by loss of a neutral ligand, and finally collapse of the outer sphere halide ion to give overall *cis* stereochemistry.[48]

$$
\begin{array}{c}
\text{OC—M} \\
\end{array}
\;+\; X_2 \;\longrightarrow\;
\left[
\begin{array}{c}
\text{OC} \cdots \text{M} \cdots \text{X} \\
\end{array}
\right]^{+}
X^{-}
\;\xrightarrow{-\text{CO}}\;
\begin{array}{c}
\text{OC} \cdots \text{M} \cdots \text{X} \\
\end{array}
$$

7.37

Exercise 7-4

Why does the *cis*-dihalo product form in equation **7.37** and not the *trans*-dihalo isomer?
[Hint: See Section **7-1-1**.]

7-2-2   Nucleophilic Oxidative Additions of R–X

Here the metal complex behaves as a nucleophile attacking an alkyl or acyl halide. Such reactions, occurring typically on 16e⁻ square planar complexes, have been well studied.[49] The first step involves nucleophilic displacement of the leaving group followed usually (but not always) by combination of the leaving group and substituted metal complex, the prototypical example of which appeared in Scheme **7.3** (p. 163) where CH_3I adds to Vaska's compound. Another excellent example of OA of CH_3I to a metal occurs as a key step in the Monsanto acetic acid synthesis (to be discussed in Chapter 9), shown in equation **7.38**.

[48]J.P. Collman and W.R. Roper, "Oxidative-Addition Reactions of d^8 Complexes," in F.G.A. Stone and R. West, Eds., *Advances in Organometallic Chemistry*, vol. 7, Academic Press, New York, 1968, 53–94.
[49]R.J. Cross, *Chem. Soc. Rev.*, **1985**, *14*, 197.

$$CH_3I \quad + \quad [Rh(I)_2(CO)_2]^- \quad \longrightarrow \quad \begin{array}{c} CH_3 \\ I_{\prime\prime\prime\prime}\!\!\underset{OC^{\nearrow}\;\;|\;\;\diagdown I}{\overset{|}{Rh}}\!\!{}^{\prime\prime\prime\prime}CO \\ I \end{array} \qquad \textbf{7.38}$$

The overall effect of the reaction results in the addition of R–X to the metal. Substrate types that will undergo this mode of OA are generally limited to benzyl, allyl, some acyl, and methyl. These are the very same substrates that are most reactive in S_N2 displacements (R = alkyl) or, when R = acyl, in nucleophilic acyl substitution. Other characteristics of these OA reactions associated with the S_N2 pathway include the following:

- Second-order kinetics (first-order with respect to metal complex and alkyl halide)
- Dependence of rate on leaving group ability:
  X = CF_3SO_3 > I > tosylate ≈ Br > Cl
- Large negative entropies of activation (-40 to -50 eu)
- Increase in rate when phosphine ligands are modified to be more electron releasing, e.g., PMe_3 or PEt_3
- Dependence of rate on solvent (polar solvents increase the rate)
- Retardation of the rate of reaction due to increasing steric bulk of ligands already attached to the complex
- Isolation, in certain instances, of ionic intermediates such as **8**,[50] the result expected after the nucleophilic displacement of the leaving group

$$IrCp(CO)(PPh_3) \quad + \quad RI \quad \longrightarrow \quad \left[\begin{array}{c} \bigoplus \\ | \\ R^{\prime\prime\prime\prime}\!\!\underset{PPh_3}{\overset{Ir}{|}}\!\!{}^{\prime\prime\prime}CO \end{array} \right]^+ \; I^- \qquad R = Me, CH_2Ph$$

8

The evidence described above is strongly suggestive of an S_N2 displacement mechanism exactly analogous to its counterpart in organic chemistry. The one piece missing from this mechanistic puzzle is an examination of the stereochemistry at the carbon atom of the substrate. Probably the most convincing

[50] J.W. Kang and P.M. Maitlis, *J. Organometal. Chem.*, **1971**, *26*, 393 and A.J. Oliver and W.A.G. Graham, *Inorg. Chem.*, **1970**, *9*, 2653.

evidence for an S_N2 pathway would be the unambiguous demonstration of the occurrence of inversion of configuration at the carbon bearing the leaving group. Until about 20 years ago, clear-cut examples of inversion of configuration taking place upon OA of R–X were lacking. Making up for this deficiency, the detailed studies by Stille, *et al.*,[51] on OA of substituted benzyl halides onto palladium complexes, provided the first unambiguous examples of inversion of configuration taking place during attack on the substrate by a metal.

Scheme **7.4** shows the reactions carried out by the Stille group. Up to the time of the study, the stereochemistry of all steps in the reaction sequences, except that of OA, had been established. Thus inversion of configuration to give **9** or **10** must have occurred in the first step, since: (1) the next step going to **11** or **12** is carbonyl insertion (to be discussed in detail in Chapter 8), known to occur with retention of configuration; and (2) the final step, called methanolysis, proceeds without disturbing a bond at the carbon stereocenter. The alternate sequence through the middle of the scheme shows a carbonylated complex, **13**, reacting with the alkyl halide giving an intermediate that already has a carbonyl ligand on the metal. Rearrangement involving carbonyl insertion resulting in retention of configuration of the migrating alkyl group followed by methanolysis gives the same product: **14**. By comparing the optical purity and absolute configuration of the starting alkyl halides to those of **14**, the change at the stereocenter was clearly shown to be inversion. These studies also found that the order of reactivity of benzyl halides was $PhCH_2Br$ > $PhCH_2Cl$ > $PhCHBrCH_3$ > $PhCHClCH_3$. This is consistent with the order expected for an S_N2 reaction, since Br is a better leaving group than Cl and primary substrates react faster than secondary. Stille, *et al.*, also found that the more electron-rich the metal, the faster the reaction—again consistent with an S_N2 pathway.

Another example of OA involving a nucleophilic mechanism concerns attack on epoxides by a square planar iridium complex, shown in Scheme **7.5**. The pathway seems to involve OA to give **17** in two steps, *via* unstable intermediate **15**.[52] The transformation is interesting because it involves an epoxide, a most useful reagent in organic synthesis, and because there are mechanistic parallels with reactions that occur in organic chemistry. The reaction begins with nucleophilic attack by the metal at the less-substituted epoxide carbon, which is analogous to attack by organic bases such as CH_3O^- or RNH_2 on unsymmetrical epoxides. Such reactions are sensitive to steric hindrance as we might expect from a process that really is S_N2 displacement. Intermediate **15** can react directly or form the metallooxetane **16**, the OA product that could in principle result from concerted, three-center addition of a C-O bond to the

[51] J.K. Stille and K.S.Y. Lau, *Accounts Chem. Res.*, **1977**, *10*, 434.
[52] D. Millstein and J.C. Calabrese, *J. Am. Chem. Soc.*, **1982**, *104*, 3773.

metal. Pushing electrons from oxygen in **16** to carbon to form the double bond forces a hydride to migrate to the electron-deficient iridium to complete the reaction, resulting in **17**. The last step is a hydride shift.

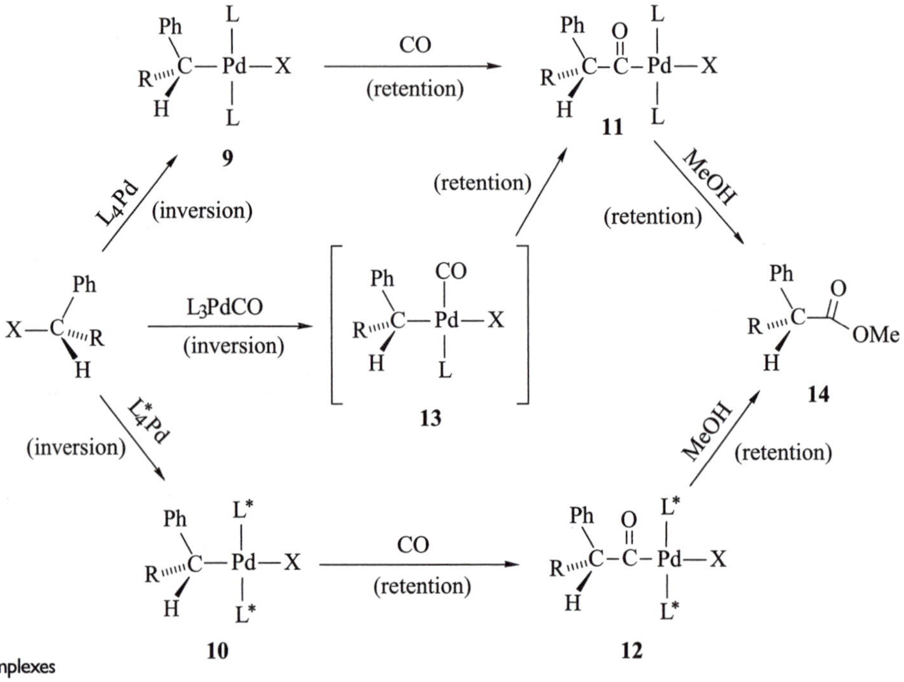

Scheme 7.4
Stereospecific
Reactions of
Palladium Complexes

$L = Ph_3P$ $L^* = Et_3P$ $X = Br$ or Cl

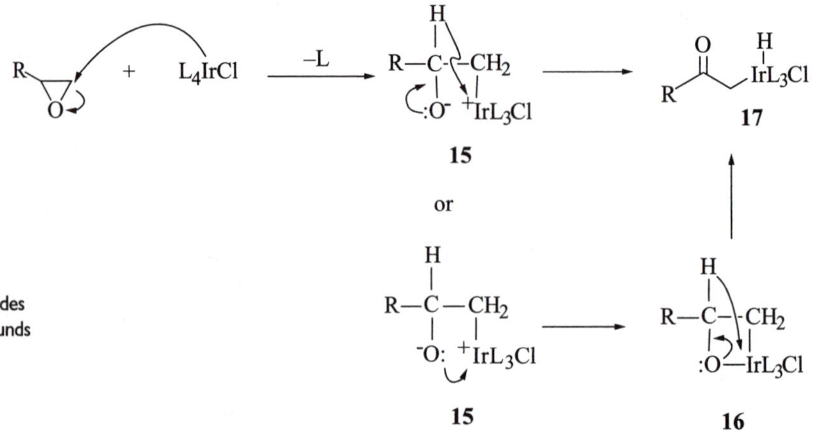

Scheme 7.5
Reaction of Epoxides
with Ir(I) Compounds

Exercise 7-5

When 1-phenyl-2,2-dideuterioethylene oxide was allowed to react with $[(PPh_3)_4Ir]Cl$, only product **C** was obtained. Explain.

C

7-2-3   Radical Pathways

Mechanistic pathways that involve radical intermediates often accompany S_N2-type oxidative additions. Under certain conditions, such pathways may even be the dominant route between reactant and product. Metals with an odd number of d electrons, including Co(II) and Rh(II), are good candidates for radical oxidative additions. For example, the binuclear complex, $(CO)_5Mn-Mn(CO)_5$, reacts with Br_2 to give $BrMn(CO)_5$ via a pathway involving radicals in which each atom of Mn undergoes a one-electron change in oxidation state. Complexes with an even number of d electrons, such as the much-studied d^8 Vaska's complex, can also react via radical mechanisms during OA of R-X.

We will consider two types of radical pathways: the chain and nonchain mechanisms. While other mechanisms are possible, these two routes have been well investigated. Our consideration of them allows us not only to exemplify OA pathways common to a number of metal complexes, but also to introduce some useful and generally applicable techniques for examining mechanisms. At this time, unfortunately, there are no real guidelines for predicting which metal complexes will undergo a particular type of pathway involving radical intermediates.

Iridium complexes, such as Vaska's compound, seem to undergo OA of R-X (where R ≠ methyl, allyl, or benzyl) by a radical chain pathway according to the equations shown in **7.39**.[53]

$$Init\cdot + Ir(I) \rightarrow Init-Ir(II)$$
$$Init-Ir(II) + R-X \rightarrow Init-IrX + R\cdot$$

7.39

$$R\cdot + Ir(I) \rightarrow R-Ir(II)$$
$$R-Ir(II) + R-X \rightarrow R-Ir-X + R\cdot$$

[53]J.A. Labinger and J.A. Osborn, *Inorg. Chem.*, **1980**, *19*, 3230 and J.A. Labinger, J.A. Osborn, and N.J. Colville, *op. cit.*, 3236.

Evidence for a radical pathway includes the observation that the reaction is accelerated by radical initiators (such as oxygen or peroxides) and the presence of UV light. Moreover, the order of reactivity for the R group is III° > II° > I°, which is inconsistent with a direct displacement mechanism, but is in accord with the stability of alkyl radicals. Radical inhibitors (such as sterically hindered phenols) retard the rate of reaction with sterically hindered alkyl halides, but not when R = methyl, allyl, and benzyl. When stereoisomerically pure alkyl halides are used, OA results in the formation of a 1:1 mixture of stereoisomeric alkyl iridium complexes, consistent with the formation of an intermediate radical R·.

Stereochemical Probes

The stereochemical outcome of a reaction can suggest much about its mechanistic pathway. If a molecular fragment, whose stereochemistry is clearly defined, can be attached to the reactant and remain attached during the course of the reaction without affecting the chemistry, then determination of product stereochemistry can be straightforward using spectroscopic means.

For example, the two primary alkyl bromides, **18** and **19**, are diastereomerically related, the former called *erythro* and the latter *threo*. The designations *erythro* and *threo* arise from the corresponding simple sugars, erythrose and threose, **20** and **21**. These bromides serve as useful probes, because the possible stereochemical outcomes of retention, inversion, or racemization at the carbon bearing the bromide may be readily determined by NMR spectroscopy without requiring the use of optically active compounds. Starting with **19** (R = D, R* = *tert*-butyl), we would expect the stereochemistry found in **22**, if retention of configuration occurred upon OA. Note that the only reasonable conformation for product requires that the bulky *tert*-butyl group be *anti* to the metal. Inversion of configuration would give **23**. We can distinguish between **22** and **23** by measuring the magnitude of the coupling constant between the two vicinal protons H_1 and H_2 and then applying the Karplus[54] relationship. Finally, if racemization occurs, the NMR spectra of product will be identical regardless of whether we start from **18** or **19**, because equal molar amounts of **22** and **23** will be present.

Variations of **18** and **19** have also been used as probes where R = F and R′ = phenyl. Fluorine, like hydrogen, has a spin of one half and couples with the vicinal proton. Fluorine-hydrogen coupling constants are significantly larger than those for vicinal hydrogens. Using the Karplus relation, it is relatively easy to discern *anti* versus *gauche* orientation of adjacent protons.

---

[54]M. Karplus, *J. Am. Chem. Soc.*, **1963**, *85*, 2870. Karplus found a relationship between the coupling constant of vicinal protons and the dihedral angle described by these adjacent nuclei. Coupling is maximal when the dihedral angle (ϕ) = 0° or 180° and minimal when ϕ = 90°.

Φ = Dihedral angle

18 *Erythro*   **19** *Threo*

20 Erythrose   **21** Threose

22   **23**

gauche orientation   *anti* orientation
of protons   of protons

Another mechanistic possibility involving radical intermediates is the non–chain mechanism as outlined in equations **7.40**, particularly applicable for zero valent members of the nickel triad (Ni, Pd, Pt).

$$ML_n \rightarrow ML_{n-1} + L$$
$$ML_{n-1} + R\text{-}X \rightarrow \cdot MXL_{n-1} + R\cdot \quad \text{(slow)}$$

7.40

$$\cdot MXL_{n-1} + R\cdot \rightarrow RMXL_{n-1} \quad \text{(fast)}$$
$$L = Ph_3P$$

Two observations suggest a mechanism involving radical intermediates. First, a stereorandomization occurs at R. For instance, if R is chiral and one starts with the *S* enantiomer, the resulting product, $RMXL_{n-1}$, is racemic. Second, the rates of reaction as a function of the structure of R decrease according to R = III° > II°> I°. Since this also corresponds to the order of stability of alkyl radicals, the observation is consistent with the presence of radical intermediates. On the other hand, we can distinguish nonchain pathways from chain mechanisms because the former are not subject to rate change due to the addition of radical scavengers or initiators.

Oxidative addition is complex in terms of the mechanistic possibilities available. Subtle variations in reaction conditions can result in a change in mechanism. Some symmetrical addenda (H$_2$) or those with non-polar bonds involving electropositive nonmetals (C-H or C-C) seem to undergo addition by a concerted three-center process; others involving electronegative nonmetals—whether symmetrical or nonsymmetrical (e.g., C-O, C-X, H-X, or X-X)—undergo OA via polar or radical pathways.

7-3   Reductive Elimination

Reductive elimination (RE) is the reverse of oxidative addition whereby oxidation state, coordination number, and electron count all decrease—usually by two units. Equation **7.41** shows an example of RE from a rhodium complex to give an alkene. The methyl group attaches to the sp^2 center on the η^1-alkene ligand with retention of configuration.

$$7.41$$

Reductive elimination is favored in complexes with bulky ligands (due to relief of steric hindrance upon ligand loss), a low electron density at the metal (high oxidation state), and the presence of groups that can stabilize the reduced metal fragment upon ligand loss [such as CO or P(OR)$_3$]. It is a reaction type that is less well studied than OA, because complexes that are susceptible to RE are usually unstable. The reaction seems particularly facile for complexes of the late transition metals in Groups 8 to 10, particularly d^8 nickel(II) and palladium(II) as well as for d^6 platinum (IV), palladium (IV), rhodium (III), and iridium (III). Copper triad (Cu, Ag, or Au) complexes also have a tendency to undergo RE; indeed, some of the first compounds observed to do so were d^8 gold (III) complexes.[55]

[55]A. Buroway, C.S. Gibson, and S. Holt, *J. Chem. Soc.*, **1935**, 1024; W.L.G. Gent and C.S. Gibson, *J. Chem. Soc.*, **1949**, 1835; and M.E. Foss and C.S. Gibson, *J. Chem. Soc.*, **1949**, 3063. A more recent discussion of RE from Au complexes may be found in A. Tamaki, S.A. Magennis, and J. Kochi, *J. Am. Chem. Soc.*, **1974**, *96*, 6140.

As chemists seek to discover new methods of producing C–C bonds, RE stands out as one of the best ways of accomplishing this task. Often the last step in a catalytic cycle involving the juncture of two different carbon fragments, RE is part of the overall process of the synthesis of several different molecules with industrial or biological significance (see Chapter 9). In addition to producing hydrocarbons, R–R′ (R = alkyl or aryl and R′= alkyl, aryl, or vinyl), RE also provides a pathway to compounds with bonds of the following types: R–H (R = alkyl, aryl), RC(=O)–H (aldehydes), R(C=O)–R′ (ketones), and R–X (alkyl halides).

7-3-1 *cis*-Elimination

The mechanistic possibilities for RE are fewer than those for oxidative addition. Most eliminations occur by a concerted three-centered process (the reverse of three-center OA) outlined in equation **7.42**.

$$
L_{\prime\prime\prime}\overset{X}{\underset{L}{\underset{|}{Pt}}}{}^{\prime\prime\prime\prime}Y \longrightarrow \left[L_{\prime\prime\prime}\overset{X}{\underset{L}{\underset{|}{Pt}}}Y \right]^{\ddagger} \longrightarrow \overset{X-Y}{\underset{L_{\prime\prime\prime}\overset{}{Pt}\,{}^{\prime\prime\prime\prime}L}{+}} \qquad \textbf{7.42}
$$

L = PR₃

$$L = PR_3$$

Such a mechanism requires that the two leaving groups be *cis* to each other, and strongly suggests, moreover, that retention of stereochemistry occurs at the leaving group atom. Theoretical investigations on RE of alkanes from gold and palladium complexes indicate that *cis*-1,1-elimination is allowed by symmetry, and takes place from either a tricoordinate Y-shaped (Au[56] or Pd[57]) activated complex, **24**, or T-shaped (Pd) intermediate, **25**, where the leaving groups are *cis* to each other. The T- or Y-shaped species forms upon prior dissociation of a ligand, typically a phosphine.

| 24 | 25 |

[56]S. Komiya, T.A. Albright, R. Hoffmann, and J. Kochi, *J. Am. Chem. Soc.*, **1976**, *98*, 7255.
[57]K. Tatsumi, R. Hoffmann, A. Yamamoto, and J.K. Stille, *Bull. Chem. Soc. Jpn.*, **1981**, *54*, 1857.

Equation **7.43** provides some experimental evidence consistent with *cis*-elimination. The *cis*-ethyl complex gives butane directly while the *trans*-isomer undergoes facile β-elimination to produce ethene and the ethylhydridopalladium complex, **26**, which finally yields ethane[58] after RE.

$$
\begin{array}{c}
\text{Et} \\
| \\
\text{L--Pd--CH}_2\text{CH}_3 \\
| \\
\text{L}
\end{array}
\longrightarrow \text{L}_2\text{Pd} + \text{CH}_3\text{--CH}_2\text{--CH}_2\text{--CH}_3
$$

7.43

$$
\begin{array}{c}
\text{L} \\
| \\
\text{Et--Pd--CH}_2\text{CH}_3 \\
| \\
\text{L}
\end{array}
\longrightarrow \text{CH}_2\text{=CH}_2 +
\begin{array}{c}
\text{Et} \\
| \\
\text{L--Pd--H} \\
| \\
\text{L} \\
\mathbf{26}
\end{array}
\longrightarrow \text{L}_2\text{Pd} + \text{CH}_3\text{--CH}_3
$$

The seminal investigations of Stille and co-workers several years ago,[59] however, provided substantial information on the mechanism of RE. These studies established that, for the palladium complexes shown in Figure **7-11**, the stereochemical relationship between the leaving groups was necessarily *cis*. The rates of reductive elimination of these compounds were measured; those rates for the *cis* series are listed in Table **7-5**. Two of the *trans* compounds, **28** and **30**, were obliged to isomerize to the *cis* isomer before reductive elimination could occur, and the overall rate of RE was retarded. Addition of Ph$_3$P retarded the rates of reaction of the *cis* complexes because Stille's group found that ligand dissociation was required before reductive elimination could occur (equation **7.44**). The order of reactivity of the *cis* complexes reflects the chelate effect[60] (**29** > **31**) and the increased stabilizing

$$
\begin{array}{ccc}
\underset{\underset{\text{L}}{\displaystyle\diagup}}{\overset{\text{L}\text{'''''}}{\text{Pd}}}\diagdown\text{CH}_3 & & \\
\end{array}
\xrightarrow{+\ \text{Solv}}
\cdots
\xrightarrow{\text{RE}}
\begin{array}{c}
\text{L} \\
| \\
\text{Pd} \\
| \\
\text{Solv}
\end{array}
+ \ \text{CH}_3\text{--CH}_3
$$

L = PR$_3$ +L

7.44

$$
\xrightarrow{+\ \text{Solv}}
\rightleftharpoons
\rightleftharpoons
\rightleftharpoons
$$

7.45

[58] F. Ozawa, T. Ito, Y. Nakamura, and A. Yamamoto, *Bull. Chem. Soc. Jpn.*, **1981**, *54*, 1868.
[59] For a good example of this work, see A. Gillie and J.K. Stille, *J. Am. Chem. Soc.*, **1980**, *102*, 4933.

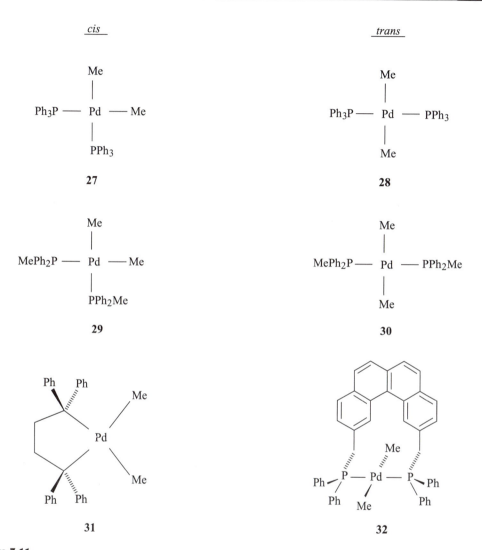

Figure 7-11
Pd(II) Complexes Used for
Reductive Elimination Studies

power of phosphine ligands upon alkyl substitution (**27** > **29**). Polar solvents seemed to enhance the rate of reductive elimination by assisting in the *trans* to *cis* isomerization and in stabilizing the reduced coordination product formed, as shown in equation **7.45**.

[60]Bidentate or chelating ligands tend to bind tightly to transition metals. A general ligand substitution reaction, involving displacement by a bidentate ligand (shown below), should have a positive ΔS, which therefore makes ΔG tend to be negative. A subsequent reaction of losing a bidentate ligand then becomes unfavorable thermodynamically. $L_nM + L^*-L^* \rightarrow L_{n-2}M(L^*-L^*) + 2 L$

Table 7-5 Relative Rates of RE of *cis*-Complexes at 60°C

cis-Complex	k (sec^{-1})
27	1.0410×10^{-3}
29	9.6250×10^{-5}
31[a]	4.778×10^{-7}

[a]at 80°C

One of the *trans* compounds, **32**, substituted with the *TRANSPHOS* ligand, did not undergo reductive elimination even at 100°, because isomerization to the *cis* isomer would be difficult. Not only does the presence of a bidentate ligand tend to retard isomerization due to the chelate effect, but also the rigidity of the *TRANSPHOS* group makes requisite first-step phosphine dissociation unlikely.

Although **32** did not undergo reductive elimination, the addition of methyl iodide immediately produced ethane. This suggested that methyl iodide added oxidatively to the Pd(II) complex to give **33**, which then underwent RE of the adjacent *cis* methyl groups. The observation that CD$_3$I reacts with **32** to give only CD$_3$-CH$_3$ lends further support to this hypothesis and to the overall notion of *cis*-elimination (via **33**, shown below):

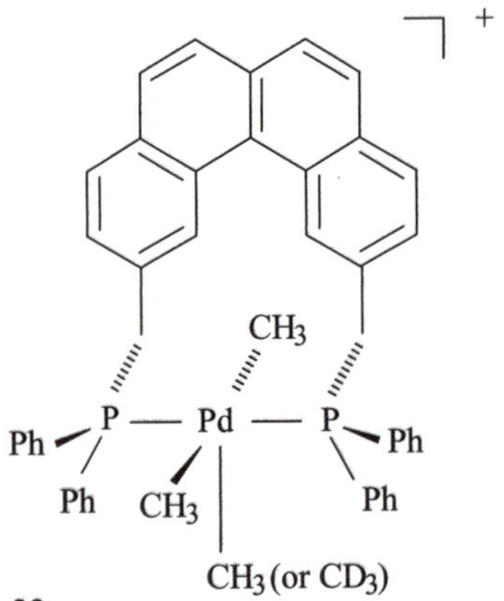

33

Crossover Experiments

Crossover experiments are helpful in determining whether a mechanistic pathway is intramolecular or intermolecular. In this type of experiment, two forms of the same reactant are used—one unlabeled and the other labeled with isotopes of atoms found in the original compound (e.g., hydrogen vs. deuterium). If the reaction is intermolecular, products should show scrambling of the label such that some products will be completely labeled, some completely unlabeled, and some partially labeled. Intramolecular pathways should show no such scrambling. Crossover experiments performed by the Stille group (equation **7.46**) showed that, when equimolar mixtures of **27**, **29**, or **31** and their hexadeuterio counterparts were allowed to react, *only ethane and* CD_3CD_3 *were produced.* The absence of the crossover product, CD_3CH_3, and the presence of products that seemingly could form only via an intramolecular pathway, give strong support to the now well-accepted view that RE is a concerted, intramolecular reaction involving leaving groups that are *cis* with respect to each other.

$$
\underset{\underset{PPh_3}{\overset{|}{}}}{\overset{\overset{CD_3}{|}}{Ph_3P-Pd-CD_3}} \quad + \quad \underset{\underset{PPh_3}{\overset{|}{}}}{\overset{\overset{CH_3}{|}}{Ph_3P-Pd-CH_3}} \quad\longrightarrow\quad CH_3-CH_3 \quad + \quad CD_3-CD_3
$$

$$\textbf{7.46}$$

A more recent example of dissociative *trans* to *cis* isomerization followed by RE occurs in a vinylic cross-coupling reaction, such as shown in equation **7.47**. This is the first extensive examination of RE from Group 8 to 10 metals involving unsaturated σ bonded ligands.[61] Here again, addition of phosphines retards the rate of elimination.

$$\textbf{7.47}$$

Although the dissociative pathway for RE seems to occur during RE from most metal complexes (certainly those where the metals are from Groups 8 to 10), in some instances, phosphine dissociation is not required for RE to occur.[62] Addition of phosphines actually accelerates the rate of RE of biaryls (Ar = Ph or CH_3Ph) from *cis*-Pt(II) complexes[63] as shown in equation **7.48**.

$$\textbf{7.48}$$

L = PPh₃ → L = PPh$_3$

34

[61]P.J. Stang and M.H. Kowalski, *J. Am. Chem. Soc.*, **1989**, *111*, 3356.

[62]See J.P. Collman, L.S. Hegedus, J.R. Norton, and R.G. Finke, *op. cit.* 329-330 for examples.

[63]P.S. Braterman, R.J. Cross, and G.B. Young, *J. Chem. Soc., Dalton*, **1977**, 1892.

The reaction is first order in Pt(II) complex (consistent with concerted, uni-molecular elimination), but the influence of the added phosphine has not been completely elucidated quantitatively. It is possible that added phosphine coordinates with the Pt to give a five-coordinate activated complex or intermediate that can rearrange to a *cis*-diaryl trigonal bipyramidal species, **34**. It has been shown that RE from such five-coordinate species is facile if the leaving groups are *cis*.[64]

The contrast between the dissociative nature of RE from Pd(II) complexes and the apparent associative pathway for Pt(II) deserves comment. Palladium seems to have less tendency to form 18-electron complexes than either Pt or Ni.[65] Therefore, 16-electron Pd(II) complexes such as **27** will usually undergo dissociation before acquiring more electrons through ligand addition.

7-3-2   Stereochemistry at the Leaving Group

When determination of stereochemistry at the leaving carbon stereocenter is possible, retention of configuration is observed during the RE step. In the example shown in Equation **7.49**, Milstein and Stille[66] found a net inversion of configuration for OA followed by RE using a Pd(II) complex. The pathway requires first, inversion of configuration of **35** during the OA step (see Section **7-2-2**) to give **36**; and second, retention of configuration during the RE step to produce **37**. Equation **7.50** depicts another example of RE with retention of configuration—this time at an sp^2 center.[67]

7.49

7.50

[64] K. Tatsumi, A. Nakamura, S. Komiya, A. Yamamoto, and T. Yamamoto, *J. Am. Chem. Soc.*, **1984**, *106*, 8181.
[65] See J.K. Stille in *The Chemistry of the Metal-Carbon Bond*, F.R. Hartley and S. Patai, Eds., Wiley, London, 1985, vol. 2, chap. 9, 764 and references therein.
[66] D. Milstein and J.K. Stille, *J. Am. Chem. Soc.*, **1979**, *101*, 4981.
[67] J.K. Stille, *Angew. Chem. Int. Ed. Eng.*, **1986**, *25*, 508.

This reaction represents the last step in a Pd-catalyzed coupling of two fragments, **38** and **39**. Coupling reactions of this sort will be discussed more fully in Chapter 11.

7-3-3   C–H Elimination

Examples of elimination of C-H are relatively rare, but they are increasingly cited in the recent literature. Research with d^8 rhodium and iridium complexes[68] shows that RE in these cases is concerted, intramolecular, and involves the loss of carbon and hydrogen that are *cis* with respect to each other. In these cases, the now-familiar rate-determining loss of phosphine ligand (a dissociative process) occurs before a new C-H bond can form. In a simple way, we can view the loss of phosphine as both effectively reducing electron density at the metal (thus forcing the need for lowering the oxidation state through RE) and providing a less-stable stereochemical configuration. RE of this unstable configuration is thus energetically favorable. Equations **7.51** and **7.52** provide two examples of C-H elimination.

$$\text{Rh(L)}_3\text{Cl} + \text{R}-\!\!\begin{smallmatrix}O\\\|\end{smallmatrix}\!\!-\text{H} \xrightarrow{\text{OA}} \mathbf{40} \rightleftharpoons \mathbf{41} \rightleftharpoons \mathbf{42} \xrightarrow{\text{RE}} \text{RhL}_2(\text{CO})\text{Cl} + \text{R-H}$$

7.51

$$\mathbf{} \longrightarrow \text{RhL}_3\text{Cl} + \text{H-CH}_2\text{-}\overset{O}{\underset{}{\|}}\text{C-R}$$

7.52

As shown in Equation **7.51**, OA of the aldehyde gives the *cis*-hydridoacylrhodium complex, **40**. Heating results in ligand dissociation to give **41**, which undergoes migratory deinsertion (to be discussed in Chapter 8) to produce **42**. Isomerization of **42** to a *cis*-hydridoalkyl complex presumably occurs before the final RE step to give the C-H elimination product. These reactions have applications in catalytic industrial process known as olefin hydroformylation (to be discussed in Chapter 9).

[68]D. Milstein, *Accounts Chem. Res.*, **1984**, *17*, 223.

Exercise 7-6

Provide a mechanism for the conversion of **42** to $RhL_2(CO)Cl$ in equation **7.51**.

The timing of the breakage of the M–C and M–H bonds and the formation of the C–H bond is of interest to chemists. In at least one case,[69] it appears that a metal-σ C–H bond complex, **43**,[70] is a true intermediate in the pathway toward alkane and reduced metal (equation **7.53** shown below).

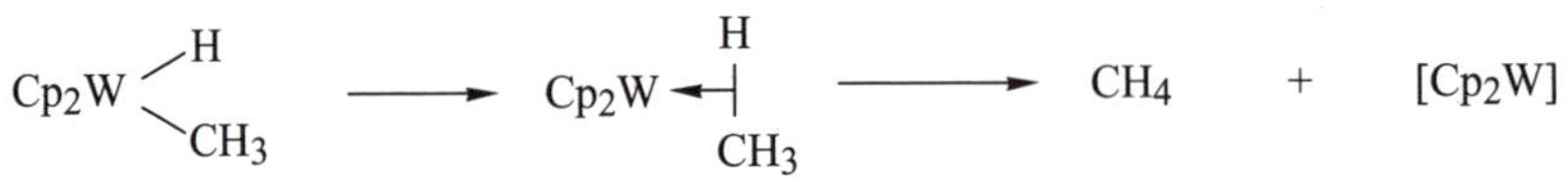

$$Cp_2W \overset{H}{\underset{CH_3}{<}} \longrightarrow Cp_2W \overset{H}{\underset{CH_3}{\dashv}} \longrightarrow CH_4 \quad + \quad [Cp_2W] \qquad \textbf{7.53}$$

43

7-3-4   Dinuclear Reductive Eliminations

Reductive elimination is also possible when two complexes react according to the equation **7.54**.

$$M\text{-}X + M'\text{-}Y \rightarrow M\text{-}M' + X\text{-}Y \qquad \textbf{7.54}$$

Several examples of this type of elimination have been discovered and a variety of mechanisms postulated. It is beyond the scope of this book to provide a detailed investigation of these reactions, but the reader should be aware that they do occur.

One well-known example (equation **7.55**) involves the formation of dihydrogen from reaction of two equivalents of $HCo(CO)_4$. A radical mechanism appears to be operative in this case involving hydrogen atom transfer.[71]

$$2\ HCo(CO)_4 \rightarrow H_2 + Co_2(CO)_8 \qquad \textbf{7.55}$$

Another interesting example, discovered by Norton, *et al.*,[72] involves reaction of two molecules of a hydridomethylosmium complex to give methane according to equation **7.56**.

$$2\ cis\text{-}Os(CO)_4(H)CH_3 \rightarrow CH_3\text{-}H + (H)(CO)_4Os\text{-}Os(CO)_4CH_3 \qquad \textbf{7.56}$$

[69]R.M. Bullock, C.E.L. Headford, K.M. Hennessy, S.E. Kegley, and J.R. Norton, *J. Am. Chem. Soc.*, **1989**, *111*, 3897.

[70]The C–H bond coordinates to the metal in a manner analogous to that of H_2.

[71]J.P. Collman, L. S. Hegedus, J.R. Norton, and R.G. Finke, *op. cit.*, 337.

[72]W.J. Carter, S.J. Okrasinski, and J.R. Norton, *Organometallics*, **1985**, *4*, 1376.

Several experiments support the idea that the reaction first involves rearrangement (Scheme **7.6**) of one equivalent of the osmium complex to an acylhydridoosmium intermediate, **44**, which is then attacked by another equivalent of $HOs(CO)_4CH_3$ (the Os–H bond is considered to be nucleophilic) to give a binuclear species, **45**. The last step involves what Norton termed an "alkyl migration directly onto the hydride ligand." Such a reaction has not been well investigated, but seems plausible, if only because alkyl migration (see Section **8-1**) to the metal to give **46** would violate the 18-electron rule (we will discuss this kind of rearrangement in Chapter 8).

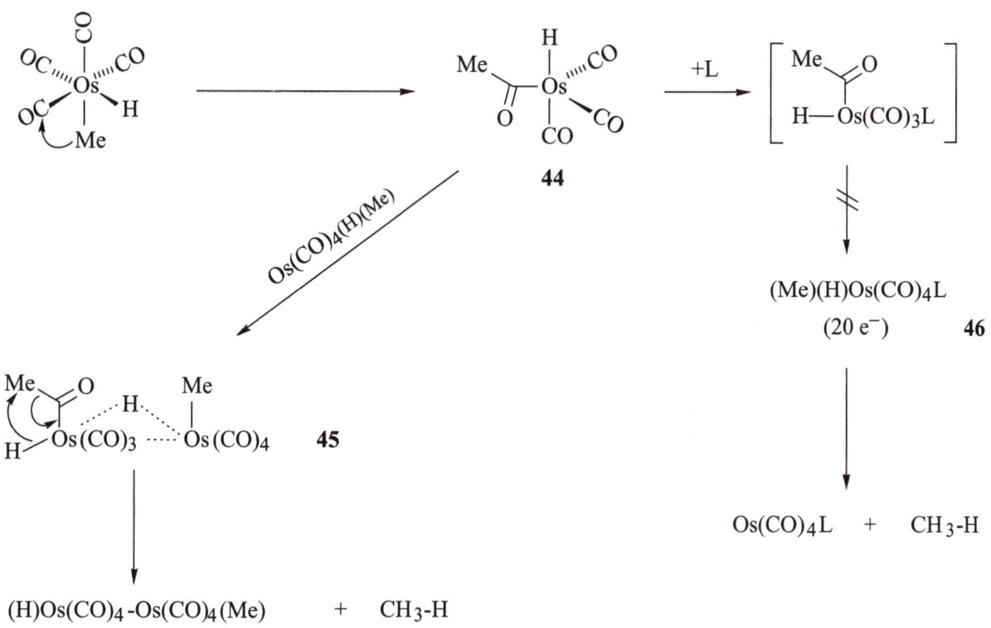

Scheme 7.6
Proposed Mechanism for Reductive
Elimination of Methane from
cis-Os(CO)$_4$(H)(Me)

Problems

7-1   A reaction $L_nML_1 + L_2 \rightarrow L_nML_2 + L_1$ has the rate law:

rate $= k[L_nML_1][L_2] + k'[L_nML_1]$

What does this rate law imply about the mechanism of this reaction?

7-2   In a series of experiments, the rate of phosphine dissociation from *cis*-Mo $(CO)_4L_2$ (L = phosphine) was determined for several phosphines. The overall reaction in each case was of the form below:

cis-Mo$(CO)_4L_2 + CO \rightarrow M(CO)_5L + L$

The rate data:	Phosphine	Rate Constant (s^{-1})
	PMe_2Ph	$< 1.0 \times 10^{-6}$
	$PMePh_2$	1.3×10^{-5}
	PPh_3	3.2×10^{-3}

Account for this trend in reaction rates.

7-3   Verify the rate law shown below and expressed in Section **7-1-4**, relating to Scheme **7.2**.

Rate $= k_1k_2[M(CO)_4diene][L]/(k_{-1} + k_2[L]) + k_3[M(CO)_4diene][L]$

Also verify that when $k_2[L]$ is large, the rate law reduces to:

Rate $= k_1[M(CO)_4diene] + k_3[M(CO)_4diene][L]$

7-4   When PEt_3 reacts with $Mo(CO)_6$, the final product is *fac*-Mo$(CO)_3(PEt_3)_3$. Substitution goes no farther than trisubstitution. Give two reasons why this might be so. Account for the stereochemistry of the product. [Hint: Why would production of the *mer*-isomer be not as likely?]

7-5   Propose a mechanism that would account for the transformations shown below:

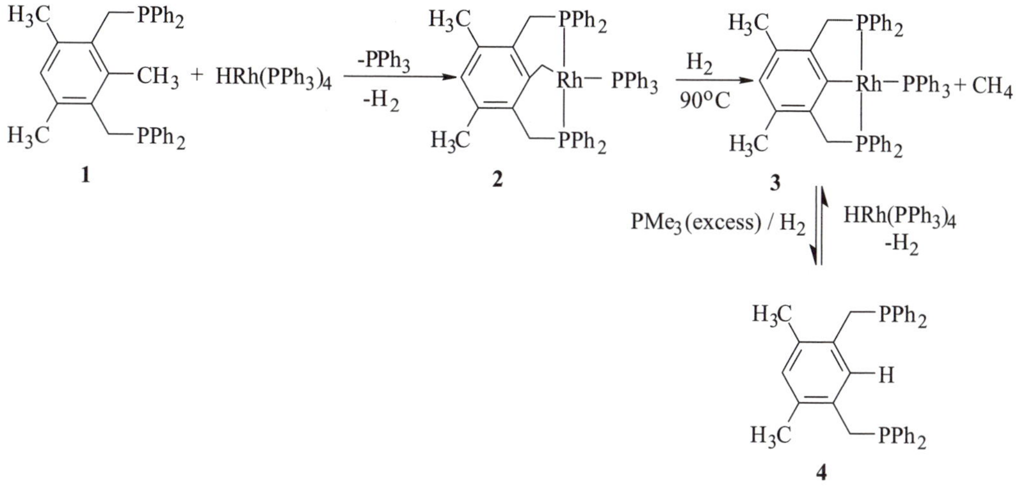

7-6   The reaction below shows a rate law that is second order, first order in substrate metal complex, and first order in phosphine. The $\Delta S^{\ddagger}$ is negative, and an intermediate was isolated and shown to be $(PPh_3)(CO)_4Mn(NO)$. What is the mechanism of the reaction? What is the nature of the bonding of the NO ligand at each stage of the reaction?

$$(CO)_4Mn(NO) + PPh_3 \rightarrow (PPh_3)(CO)_3Mn(NO)$$

7-7   The chiral Mn complex below undergoes racemization according to the equation shown:

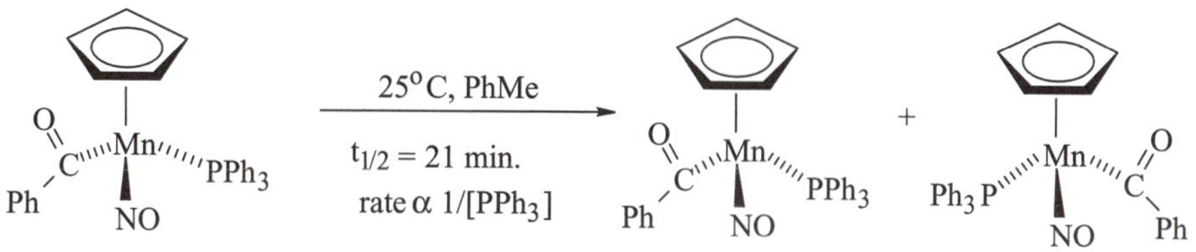

Propose a mechanism that is consistent with the rate being affected inversely by the concentration of PPh_3.

7-8   Bidentate phosphine analogs of Vaska's compound react with H_2 to give two
             diastereomers as shown below:

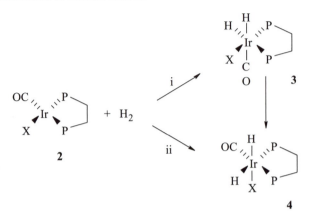

Isomer **3** is formed faster, but isomer **4** is more stable. Two mechanisms were
found to operate at 25° and 1 atm of H_2 for isomerization of **3** to **4**. Mechanism
i is first order in Ir complex, while mechanism ii is second order in Ir complex.
When the pressure of H_2 drops significantly below 1 atm, the second mechanism predominates. Propose mechanistic explanations for the two different
pathways.[73]

[73]See W.D. Jones and R. Eisenberg, et al., *J. Am. Chem. Soc.*, **1987**, *109*, 2963.

8

Organometallic Reactions II

Reactions Involving Modification of Ligands

Once a ligand is attached to a metal, several events may occur. In this chapter we will consider a few of the most common possibilities for reactions involving changes in the ligand. Such transformations are often integral parts of catalytic cycles, which we will examine in Chapter 9. Reactions on ligands bound to a transition metal are also of great interest to the organic chemist, for transformations are now possible that would be difficult or impossible to do using conventional carbon-based chemistry. We will explore the basis for these new possibilities in this chapter, and then apply what we learn here in Chapter 11. We can divide our discussion into two major parts: (1) insertion or deinsertion of a ligand with respect to a bond between a metal and another ligand, and (2) attack on a ligand by a nucleophile or electrophile.

8-1   Insertion and Deinsertion

Equations **8.1** and **8.2** (on p. 192) describe the general process of insertion of a ligand into an M-Y bond. The former shows a process known as 1,1-insertion and the latter its 1,2 counterpart. The reverse reactions are known interchangeably as deinsertion, extrusion, or elimination.

$$\underset{\underset{1}{\overset{\overset{\displaystyle Y}{\displaystyle |}}{L_nM-X=Z}}}{} \longrightarrow \underset{1}{\overset{\overset{\displaystyle Z}{\displaystyle ||}}{L_nM-X-Y}} \qquad\qquad \textbf{8.1}$$

1,1-Insertion

$$\underset{\underset{1}{\overset{\overset{\displaystyle Y}{\displaystyle |}}{L_nM-\overset{\overset{\displaystyle }{\displaystyle ||}}{X}}}}{\overset{\displaystyle Z}{}}{}^2 \longrightarrow \underset{1\quad 2}{L_nM-X-Z-Y} \qquad\qquad \textbf{8.2}$$

1,2-Insertion

8-1-1   1,1-Insertion: Carbonyl Insertion (Alkyl Migration) Reactions

Perhaps the most well-studied example of the insertion–deinsertion reaction is that of the 1,1-insertion of carbon monoxide into a metal-alkyl (or aryl) bond. This reaction figures prominently as a key step in a number of catalytic processes. The numbering designation of 1,1 refers to the numbering on the CO ligand. If we designate the carbon of the CO ligand to be the 1-position, the insertion occurs such that both the metal and the original alkyl ligand are bonded to that carbon when insertion is complete. Equation **8.3** shows insertion of a carbonyl ligand into a Pd-C bond. (See Chapter 7, Scheme **7.4**, p. 174.)

$$\qquad\qquad\qquad\qquad\qquad\qquad\qquad\qquad\qquad\qquad\qquad\qquad\qquad\qquad\qquad\qquad \textbf{8.3}$$

The prototypical example of carbonyl insertion, and the one most thoroughly investigated, involves reaction of CO with $(CH_3)Mn(CO)_5$, the stoichiometry of which is shown in Equation **8.4**.

$$H_3C-Mn(CO)_5 + CO \rightarrow H_3C-\overset{\overset{\displaystyle O}{\displaystyle ||}}{C}-Mn(CO)_5 \qquad\qquad \textbf{8.4}$$

From the net equation we might expect that the CO inserts directly into the Mn-CH$_3$ bond; were such the case, the label "CO insertion" would be entirely appropriate for this reaction. Other mechanisms, however, are possible that would give the overall reaction stoichiometry while involving steps other than insertion of an incoming CO. The following are three plausible mechanisms that have been suggested for this reaction:

Mechanism 1: CO Insertion
Direct insertion of CO into metal-carbon bond.

Mechanism 2: CO Migration
Migration of CO to give *intramolecular* CO insertion. This would give rise to a five-coordinate intermediate, with a vacant site available for attachment of an incoming CO.

Mechanism 3: Alkyl Migration
In this case the alkyl group would migrate—rather than the CO—and attach itself to a CO *cis* to the alkyl. This would also give a five-coordinate intermediate with a vacant site available for an incoming CO.

These mechanisms are described schematically in Figure **8-1**. In both mechanisms **2** and **3** the intramolecular migration is considered most likely to occur to the migrating group's nearest neighbors, located in *cis* positions.

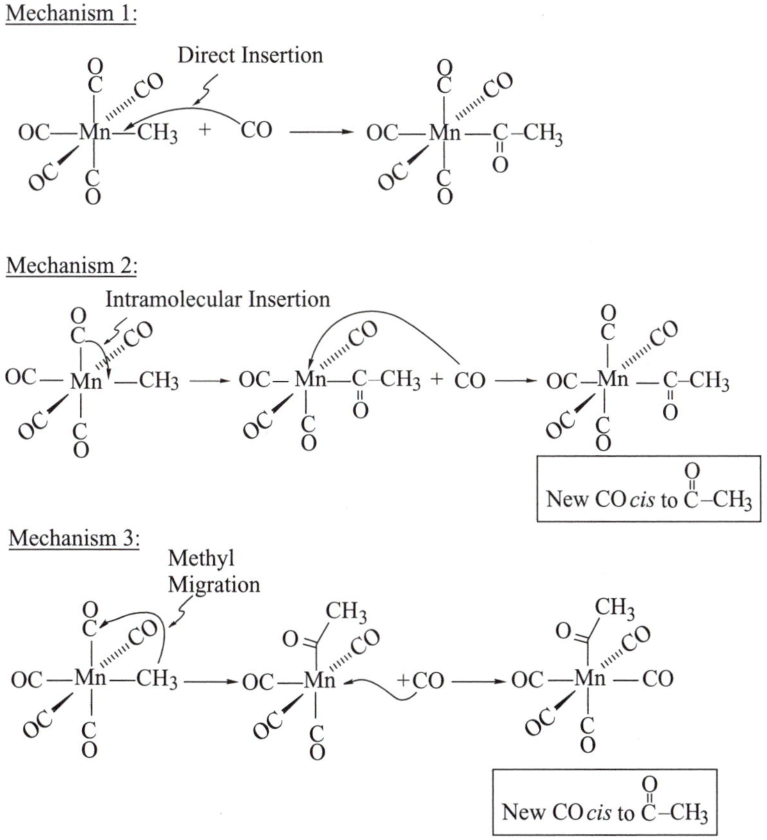

Mechanism 1:

Mechanism 2:

Mechanism 3:

Figure 8-1
Possible Mechanisms for
CO Insertion Reactions

Experimental evidence that may be applied to evaluating these mechanisms includes the following:[1]

I.   Reaction of $CH_3Mn(CO)_5$ with ^{13}CO gives a product with the labeled CO in carbonyl ligands only—*none* is found in the acyl [$CH_3C(=O)$] position.

II.   When the reverse reaction (which occurs readily on heating $CH_3C(=O)Mn(CO)_5$) is carried out with ^{13}C in the acyl position, the $CH_3Mn(CO)_5$ formed has the labeled CO entirely *cis* to CH_3. No labeled CO is lost in this reaction.

$$H_3C-\overset{\overset{\displaystyle O}{\|}}{C}-Mn(CO)_5 \rightarrow H_3C-Mn(CO)_5 + CO$$

III. When the reverse reaction is carried out with a ^{13}CO *cis* to the acyl group, the product has a 2:1 ratio of *cis* to *trans* (*cis* and *trans* referring to the position of labeled CO relative to CH_3 in the product). Some labeled CO is also lost in this reaction.

The mechanisms can now be evaluated on the basis of these data. First, mechanism **1** is definitely ruled out by experiment **I**. Direct insertion of ^{13}C must result in ^{13}C in the acyl ligand; since none is found, the mechanism cannot be a direct insertion. Mechanisms **2** and **3**, on the other hand, are both compatible with the results of this experiment.

The principle of microscopic reversibility requires that any reversible reaction must have identical pathways for the forward and reverse reactions, simply proceeding in opposite directions. (This principle is similar to the idea that the lowest pathway over a mountain chain must be the same regardless of the direction of travel.) If the forward reaction is carbonyl migration (mechanism **2**), the reverse reaction must proceed by loss of a CO ligand from the acyl compound, followed by migration of CO from the acyl ligand to the empty site. Since this migration is unlikely to occur to a trans position, all the product should be cis. If the mechanism is alkyl migration (mechanism **3**), the reverse reaction must proceed by loss of a CO ligand from the acyl compound, followed by migration of the alkyl portion of the acyl ligand to the vacant site. Again, all the product should be cis. Both mechanisms **2** and **3** would transfer labeled CO in the acyl group to a cis position and are therefore consistent with the experimental data for the second experiment (Figure **8-2**).

[1]T. C. Flood, J. E. Jensen, and J. A. Statler, *J. Am. Chem. Soc.*, **1981**, *103*, 4410 and references cited therein.

Mechanism 2 vs. Mechanism 3:

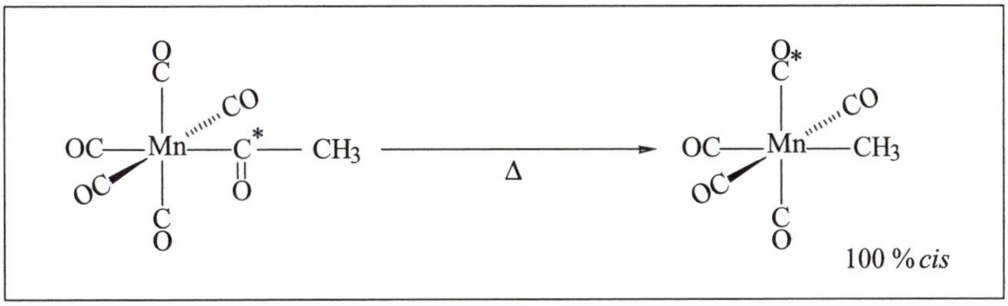

Mechanism 2:

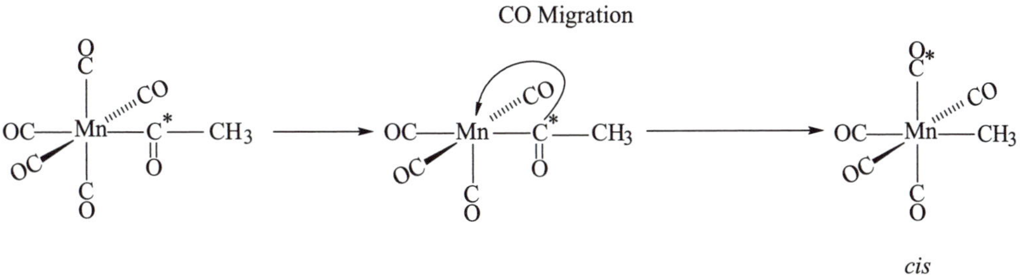

Mechanism 3:

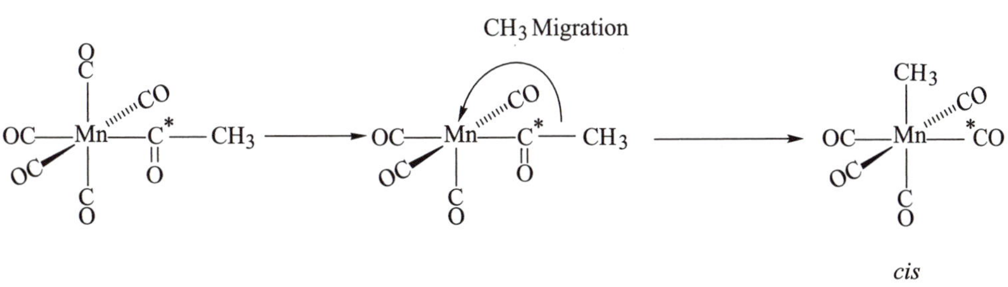

Figure 8-2
Mechanisms of Reverse Reactions for
CO Migration and Alkyl Migration (1)
(Key: * Indicates Location of ^{13}C)

Exercise 8-1

Show that heating of $CH_3-^{13}\overset{O}{\overset{\|}{C}}-Mn(CO)_5$ would not be expected to give the *cis* product by mechanism **1**.

The third experiment (Figure **8-3**) allows a choice between mechanisms **2** and **3**. The CO migration of mechanism **2**, starting with ^{13}CO cis to the acyl ligand, requires migration of CO from the acyl ligand to the vacant site. As a result, 25% of the product should have no ^{13}CO label, and 75% should have the labeled CO cis to the alkyl, as shown in Figure **8-3**. On the other hand, alkyl migration (mechanism **3**) should yield 25% with no label, 50% with the label cis to the alkyl, and 25% with the label trans to the alkyl. Since this is the ratio of cis to trans found in the experiment, the evidence supports mechanism **3**, which is the accepted pathway for this reaction.

One additional point about the mechanism of these reactions should be made. In the discussion of mechanisms **2** and **3**, it was assumed that the intermediate was a square pyramid and that no rearrangement to other geometries (such as trigonal bipyramidal) occurred. Other labeling studies, involving reactions of labeled $CH_3Mn(CO)_5$ with phosphines, have supported a square pyramidal intermediate.[2]

Exercise 8-2

Predict the product distribution for the reaction of *cis*-$CH_3Mn(CO)_4(^{13}CO)$ with PR_3 ($R = C_2H_5$). (You may check your answer by consulting reference 1.)

Pseudotetrahedral iron complexes shown in Scheme **8.1**[3] (p. 198) serve as useful stereochemical probes for examining the mechanism of carbonyl insertion–deinsertion.[4] If the alkyl group migrates, we would expect **1** to form. If, on the other hand, the carbonyl group migrates, the enantiomeric complex, **2**, should be produced. Note that formation of **1** requires inversion of configuration at the metal, while **2** exhibits retention at the iron center. Under appropriate thermal or photochemical conditions, product **1** forms predominantly, thus adding support to the information provided by studies on manganese-alkyl complexes.

---

[2] Flood, Jenson, and Statler, *op. cit.*

[3] A comment on notation is in order here. A box (□) attached to a metal signifies an open coordination site available for ligand coordination.

[4] T.C. Flood in *Topics in Organic and Organometallic Stereochemistry*, G.L. Geoffroy, Ed., *Topics in Stereochemistry*, vol. 12, Wiley, New York, 1981, 83-90.

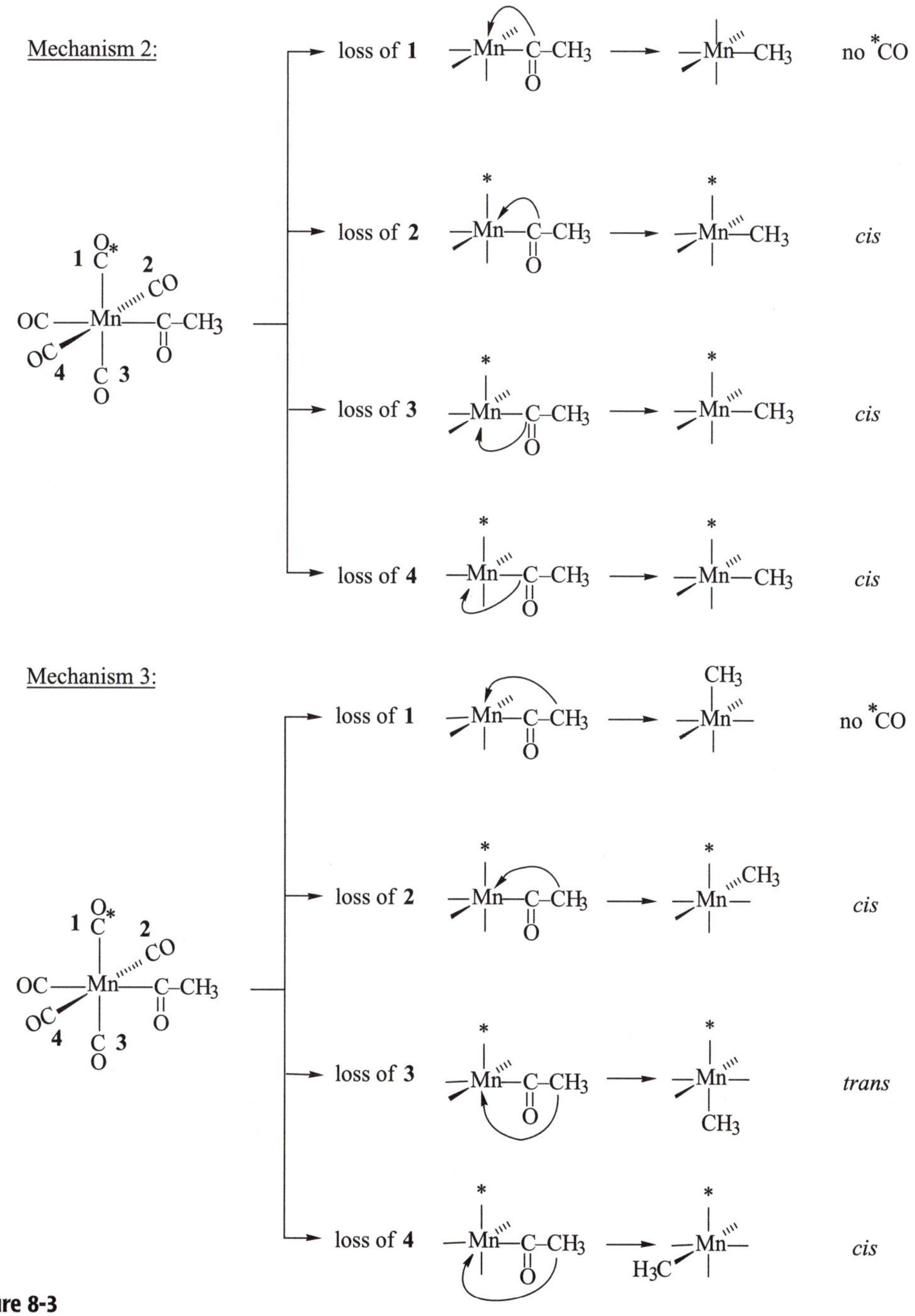

Figure 8-3
Mechanisms of Reverse Reactions for
CO Migration and Alkyl Migration (2).
(Key=*Indicates Location of ^{13}C.)

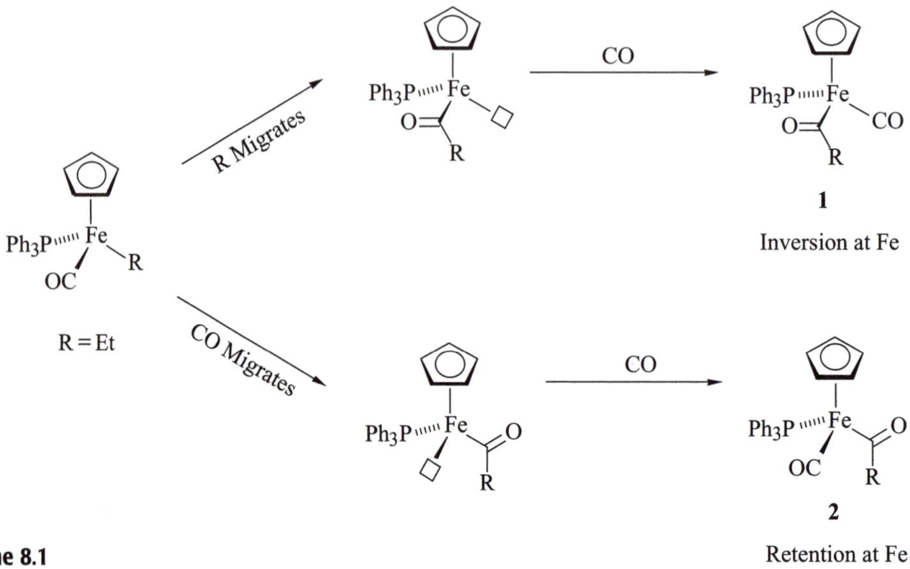

Scheme 8.1
Investigation of Carbonyl Insertion
on Chiral Iron Complexes

The result is that a reaction that initially appears to involve CO insertion, and is often so designated, does not involve CO insertion at all! In fact, it is really analogous to a 1,2-alkyl shift in organic chemistry where a relatively electron-rich alkyl-carbon migrates (with retention of configuration) to a relatively electron-poor center in order to produce a more stable carbocation (equation **8.5**). It is not uncommon for reactions, on close study, to be found to differ substantially from how they might at first appear; the "carbonyl insertion" reaction may, in fact, be substantially more complicated than described here. In this reaction, as in all chemical reactions, it is extremely important for chemists to undertake mechanistic studies and to keep an open mind on possible alternative mechanisms. No mechanism can be proven; it is always possible to suggest alternatives consistent with the known data. It does appear, however, that for the vast majority of carbonyl insertions, the reaction involves migration of the alkyl group to the carbonyl with retention of configuration of the migrating center and inversion of configuration at the metal.

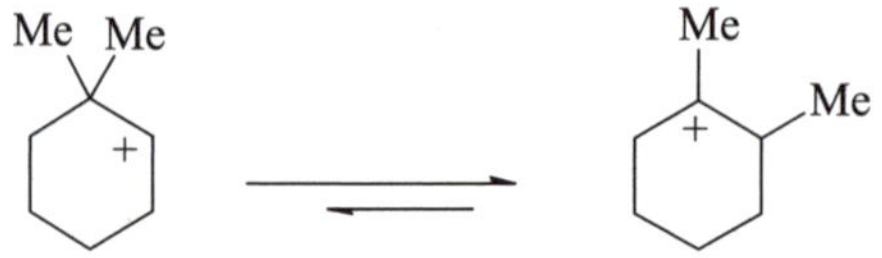

8.5

Exercise 8-3

Complex **A** [ν(CO) = 1670 cm⁻¹] rearranges cleanly to the isomeric compound **B** [ν(CO) = 2040 cm⁻¹] at 30° in benzene. Draw a possible structure for **B**.

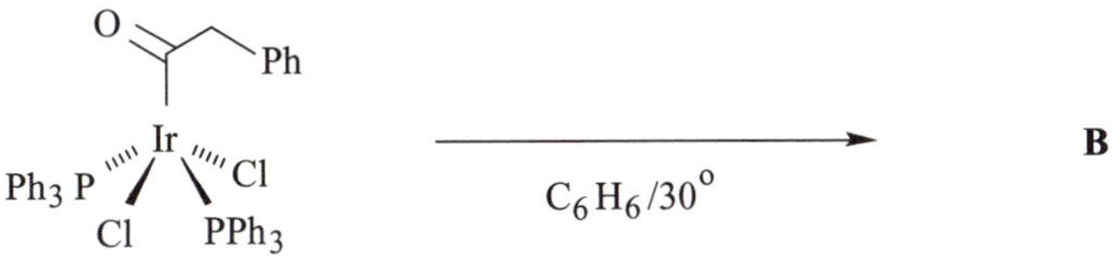

A

8-1-2   1,2-Insertion and Deinsertion (β-Elimination)

Alkenes and alkynes can undergo 1,2-insertion into M-H bonds. The numbering again refers to the numbers for the carbon atoms on the double or triple bond. A schematic insertion reaction is outlined below:

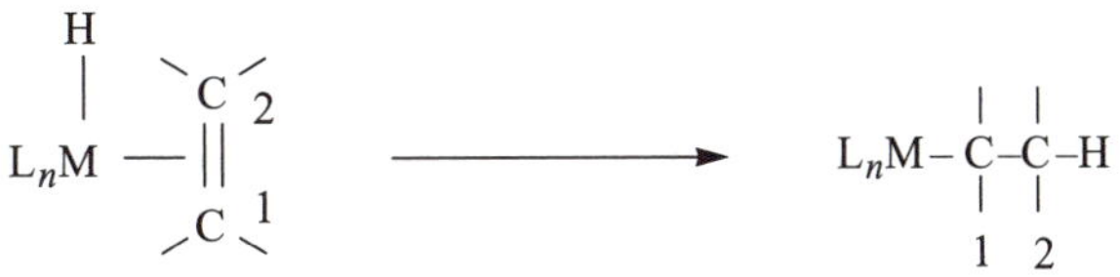

or

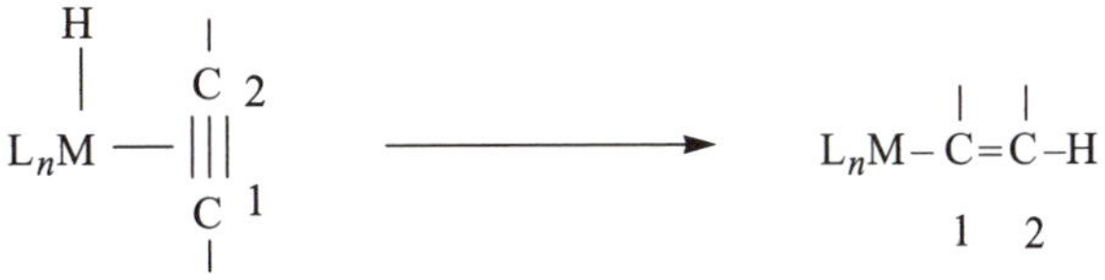

A useful 1,2-insertion reaction encountered in organic chemistry is hydroboration of an alkene to form an alkylborane (equation **8.6**), which subsequently undergoes treatment with alkaline H_2O_2 to yield an alcohol.

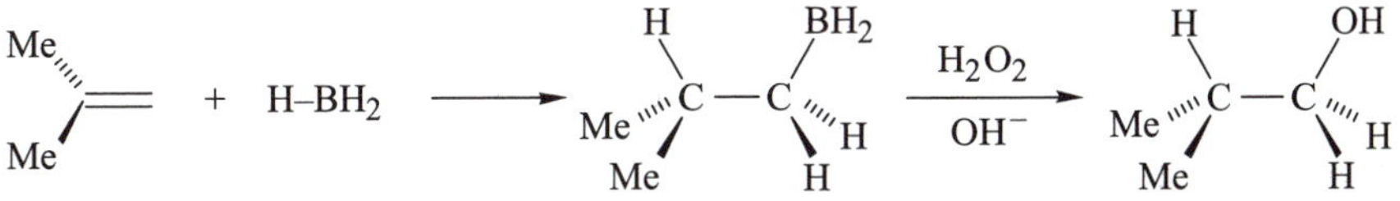

From the organic chemist's standpoint, hydroboration is considered an addition of a B-H bond across a C=C or C≡C bond. Categorizing the reaction as a 1,2-insertion is, however, equally valid.

We saw in Chapter 6 that a reaction that is the reverse of 1,2-insertion—
called 1,2- or β-elimination—may also occur with metal alkyls, the product of
insertion.[5] Equation **8.7** details the insertion–elimination reaction that yields a
vacant site for coordination in the forward direction (1,2-insertion) and
requires a vacant site for the reverse (β-elimination) pathway.

$$
\underset{\text{H}}{\overset{H_2C=CH_2}{L_nM}} \rightleftharpoons
\left[\underset{L_nM \cdots H}{\overset{H_2C = CH_2}{}} \right] \rightleftharpoons
\underset{H}{L_nM} \overset{\overset{\alpha}{CH_2}}{\underset{CH_2}{\diamond}} \overset{\beta}{} \rightleftharpoons
L_nM\text{--}CH_2CH_3
$$

$$
\xrightarrow{\text{Insertion}}
$$
$$
\xleftarrow{\text{β–Elimination}}
$$

                                                                                                    8.7

Experimental evidence points to a mechanism in which the alkene (or alkyne)
carbons and the M-H bond must be nearly coplanar to react. Once the metal
alkene complex has achieved such geometry, 1,2-insertion can occur. During
insertion the reactant proceeds through a four-center transition state. The reaction involves simultaneous breakage of the M-H and C-C π bonds as well as
formation of an M-C σ bond and a C-H bond at the two-position of the
alkene (or alkyne). The result is a linear compound, $L_nM(CH_2CH_3)$. The
reverse reaction, β-elimination, follows the same pathway starting from a
metal-alkyl complex with an open coordination site. Notice that the hapticity
of the organic ligand decreases by one (η^2 to η^1) with insertion and increases
by one (η^1 to η^2) during β-elimination.

    Insertion is slightly exothermic according to the following analysis using
the estimated energies of bonds made and broken in the schematic equation **8.8**:

$$
\underset{M}{\overset{H}{|}} + C{=}C \rightarrow M\text{-}\overset{\square}{\underset{|}{C}}\text{-}C\text{-}H
$$
                                                                                                    8.8

Bond Energies:

Broken		Formed	
M–H:	ca.60 kcal/mole	M–C:	35–40 kcal/mole
π-C=C:	64 kcal/mole	C–H:	98 kcal/mole
total:	124 kcal/mole	total:	133–138 kcal/mole

    ΔH = -9 to -14 kcal/mole

[5]The Greek letter α is used to designate the carbon atom (or other non hydrogen atom) directly
attached to the metal. The next position beyond the α position is designated β and the next one
beyond that γ, etc.

Transition metal alkyl complexes tend to be kinetically unstable and rapidly undergo β-elimination. Equation **8.9** shows the decomposition of a dibutylplatinum complex first through β-elimination and then through the already familiar mode of reductive elimination:

$$8.9$$

There are several ways to prevent or at least lessen the propensity for β-elimination to occur (especially after 1,2-insertion has occurred). Obviously, if there is no β-hydrogen atom, elimination cannot occur. Electron withdrawing atoms, such as fluorine or groups such as CF_3 placed at the α- or β-alkyl positions (equation **8.10**), will also lessen the tendency for β-elimination by strengthening the M–C σ bond:

$$8.10$$

Exercise 8-4

Why does the presence of C–F bonds at the α or β positions of an alkyl ligand tend to strengthen the M–C bond of the metal alkyl complex?

Coordinatively saturated metal-alkyl complexes tend not to undergo β-elimination, especially if they are in solution in the presence of an excess of L-type ligand, such as CO or PR_3, or if ligands also bound to the complex, such as Cp[6] or bidentate phosphines, are reluctant to leave and provide an open coordination site.

---

[6]Cp is the abbreviation for the cyclopentadienyl ligand (C_5H_5) and Cp* corresponds to the pentamethyl derivative (C_5Me_5).

Metal alkyls in which the metal, the C–C bond, and the β-hydrogen cannot become coplanar either do not undergo elimination or do so at a severely retarded rate. For example, the platinacyclopentane,[7] **3**, is kinetically more stable than the corresponding dibutyl complex by a factor of 10^4 as shown below.

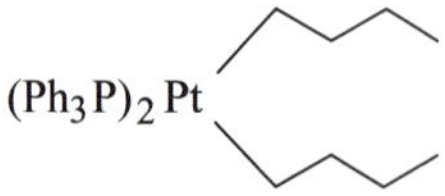

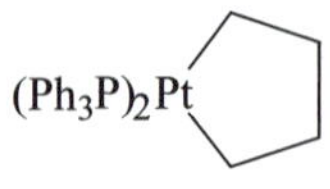

$$(Ph_3P)_2\,Pt \qquad\qquad\qquad (Ph_3P)_2Pt$$

3

Equation **8.11** shows β-elimination from a rhodium complex, a reaction preceded by deinsertion of CO. The reaction is significant in demonstrating that the stereochemistry of the elimination is typically *syn* in which the dihedral angle (**Φ**) defined by the C(β)–H bond and the M–C(α) bond is 0°. The stereochemistry of β-elimination is analogous to that observed in organic chemistry with the Cope reaction shown in equation **8.12**. The success of this unimolecular, concerted reaction requires that the amine oxide group and the β-hydrogen be coplanar (*syn*):

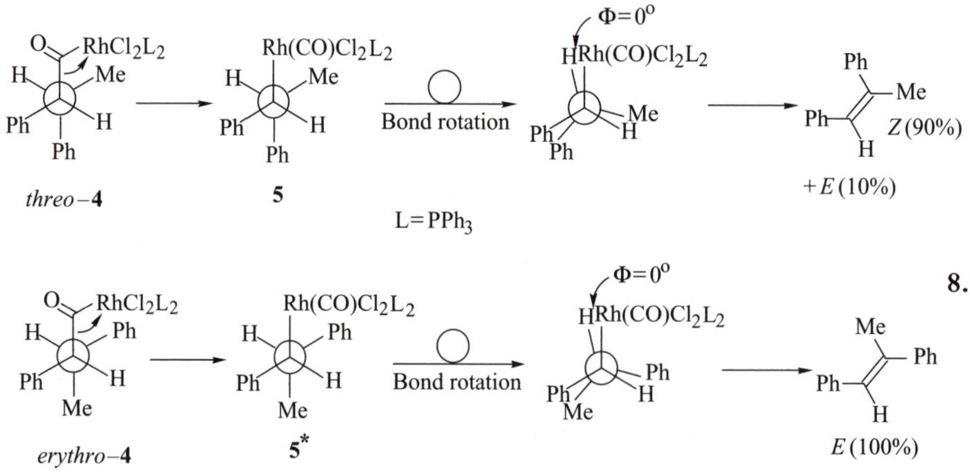

threo–**4**　　　　　　**5**　　　　　　　　　　　Bond rotation　　　　　　　　　　　　　　　　　Z (90%)

L = PPh₃　　　　　　　　　　　　　　　　　　　　　　　　　　　　+ E (10%)

8.11

erythro–**4**　　　　　　**5***　　　　　Bond rotation　　　　　　　　　　　　　　　　E (100%)

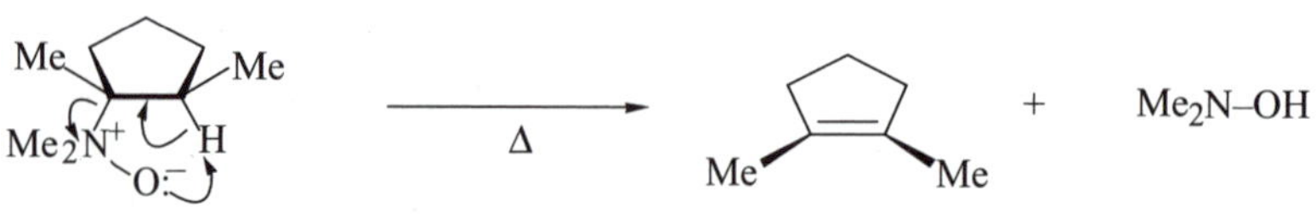

8.12

[7]J.X. McDermott, J.F. White, and G.M. Whitesides, *J. Am. Chem. Soc.*, **1976**, *98*, 6521.

The elimination from the rhodium complex is another example of the use of stereochemical probes to demonstrate in this case *syn* orientation of the β-hydrogen and the metal-carbon bond. Starting with the *threo* acylrhodium complex, **4**, the reaction first proceeds through 1,1-deinsertion of the CO with retention of configuration to give the *threo* alkylrhodium, **5**. β-elimination yields (*Z*)-1-methyl-1,2-diphenylethene as 90% of the product (10% is the corresponding *E*-isomer). Starting with the *erythro* isomer, 1,1-deinsertion of CO (to give **5***) and elimination provide exclusively the *E*-alkene.[8] The results are entirely consistent with *syn* stereochemistry for β-elimination and also, by virtue of the principle of microscopic reversibility, correspond to the *syn* geometry for 1,2-insertion. The metal complex thus serves as a template allowing stereospecific insertion or deinsertion to occur.

Equation **8.13** shows how the equilibrium between insertion and elimination can be shifted in favor of the metal alkyl by adding an electron-rich trialkyl-phosphine ligand to a Ru(I) complex.[9] The alkyl complex is positively charged, but electron-rich phosphines stablilize it. This means that loss of one of the phosphines to create an open coordination site necessary for β-elimination is less likely than in a neutral complex.

One of the most synthetically useful 1,2-insertions, hydrozirconation, is shown in equation **8.14**. The starting material, known as Schwartz's reagent,[10] readily reacts with 1-alkenes to give the corresponding alkyl zirconium complex, **6**. This d^0 Zr(IV) complex is stable and undergoes relatively little useful synthetic

[8]J.K. Stille, F. Huang, and M.T. Regan, *J. Am. Chem. Soc.*, **1974**, *96*, 1518.
[9]H. Werner and R. Werner, *J. Organometal. Chem.*, **1979**, *174*, C63.
[10]J. Schwartz and J.A. Labinger, *Angew. Chem. Int. Ed. Eng.*, **1976**, *15*, 333.

chemistry on its own, but ultimately it is a synthetically valuable intermediate (as we shall see in Chapter 11). Lack of *d* electrons on the metal that could back donate into a β-C-H σ* orbital[11] serves as a major reason for the stability of **6**. Efficient back donation into this orbital often results in rapid β-elimination.

Hydrozirconation occurs with *syn*-addition of the Zr-H bond across a C=C or C≡C bond (equation **8.15**). Due to lower steric hindrance, the addition also tends to be regiospecific with the zirconium attached to the less-substituted position (just as in hydroboration). Internal alkenes and alkynes isomerize to 1-alkyls and 1-alkenyl complexes, respectively—presumably by alternating reactions of insertion and deinsertion—until the complex with the least steric hindrance is formed.

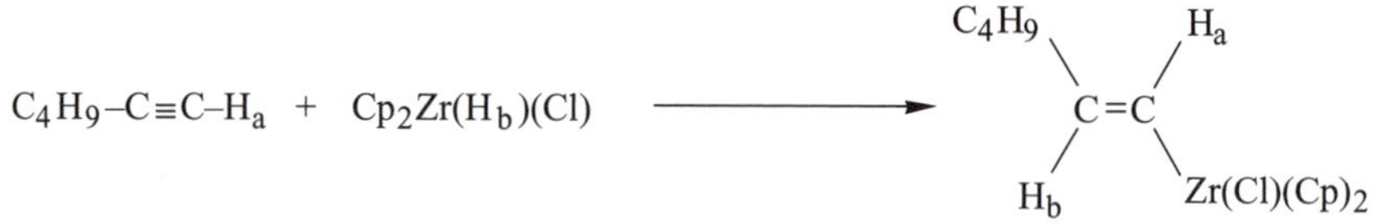

$C_4H_9-C\equiv C-H_a$ + $Cp_2Zr(H_b)(Cl)$
C_4H_9
H_a
C=C
H_b
$Zr(Cl)(Cp)_2$

8.15

Exercise 8-5

Provide a mechanism for the following reaction:
$L_nZr-H + E-2\text{-butene} \rightarrow L_nZr-CH_2CH_2CH_2CH_3$

Equations **8.16**[12] and **8.17** show that zirconium alkyls do undergo cleavage by electrophiles such as protic acids or halogens (X_2) to give the corresponding hydrocarbon or alkyl (alkenyl) halide. Note the retention of configuration at the carbon attached to the metal.

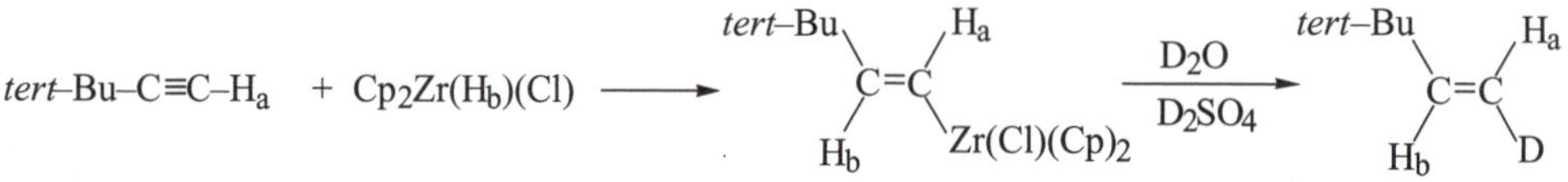

tert-Bu-C≡C-H_a + $Cp_2Zr(H_b)(Cl)$
tert–Bu
H_a
C=C
H_b
$Zr(Cl)(Cp)_2$
D_2O
D_2SO_4
tert–Bu
H_a
C=C
H_b
D

8.16

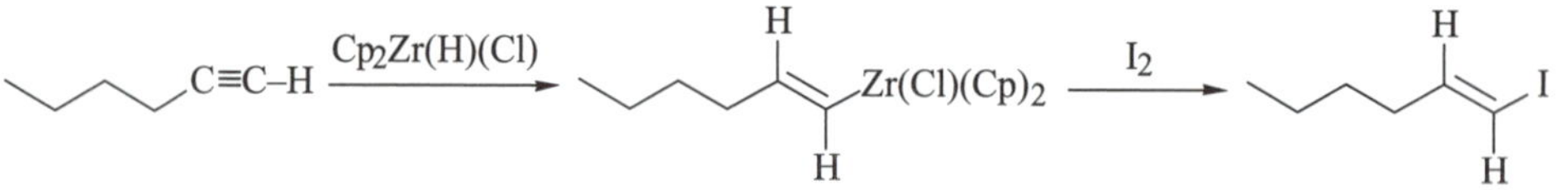

C≡C–H
$Cp_2Zr(H)(Cl)$
H
$Zr(Cl)(Cp)_2$
H
I_2
H
I
H

8.17

[11]One of the driving forces for β-elimination is thought to be the ability of the metal to donate *d* electrons to an antibonding σ orbital located mainly at the β C-H bond. Placing electrons in this orbital weakens the C-H bond, an event necessary for elimination to occur. High-valent, early transition metals, such as Zr(IV), lack *d* electrons and are thus unable to donate electron density to the σ* orbital.

[12]J.A. Labinger, D.W. Hart, W.E. Seibert, III, and J. Schwartz, *J. Am. Chem. Soc.*, **1975**, *97*, 3851.

1,2-Insertion of a C=C Bond into M–C

The insertion of a C=C bond between a metal and a carbon atom is of great importance in building carbon chains. When the process occurs according to equation **8.18**, giant molecules with masses well over 10,000 daltons[13] can result.

$$L_nM–R \ + \ CH_2=CH_2 \quad\longrightarrow\quad L_nM–CH_2–CH_2–R$$

8.18

where R = Alkyl

The molecules are then processed into plastics such as polypropylene or high-density polyethylene. Some chemists feel that a key step in such a polymerization involves a "classical" insertion of C=C into an M-C bond analogous to 1,2 M-H insertion. There seem, unfortunately, to be no clear-cut, well-behaved examples of the process described in **8.18**, where the alkyl group and the alkene are simple acyclic, inactivated ligands.[14] While thermodynamically feasible, M-C insertion appears to be unfavorable kinetically. Rapid β-elimination of the insertion product, moreover, would normally preclude its direct observation.

A key experiment reported by Bergman[15] lends credence to the possibility that C=C insertion can occur in the classical manner. In Scheme **8.2** labeled cobalt complex **7** undergoes insertion of ethylene between the metal and CD_3 group to give the intermediate propyl-cobalt complex **8**. β-elimination and subsequent reductive elimination ultimately provide $CD_3–CH=CH_2$ and CHD_3, products that seem reasonable only as the result of direct insertion by the classical pathway. An alternative mechanism, also shown in Scheme **8.2** without deuterium labeling, involves an early α-elimination step, and would give instead CD_4 and $CD_2=CH–CH_3$ starting from the same deuterated Co complex.

Exercise 8-6

Show how the alternative mechanism in Scheme **8.2** would give CD_4 and $CD_2=CH–CH_3$ starting with the hexadeuterodimethyl-cobalt complex and ethene.

If the mechanism of C=C insertion into an M-C bond is truly analogous to that for M-H bonds, then the stereochemistry should be the same, i.e., *syn* addition of M-C to the double bond. If the R group (equation **8.18**)

---

[13]A dalton is the same as the atomic mass unit, one-twelfth the mass of a ^{12}C atom.
[14]T.C. Flood, *op. cit.*, 71.
[15]E.R. Evitt and R.G. Bergman, *J. Am. Chem. Soc.*, **1979**, *101*, 3973.

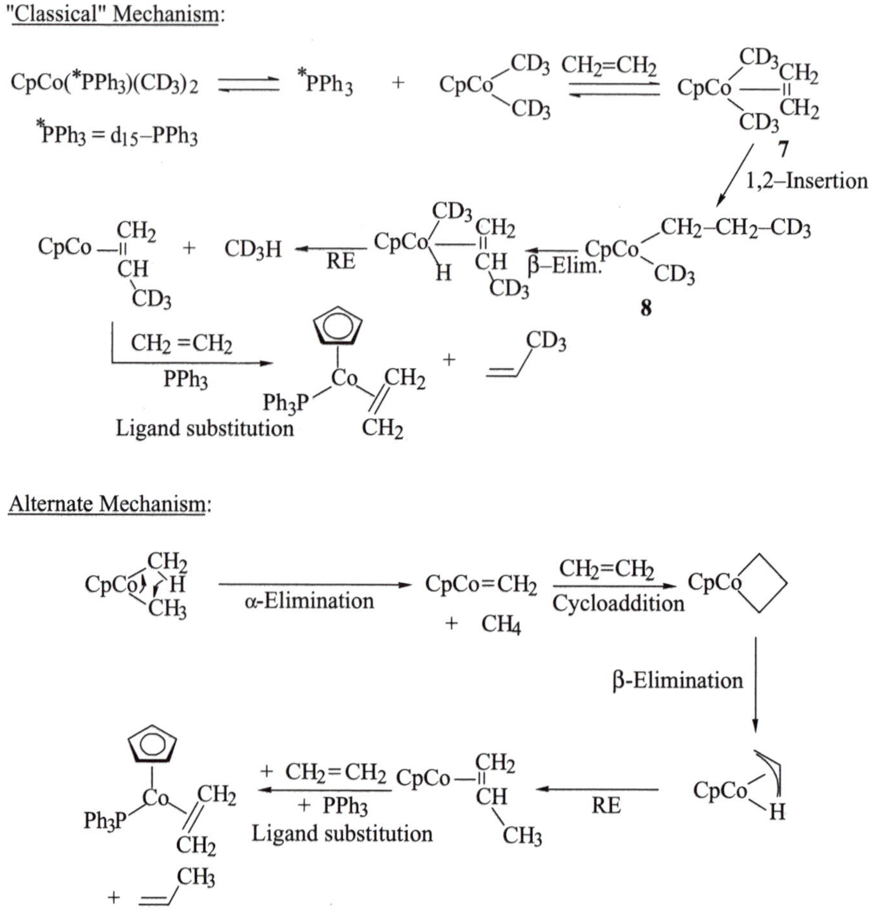

"Classical" Mechanism:

Scheme 8.2
C=C Insertion into an M−C Bond

constitutes a stereogenic center, moreover, retention of stereochemistry should be observed. Equation **8.19** demonstrates one of the few examples of stereospecific C=C insertion where the metal and R group have no real alternative but to undergo *syn* addition with retention of configuration at R.[16]

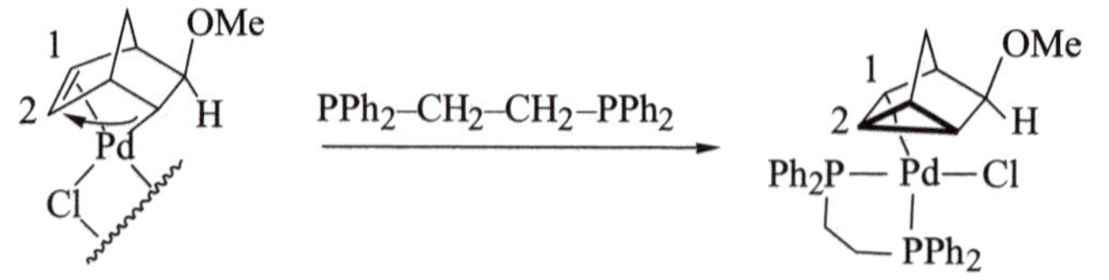

8.19

[16]D.R. Coulson, *J. Am. Chem. Soc.*, **1969**, *91*, 200.

The sequence of reactions outlined in Scheme **8.3**[17] shows a high degree of stereospecificity for C=C insertion.[18] (Z)-1-Phenyl-1-propene gives finally (Z)-1,2-diphenylpropene as 90% of the products collected. Starting with (E)-1-phenyl-1-propene, the corresponding (E)-1,2-diphenyl-1-propene was obtained as virtually the only product. If one assumes that the last step, which is β-elimination, proceeds with *syn* stereochemistry only, then C=C insertion must also be *syn*.

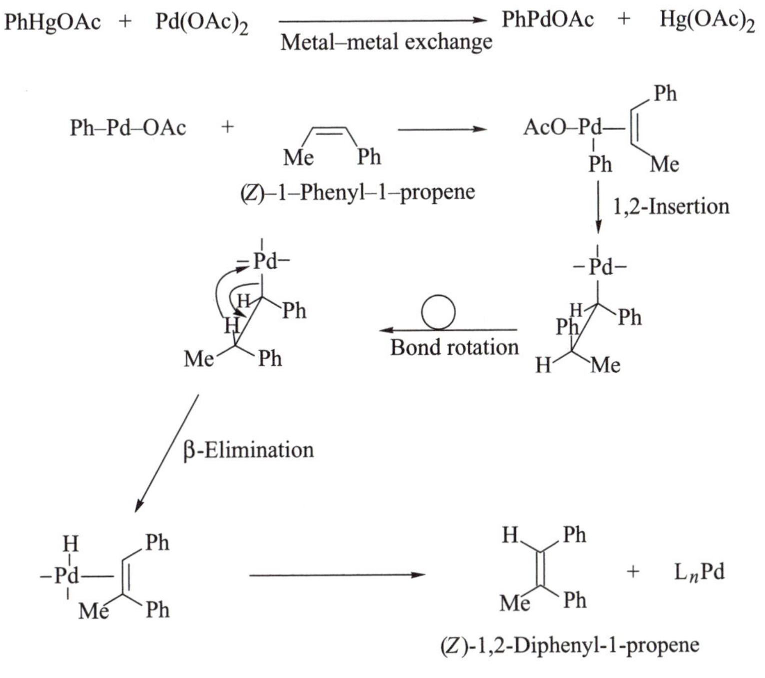

$$PhHgOAc \ + \ Pd(OAc)_2 \ \xrightarrow[\text{Metal–metal exchange}]{} \ PhPdOAc \ + \ Hg(OAc)_2$$

Scheme 8.3
C=C Insertion into a Pd–Ph Bond

Exercise 8-7

Start with (E)-1-phenyl-1-propene instead of the Z-isomer in Scheme **8.3**, and show that the corresponding E-diphenylpropene forms as the result of *syn* C=C insertion and *syn* β-elimination.

[17]The first step in Scheme **8.3** shows a metal-metal exchange, a reaction common in organometallic chemistry. Such exchanges will be discussed in Chapter 11. The exchange shown in the scheme is reminiscent of the metal-metal exchange that occurs in the formation of Gilman reagents, e.g., 2 R–Li + CuI → [R–Cu–R]⁻ Li⁺ + LiI
[18]R. F. Heck, *J. Am. Chem. Soc.*, **1969**, *91*, 6707.

Efforts to completely elucidate the mechanism of C=C insertion into a M-C bond remain an active and controversial area of research in organometallic chemistry. We shall again consider this reaction when we discuss metal-catalyzed polymerization in Chapter 10.

1,1- and 1,2-Insertion of SO_2

SO_2 may insert into an M-C bond in a variety of ways giving structures **9** to **12**, as shown below. Only the S-sulfinate, **9**, and O-sulfinate, **10**, are observed, however. Whether the sulfur or the oxygen binds to the metal depends upon its softness, with softer metals binding to sulfur and harder to oxygen (see Section **7-1-2**). For most of the transition metals, the S-sulfinate form predominates. Complexes of titanium and zirconium, on the other hand, often give the O-sulfinate insertion product, owing to the harder nature of the early transition metals. A reaction to form **9** constitutes a 1,1-insertion, while overall 1,2-insertion leads to **10**.

The mechanism of SO_2 insertion has been investigated extensively for

$$
\begin{array}{cccc}
\overset{\displaystyle O}{\overset{\|}{\underset{\underset{\displaystyle O}{\|}}{M-S-R}}} & \overset{\displaystyle O}{\overset{\|}{M-O-S-R}} & \overset{\displaystyle O}{\overset{\|}{M-S-O-R}} & M\underset{O}{\overset{O}{\diamond}}S-R \\[2em]
\mathbf{9} & \mathbf{10} & \mathbf{11} & \mathbf{12}
\end{array}
$$

18-electron iron complexes.[19] Scheme **8.4** outlines the general mechanism for insertion. Note that SO_2 is a Lewis acid (electrophile), and that it does not attack the metal.

The reaction is considered to proceed first via an S_E2[20] pathway where, like its nucleophilic counterpart in organic chemistry, inversion of configuration is observed at the carbon attached to the metal (the α-carbon). After electrophilic attack, a tight or intimate ion pair, **13**, forms that retains stereochemical integrity at the carbon. Collapse of **13** leads to the O-sulfinate. The O-sulfinate seems to be the kinetically favored product that ultimately rearranges to the more thermodynamically stable S-sulfinate. The rate of the reaction is sensitive to the nature of substituents attached to the metal-bound carbon. Bulky alkyl groups decrease the rate, consistent with a bimolecular, concerted step involving a sterically congested transition state. Electron withdrawing substituents attached to the α-carbon also result in a rate retardation by making the carbon less nucleophilic.

Equations **8.20** and **8.21** show the results of some studies designed to elucidate the stereochemistry involved in SO_2 insertion into iron complexes. In **8.20** the *threo* Fe complex undergoes insertion to give the corresponding *erythro*

---

[19] A. Wojcicki, "Insertion Reactions of Transition Metal-Carbon σ Bonded Compounds II. Sulfur Dioxide and Other Molecules," in F.G.A. Stone and R. West, Eds., *Advances in Organometallic Chemistry*, vol. 12. Academic Press, New York, 1974, 31-81 and references therein.
[20] S_E2 stands for "substitution-bimolecular-electrophilic."

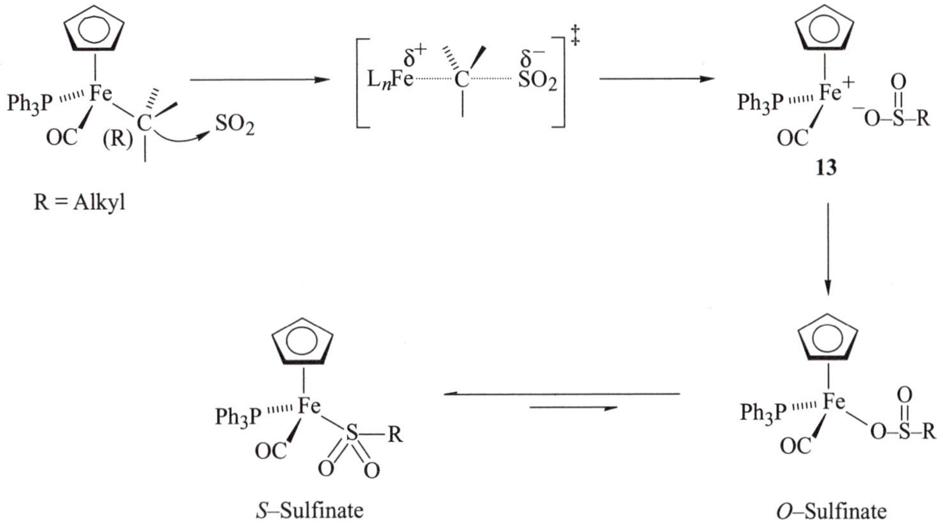

Scheme 8.4
Mechanism of SO$_2$ Insertion

S-sulfinate complex with a high degree of stereospecificity. Unlike CO insertion, retention of configuration is observed at the metal center. Equation **8.21** shows an experiment designed to correlate stereochemistry of reactant and insertion product. The stereospecificity of the insertion is greater than 90% retention at the metal center.[21]

8.20

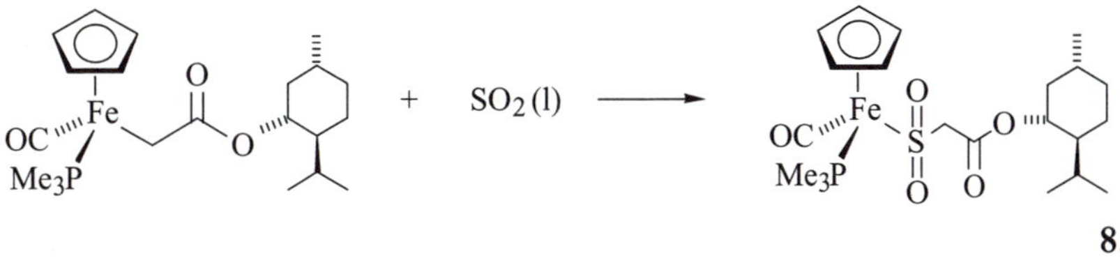

8.21

[21]T.C. Flood and D.L. Miles, *J. Am. Chem. Soc.*, **1973**, *95*, 6460.

Insertion of SO_2 occurs with a wide variety of transition metal complexes, almost as commonly as CO insertion. Unlike CO, however, deinsertion of SO_2 (desulfination) is not common. Since SO_2 reacts with 18-electron complexes, the resulting coordinatively saturated insertion complex must lose a ligand to provide an open site for alkyl migration to occur. Apparently this is a difficult process, and when it does occur, is accompanied by substantial decomposition.

Several other small molecules undergo "insertion" into M–C bonds. Examples of these are R–NC (1,1- and 1,2-insertion), NO (1,1-insertion), and CO_2 (1,2-insertion). Discussion of these and other insertions is beyond the scope of this text.[22]

8-2   Nucleophilic Addition to the Ligand

A nucleophile may attack an organometallic complex—substituted with at least one unsaturated ligand (e.g., CO, alkene, polyene, arene)[23]—in a variety of ways, as shown in Scheme **8.5**. Path **a** involves nucleophilic attack at the metal, and has already been discussed to some extent in Chapter 7.

Nucleophilic attack at the metal is also of importance with regard to transmetallation reactions in which alkyl, alkenyl, alkynyl, and aryl groups may be transferred from either ionically bonded main group organometallic compounds such as Grignard reagents or from more covalent metal-C compounds such as R-Hg-Cl. These reactions were discussed previously in Chapter 6, Section **6-1**; we will see how they may be applied to organic synthesis in Chapter 11.

Path **b** also involves nucleophilic attack at the metal, but instead of displacement, a *syn* insertion of an unsaturated ligand between the metal and the nucleophilic ligand can occur, some of the details of which we discussed earlier in this chapter. In contrast to the first two paths, path **c** shows the nucleophile attacking the ligand. It is this route that we shall consider in our discussion of nucleophilic reactions, for such a pathway offers the chemist numerous possibilities for accomplishing synthetic transformations that would be impossible without the presence of the metal.

Path **c** demonstrates a phenomenon that often occurs in organometallic chemistry—*umpolung. Umpolung* is a German word that translates into English

---

[22]See C.M. Lukehart, *Fundamental Transition Metal Organometallic Chemistry*, Brooks-Cole, Pacific Grove, CA, 1985, 234-245 and A.Yamamoto, *Organotransition Metal Chemistry*, Wiley-Interscience, New York, 1986, 257-260.

[23]Also known as π ligands.

Scheme 8.5

Possible Reactions of Nucleophiles

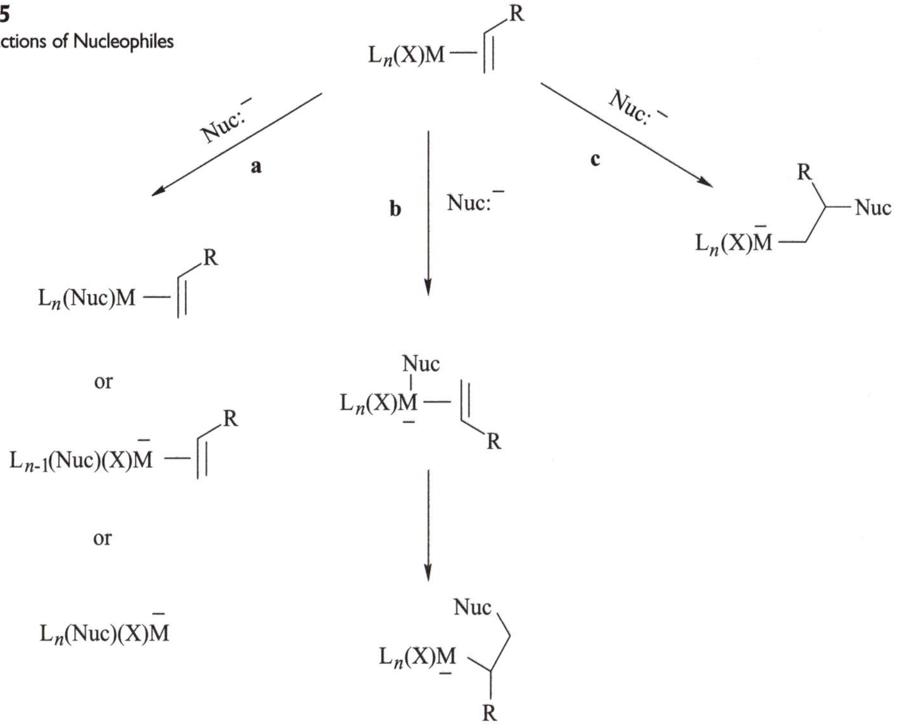

approximately to mean "reversal of polarity." One example of *umpolung* from the realm of organic chemistry is shown in equation **8.22** where an electrophile, CH$_3$I, adds to a dithiane to give a ketone.

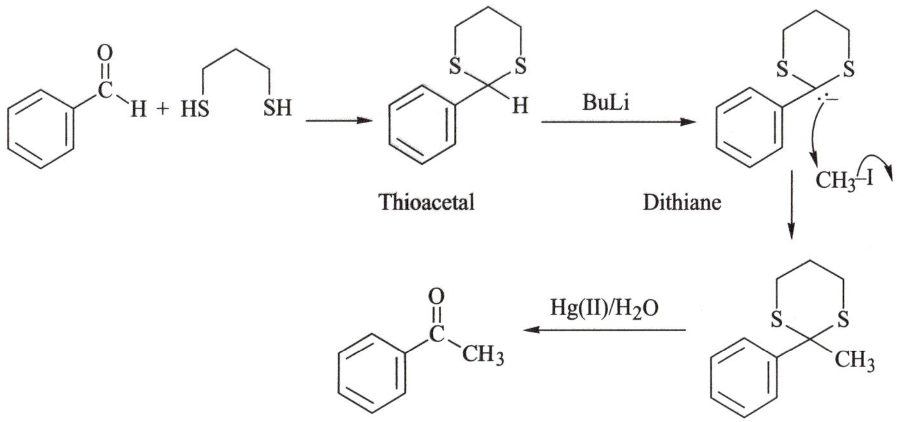

8.22

The overall transformation represents formally the attack of an electrophile at the carbon of a carbonyl group. Normally only nucleophiles attack carbonyl carbon atoms because these carbons represent electron-deficient sites. By converting the carbonyl to a thioacetal and pulling off the now relatively acidic proton with base, the carbonyl carbon becomes nucleophilic instead of electrophilic.

Alkenes, polyenes, arenes, and CO normally do not react with nucleophiles because these species are already electron rich. When these π ligands complex with a metal, however, (especially if the metal is electron-deficient due to the presence of other electron withdrawing ligands or due to a relatively high oxidation state), they are forced to give up some of their electron density to the attached metal complex fragment. The complexed ligands are now electron deficient compared to their free state. Nucleophilic attack on them is now possible, giving a substituted ligand that can either stay attached to the metal or be released from the metal by numerous methods such as β-elimination or oxidative cleavage.

The tendency for a nucleophile to attack an unsaturated ligand directly is a function of the following: the electron density on the metal (i.e., complexes with a formal positive charge on the metal are more reactive than neutral ones); the degree of coordinative saturation of the metal (unsaturated metals have a higher probability of attack directly at the metal); and the presence of π electron-attracting ligands (such as CO) that can absorb some of the increased electron density on the metal after attack has occurred.

8-2-1 Addition at CO and Carbene Ligands

Equations **8.23** to **8.25** show that carbonyl and carbene ligands can undergo nucleophilic reactions. In equation **8.23**, $NaBH_4$ delivers a hydride to one of the carbonyl ligands to yield the corresponding formyl complex.[24] Attack by OH^- on one of the CO ligands in $Fe(CO)_5$ liberates CO_2 in equation **8.24**.[25] This is a key step in the iron-catalyzed[26] water-gas "shift reaction," a reaction used to produce more H_2 from a mixture of CO and H_2O. Equation **8.25** shows a Fischer carbene complex[27] undergoing attack by a I° amine to give a product that requires replacement of OMe by RNH. This is analogous to a reaction in organic chemistry known as aminolysis (shown also for comparison in **8.25**) where the amine nucleophile attacks the carbon to give a tetrahedral

[24]J. R. Sweet and W.A.G. Graham, *J. Am. Chem. Soc.*, **1982**, *104*, 2811.

[25]This reaction is analogous to the loss of CO_2 under thermal conditions from a β-carbonyl carboxylic acid such as malonic acid, $CH_2(COOH)_2$.

[26]J. W. Reppe, *Annalen*, **1953**, *582*, 121.

[27]H. Werner, E.O. Fischer, B. Heckl, and C.G. Kreiter, *J. Organomet. Chem.*, **1971**, *28*, 367.

intermediate. Breakdown of the intermediate gives the substituted product plus alcohol (or alkoxide) as the leaving group. The carbene complex in this case actually reacts as if it were a carboxylic acid derivative.

$$8.23$$

$$8.24$$

$$8.25$$

8-2-2 Addition at Pi Ligands

There are numerous examples of nucleophilic addition to π hydrocarbon ligands.[28] The regiochemistry of these additions to cationic metal complexes has been well studied and summarized by Davies, Green, and Mingos[29] (DGM) as a set of rules. These rules can be used to predict where the nucleophile will add to a variety of π ligands when the reaction is under kinetic control. A summary of the rules follows:

1. Nucleophilic attack occurs preferentially at *even* instead of *odd* coordinated polyenes.

2. Nucleophilic addition to *open* coordinated polyenes is preferred over addition to *closed* polyenes.

---

[28]L.S. Hegedus, "Nucleophilic Attack on Transition Metal Organometallic Compounds," in F.R. Hartley and S..R. Hartley and S. Patai, Eds., *The Chemistry of the Metal-Carbon Bond*, Wiley-Interscience, New York, 1985, 401–512.

[29]S.G. Davies, M.L.H. Green, and D.M.P. Mingos, *Tetrahedron*, **1978**, *34*, 3047; a simplified explanation of the rules and several examples are provided in S.G. Davies, *Organotransition Metal Chemistry: Applications to Organic Synthesis*, Pergamon Press, Oxford, 1982, 116–186.

3.   For *even, open* polyenes, nucleophilic addition occurs preferentially at the terminal position; for *odd, open* polyenes, attack occurs at the terminal positions only if the metal fragment is strongly electron withdrawing.

Figure **8-4** shows some examples of ligands attached to the generic metal, M, according to their classification under the DGM scheme. Take note that DGM system classifies a ligand as *even* or *odd* depending on its hapticity and electron count, assuming that it is a neutral ligand. Thus η^3-allyl (3 e$^-$) is an *odd, open* ligand, and η^4-cyclohexadienyl (4 e$^-$) is *even* and also *open*.[30] The first two rules may be simplified further, as in the following:

1.   *Even* before *odd*.
2.   *Open* before *closed*.

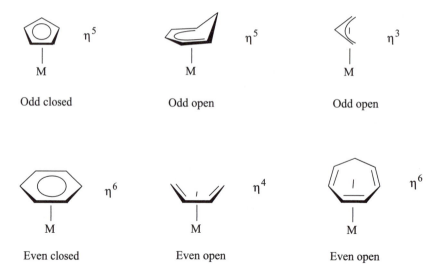

Figure 8-4
Classification of Pi Ligands

Figure **8-5** expresses rules 1 and 2 another way by ranking the various ligands in terms of their relative reactivity to nucleophilic attack.

---

[30]Note that a cyclic conjugated system such as benzene or Cp is considered *closed*.

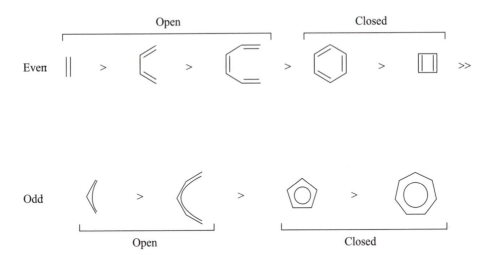

Figure 8-5
Relative Reactivity of Pi Ligands

Equations **8.26** to **8.29** provide some examples of some cationic complexes, polysubstituted with π ligands, that seem to obey the rules.

8.26

8.27

+   MeS⁻ ⟶

Mo⁺   Mo — SMe

8.28

NC

Co⁺   +   CN⁻ ⟶   Co

8.29

DGM and others have attempted to explain the basis for these rules. Rule 1 may be explained on the basis of the reactions being under charge control. Consider the orbitals of the π ligand once attached to the metal (Figure **8-6**). For *even* ligands the HOMO is doubly occupied.[31] If the metal is quite electron deficient, up to two electrons could transfer from the HOMO of the ligand to the metal. The net charge on the ligand could then be as high as +2. *Odd* π ligands are singly occupied in the HOMO. With *odd* ligands, bonding electrons are contributed by both the ligand and the metal. At most, one electron may be transferred from the ligand to the metal, and thus the net charge on the ligand will be maximally +1. In a complex with both an *even* and an *odd* π ligand, the *even* ligand will be more positive than the *odd* by up to one unit of positive charge.

Rule 2 states that *open* ligands will undergo nucleophilic attack before *closed*. Although not explicitly proven by DGM, it seems reasonable that, for cyclic, *closed* polyenes, the positive charge on the ring is more evenly distributed due to symmetry. As a consequence, any site on a *closed* ligand will have less positive charge than at least some sites on an *open* ligand with the same number of π electrons.

---

[31]One exception to this statement is cyclobutadiene. This molecule has two HOMOs of equal energy (a situation termed *degenerate*), each of which is singly occupied with an electron. Nucleophilic attack to other *even* polyenes is preferred according to Rule 1. Cyclobutadiene is, however, attacked in preference to *odd* polyenyl ligands.

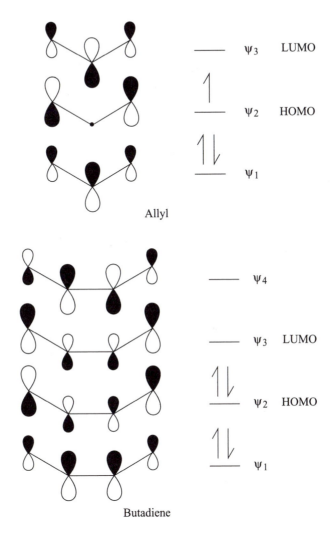

Allyl

Butadiene

Figure 8-6
Pi MOs of Allyl and Butadiene

There does, however, seem to be an empirical basis for the increased reactivity of *open* over *closed* ligands when the reactivity of ethene and benzene ligands are compared. Work reported after DGM's initial investigations by Bush and Angelici[32] indicates that there is a relation between the tendency of a ligand to undergo nucleophilic attack and the effective force constant,[33] $k_{CO}{}^*$, when reactions are under kinetic control. The reasoning goes as follows:

[32]R.C. Bush and R.J. Angelici, *J. Am. Chem. Soc.*, **1986**, *108*, 2735.
[33]If we assume that bonds are analogous to springs with varying degrees of stiffness, the force constant, *k*, measures the stiffness of the bond; see Chapter 4, Section **4-1**, for more details.

The electron withdrawing nature of a positive metal center should have the same effect on a π ligand as it would on a carbonyl group. The IR stretching frequency of a carbonyl (more accurately the force constant, k) is a reflection of the electronic interaction of the carbonyl group and the metal. Electron-poor metals (or metals also coordinated to π acids) do not back bond effectively to the carbonyl. Thus the Lewis structure **14** makes a substantial contribution to the actual structure of the complex.

$$\delta^+$$
$$L_nM\text{--}CO$$

14

The more electron withdrawing the metal fragment, the more complete is the σ donation to the metal. This will increase the triple-bond character of the C-O bond and thus increase the force constant. Bush and Angelici analyzed a number of cationic complexes containing π ligands for their tendency to undergo addition with a variety of nucleophiles. The force constant, $k_{CO}{}^*$, represents the effective force constant if the π ligand were replaced by the equivalent number of carbonyl groups. For instance, $[(C_2H_4)Rh(PMe_3)_2Cp]^{2+}$ would be equivalent to $[(CO)Rh(PMe_3)_2Cp]^{2+}$ and $[(\eta^6\text{-benzene})Mn(CO)_3]^+$ would correspond to three CO ligands grouped in a *facial* arrangement on $[Mn(CO)_6]^+$. Bush and Angelici found, in general, that addition of a particular kind of nucleophile such as an amine or a phosphine would not occur in complexes when $k_{CO}{}^*$ was below a certain threshold value. One particularly interesting result from this analysis was that the threshold value of $k_{CO}{}^*$ for addition of a particular kind of nucleophile—phosphines, for instance—was lower when ethene (an *open* ligand) was coordinated to the metal than benzene (a *closed* ligand). In other words, the ethene ligand, attached to a number of different metals, seemed to be more susceptible to attack by a variety of nucleophiles than the benzene ligand. We must be careful not to extrapolate this analysis too far since nucleophilic addition could be influenced by both kinetic *and* thermodynamic factors. In this limited case, however, there does seem to be a way of predicting whether or not a complex containing either a benzene or ethene ligand (or both) will undergo addition. If addition does occur, the ethene reacts preferentially to benzene in accordance with Rule 2.

Rule 3 considers the regiochemistry of attack of the nucleophile. For *even, open* ligands, the nucleophile attacks at the terminal position of the ligand. Consider butadiene as a typical example. The LUMO of butadiene, (Ψ_3)[34] has larger lobes at the termini of the orbital than in the middle (Figure **8-6**). The incoming nucleophile (if it is reasonably soft) will be attracted to that part of the orbital with the biggest lobe.

---

[34]Assuming that there is still electron occupation of Ψ_2.

The picture is more complicated when we apply Rule 3 to *odd, open* ligands. Attack may occur at the termini or in the middle depending on the electron-richness of the attached metal fragment. Usually, attack occurs at the terminal carbons of η^3-allyl ligands, but for relatively electron-rich metal complexes such as $[Cp_2Mo(\eta^3\text{-allyl})]^+$ (equation **8.30**[35]) and $[Cp^*(PMe_3)Rh(\eta^3\text{-allyl})]^+$ (equation **8.31**[36]), addition occurs at C-2 to form metallacyclic products.

8.30

8.31

We can understand this difference in regiochemistry if we consider the molecular orbitals of the allyl system in Figure **8-6**. For electron-rich metal fragments, the LUMO becomes ψ_3, which has a large lobe (or high lack of electron density) at C-2. Nucleophiles thus would be attracted to the middle carbon of the allyl ligand in these cases. Electron-poor metal fragments, on the other hand, pull electron density out of the allyl ligand, and the LUMO becomes ψ_2—an MO with the largest lobes at the ends of the system. Incoming nucleophiles thus tend to attack at these positions.

The stereochemistry of addition in equation **8.31** is interesting. If LiEt$_3$BD is used instead of its protium analog, the deuterium ends up at the C-2 of the allyl ligand *syn* to the Cp*. As we shall see in the examples to follow, nucleophilic addition onto π ligands almost always occurs *anti* to the metal fragment. This seems reasonable since *anti* addition involves a pathway with less steric hindrance than *syn*.

[35]M. Ephretikine, B.R. Francis, M.L.H. Green, R.E. Mackenzie and M.J. Smith, *J. Chem. Soc., Dalton Trans.*, **1977**, 1131.
[36]R.A. Periana and R.G. Bergman, *J. Am. Chem. Soc.*, **1984**, *106*, 7272.

Exercise 8-8

Propose a mechanism that might explain why deuterium adds syn to the Cp* ligand in equation **8.31**.

$\eta^2 \pi$ Ligands

Addition of a variety of nucleophiles to η^2-alkene ligands complexed to iron has been well-studied,[37] and is typical of nucleophilic attack in general. Careful investigation of the reaction of amines with (E)- and (Z)-2-butenyl iron complexes showed the stereochemistry to be clearly *anti* (equation **8.32**).

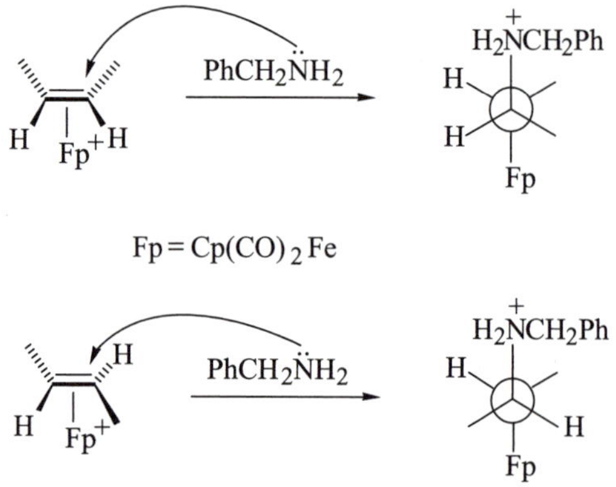

$$Fp = Cp(CO)_2\,Fe$$

8.32

The study also showed that the regioselectivity was low for alkenes substituted with alkyl groups, but was high when styrene was the ligand (equation **8.33**).

Addition to unsymmetrical η^2-alkenes seems to occur generally at the more-substituted carbon, analogous to a reaction from the realm of organic

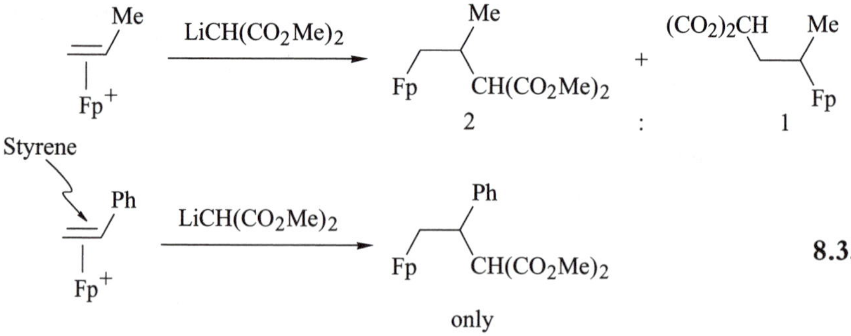

8.33

---

[37]P. Lennon, A.M. Rosan, and M. Rosenblum, *J. Am. Chem. Soc.*, **1977**, *99*, 8426.

chemistry known as *solvomercuration* (equation **8.34**). Theoretical analysis[38] of attack at alkene ligands has shown that a "slippage" occurs from the η^2- (**15**) to the η^1- complex (**16**) along the path to product. Calculations show that, for donor substituents such as alkyl groups, the energies of the pathways leading to attack at the more-substituted position *vis-a-vis* the less-substituted do not differ greatly. A phenyl group, on the other hand, is slightly electron withdrawing yet can effectively stabilize, through resonance, the positive charge that would develop in the η^1-complex (**17**). Attack, therefore, occurs exclusively at the benzylic position.

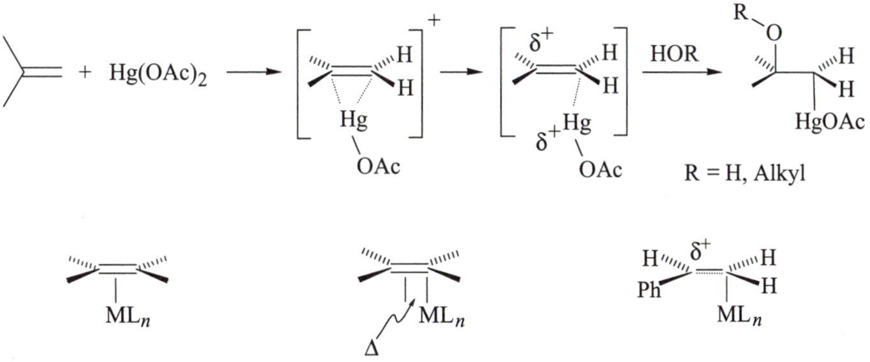

8.34

η^3-Allyl Ligands

Equation **8.35** shows an example of addition to a η^3-allyl complex. Desirable nucleophiles typically include stabilized carbanions such as $^-CH(COOR)_2$ or I° and II° amines. Unstablilized nucleophiles such as MeMgBr or MeLi may attack the metal and not the ligand. Additions to allyl ligands are extremely useful in organic synthesis, and many applications of this reaction have appeared in the literature.[39] We will discuss these additions more extensively in Chapter 11.

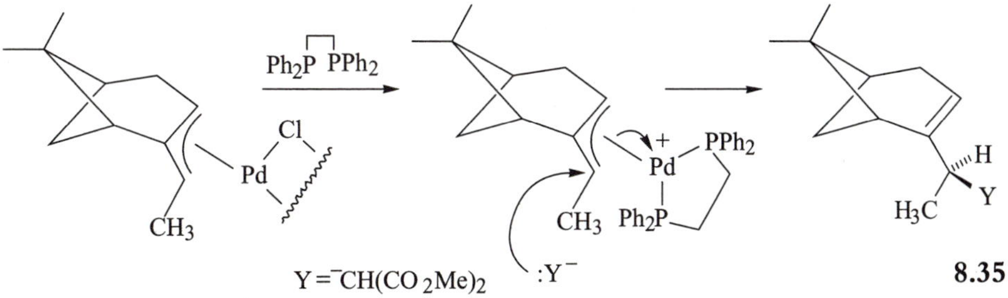

8.35

---

[38]O. Eisenstein and R. Hoffmann, *J. Am. Chem. Soc.*, **1981**, *103*, 4308.
[39]See L.S. Hegedus, *op. cit.*, 401-512.

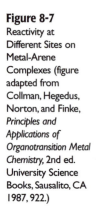

Figure 8-7
Reactivity at Different Sites on Metal-Arene Complexes (figure adapted from Collman, Hegedus, Norton, and Finke, *Principles and Applications of Organotransition Metal Chemistry,* 2nd ed. University Science Books, Sausalito, CA 1987, 922.)

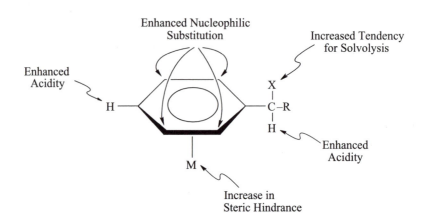

Pi-Arene Ligands

Pi-Arene complexes offer interesting possibilities for demonstrating the utility of *umpolung.* Group 6 (especially Cr) metal-arene compounds[40] have been particularly well investigated and exploited for their synthetic utility. Complexation of a transition metal to the arene confers enhanced reactivity to not only the metal but also to the arene and its attached carbon centers. Figure **8-7** shows what changes in reactivity metal complexation causes.

When M = Cr, profound effects on the acidity of arene and neighboring protons, electron density of the arene, and steric hindrance at the metal occur. As Table **8-1** shows, the metal exerts an electron withdrawing effect on the arene comparable to that of a *para*-NO_2 group on the acidity of a carboxylic acid. Metallation of the ring using strong bases such as butyllithium occurs at even -78°, a much lower temperature than would be required in the uncomplexed arene. The electron withdrawing effect of the metal fragment is also manifested in the enhanced ability of complexed arenes to undergo nucleophilic aromatic substitution, a reaction normally requiring strong nucleophiles in uncomplexed arenes. The reaction of methoxide with the chloroarene

Table 8-1 Acidities of Chromium-Aryl Carboxylic Acids Compared to Uncomplexed Analogs

Acid	pK$_a$	Acid	pK$_a$
PhCOOH	5.68	PhCH$_2$COOH	5.54
(PhCOOH)Cr(CO)$_3$	4.77	(PhCH$_2$COOH)Cr(CO)$_3$	5.02
p-NO$_2$-C$_6$H$_4$-COOH	4.48	*p*-NO$_2$-C$_6$H$_4$-CH$_2$COOH	5.01

[40]For a recent review on Cr(0)-arene complexes see S. Maiorana, C. Baldoli, and E. Licandro, "Carbonylchromium(0) Complexes in Organic Synthesis" in A.F. Noels, et al., Eds., *Metal Promoted Selectivity in Organic Synthesis,* Kluwer Academic Publishers, The Netherlands, 1991, 261-286.

chromium complex shown in equation **8.36** gives the corresponding methoxy-yarene complex at a rate comparable to that for substitution by methoxide on uncomplexed 1-nitro-4-chlorobenzene. As with their uncomplexed counterparts in organic chemistry, nucleophilic substitutions on coordinated arenes go through an intermediate σ complex, **18**, known as a Meisenheimer complex.[41]

$$\underset{\text{Cr(CO)}_3}{\text{⬡}}-\text{Cl} \quad \xrightarrow{\text{CH}_3\text{O}^-} \quad \left[\underset{\text{Cr(CO)}_3}{\underset{\textbf{18}}{\text{⬡}}}\overset{\text{OCH}_3}{\underset{}{\text{Cl}}}\right]^- \quad \longrightarrow \quad \underset{\text{Cr(CO)}_3}{\text{⬡}}-\text{OCH}_3 \;+\; \text{Cl}^-$$

8.36

When a substituent is present that is not a leaving group, the directing effects of that substituent tend to be the reverse of what they would be for electrophilic substitution on the uncomplexed arene. For instance, in equation **8.37** the electron donating OCH_3 directs primarily to the *meta* position, whereas in **8.38**, the electron withdrawing CF_3 directs to the *para* position primarily. The last step in these reactions involves oxidative cleavage (using I_2) of the arene-metal bonds.

$$\underset{\text{Cr(CO)}_3}{\text{⬡}}-\text{OCH}_3 \quad \xrightarrow[\text{2) I}_2]{\text{1) Li–CH}_2\text{CO}_2\text{R}} \quad \underset{\text{RO}_2\text{C–H}_2\text{C}}{\text{⬡}}-\text{OMe}$$

$o:m:p = 4:96:0$ **8.37**

$$\underset{\text{Cr(CO)}_3}{\text{⬡}}-\text{CF}_3 \quad \xrightarrow[\text{2) I}_2]{\text{1) LiY}} \quad \underset{\text{Y}}{\text{⬡}}-\text{CF}_3$$

$o:m:p = 0:30:70$

$$Y = {}^-\!:\overset{\overset{\displaystyle CN}{|}}{\underset{\underset{\displaystyle Me}{|}}{C}}-O-C\overset{O\diagdown}{\underset{\diagdown Me}{}}$$

8.38

Again, the stereochemistry of attack is such that the nucleophile adds *anti* to the metal.

---

[41]See J. McMurry, *Organic Chemistry*, 4th ed. Brooks-Cole, Pacific Grove, CA, 1996, 597-602 for a discussion of nucleophilic aromatic substitution.

The rationalization for the regiochemistry of addition when a directing group is present and when the reaction is under kinetic control is based on inspection of the LUMO of the metal arene complex.[42] If the LUMO of the arene complex and the HOMO of the base are reasonably close in energy, then inspection of the coefficients on the LUMO offers some insight into the orientation of attack on the aromatic ring. The LUMO, when the substituent is electron donating, is given by **19** (below) and shows large coefficients at the *ortho* and *meta* positions. When the substituent is electron withdrawing, however, the coefficients are large mainly at the 1 and 4 positions as shown in **20**. We would expect an incoming nucleophile to attack at the positions where the coefficients are large since this would provide the best overlap with the lobe(s) of the nucleophile's HOMO. This analysis is reasonable, however, only if the reaction is under kinetic control and if considerations of steric hindrance are unimportant.

$\overline{e}$ -Donating substituent   Top view $\overline{e}$ -Withdrawing substituent

19   **20**

Equations **8.39** to **8.41** show other examples of nucleophilic addition to π ligands. The cationic manganese arene complex in **8.39**,[43] produced by ligand substitution in the presence of a Lewis acid, undergoes nucleophilic substitution by a variety of alkoxide and amine nucleophiles. In equation **8.40** the molybdenum η^4-diene complex undergoes attack by methyllithium to give an η^3-allyl intermediate.[44] Hydride abstraction with trityl cation, Ph_3C^+, gives the product with the methyl group *anti* to the metal fragment. Reaction of the η^6-trienyl Mn complex with a stabilized carbanion such as malonate results in an η^5-pentadienyl complex—again showing the nucleophile *anti* to the metal (equation **8.41**).[45]

[42]M. F. Semmelhack, G.R. Clark, R. Farina, and M. Saeman, *J. Am. Chem. Soc.*, **1979**, *101*, 217; for a more recent discussion of orientation effects based on kinetic and thermodynamic control, see E.P. Kunding, V. Desobry, D.P. Simmons, and E. Wenger, *J. Am. Chem. Soc.*, **1989**, *111*, 1804.

[43]P.L. Pauson and J.A. Segal, *J. Chem. Soc., Dalton*, **1975**, 1677, 1683.

[44]J.W. Faller, H.H. Murray, D.L. White, and K.H. Chao, *Organometallics*, **1983**, *2*, 400.

[45]A.J. Pearson, P. Bruhn, and I.C. Richards, *Tetrahedron Lett.*, **1984**, *25*, 387.

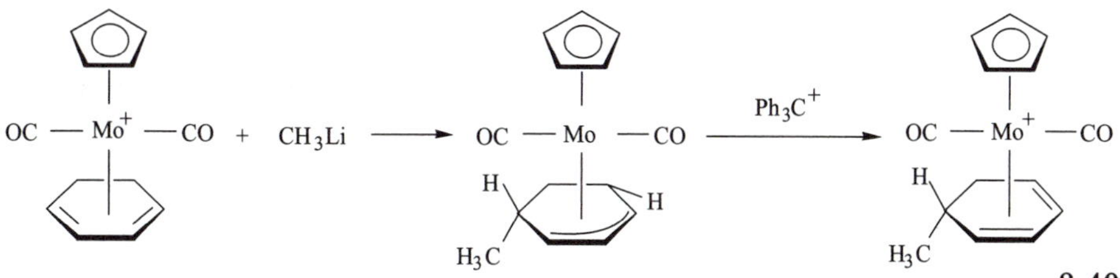

8.40

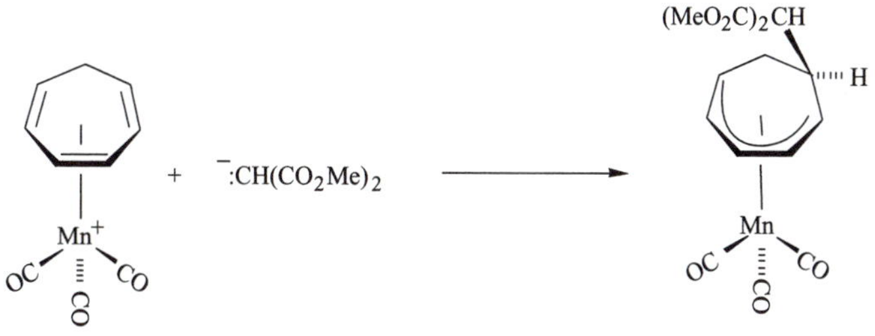

8.41

8-3 Nucleophilic Abstraction

Instead of simply adding to a ligand, the possibility exists for a nucleophile to attack the ligand in such a way that part or all of the original ligand is removed along with the nucleophile. We call this reaction *nucleophilic abstraction*. Equation **8.42** demonstrates an example of this type of reaction. Here, electron-rich, zero-valent Pd attacks the chiral Pd(II) σ-alkyl complex, resulting in inversion of configuration at the stereogenic center of the alkyl ligand and complete abstraction of the alkyl group. The Pd(0) by-product will then attack the inverted alkyl-palladium complex, and inversion of configuration will occur again. This reaction probably accounts for the observation of partial racemization during studies on the stereochemical course of oxidative addition (see Chapter 7, Scheme **7.4**, p. 174).

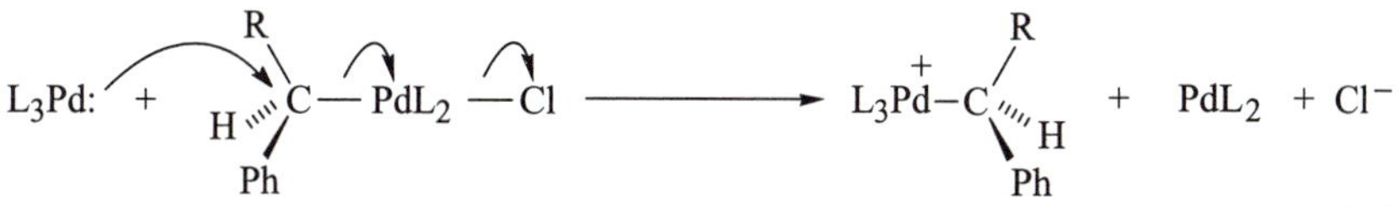

$$L \quad Ph_3P; \ R = D, CF_3$$

8.42

Although self-exchange reactions such as **8.42** are probably relatively common because they are thermoneutral,[46] abstraction of alkyl groups by nucleophiles other than low-valent metals is not as common, since reduced

---

[46]A thermoneutral reaction requires that $\Delta H_{rxn} \approx 0$; L.S. Hegedus, *op. cit.*, 427.

metals are generally poor leaving groups. If the metal fragment is first oxidized, however, nucleophilic attack on alkyl ligands at the carbon attached to the metal becomes much more feasible. Oxidation makes the leaving metal fragment less basic and also weakens the M–C bond. The halogens Br_2 and I_2 serve effectively as M–C cleaving agents; an example of oxidatively driven, nucleophilic abstraction is shown in equation **8.43**.[47]

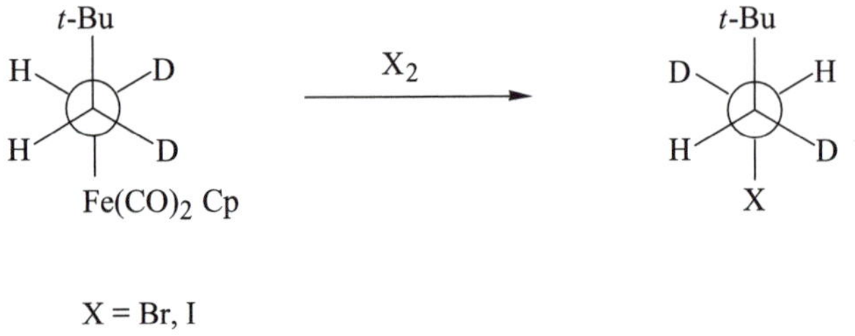

$$X = Br, I$$

(8.43)

Note that the reaction proceeds with inversion of configuration at the stereogenic center attached to the metal. However, it is not clear exactly what the initial function of the halogen is in the reaction. Scheme **8.6** presents two possible scenarios for the reaction to occur. In path **a** the halogen oxidatively adds to the metal to give a cationic complex. The remaining nucleophilic halide ion attacks the complex at the α-carbon of the alkyl group, giving inversion of configuration. Path **b**, on the other hand, shows oxidation of the metal without addition, followed by nucleophilic attack.

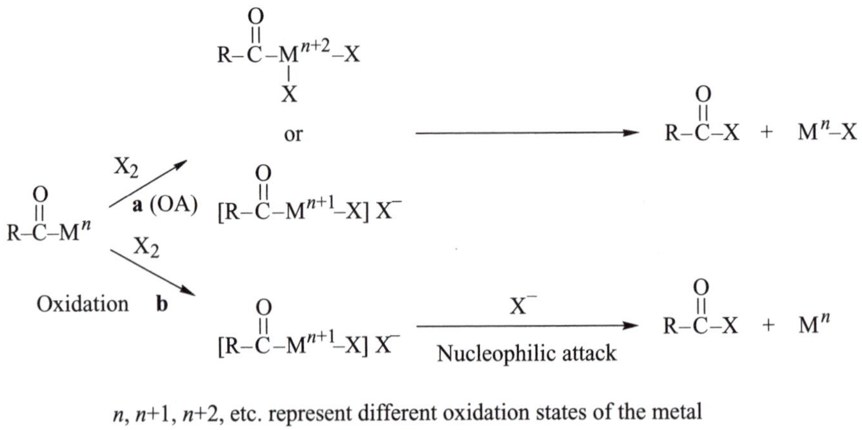

$$n, n+1, n+2, \text{etc. represent different oxidation states of the metal}$$

Scheme 8.6
Pathways for Oxidatively
Driven Nucleophilic Abstraction

[47]G.M. Whitesides and D.J. Boschetto, *J. Am. Chem. Soc.*, **1971**, *93*, 1529.

We can also envision the overall process of halogen cleavage as an *electrophilic* process if the first step is the addition of X^+ (from X_2) to the metal—analogous to the addition of halogen to a C=C bond in organic chemistry. Addition of halogen to alkenes is a two-step process where attack of the electrophile occurs first to form an intermediate halonium ion (equation **8.44**), followed by attack by the nucleophile, X^-. We see that sometimes the designation "nucleophilic" or "electrophilic" can be rather arbitrary, and it depends upon one's perspective in describing the reaction. We shall consider electrophilic processes in more detail in Section **8-4**.

$$ \diagup\kern-1em\diagup\ +\ Br_2\ \longrightarrow\ \underset{H}{\overset{+\ H}{Br}}\ \xrightarrow{Br^-}\ \underset{Br}{\overset{Br}{H\diagdown\kern-1em\diagup H}}\qquad \textbf{8.44}$$

Nucleophilic abstraction of σ-acyl-metal complexes is quite useful synthetically, especially when the metal is Pd. Scheme **8.7** shows a sequence of reactions that may be either stoichiometric or catalytic in that metal.[48]

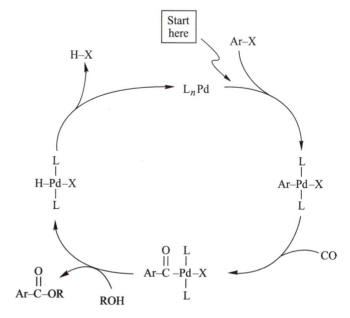

Scheme 8.7
Nucleophilic Abstraction of
Sigma-Acyl-Pd Complexes   R = Alkyl   X= Cl, Br, I

[48]R. F. Heck, *Pure Appl. Chem.*, **1978**, *50*, 691.

Palladium is the metal of choice here because it readily forms σ–M–C complexes, which then undergo facile CO insertion and subsequent σ–acyl–Pd bond cleavage with a variety of nucleophiles. The first step involves OA of a vinyl or aryl halide to a Pd(0) complex (shown as L_nPd in Scheme **8.7**). Migratory insertion of CO (Section **8-1**) gives the σ–acyl complex, which undergoes nucleophilic abstraction with MeOH to give the corresponding methyl ester (see Scheme **7.4**, p. 174). This sequence of reactions has wide synthetic applicability, because intermolecular nucleophilic abstraction produces a variety of linear acyl derivatives and intramolecular attack provides cyclic compounds such as lactones or lactams.

Iron in the form of $Na_2[Fe(CO)_4]$ (known as Collman's[49] reagent) is another metal involved in synthetically useful nucleophilic abstractions. Scheme **8.8** shows how Collman's reagent may be used to produce esters. Note that an oxidizing agent such as Br_2 is used to make the metal fragment a better leaving group. We can again postulate two different mechanisms involving the oxidizing agent analogous to those presented in Scheme **8.6**. We shall encounter again the use of Pd and Fe in organic synthesis in Chapter 11.

Scheme 8.8
The Use of Collman's
Reagent in Nucleophilic
Abstractions

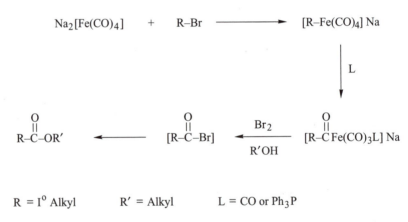

$$Na_2[Fe(CO)_4] \quad + \quad R–Br \longrightarrow [R–Fe(CO)_4]\, Na$$

$R = 1°$ Alkyl $R' = $ Alkyl $L = $ CO or Ph_3P

Equation **8.45** (on the next page) can be considered to be a nucleophilic abstraction of H^+ from the arene ligand by a strong nucleophile. An alternate interpretation of **8.45** is that it represents a Brønsted-Lowry acid-base reaction. Regardless of the viewpoint of the observer, the reaction is facile mainly because of the electron withdrawing nature of the metal fragment. Oxidative cleavage of the arene-Cr bonds with I_2 liberates iodobenzene.

---

[49]J.P. Collman, *Accounts. Chem. Res.*, **1975**, *8*, 342.

$$8.45$$

8-4   Electrophilic Reactions

Reactions involving a metal complex and an electron-deficient species (the electrophile, E^+) are common in organometallic chemistry and often useful synthetically. There are several possibilities for interaction of a metal complex and E^+, some of which we have already encountered. For example, oxidative addition of R–X to a metal (shown in Chapter 7, Section **7-2-2**, equation **7.38**, p. 172) can be considered to be the attack of "R^+" on the metal followed by addition of X^-. The insertion of SO_2 (equation **8.20**), electron deficient at sulfur, begins with attack by the electrophile at the carbon attached to the metal.

$$7.38$$

$$8.20$$

threo   *erythro*

Just where and with what stereochemistry the electrophile will attack a metal complex depends upon a number of factors. If the reaction is controlled by frontier orbital interactions, then the point of attack on the complex is dictated by the electron density of the complex's HOMO. Major lobes of the HOMO could be on the ligand or the metal or both. The stereochemistry of attack may depend on such factors as the steric hindrance of the metal fragment or even on the nature of the solvent. Extensive study of electrophilic attack on specific metal complexes has uncovered some general trends for use as a basis in predicting how an electrophile will react. This area of organometallic chemistry, nevertheless, remains enigmatic. We will limit our treatment of electrophilic attack to a few well-defined examples that illustrate the major pathways of a reaction where E^+ either directly or indirectly interacts with a coordinated ligand. Again, we distinguish between *addition* reactions, where all or part of the electrophilic species adds to the coordinated ligand, and *abstractions*, in which E^+ attacks in such a manner that all or part of the ligand is removed from the metal complex (see Chapter 7, Table **7-1**, p. 139).

Electrophiles

The term electrophile is synonymous with Lewis acid. Classifying electrophiles on the basis of their "strength" is virtually impossible (see Section **7-1-2** in Chapter 7). It is useful, however, to point out that we can classify Lewis acids according to three types as described below:[50]

1. *Metallic* electrophiles, such as Hg(II), Tl(I), Ag(I), Ce(IV), Pt(IV), Au(III), or Ir(IV)—most of which also have accessible oxidation states one or two units lower in value from the given states

2. *Organic* electrophiles, such as R_3O^+, Ph_3C^+, R–X, or $(NC)_2C=C(CN)_2$

3. *Non-metallic* electrophiles, such as X_2, SO_x, CO_2, NO_x^+, or H^+

Each of these classes of electrophiles may attack transition metal complexes. We will encounter examples involving all three classes during the remainder of our discussion.

8-4-1 Metal-Carbon Sigma Bond Cleavage

The use of an electrophile, e.g., H^+ or X_2, to remove a σ bonded carbon ligand from a metal is common in organometallic chemistry. Such a reaction constitutes an electrophilic abstraction of the entire ligand. Scheme **8.9** diagrams some of the many pathways available for this type of abstraction. Stereochemical outcomes include retention, inversion, and racemization, and depend upon such factors as the nature of the metal, steric hindrance, the nature of E^+, where the HOMO is located on the complex, and characteristics of the solvent. Pathways **a** and **b** represent a mechanism called S_E2, which we encountered earlier in the discussion on SO_2 insertion. Inversion[51] (path **a**) or retention (path **b**) of stereochemistry may occur at the stereogenic center[52] attached to the metal. Pathway **c** is oxidative addition to give intermediates, **21** and **22**, which may suffer several fates. Pathway **d**, emanating from **21**, is ligand dissociation to give a stable carbocation followed by nucleophilic capture to

---

[50]See M.D. Johnson, "Electrophilic Attack on Transition Metal η^1-Organometallic Compounds," in F.R. Hartley and S. Patai, Eds., *The Chemistry of the Metal-Carbon Bond*, Wiley, London, 1985, vol. 2. chap. 7, 515.

[51]Inversion of configuration means transformation of *R* to *S* (or vice versa) or *erythro* to *threo* (or vice versa) if more than one stereocenter is present in the σ bonded ligand.

[52]The term "chiral center" has been replaced more recently by the label "stereogenic center" or, simply, "stereocenter," deemed to be more correct semantically. A stereogenic atom is defined as one bonded to several groups of such nature that interchange of any two groups produces a stereoisomer. See K. Mislow and J. Siegel, *J. Am. Chem. Soc.*, **1984**, *106*, 3319.

give racemic[53] product. Also originating from **21** is **e**, reductive elimination, from which we would expect retention of configuration. Reductive elimination with retention (path **f**) could also lead from **22**; however, direct nucleophilic S_N2 displacement (path **g**) would yield product with inversion of configuration (see equation **8.43**). Path **h** represents a general oxidation process of the metal that may occur by a single electron mechanism. It is likely that some electrophilic cleavages occur under these conditions, but experimental evidence is difficult to obtain. Discussion of such mechanisms is beyond the scope of this text.[54] Table **8-2** summarizes the mechanistic type and stereochemical characteristics of each path.

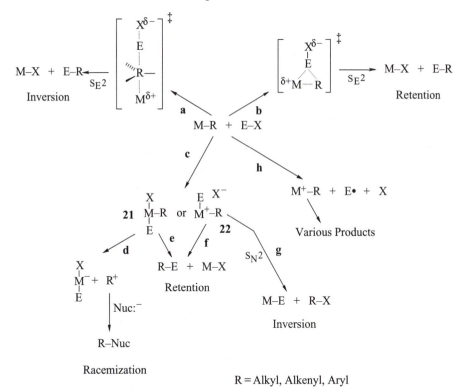

Scheme 8.9
Pathways for Electrophilic Abstraction of Sigma-Bonded M-C Ligands

[53]The term "racemic" applies strictly if only one stereogenic center is present in the σ bonded ligand. A more inclusive term for the stereochemical result upon formation of R^+ would be "stereorandomization." For example, if the σ bonded ligand had 2 stereocenters, giving it the stereochemical designation *threo*, stereorandomization would provide roughly equal amounts of two diastereomers, *threo* and *erythro*. In the subsequent discussion we will use the terms "racemic" and "racemization" to indicate that stereorandomization has occurred.
[54]See J.K. Kochi, *Organometallic Mechanisms and Catalysis*, Academic Press, New York, 1978, chap. 18.

Table 8-2   Characteristics of Electrophilic Abstraction Pathways

Pathway	Reaction Type	Stereochemistry
a	S_E2	Inversion
b	S_E2	Retention
c	OA	Different stereochemistries possible
d	Ligand dissociation followed by nucleophilic capture	Racemization
e	RE	Retention
f	RE	Retention
g	S_N2	Inversion
h	Radical	Several possible

Cleavage by H⁺

Protonolysis, or the use of Brønsted-Lowry acids to cleave σ M–C bonds, is an often-used method of M–C bond cleavage. The stereochemistry observed invariably seems to be retention of configuration at the stereocenter attached to the metal. Equations **8.46**[55] and **8.47**[56] provide two examples of protolytic cleavage using deuterated acids to provide an indication of stereochemistry. The reactions probably occur via paths **c** and then **f**, that is, protonation of the metal followed by reductive elimination.

8.46

8.47

$$L = PPh_3$$

Unlike the metals in equations **8.46** and **8.47**, the metal in equation **8.48**[57] is an early d^0-transition metal. The OA-RE pathway is not possible from such a complex, since it is difficult to remove electrons from a d^0 metal, yet protonolysis occurs again with retention. The mechanism for this reaction probably involves path **b**, a concerted S_E2 reaction involving a four-centered transition state, **23**. *This mechanism seems to be general for electrophilic*

[55] W.N. Rogers and M.C. Baird, *J. Organomet. Chem.*, **1979**, *182*, C65.
[56] D. Dodd and M.D. Johnson, *J. Organomet. Chem.*, **1973**, *52*, 1.
[57] J.S. Labinger, D.W. Hart, W.E. Seibert, III, and J. Schwartz, *op. cit.*

cleavages involving the early transition metals because the HOMO consists primarily of the M-C σ bond and oxidative pathways are not possible.

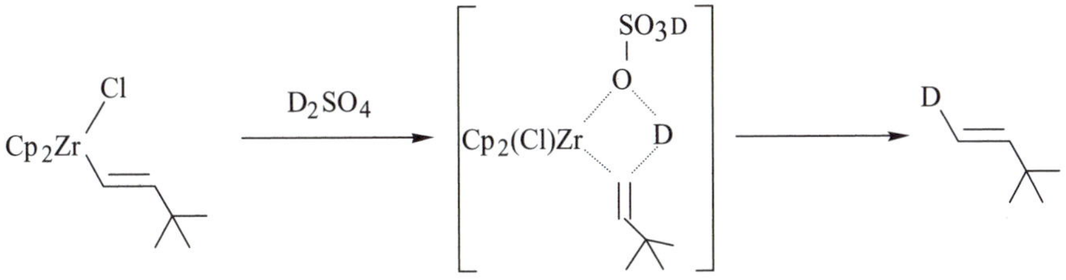

23   **8.48**

Cleavage by Halogens

Cleavage by halogens is a second important method of electrophilic cleavage. We have already encountered an example of this reaction (equation **8.43**). Another example is shown in equation **8.49** (below):

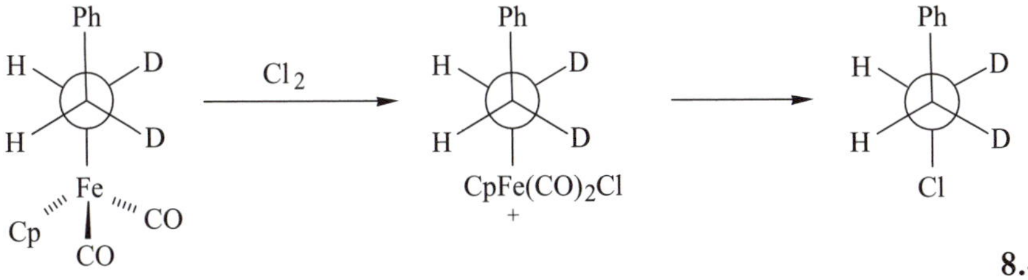

8.49

In these reactions, rather different stereochemistry results depending upon the group attached to the β-carbon of the σ-alkyl ligand. When the β-group is Ph, retention of configuration is observed. Change the group to *tert*-butyl, and cleavage occurs with inversion of configuration at the carbon originally attached to the metal. Scheme **8.10** details the stereochemistry involved in the two reactions.

The reaction begins by the addition of X^+ to the *threo*-Fe complex to give **24**. If R = *tert*-butyl, then X^- attacks at the α carbon to produce the *erythro*-alkyl halide (path **g**, Scheme **8.9**), **25**. A change in R to Ph results in attack by X^- with retention of configuration to yield *threo*-alkyl halide, **27**. There are actually two pathways to consider when explaining how retention of configuration could occur when R = Ph. Direct reductive elimination (path **e**, Scheme **8.9**) would provide **27**, since we know that RE proceeds with retention of configuration. The other possibility proceeds through a symmetrical, bridged carbocation, **26**, known generally as a phenonium ion. Attack by X^- at either alkyl carbon of **26** produces *threo*-alkyl halide. The steric bulk of the bridged phenyl group forces attack from the same side as the metal fragment was originally attached, thus ensuring retention of configuration.

Exercise 8-9

What is the stereochemical relationship between the alkyl halides produced when X⁻ attacks at the two alkyl positions in **26**? Note, only one isomer is shown in Scheme **8.10**.

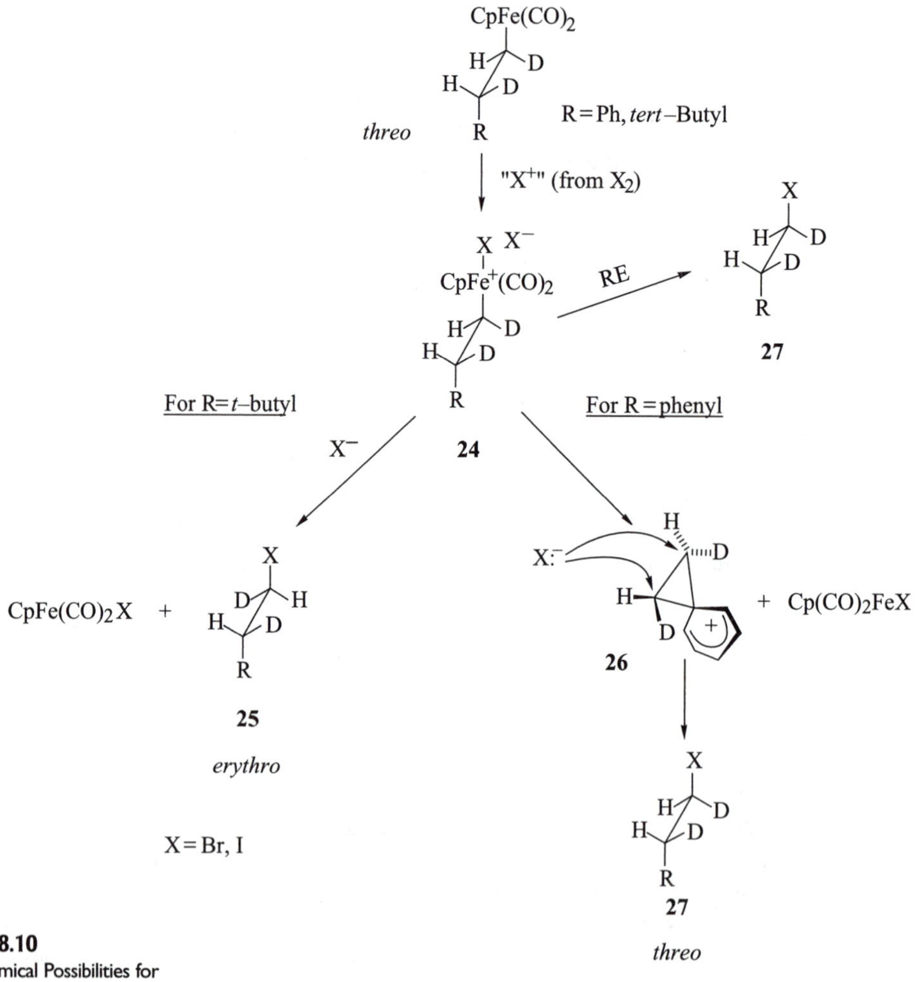

Scheme 8.10
Stereochemical Possibilities for
Metal-Halogen Cleavage

There is much evidence for phenonium ions in the realm of organic chemistry.[58] In the case described by equation **8.49**, there is experimental support for the presence of **26** based on the following experiment: Cleavage by X_2 of $CpFe(CO)_2CD_2CH_2Ph$ led to the formation of a $1:1$ mixture of XCD_2CH_2Ph and XCH_2CD_2Ph.[59] Had the reaction occurred by concerted, one-step RE, only one isomer would have been observed. With this example, we see that an apparent concerted reductive elimination with retention probably did not occur, but instead the reaction proceeded through a free carbocation to give the same stereochemical result as we would have expected from RE. This case, moreover, again points out the need for careful experimentation any time a mechanism is proposed.

Exercise 8-10

Show that halogen cleavage of $CpFe(CO)_2CD_2CH_2Ph$ will give two different constitutional isomers in equal amounts if **26** is an intermediate in the reaction.

The division between retention and inversion during halogenolysis occurs with a number of σ metal-alkyl and metal-alkenyl complexes. The factors mentioned earlier, such as solvent polarity, steric hindrance, the nature of the HOMO, etc., are influential in directing the stereochemical outcomes. It is only with the early transition metals that rather consistent stereochemical outcomes occur. As in protonolysis, it appears that halogen cleavage occurs by an S_E2 mechanism with retention of configuration. Equations **8.50**[60] and **8.51**[61] demonstrate this pathway for σ-alkyl and σ-alkenyl Zr complexes, respectively.

$$\text{8.50}$$

$$\text{8.51}$$

[58]See T.H. Lowry and K.S. Richardson, *Mechanism and Theory in Organic Chemistry*, 3rd ed. Harper and Row, New York, 1987, 434–448, for a good discussion on the evidence for the existence of phenonium ions.
[59]T.C. Flood and F.J. DiSanti, *J. Chem. Soc., Chem. Commun.*, **1975**, 18.
[60]J.A. Labinger, D.W. Hart, W.E. Seifert, III, and J. Schwartz, *op. cit.*
[61]D.W. Hart, T.F. Blackburn, and J. Schwartz, *J. Am. Chem. Soc.*, **1975**, *97*, 679.

Cleavage by Metal Ions

The third and last method of σ M–C cleavage we will consider is that with metal ions. The metal ion most thoroughly studied is Hg(II), which is well known for its ability to reversibly transfer alkyl, alkenyl, and aryl groups to a variety of transition metals. Investigations on the stereochemistry attendant cleavage of σ M–C bonds have been performed on a number of complexes involving Fe, Mn, W, Mo, and Co. There seems to be no single pathway preferred for cleavage, based on stereochemical results, though a pathway involving retention of configuration is common. Our discussion will focus on a few of these investigations.

In Scheme **8.11** the products formed upon treatment of $CpFe(CO)_2R$ with HgX_2 (X = Cl, Br, or I) seem to depend on the nature of R. The stereochemistry of the liberated organic group, moreover, also depends on the structure of R. The scheme takes into account the involvement of HgX_2 in the first two steps of the mechanism, giving an observed third-order rate law shown in the following:

$$Rate = k[L_nFeR][HgX_2]^2$$

When R = I° alkyl, the reaction proceeds with retention of configuration at R to produce $RHgX$ and L_nFeX. When R = III° or benzyl, racemization is the stereochemical outcome and the products are RX and L_nFeHgX. Apparently two pathways are operative in this example. Path **e** (Scheme **8.9**) gives retention by a reductive elimination pathway. As R becomes substituted with groups that can stabilize a carbocation, path **d** (Scheme **8.9**) takes over to produce free R^+ that picks up X^- from HgX_3^-.

The equations that follow show actual examples of electrophilic cleavage due to Hg(II) salts. Cleavage of the two iron complexes (R = *tert*-butyl[62] and R = phenyl[63]) with retention of stereochemistry is shown in equation **8.52**. When R = phenyl, a phenonium ion could also form to give retention of configuration as was demonstrated in Scheme **8.10**. The rate law showing second-order involvement of Hg(II), however, argues for the mechanism involving path **e**. The *trans*-$CpW(CO)_2(PEt)_3$(alkyl) complex in equation **8.53** also undergoes cleavage with retention[64]:

$$R = tert\text{-}Bu, Ph$$

[62] G.M. Whitesides and D.J. Boschetto, *op. cit.*
[63] D. Dong, D.A. Slack, and M.C. Baird, *Inorg. Chem.*, **1979**, *18*, 188.
[64] D. Dong, *et al.*, *op. cit.*

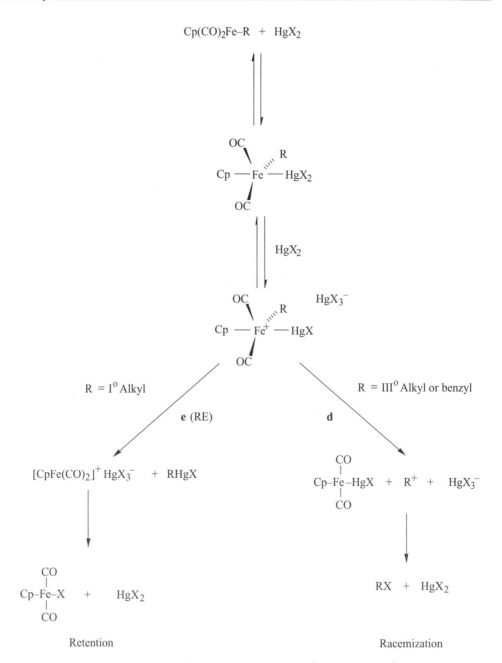

Scheme 8.11
Hg(II) Cleavage of Iron Complexes

$$ (8.53) $$

Hg(II) cleavage of the Mn[65] (equation **8.54**) or the Co[66] complex (equation **8.55**), on the other hand, results in inversion of configuration, probably via an S_E2 mechanism (path **a**, Scheme **8.9**). Steric hindrance about the metal—especially in the case of the Co complex with its large, macrocyclic ligand—seems to be the reason that Hg(II) cleavage occurs with inversion of configuration in these cases. The investigators proposed a transition state, **28**, as that involved for the Co complex.

While electrophilic cleavage of σ-bonded carbon ligands constitutes an important method for removing those ligands from the metal, we now see that the pathways available for this process are diverse indeed. A few generalizations regarding the stereochemical course of these abstractions can be made, such as retention of configuration with all electrophiles when early transition metals are involved or retention during protonolysis. However, the stereochemistry of most cleavages is difficult to predict and depends on a balance of a number of interrelated factors.

8-4-2   Addition and Abstraction on Ligands with Pi Bonds

In Section **8-2-2** much of the discussion centered on the attack by nucleophiles on π ligands, such as monoolefins or arenes. In these reactions, hapticity ($\eta^n \to \eta^{n-1}$) and the overall charge ($+n \to +n-1$) tended to change by one unit. Similar events can occur with electrophiles, and sometimes the changes in hapticity and overall charge are just the opposite of those found in nucleophilic reactions. While reports in the chemical literature of electrophilic attack on ligands containing π bonds are fewer than for nucleophilic counterparts, many examples of this kind of reaction do exist.

[65]D. Dong, *et al.*, *op. cit.*
[66]H.L Fritz, J.H. Espenson, D.A. Williams, and G.A. Molander, *J. Am. Chem. Soc.*, **1974**, *96*, 2378.

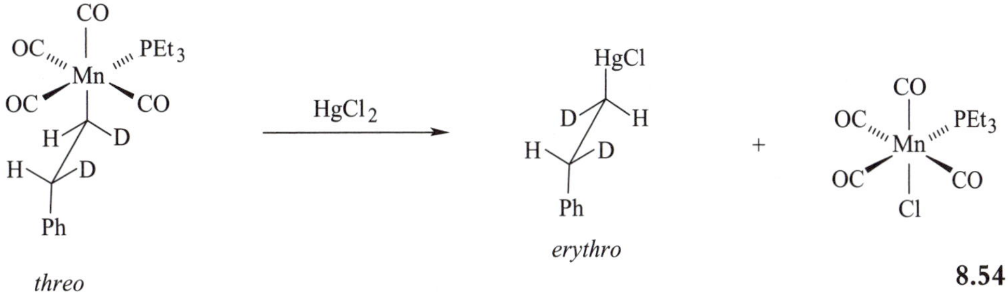

threo

erythro

8.54

28

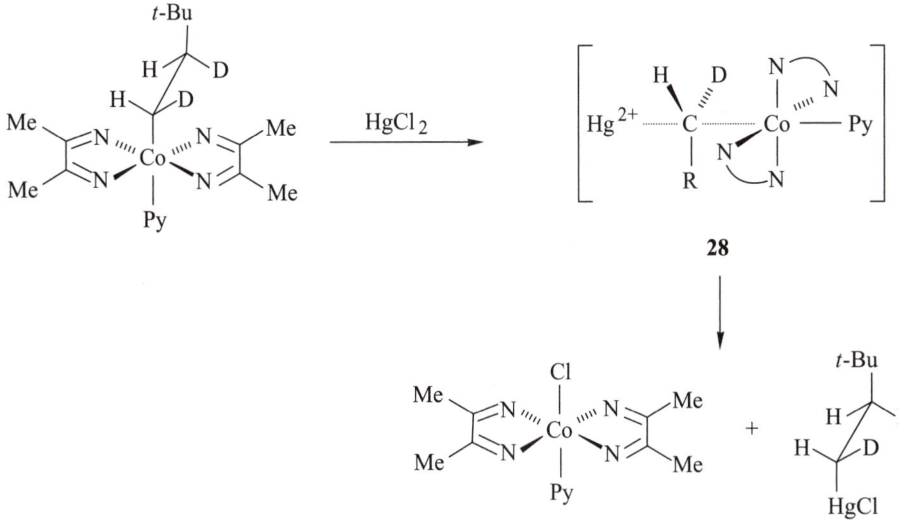

8.55

Addition

In equation **8.56**, attack of the electrophile occurs at the β position to give a carbene complex as the intermediate; subsequent loss of CH₃I produces the acyl-Re complex as the final product.[67]

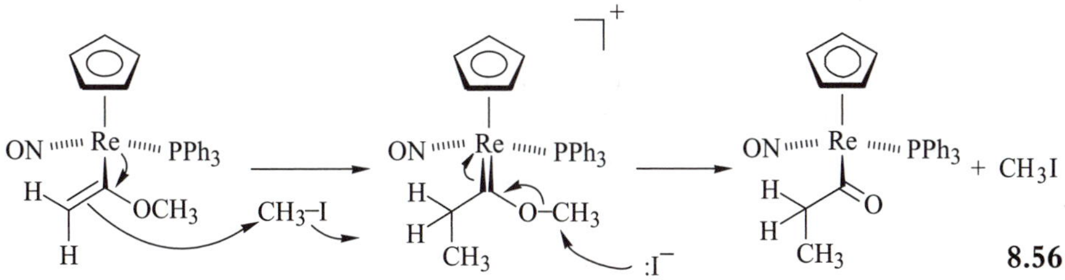

8.56

[67]D.E. Smith and J.A. Gladysz, *Organometallics*, **1985**, *4*, 1480.

Equation **8.57** demonstrates addition to the γ position of the η^1-allyl complex and represents simply the reverse of nucleophilic addition to [Fe(alkene)]$^+$ reagents (equations **8.32** and **8.33**).[68] The synthetic utility of this reaction is illustrated in equation **8.58** where the iron complex undergoes conjugate addition to the α,β-unsaturated diester. The resulting carbanion then attacks intramolecularly the η^2-π ligand to give a substituted cyclopentane that may be converted to a number of different compounds.[69]

$$Fp = Cp(CO)_2Fe \qquad\qquad\qquad\qquad\qquad\qquad\qquad \textbf{8.57}$$

$$\textbf{8.58}$$

Equation **8.59** shows attack of the electrophile ("Et$^+$") at the carbonyl oxygen, the most electron-rich site on the η^4-dienone ligand, to yield a product with η^5 hapticity.[70]

$$\textbf{8.59}$$

Abstraction

Electrophilic abstraction of a portion of a σ-bound alkyl or a coordinated π ligand may occur. The trityl cation, Ph$_3$C$^+$, is by far the most commonly used electrophile for this task. Equation **8.60** provides an example of abstraction at the β position of an alkyl ligand to provide an η^2-alkene complex,[71] a much-used route for the synthesis of these compounds.

[68]For several examples of this type of electrophilic addition, see S.G. Davies, *op. cit.*, 189-192 and references therein.
[69]T.S. Abram, R. Baker, and C.M. Exon, *Tetrahedron Lett.*, **1979**, 4103.
[70]A.J. Birch and I.D. Jenkins, *Tetrahedron Lett.*, **1975**, 119.
[71]D. Slack and M.C. Baird, *J. Chem. Soc., Chem. Commun.*, **1974**, 701.

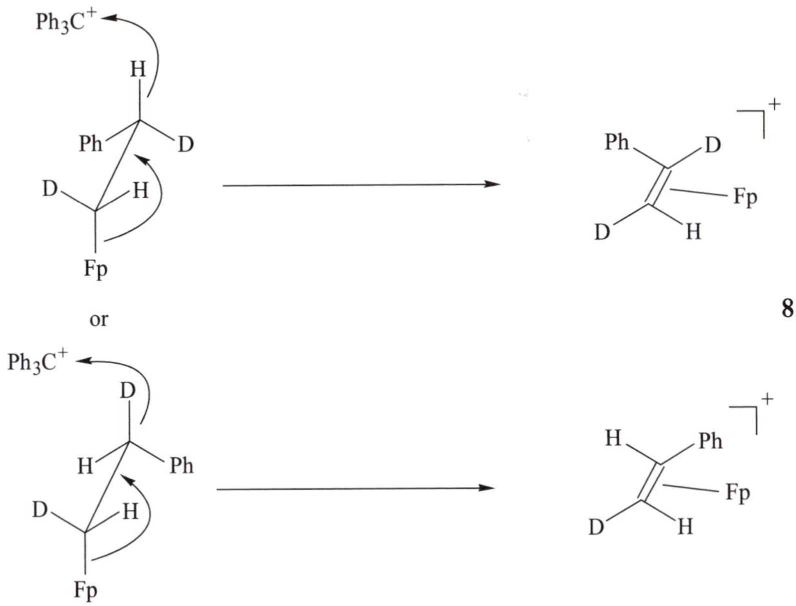

8.60

The reaction is stereospecific, as shown by the use of ligand deuterated at the β position. Formation of these products requires *anti*-elimination in a manner analogous to the E₂ pathway in organic chemistry. The difference here, of course, is that abstraction occurs with an electrophile and not with a base as we find in the organic analog. It is not surprising that trityl cation serves as the premier hydride-abstracting agent in reactions such as **8.60**. The cation is rather stable (trityl salts are commercially available) and easy to handle. Because of its steric bulk, it does not readily attack the metal directly. Finally, the reaction is thermodynamically feasible in terms of enthalpy change.

Hydride abstraction by Ph_3C^+ also occurs on the η^3-allyl-Mo complex,[72] shown in equation **8.61**, again causing an increase in hapticity of one unit.

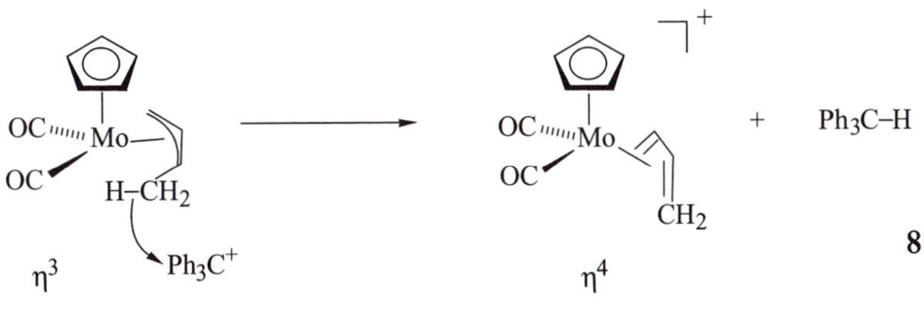

8.61

---

[72]. W. Faller and A.M. Rosan, *J. Am. Chem. Soc.*, **1977**, *99*, 4858. For references to other hydride abstractions using Ph_3C^+, see D. Mandon, L. Toupet, and D. Astruc, *J. Am. Chem. Soc.*, **1986**, *108*, 1320.

Although we will encounter other types of reactions in later chapters, ligand substitution, oxidative addition, reductive elimination, insertion, elimination, and nucleophilic and electrophilic attack comprise the fundamental processes that encompass all of organometallic chemistry. Your understanding of these reactions will enable you to appreciate the many applications of organotransition metal chemistry to other areas of the chemical sciences. In the next chapter, we consider the role of organometallic compounds as catalysts in pathways leading to industrially useful molecules. The individual steps involved in these transformations are typically the basic reaction types we have just discussed in this chapter and in Chapter 7.

Problems

8-1   Predict the products:

     a. cis-$Re(CH_3)(PR_3)(CO)_4$ + ^{13}CO → [show structures of all products]

     b. $trans$-$Ir(CO)Cl(PPh_3)_2$ + CH_3I →

     c. $CH_3Mn(CO)_5$ + SO_2 → [no gases are evolved]

     d. $(\eta^5\text{-}C_5H_5)Fe(CO)_2(CH_3)$ + PPh_3 →

     e. $(\eta^5\text{-}C_5H_5)Mo(CO)_3C(=O)CH_3$ + heat →

8-2   Predict the structure of the product of the following equilibrium (note: the product obeys the 18-electron rule):

$$[(\eta^5\text{-}Cp)Rh(CH_2CH_3)(PMe_3)(S)]^+ \rightleftharpoons \quad + \ S$$

(S = weakly coordinating solvent such as THF)

8-3   The complex shown at right loses carbon monoxide on heating. Would you expect this to be ^{12}CO, ^{13}CO, or a mixture of both? Why?

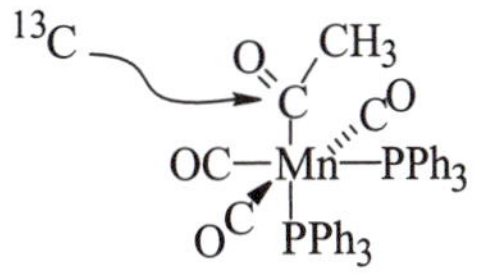

8-4   Predict the products of the following reaction, clearly showing the structure of each and the expected relative distribution of products.

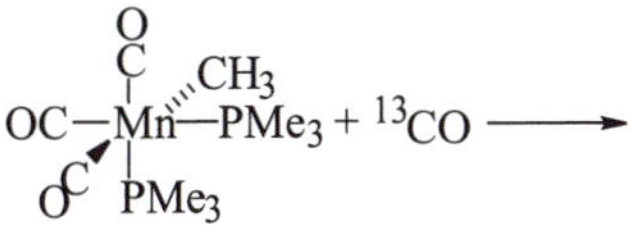

8-5   The molecule shown at right loses CO on heating. Predict the products of this reaction, assuming that the mechanism is the reverse of the following:

     a.   Direct CO insertion

     b.   Intramolecular CO migration

     c.   Intramolecular CH_3 migration

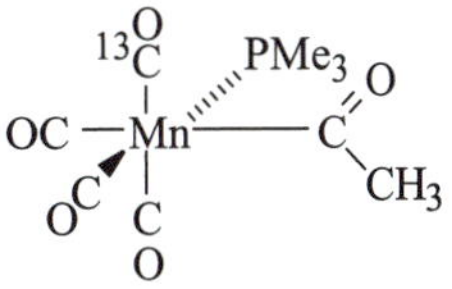

8-6   The complex $[(\eta^6\text{-}C_6H_6)Mn(CO)_3]^+$ has carbonyl bands at 2026 and 2080 cm^{-1} and a single ^1H NMR resonance at δ 6.90. This complex reacts with PBu_3 to give product **M**, which has infrared bands at 1950 and 2028 cm^{-1} and ^1H NMR signals at δ 6.30 (relative area = 1), 5.50 (2), 4.40 (1), 3.40 (2), and signals corresponding to butyl groups. Complex **M** converts photochemically into **N**, which has infrared bands at 1950 and 1997 cm^{-1} and only one signal, at δ 6.42, other than signals for the butyl groups. Suggest structures for **M** and **N**.

8-7   Propose mechanisms for the two transformations shown.

a.

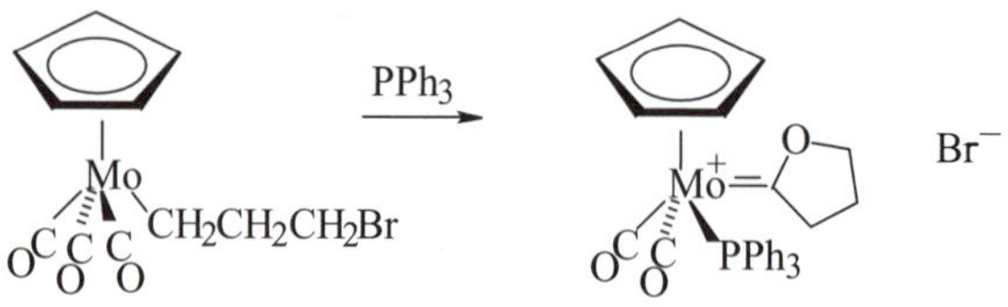

b.

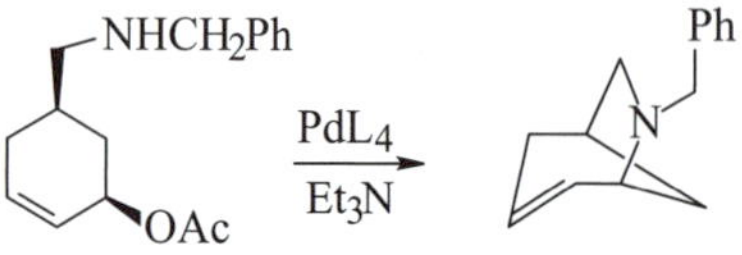

8-8   Follow the step-wise reaction shown below. For each step, indicate what kind of fundamental organometallic reaction is taking place.

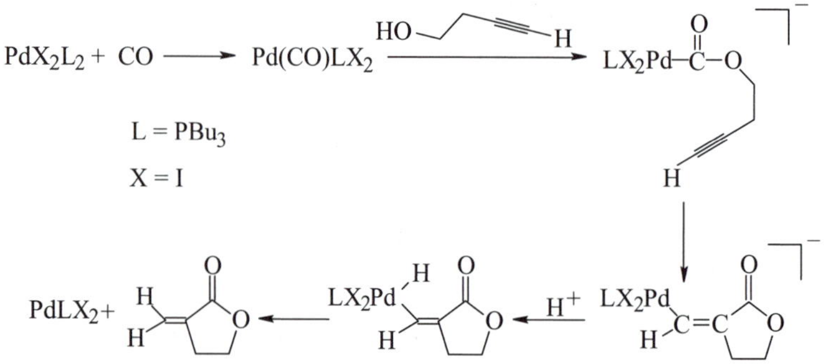

$$L = PBu_3$$

$$X = I$$

8-9   Propose a mechanism for the following multi-step reaction that involves nucleophilic participation of a remote π bond. Assume that in the first step **2** acts as a nucleophile.   [Fp = CpFe(CO)$_2$]

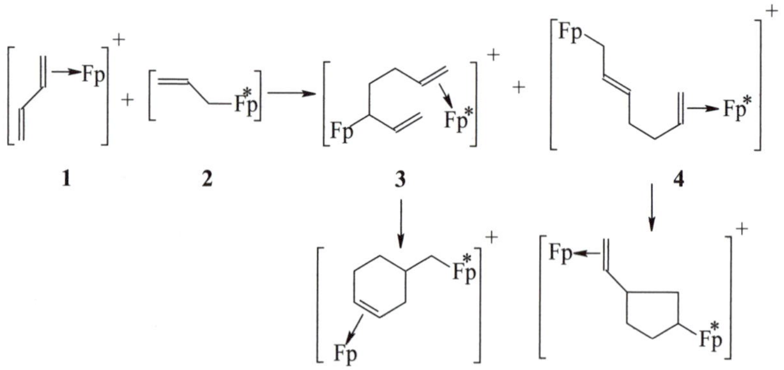

9

Homogeneous Catalysis

Use of Transition Metal Complexes in Catalytic Cycles

There are many reactions in chemistry that are favorable thermodynamically, yet occur at extremely slow rates at room temperature. A few of these are shown in equations **9.1** to **9.3**:

Water gas shift reaction:
$$H_2O \text{ (g)} + CO \text{ (g)} \rightarrow H_2 \text{ (g)} + CO_2 \text{ (g)} \qquad \textbf{9.1}$$
$$\Delta G^\circ_{rxn} = -6.9 \text{ kcal/mole}$$

Alkene hydrogenation:
$$CH_3CH=CH_2 \text{ (g)} + H_2 \text{ (g)} \rightarrow CH_3CH_2CH_3 \text{ (g)} \qquad \textbf{9.2}$$
$$\Delta G^\circ_{rxn} = -20.6 \text{ kcal/mole}$$

Glucose metabolism:
$$C_6H_{12}O_6 \text{ (glucose)} + 6 \text{ } O_2 \text{ (g)} \rightarrow 6 \text{ } H_2O \text{ (l)} + 6 \text{ } CO_2 \text{ (g)} \qquad \textbf{9.3}$$
$$\Delta G^\circ_{rxn} = -688 \text{ kcal/mole}$$

The presence of a *catalyst* dramatically increases the rate of the reactions shown and countless others, even those that may not have negative free energies of reaction. This chapter will consider the role transition metal complexes play as catalysts for several transformations, many of which are important industrially. As catalysts, transition metal complexes undergo most of the reactions we have just discussed in Chapters 7 and 8. We will encounter a number of different catalytic cycles in Chapter 9; other catalytic processes involving transition metals will be described in Chapters 10 and 11.

9-1   Fundamental Concepts of Homogeneous Catalysis

The phenomenon of catalysis was recognized over 150 years ago by Berzelius who referred to the "catalytic power of substances" that were able to "awake affinities that are asleep at this temperature by their mere presence and not by their own affinity."[1] Once the principles of thermodynamics were developed and the concept of equilibrium established by the turn of the twentieth century, scientists realized that catalysts are species that change the rate of a reaction without affecting the equilibrium distribution of reactants and products. Exactly how catalysts are able to speed up reactions without changing the free energy of either reactants or products remained a mystery for many years. Only in the last few decades has it been possible to elucidate mechanistic pathways

Figure 9-1
Reaction Coordinates
for a Catalyzed and
Uncatalyzed Chemical
Reaction

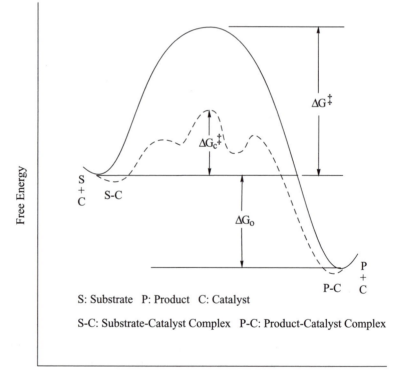

S: Substrate   P: Product   C: Catalyst

S-C: Substrate-Catalyst Complex   P-C: Product-Catalyst Complex

Reaction Coordinate

[1]Taken from a quotation in G.C. Bond, *Homogeneous Catalysis*, Oxford University Press, Oxford, 1987, 2.

involving catalysts by using such tools as kinetics, stereochemical studies, and spectroscopy. It is clear now that catalysts interact with reactants to provide a reaction pathway with a significantly lower free energy of activation than the corresponding uncatalyzed pathway. Figure **9-1** depicts this phenomenon with the solid line representing the uncatalyzed reaction and the dashed line the catalyzed reaction path.

Typically, catalysts are intimately involved with reactants (often called substrates) in a cyclic series of associative (binding), bond making and/or breaking, and dissociative steps (Scheme **9.1**). During each cycle, the catalyst is regenerated so that it may go through another cycle. Each cycle is called a *turnover*, and an effective catalyst may undergo hundreds, even thousands of turnovers before decomposing—each cycle producing a molecule of product. In a stoichiometric reaction, on the other hand, the "catalyst" (actually reagent) undergoes only *one* turnover per molecule of product produced.

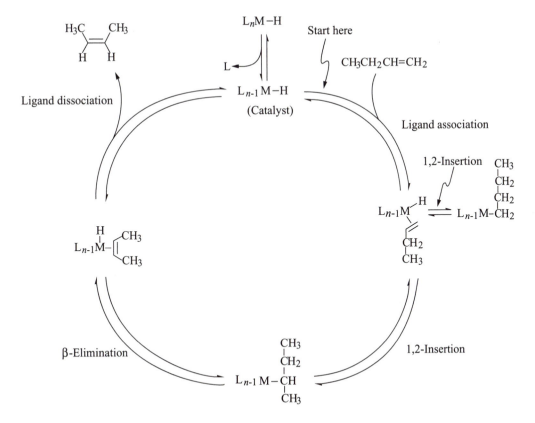

Scheme 9-1
A Schematic Picture of a Catalytic Cycle
Showing Double Bond Isomerization

9-1-1   Selectivity

The laws of thermodynamics dictate that the product distribution resulting from a catalyzed or uncatalyzed reaction must be the same if enough time is allowed for the transformation to come to equilibrium. The presence of a catalyst, however, can influence initial product distributions, allowing preferential formation of a product that may be less stable thermodynamically than another. This is a phenomenon known as *selectivity*. There are several kinds of selectivity demonstrated by catalysts—chemoselectivity, regioselectivity, and stereoselectivity.

Equation **9.4** shows an example of chemoselectivity in which hydrogenation occurs selectively at only one functional group in the reactant. In principle, hydrogenation could occur at the phenyl ring, the C=C bond, or the nitro group. In the presence of the Rh catalyst, however, H_2 adds only to the alkene group:

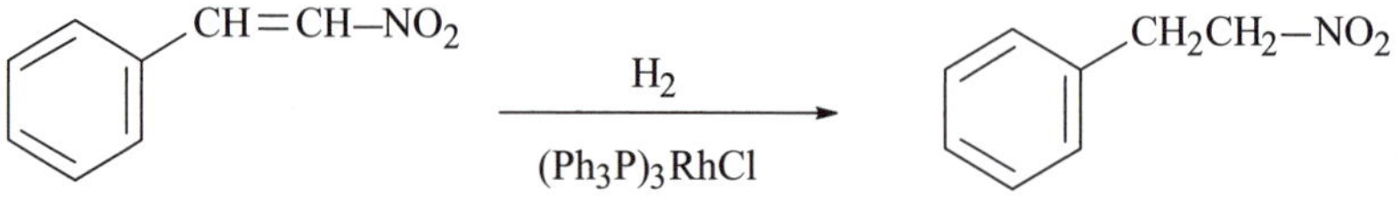

$$\text{C}_6\text{H}_5\text{—CH=CH—NO}_2 \quad \xrightarrow[\text{(Ph}_3\text{P)}_3\text{RhCl}]{\text{H}_2} \quad \text{C}_6\text{H}_5\text{—CH}_2\text{CH}_2\text{—NO}_2 \qquad \textbf{9.4}$$

The hydroformylation reaction (Section **9-2**), shown below in equation **9.5**, demonstrates regioselectivity whereby one regioisomer forms preferentially over another. The Co catalyst may be modified to give linear (anti-Markovnikov) over branched (Markovnikov) product:

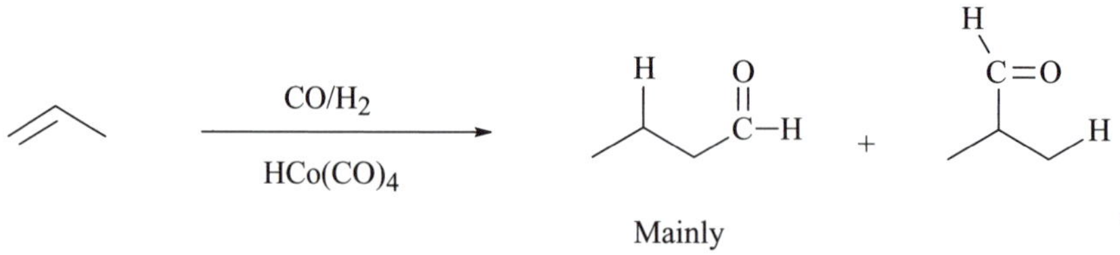

$$\text{CH}_2\text{=CH—CH}_3 \quad \xrightarrow[\text{HCo(CO)}_4]{\text{CO/H}_2} \quad \underset{\text{Mainly}}{\text{CH}_3\text{CH}_2\text{CH}_2\text{—C(=O)H}} \quad + \quad \text{(CH}_3)_2\text{CH—C(=O)H} \qquad \textbf{9.5}$$

The hydrogenation of disubstituted alkynes with Lindlar catalyst (equation **9.6**) represents an example of the use of a stereoselective catalyst. Such catalysts promote the formation of one stereoisomer in preference to one or more others. In this particular example, the *Z*-alkene is formed in great preference to the corresponding *E*-isomer.

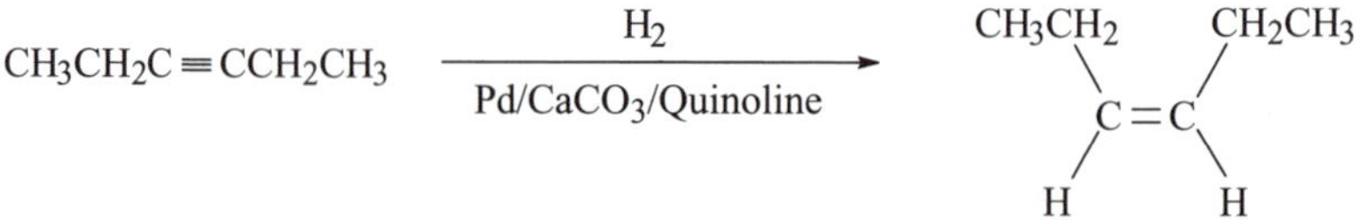

$$\text{CH}_3\text{CH}_2\text{C} \equiv \text{CCH}_2\text{CH}_3 \quad \xrightarrow[\text{Pd/CaCO}_3\text{/Quinoline}]{\text{H}_2} \quad \underset{\text{H}}{\overset{\text{CH}_3\text{CH}_2}{\diagdown}}\text{C=C}\underset{\text{H}}{\overset{\text{CH}_2\text{CH}_3}{\diagup}} \qquad \textbf{9.6}$$

Catalysts also exhibit selectivity in their initial binding to reactants. Enzymes are well known for their ability to bind selectively to only one member of a pair of stereoisomers. The bound stereoisomer will undergo reaction, and the remaining isomer is inert to the reaction conditions. For example, the addition of water to fumaric acid to give malic acid (equation **9.7**) is catalyzed by an enzyme called fumarase. Isomeric maleic acid (the *Z*-isomer) fails to react in the presence of fumarase.

$$
\underset{\substack{\text{H}}}{\overset{\substack{\text{HO}_2\text{C}}}{}}\text{C}=\text{C}\underset{\substack{\text{CO}_2\text{H}}}{\overset{\substack{\text{H}}}{}}\quad\xrightarrow[\text{Fumarase}]{\text{H}_2\text{O}}\quad\text{HO}_2\text{C}-\underset{\substack{\text{H}\quad\text{OH}}}{\text{C}}-\text{CH}_2-\text{CO}_2\text{H}\qquad\textbf{9.7}
$$

(*S*)-Malic Acid

9-1-2   Homogeneous vs. Heterogeneous Catalysts

In your study of chemistry thus far you have probably encountered a number of catalyzed reactions. Those who are familiar with organic chemistry will recognize equation **9.2** (on p. 245) as an alkene hydrogenation reaction that occurs at a reasonable rate only in the presence of a catalytic amount of Pd or Pt deposited on an inert solid. The synthesis of esters from alcohols and carboxylic acids (equation **9.8**) is catalyzed by mineral acids such as H_2SO_4 or HCl:

$$
\underset{}{\overset{\overset{\text{O}}{\|}}{\text{C}}}-\text{OH} \quad + \quad \text{CH}_3\text{CH}_2\text{OH} \quad \underset{\text{H}^+}{\rightleftharpoons} \quad \underset{}{\overset{\overset{\text{O}}{\|}}{\text{C}}}-\text{O}-\text{CH}_2\text{CH}_3 \quad + \quad \text{H}_2\text{O} \qquad \textbf{9.8}
$$

The catalyst involved in the former reaction is known as a *heterogeneous* catalyst. Equation **9.8**, on the other hand, demonstrates the use of a *homogeneous* catalyst. A heterogeneous catalyst exists as another phase in the reaction medium, typically as a solid in the presence of a liquid or gaseous solution of reactants. A homogeneous catalyst is *dissolved* in the reaction medium along with the reactants. Homogeneous catalysts that are transition metal complexes find increasing importance in the chemical industry where the use of heterogeneous catalysts has historically been predominant.

Table **9–1** summarizes the major differences between homogeneous and hetereogeneous catalysts. The subsequent discussion assumes that the homogeneous catalysts being discussed contain transition metals.

Table 9-1 Major Differences between Homogeneous and Heterogeneous Catalysts

Characteristic	Homogeneous	Heterogenous
1. Catalyst composition and nature of active site	Discrete molecules with well-defined active site	Nondiscrete molecular entities; active site not well-defined
2. Determination of reaction mechanism	Relatively straightforward using standard techniques	Very difficult
3. Catalyst properties	Easily modified, often highly selective, poor thermal stability, mild reaction conditions	Difficult to modify, relatively unselective, thermally robust, vigorous reaction conditions
4. Ease of separation from product	Often difficult	Relatively easy

Catalyst Composition and Nature of the Active Site

The active sites on a heterogeneous catalyst are difficult to characterize because they are not discrete molecular entities. Instead, the active sites may be aggregations of solid support material (e.g., silica gel or zeolites) coated with deposited metal atoms. Not all sites on the surface of the catalyst have the same activity and physical or chemical characteristics. Analytical techniques (e.g., Auger and ESCA spectroscopy) as well as scanning tunneling microscopy have been applied recently in attempts to determine the nature of the catalytic surface. Much progress has been made; however, much more information must be obtained before a complete and cogent understanding of heterogeneous catalysis can emerge.

Homogeneous catalysts, on the other hand, are discrete molecules that are relatively easy to characterize by standard spectroscopic techniques, such as NMR, IR, etc. The active site consists of the metal center and adjoining ligands.

Determination of the Reaction Mechanism

Because the constitution of the active site of a heterogeneous catalyst is difficult to determine, the elucidation of reaction mechanisms involving these catalysts can be troublesome indeed. The field of homogeneous catalysis, in contrast, has advanced rapidly over the last few decades because chemists have developed many techniques useful for studying reaction mechanisms. Elucidating the mechanism of a homogeneously catalyzed reaction means studying the mechanism of each individual step in a series of relatively elementary chemical reactions by conventional methods. Each step must be shown to be kinetically and thermodynamically reasonable.[2] While this can be a daunting task for a reaction involving several catalytic steps, it is not, at present, as difficult as determining exactly what goes on during a heterogeneous catalysis.

Catalyst Properties: Ease of Modification, Selectivity, Thermal Stability, and Reaction Conditions

Because homogeneous catalysts are typically organotransition metal complexes, it is relatively easy to modify these compounds in order to increase selectivity. As we shall see, phosphines appear commonly as ligands in homogeneous catalysts. Phosphines, of course, offer a wide variety of stereoelectronic properties that can substantially influence the course of a catalyzed reaction.

Homogeneous catalysts are often much less thermally stable than their heterogeneous counterparts. The use of a homogeneous catalyst rather than a heterogeneous one requires milder conditions of temperature and pressure. If a sufficiently active homogeneous catalyst can be found that can do the job of a heterogeneous one, substantial savings in energy and initial capital cost (plants that run at high temperature and pressure are very expensive to build) accrue to the manufacturer that chooses to employ a homogeneously catalyzed process.

Ease of Separation from Reaction Products

Homogeneous catalysts suffer from one key disadvantage when compared to their heterogeneous counterparts. They are often quite difficult to separate from reaction products. Catalyst recovery is clearly important, not only in ensuring product purity but also in conserving often-used precious metals such as palladium and rhodium.

9-1-3   Enzymes: Homogeneous or Heterogeneous Catalysts?

A discussion of catalysis would not be complete without comparison of the catalysts we normally encounter in the laboratory or in industry with those that occur naturally, i.e., enzymes. Enzymes are proteins that are either soluble or attached to a cellular membrane. Soluble enzymes resemble homogeneous transition metal catalysts in ways other than solubility characteristics. Enzymes have at least one region that serves as an active site (see Figure **9-2**) for substrate(s) to bind[3] and thence to undergo transformation into product(s). Modification of the enzyme's amino acid sequence can drastically alter the catalytic efficiency of the enzyme. Often enzymes have sites remote from the

---

[2]It should be pointed out that an individual step may be endothermic even if the overall reaction is exothermic.

[3]The binding energies involved in enzyme-substrate interactions are much less than ordinary chemical bonds. The cumulative effect, however, of several weak binding interactions is significant.

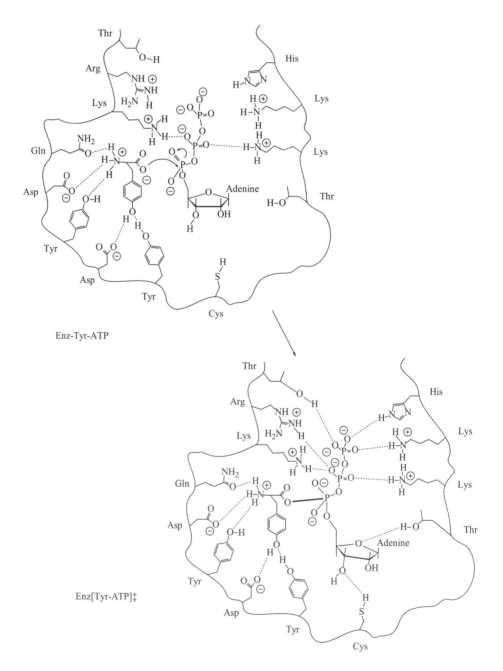

Enz-Tyr-ATP

Enz[Tyr-ATP]‡

Figure 9-2
A Schematic View of the Active Site of Tryrosyl-tRNA
Synthetase (Adapted from L.A. Moran, et al., *Biochemistry*,
2nd ed., Prentice-Hall, Englewood Cliffs, NJ, 1994, 7-30.)

active site where ligands can bind. This binding can alter the nature of the active site by causing a conformational change in the structure of the protein. Binding at these remote sites may be a necessary precursor for catalytic activity. Similarly, transition metal complexes may change significantly in catalytic activity when modifications are made of ligands that are not directly involved in the overall chemical transformation (i.e., phosphines).

Enzymes are analogous to heterogeneous catalysts in that they are large molecules with no discrete molecular active site. The active site is comprised of functional groups from several non-neighboring amino acids. These functional groups become neighboring only because of the folding of the polymeric chain of amino acids. Enzymes are large molecules that have a vast surface area compared to transition metal complexes, and as such, may have complex interactions with other cellular entities.

Membrane bound enzymes represent loosely a hybrid between homogeneous and heterogeneous catalysts. The membrane represents a large surface area that supports catalytically active entities (the enzymes). The enzymes, however, are discrete molecules (though giant ones) in contrast to metal crystallites found in most heterogeneous catalysts.

Interestingly, one of the most active areas in transition metal catalysis concerns the use of a hybrid catalyst that resembles a membrane bound enzyme. Much work has been reported in which transition metal complexes are bound to solid supports such as silica or alumina. These catalytic systems have some of the advantages of both types of catalysts. The transition metal complexes may be easily modified to increase selectivity based on studies in homogeneous media. The complex, on the other hand, may be readily separated from product because it is bound to a solid support—a useful property indeed as discussed above.

9-1-4   Unique Fitness of the Transition Metals

While there are many species that serve as useful homogeneous catalysts (e.g., H^+, OH^-, Al^{3+}, imidazole), complexes of transition metals are far and away the most selective of these. Why is this so? In the next paragraphs we attempt to shed some light on this very important question—in part by reminding readers of several concepts that we have already encountered.[4]

Variety of Ligands Will Bind to Transition Metals

Chapters 4 to 6 have shown that a host of different ligands will bind to metals and that these ligands may be classified as X- or L-type. In fact, transition metals will bind to virtually any other element in the Periodic Table and almost all organic molecules. Ligands may either be directly involved in the catalytic

---

[4]An excellent discussion on the fitness of the transition metals for catalytic activity is found in C. Masters, *Homogeneous Transition Metal Catalysis—A Gentle Art*, Chapman and Hall, London, 1981, 5-20.

process or indirectly affect catalysis by exerting steric and/or electronic effects on the complex.

Transition Metals Have the Ability to Bind to Ligands in a Number of Ways

In Chapters 4 to 6 we described how various ligands can bind to transition metals. The availability of d as well as s and p orbitals on the metal allows for the formation of σ and π bonds from metal to ligand. For example, a single ligand, such as an alkene, may bind in an η^1 or η^2 manner. The hapticity of such a ligand may change during a catalytic cycle, and the ease of this change facilitates ligand modification. Carbonyl ligands attach to a metal in a terminal or bridging bonding mode. Groups such as methyl or hydride may be considered to be anionic, neutral (radical), or cationic depending on the electron density of the metal. Finally, the strength of metal-ligand bonds is moderate (30 to 80 kcal/mole), allowing bonds to form or break relatively easily—a necessity in order for the catalytic cycle to proceed.

Variety of Oxidation States is Available

As ligands are added to or removed from the metal by processes such as oxidative addition (OA) and reductive elimination (RE), the oxidation state of the metal changes. Transition metals, with their d valence electrons, usually have a rather large number of oxidation states available—particularly in comparison to main group metals. The elements in groups 8 to 10 especially possess a tendency for rapid, reversible two-electron change (such as from 18 e⁻ to 16 e⁻ and back), and thus it is not surprising that they are often involved in homogeneous catalysis.

Transition Metal Complexes Exhibit Several Different Geometries

Depending on the coordination number, a variety of structural possibilities exists for transition metal complexes. Geometries such as square planar, octahedral, tetrahedral, square pyramidal, and trigonal bipyramidal are common in complexes involving the transition metals. Much is already known about the behavior of ligands attached to metals in these geometries. For example, in square planar complexes, a ligand *trans* to another may cause the latter ligand to be quite labile and thus happy to depart from the metal. If loss of the latter ligand is required for effective catalysis, then it is desirable to design the catalytic cycle in such a way that a square planar complex is one of the intermediates along the path to the product and to have the directing group with high *trans* effect positioned *trans* to the leaving group. Another example might involve a process where a key step is the reductive elimination of two ligands. In Chapter 7 we saw that RE requires the two leaving groups to be *cis* with respect to each other before reaction may occur. The catalysis will be successful in this case if an intermediate forms such that the two leaving groups are *cis*.

Transition metal complexes, with their well-defined geometries, serve as "templates" for the occurrence of reliable stereospecific or stereoselective ligand interaction. We know that alkyl migration to a carbonyl group occurs with retention of configuration and that reductive elimination involves retention of configuration in the two leaving groups. In this way, transition metal complexes mimic the stereospecific reactions often catalyzed by enzymes.

Transition Metal Complexes Possess the "Correct" Stability

By varying metal and ligands, transition metal complexes are readily available to serve as intermediates that are not too reactive or too unreactive. In order for a catalytic turnover to occur, each intermediate in the cycle must be reactive enough to proceed to the next stage, yet not so reactive that other pathways, e.g., decomposition or a different bonding mode, become feasible.

We have just seen several reasons for the unique fitness of transition metals and their complexes as catalysts. The take-home lesson from all of this discussion is that the transition metals and the complexes derived from them are **versatile**. The stage is now set to examine several examples of homogeneous transition metal catalysis.

9-2   Hydroformylation Reaction

Among all of the homogeneous processes catalyzed by transition metals, hydroformylation stands out in two respects: It is the oldest process still in use today; and it is responsible for producing the largest amount of material resulting from a homogeneous transition metal-catalyzed reaction. The hydroformylation reaction is outlined below.

Discovered in 1938 by Otto Roelen[5] (of Ruhrchemie in Germany), cobalt-catalyzed hydroformylation (also known as the *oxo*[6] reaction) was used successfully during World War II by the Germans to produce aldehydes from alkenes, H_2, and CO. In 1945, American and British chemical engineers obtained details of the process, and since then, hydroformylation plants have been built worldwide to produce millions of tons of alcohols according to reactions shown in Scheme **9.2**. Because of its economic importance and because it serves as a prototypical example of homogeneous transition metal catalysis, we will consider hydroformylation in more detail than some other catalytic processes that appear later in this chapter.

[5]J. Falbe, *Carbon Monoxide in Organic Synthesis*, Springer Verlag, New York, 1970; O. Roelen, Ger. Patent 949 548, 1938.

[6]The name *oxo* was given to the reaction because in German, *oxo* means carbonyl, and carbonyl compounds (aldehydes) are the initial products of hydroformylation processes.

Propene is the most common feedstock for hydroformylation and is transformed into butanal and 2-methylpropanal, the former being the much more valuable product. Butanal may be hydrogenated in a separate step using a heterogeneous catalyst to give 1-butanol, a useful solvent, or it may undergo aldol condensation followed by hydrogenation to give 2-ethyl-1-hexanol. The hexanol is used to make a diester of phthalic acid, which then serves as an agent (called a plasticizer) for making the normally rigid plastic, polyvinyl chloride (PVC), flexible.[7] Hydroformylation converts C_7-C_9 alkenes to aldehydes with one more carbon, which are then hydrogenated to give straight-chain alcohols, also useful as plasticizers. Alcohols containing 12 to 16 carbons result from hydroformylation-hydrogenation of corresponding alkenes with one less carbon atom. These alcohols serve as the basis for surfactant (detergent) compounds with several domestic and industrial uses.

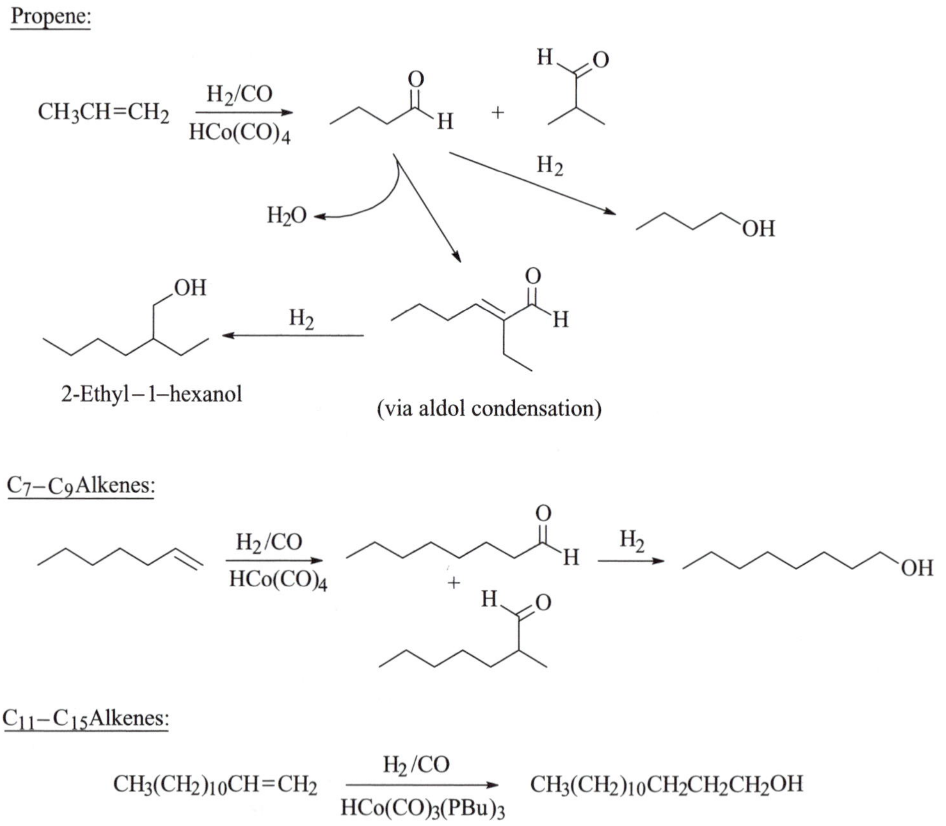
Propene:
CH3CH=CH2
H2/CO
HCo(CO)4
O
H
+
H
O
H2
H2O
OH
OH
H2
O
H
2-Ethyl–1–hexanol
(via aldol condensation)
C7–C9Alkenes:
H2/CO
HCo(CO)4
O
H
H2
OH
+
H
O
C11–C15Alkenes:
CH3(CH2)10CH=CH2
H2/CO
HCo(CO)3(PBu)3
CH3(CH2)10CH2CH2CH2OH

Scheme 9.2
Industrially Useful Compounds Produced by Hydroformylation

[7]Tygon tubing, commonly used in laboratories, is an excellent example of plasticized PVC.

9-2-1   Cobalt-Catalyzed Hydroformylation

Scheme **9.3** shows the cycle of catalytic steps proposed by Heck and Breslow[8] for cobalt-catalyzed hydroformylation. The mechanism they proposed resulted from studies on model organo-cobalt carbonyl complexes. The cycle shown is an excellent example of transition metal catalysis because of the following: it contains steps converting a precatalyst to the active complex; it consists of several simple mono- and bimolecular steps involving the fundamental reactions of organometallic chemistry; it demonstrates the validity of the 16- and 18-electron rules; and it shows intermediates with plausible geometries. It must be pointed out, however, that the mechanism for cobalt-catalyzed

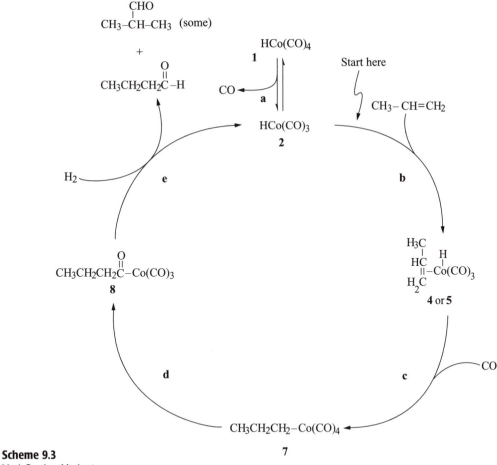

Scheme 9.3
Heck-Breslow Mechanism
for Hydroformylation

[8]R.F. Heck and D.S. Breslow, *J. Am. Chem. Soc.*, **1961**, *83*, 1961.

hydroformylation is not completely understood with regard to kinetic and thermodynamic parameters for each step.[9] We will look at each step, commenting on its relation to the catalytic cycle and to material you have already encountered in this text, and we will also point out some of the more recent work aimed at elucidating the mechanism of hydroformylation.

Step a

The true catalytically active species is probably $HCo(CO)_3$, a 16-electron species. This complex derives from 18-electron $HCo(CO)_4$, **1**, which in turn ultimately comes from Co(0) or Co(II), via $Co_2(CO)_8$, in the presence of a 1:1 mixture of CO and H_2[10] (synthesis gas). Sometimes **1** is prepared in a separate step and introduced to the alkene in the presence of synthesis gas; this allows the subsequent hydroformylation to be run at a lower temperature (90° to 120° rather than the usual 120° to 170°). The dissociation step to form the active catalyst occurs with a relatively high activation energy, and is, of course, inhibited by a high concentration of CO. Calculations indicate that the preferred geometry for $HCo(CO)_3$ is more likely to be **2** instead of **3**.[11] The reaction is run, however, under very high pressure (200 to 300 atm) in order to stabilize $HCo(CO)_3$ and later intermediates in the catalytic cycle. This demonstrates a balance in reaction conditions between formation of sufficient $HCo(CO)_3$ for hydroformylation to occur at a reasonable rate and the enhancement of the stability of catalytic intermediates.[12]

Step b

The next step involves binding the alkene to **2**, forming an 18-electron hydrido–alkene complex that can have two structures, **4** or **5**. Although **4** is a

[9]J.P. Collman, *et al., op. cit.*, 623.

[10]Mixtures of CO and H_2 are called *synthesis gas* (sometimes referred to as "water gas"). Synthesis gas results from reaction of natural gas (or petroleum distillates) and steam over a nickel catalyst, and it is the basis for numerous organic molecules produced in large scale by the petrochemical industry.

[11]T. Ziegler and L. Versluis, "The Tricarbonylhydridocobalt-Based Hydroformylation Reaction," in W.R. Moser and D.W. Slocum, Eds., *Homogeneous Transition Metal-Catalyzed Reactions*, American Chemical Society, Washington, D.C., 1992, 75-93.

[12]For a good discussion of hydroformylation reaction conditions, see G. W. Parshall and S.D. Ittel, *Homogeneous Catalysis*, 2nd ed. Wiley, New York, 1992, 106-111.

little more stable than **5** according to calculations (probably due to less steric hindrance), only **5** has the requisite geometry (Co, H, and the double bond carbon atoms all coplanar) for the next 1,2-insertion step (step c).[13] Overall, therefore, steps a and b constitute a ligand substitution via a D mechanism.

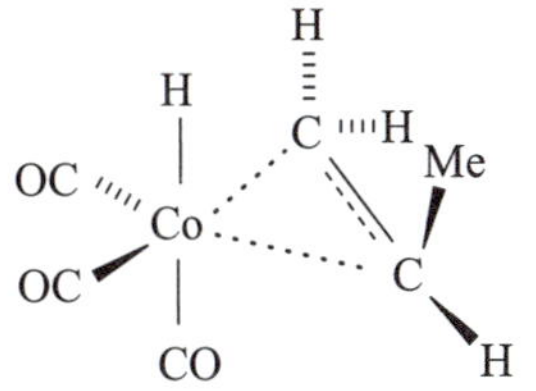

4

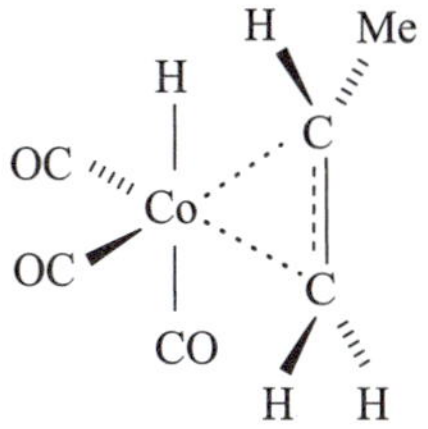

5

Step c

Once complexation of the alkenes occurs, 1,2-insertion is facile. Insertion initially gives a 16-electron complex, **6**. The high concentration of CO present quickly drives step c to completion, resulting in 18-electron intermediate, **7**. Although β-elimination is possible now, the high partial pressure of CO present in the reaction vessel tends to stabilize **7** and prevent loss of CO that would generate the vacant site necessary for elimination to occur. Structures **6** and **7** show the result of "anti-Markovnikov" insertion from which the linear aldehyde, rather than the branched isomer, will result.

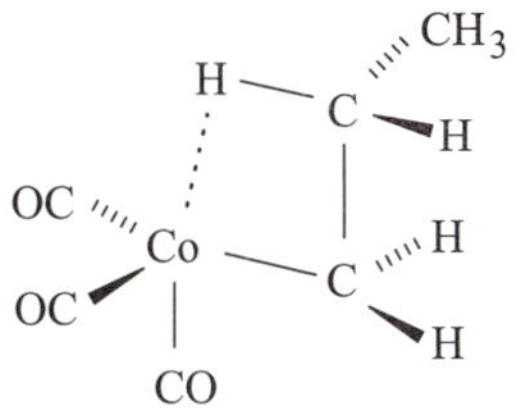

6

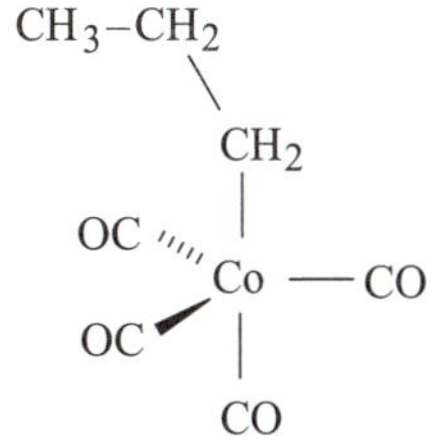

7

Step d

Step d involves CO insertion, which really is, as we saw in Chapter 8, a 1,2-migration of the alkyl group to CO (equation **9.9**). Calculations indicate that the rearrangement has a low activation energy and is endothermic.[14] Under the reaction conditions of high CO partial pressure, the 16-electron alkyl

[13]Ziegler and Versluis, *op. cit.*
[14]Ziegler and Versluis, *op. cit.*

carbonyl complex, **8**, must reversibly add CO to form $(RCO)Co(CO)_4$, which is the only detectable intermediate observed when the reaction has been followed using IR spectroscopy.[15] The stage is now set for the addition of hydrogen to give the aldehyde mixture as the final products of the catalytic cycle.

$$9.9$$

Step e

Two pathways have been proposed for step e, which may be the rate-limiting stage in the cycle. Equation **9.10** shows an oxidative addition followed by reductive elimination sequence, as originally proposed. Since Heck and Breslow's work, a bimolecular process, described by equation **9.11** below, has been suggested.[16]

$$9.10$$

$$8 + HCo(CO)_4 \longrightarrow CH_3-CH_2-CH_2-\overset{\overset{\displaystyle O}{\|}}{C}-H + Co_2(CO)_7 \longrightarrow HCo(CO)_4 + HCo(CO)_3$$

$$9.11$$

Although the bimolecular process involving reaction of the acylcobalt complex with $HCo(CO)_4$ occurs readily under stoichiometric conditions, the low concentration of the two required cobalt complexes under catalytic conditions probably argues against **9.11** as a catalytic step. Whether the OA-RE process occurs (equation **9.10**) in the usual manner is still a matter of question since calculations show that a process involving addition of H_2 to **8** to give dihydride **9** occurs with a rather high activation energy. A lower energy process seems to be possible according to calculations if an η^2-dihydrogen complex, **10**, forms instead, followed by four-center rearrangement to $HCo(CO)_3$ and aldehyde

[15]R.J.Whyman, *J. Organomet. Chem.*, **1974**, *66*, C23 and **1974**, *81*, 97.
[16]N.H. Alemaroglu, J.M.L. Renninger, and E. Oltay, *Monatsh. Chem.*, **1976**, *107*, 1043.

(equation **9.12**).[17] The determination of the actual mechanism for step e awaits further investigation.

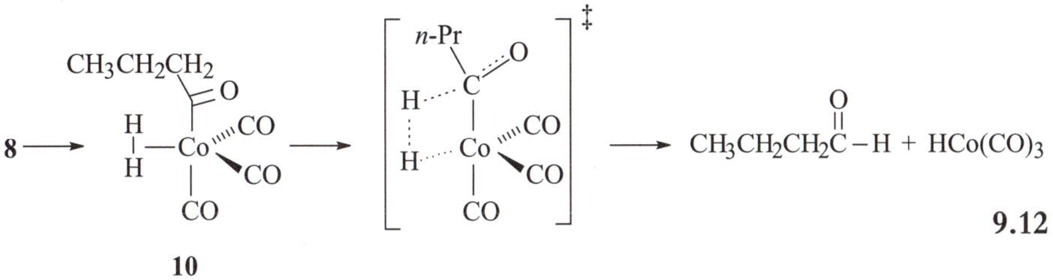

$$8 \longrightarrow 10 \longrightarrow [\ldots]^{\ddagger} \longrightarrow CH_3CH_2CH_2\overset{O}{\underset{\|}{C}}-H + HCo(CO)_3$$

9.12

The original hydroformylation process described above, while still widely used, is not without problems in actual practice. Some of these problems include the following:

1.   The ratio of linear to branched aldehydes is at best only ca. 4 to 1, and since the linear isomer is far more valuable than its branched co-product, improvement in this regard would be important from an economic standpoint.

2.   The active catalyst is unstable, and its separation and recovery are difficult.

3.   The high partial pressure of CO required for the process means that plants are expensive to build and operate.

9-2-2   Phosphine-Modified Hydroformylation

In 1968 Slaugh and Mullineaux[18] (Shell Oil Company) reported that adding tertiary phosphines, such as PBu_3, resulted in hydroformylation taking place at less than 100 atm (vs. the 200 to 300 atm normally required). Although the phosphine-modified catalyst was not as active as $HCo(CO)_3$ toward hydroformylation, it was a better hydrogenation catalyst. Thus the two stages of hydroformylation and hydrogenation could be combined into one step. The ratio of linear to branched product, moreover, was as high as 9 to 1. Finally, because the modified catalyst, $HCo(CO)_3(PBu_3)$, turned out to be more stable than the original, it was easier to separate it from product alcohols. The advantages of the Shell process are diminished somewhat because hydroformylation must occur at higher temperature (160° to 200°) and some alkene (ca. 15%) is converted directly to the corresponding alkane, owing to the effectiveness of the catalyst in promoting hydrogenation.

[17]Ziegler and Versluis, *op. cit.*

[18]L.H. Slaugh and R.D. Mullineaux, *J. Organomet. Chem.*, **1968**, *13*, 469.

The higher selectivity[19] of the modified catalyst in producing anti-Markovnikov product vs. branched probably is due mainly to steric factors. The large steric bulk of PBu_3 (compared to that of CO), as measured by its cone angle (see Chapter 6, Section **6-3**), must influence the course of the insertion of the alkene into the Co-H bond as shown in transition state structures **11** and **12** below. Structure **11** shows 1,2-insertion occurring that would ultimately lead to linear aldehyde or alcohol. The phosphine ligand points away from the R group on the alkene. Much more steric hindrance occurs with the other orientation, shown by **12**.

11   **12**

Studies with a variety of phosphine ligands have indicated that electronic factors may also play a role in the rate and orientation of phosphine-modified hydroformylation. As σ donors, phosphines undoubtedly donate electron density to the Co center that can then be taken up by the CO ligands, thus providing overall stability to the catalyst complex.[20]

Investigations of phosphine-modified hydroformylation have attempted to identify key intermediates in the catalytic cycle. Instead of acyl- and alkyl-cobalt complexes typical of the unmodified process, only phosphine-substituted cobalt carbonyl species have been observed.[21] We can only assume that the steps in the catalytic cycle are analogous to those for the unmodified process (Scheme **9.3**), but that the rate-determining step occurs at a different stage, perhaps during alkene insertion.

---

[19]An interesting report recently showed that hydroformylation, catalyzed by $HCo(CO)_4$, can be run at low temperature (80°) with high selectivity toward the linear aldehyde if the solvent is supercritical CO_2. Gases such as H_2 and CO as well as the catalyst are highly soluble in this solvent. See J.W. Rathke, R.J. Klinger, and T.R. Krause, *Organometallics*, **1991**, *10*, 1350.

[20]For a good discussion of the modified hydroformylation process including an analysis of phosphine ligand effects, see C. Masters, *op. cit.*, 114-120.

[21]F. Piacenti, M. Bianchi, and M. Bendetti, *Chim. Ind. Milan*, **1969**, *49*, 245.

9-2-3   Rhodium-Catalyzed Hydroformylation

As development of suitable cobalt hydroformylation catalysts occurred, work was also carried out to create corresponding Rh complexes that could also serve as suitable catalysts. Under appropriate conditions in the presence of H_2 and CO, $HRh(CO)_4$ forms from Rh-carbonyl cluster compounds. The hydrido-Rh complex catalyzes hydroformylation but also serves as a good alkene hydrogenation catalyst. Ratios of linear to branched hydroformylation products also tend to be low. In the 1960s, several groups[22] discovered that the addition of phosphines allowed hydroformylation to occur even at atmospheric pressure and at relatively low temperature. With the appropriate phosphine and by adjusting other reaction parameters, high linear to branched aldehyde product ratios were achieved—without extensive hydrogenation of either the alkene starting material or the aldehyde (to give alcohol). In 1976 Union Carbide first commercialized the use of phosphine-modified rhodium catalysts in hydroformylation.

The use of rhodium-phosphine catalysts offers many advantages over cobalt systems. Rhodium complexes are 100 to 1000 times more active than Co complexes, so much less rhodium needs to be present in the reactor.[23] The pressure (15 to 25 atm) and temperature (80° to 120°) required for reaction are significantly less than for either Co-based process, meaning that initial costs for plant construction are relatively low, as are energy expenses to keep the plant running. If the hydrocarbon starting material consists of only 1-alkenes, linear to branched aldehyde product ratios as high as 14 to 1 are obtained. In spite of the need to carefully recover expensive catalyst, rhodium-phosphine catalyzed hydroformylation is clearly the optimal process if linear aldehydes are the desired products. If alcohols are the goal, then the use of a phosphine-modified Co catalyst is preferred, since hydroformylation and hydrogenation occur in the same reaction vessel.[24]

Scheme **9.4** shows the steps involved in the phosphine-rhodium catalytic cycle. The immediate precursor to the active species is either $HRh(CO)_2(PR_3)_2$ or $HRh(CO)(PR_3)_3$[25] depending on the concentration of CO with respect to phosphine.

In contrast to PBu_3, used in the phosphine-modified Co process, the optimal phosphine for Rh-based catalysis appears to be PPh_3. Apparently, PBu_3

---

[22]J. A. Osborne, J. F. Young, and G. Wilkinson, *J. Chem. Soc., Chem. Commun.*, **1965**, 17; C.K. Brown and G. Wilkinson, *J. Chem. Soc. A*, **1970**, 2753; R.L. Pruett and J.A. Smith, *J. Org. Chem.*, **1969**, *34*, 327; L.H. Slaugh and R.D. Mullineaux, U.S. Patent 3 239 566, 1966.

[23]This is fortunate because the cost per gram of Rh is over 1000 times higher than that of Co.

[24]C. Masters, *op. cit.*, 127–128.

[25]It is interesting to note that this complex is an efficient alkene hydrogenation catalyst. It also resembles in reactivity the very effective hydrogenation catalyst, $ClRh(PPh_3)_3$ (Wilkinson's catalyst, Section **9-4-2**). Apparently, when the partial pressure of CO exceeds 10 atm, the capability of the complex to hydrogenate alkenes is suppressed.

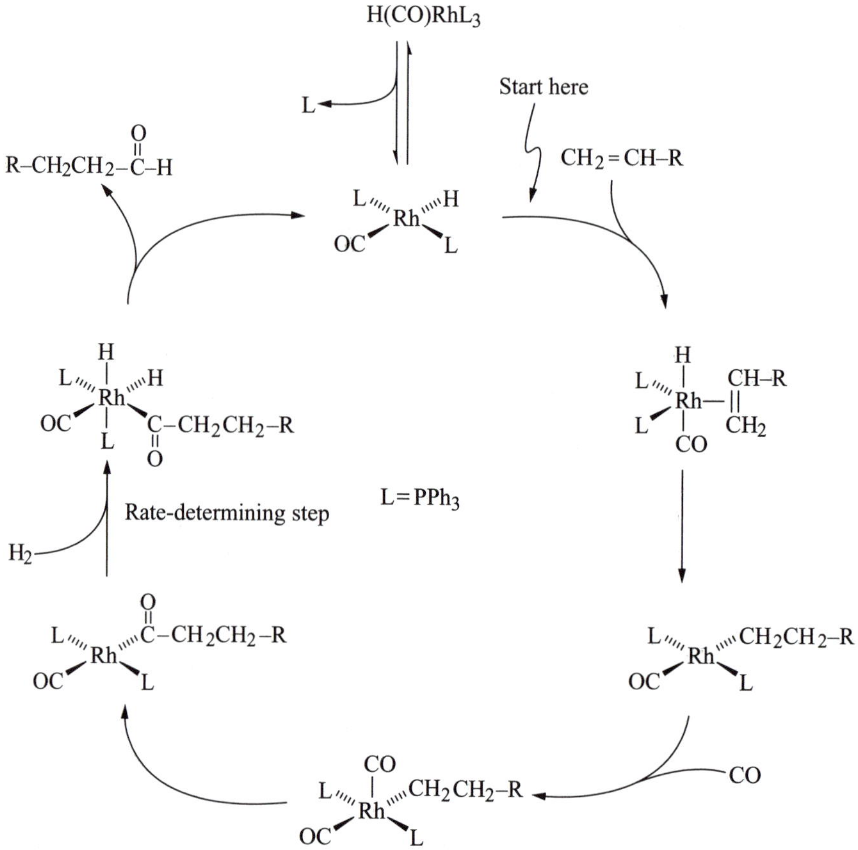

H(CO)RhL$_3$

Start here

L

R–CH$_2$CH$_2$–$\overset{\overset{\text{O}}{\|}}{\text{C}}$–H

CH$_2$=CH–R

L = PPh$_3$

Rate-determining step

H$_2$

Scheme 9.4
Phosphine-Rhodium-Catalyzed Hydroformylation

is too good at donating electron density to the metal in the case of Rh, and the intermediate catalytic species are too stable for effective catalysis. The triaryl phosphine seems to have the right combination of steric (to induce the formation of linear product at the 1,2-insertion stage) and electronic (to donate electron density to metal in order to stabilize CO ligands) properties. Studies indicate that the rate-determining step is likely to be hydrogenation of the acylrhodium intermediate (as with unmodified Co hydroformylation), but the mechanism of this apparent OA-RE step is not completely understood.[26]

Exercise 9-1

Increasing the concentration of phosphine in the phosphine-rhodium cycle slows the reaction rate, but also raises the linear/branched product ratio. Explain.

Other metals (e.g., Ru, Mn, and Fe) may serve as the basis for hydroformylation catalysts. The overall effectiveness of these metals compared to Co and Rh is as follows:

$$Rh > Co > Ru > Mn > Fe > Cr, \ Mo, \ W, \ Ni$$

Rel. Reactivity: $10^2\text{-}10^3$ 1 10^{-2} 10^{-4} 10^{-6} 0

Although hydroformylation catalyzed by Ru is significantly slower than by Co or Rh, anionic and cluster (Ru$_3$) complexes of this metal catalyze the formation of linear instead of branched aldehydes with as much as 97% or more selectivity.[27]

9-3   Wacker-Smidt Synthesis of Acetaldehyde

Ethanal (more commonly known as acetaldehyde) is another aldehyde of commercial importance. Oxidation to acetic acid and possible subsequent dehydration to form acetic anhydride are the typical fates of this compound. Acetaldehyde was originally prepared by hydration of acetylene according to equation **9.13**—a facile and high-yield route.

$$HC \equiv CH \ + \ H_2O \ \xrightarrow[\text{Hg(II)}]{H^+} \ H_2C = CHOH \ \rightleftharpoons \ \underset{\displaystyle CH_3\overset{\displaystyle O}{\overset{\displaystyle \|}{C}}-H}{} \qquad \textbf{9.13}$$

[26]A.A. Oswald, D.E. Hendriksen, R.V. Kastrup, and E.J. Mozeleski, "Electronic Effects on the Synthesis, Structure, Reactivity, and Selectivity of Rhodium Hydroformylation Catalysts," in W.R. Moser and D.W. Slocum, Eds., *Homogeneous Transition Metal-Catalyzed Reactions*, American Chemical Society, Washington, D.C., 1992, 395-417; C. Masters, *op. cit.*, 120-127; J.P. Collman, *et al.*, *op. cit.*, 625-630.
[27]For a review of recent work on Ru-catalyzed hydroformylation, see P. Kalck, Y. Peres, and J. Jenck, *Adv. Organomet. Chem.*, **1992**, *32*, 121.

This synthesis is now obsolete because of problems associated with the starting material, acetylene. Acetylene must be produced by heating a hydrocarbon gas stream to high temperature, sometimes in the presence of an electric arc. All processes for producing it require large amounts of energy. Acetylene is also thermodynamically unstable, and it must be handled with extreme care in order to prevent explosion.

The incentive, therefore, existed to develop a process for producing acetaldehyde from a cheaper and less hazardous starting material. It had long been known that acetaldehyde formed directly from ethylene and water in the presence of a stoichiometric amount of $PdCl_2$ according to equation **9.14**.[28] It was not until the 1950s, however, that a commercially feasible process was developed by Smidt at Wacker Chemie in Germany.[29] The Wacker–Smidt synthesis of acetaldehyde combines equation **9.14** with steps (equations **9.15** and **9.16**), that allow deployment of Pd in *catalytic* amount because much less expensive reagents—$CuCl_2$, HCl, and O_2—are used to keep Pd in the proper oxidation state.

The overall reaction is equivalent to direct oxidation of ethylene by O_2. The process is run in one or two stages depending on whether the catalyst is regenerated *in situ* or in a separate reactor. The former process uses pure oxygen in the presence of ethylene, Pd(II), $CuCl_2$, and HCl. The latter process uses the chemistry described in equation **9.14** in one reactor and that described in **9.15** and **9.16** in another separate reaction vessel with air as the oxygen source. Both processes have their advantages and disadvantages and

$$CH_2=CH_2 \ + \ PdCl_2 \ + \ H_2O \longrightarrow CH_3\overset{\overset{\textstyle O}{\|}}{C}-H \ + \ Pd(0) \ + \ 2\,HCl \qquad \textbf{9.14}$$

$$Pd(0) \ + \ 2\,CuCl_2 \longrightarrow PdCl_2 \ + \ 2\,CuCl \qquad \textbf{9.15}$$

$$2\,CuCl \ + \ 2\,HCl \ + \ \tfrac{1}{2}\,O_2 \longrightarrow 2\,CuCl_2 \ + \ H_2O \qquad \textbf{9.16}$$

$$CH_2=CH_2 \ + \ \tfrac{1}{2}\,O_2 \longrightarrow CH_3\overset{\overset{\textstyle O}{\|}}{C}-H$$

[28]F.C. Phillips, *Am. Chem. J.*, **1894**, *16*, 255.
[29]J. Smidt, W. Hafner, R. Jira, J. Sedlmeier, R. Sieber, R. Rütlinger, and H. Kojer, *Angew. Chem.*, **1959**, *71*, 176.

both are used commercially in the U.S. and Europe, producing acetaldehyde at about the same cost. In terms of amount of product produced, the Wacker-Smidt process represents an economically significant example of the use of transition metal homogeneous catalysis. The technology has also been used to produce vinyl acetate (equation **9.17**), a monomer that is subsequently polymerized to give polyvinylacetate films.[30]

$$CH_2 = CH_2 \ + \ Cu(OAc)_2 \ + \ KCl \ + \ KOAc \ \xrightarrow[\text{PdCl}_2]{\text{O}_2} \ CH_2 = CH\text{–OAc} \ + \ H_2O$$

9.17

The Wacker-Smidt process, leading ultimately to acetic acid and acetic anhydride, has enjoyed considerable success, yet its use has declined over the last ten years for at least two reasons. First, manufacturing plants are expensive to build and maintain because they must be constructed to withstand a corrosive environment. Second, another procedure, which yields acetic acid directly from synthesis gas, has been developed and now supplants the Wacker-Smidt process. This newer route also uses homogeneous catalysis involving Rh complexes and will be described in section **9-5**.

While the Pd(II)–catalyzed route to acetaldehyde is no longer as important to the chemical industry as it was 20 to 30 years ago, the mechanistic basis of the process is interesting because it involves a number of fundamental kinds of organometallic reactions, such as ligand substitution, β-elimination, and 1,2-insertion. Scheme **9.5** shows an outline of the catalytic cycle involving organopalladium complexes. It also shows how Cu(II) and O_2 couple with the Pd cycle in order to regenerate Pd(II). The scheme is consistent with the rate law[31] observed for Wacker-Smidt oxidation:

$$\text{Rate} = \frac{k[CH_2{=}CH_2][PdCl_4{}^{2-}]}{[H^+][Cl^-]^2}$$

[30]For a good discussion of the Wacker-Smidt and related processes, see P. Wiseman, *An Introduction to Industrial Organic Chemistry*, Wiley, New York, 1976, 97–103. A process using palladium-impregnated silica, developed more recently, is superior to the homogeneous-catalyzed process described in equation **9.17**.

The exact pathway for acetaldehyde formation is still not completely clear. One difficulty rests with the instability of Pd-alkene and Pd-alkyl intermediates (which goes hand-in-hand with Pd(II) being such an effective catalyst). Consequently, platinum analogs, which are much more stable and not catalytically active, have been studied to obtain information regarding the Pd-catalyzed pathway. We shall also see that experiments have been reported under conditions that do not exactly duplicate those for the Wacker-Smidt process. Nevertheless, these experiments have provided useful, though indirect, information.

Step a

The first major transformation in the pathway probably consists of two steps in which two Cl$^-$ ions are displaced and substituted with alkene and H$_2$O. These set the stage for nucleophilic attack on the complexed alkene.

Step b

The mechanism of attack of the nucleophile (H$_2$O or OH$^-$) on the coordinated alkene is still unclear. Does the nucleophile attack externally in a manner *trans* to Pd to give the hydroxyalkylpalladium complex (path b, Scheme **9.5** on page 269), or does intramolecular 1,2-insertion of the alkene between the metal and a coordinated OH group occur with OH *cis* to Pd (path b′, Scheme **9.5**)? The observed rate law is consistent with either mechanism.

In order to answer this question, researchers designed rather elegant experiments. The first, reported by Stille and Divakaruni (equation **9.18**),[32] showed *cis*-dideuterioethylene undergoing hydroxypalladation in the presence of CuCl$_2$ followed by CO insertion to give lactone **13**. The deuterium atoms in **13** ended up *trans* to one another, meaning that inversion of configuration occurred. Since CO insertion is known to occur with retention of configuration (Chapter 8, Section **8-1**), inversion must have taken place during attack by H$_2$O, exactly the result expected when nucleophiles attack π ligands (Chapter 8, Section **8-2-1**).

Another investigation (equation **9.19**), reported by Bäckvall and Åkermark,[33] described *trans* attack on a *trans*-dideuterioethylene Pd complex. In the presence of high concentrations of Cu(II) and Cl$^-$, the final product was the chlorohydrin, **14**, rather than the aldehyde. Treatment of the chlorohydrin with base gave dideuterioethylene oxide, **15**, in which the two deuterium atoms were *cis*. This result is again consistent with attack by an external nucleophile.

[31]J.E. Bäckvall, B. Åkermark, and S.O. Ljunggren, *J. Am. Chem. Soc.*, **1979**, *101*, 2411 and references therein.
[32]J.K. Stille and R. Divakaruni, *J. Am. Chem. Soc.*, **1978**, *100*, 1303.

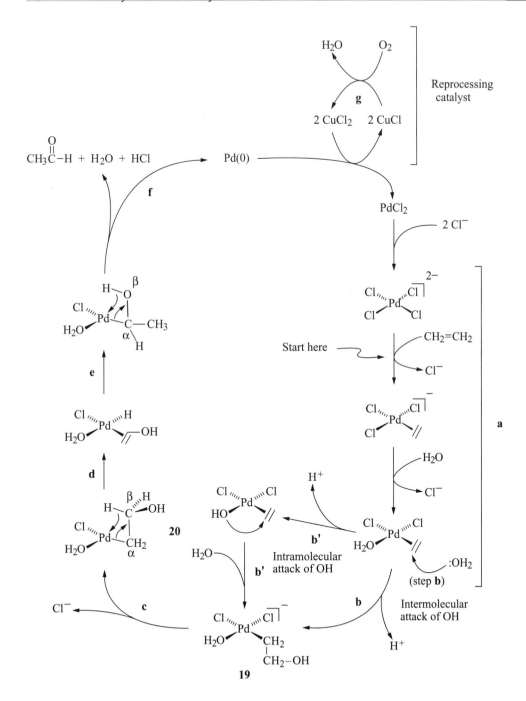

Scheme 9.5
Wacker-Smidt Synthesis of Acetaldehyde

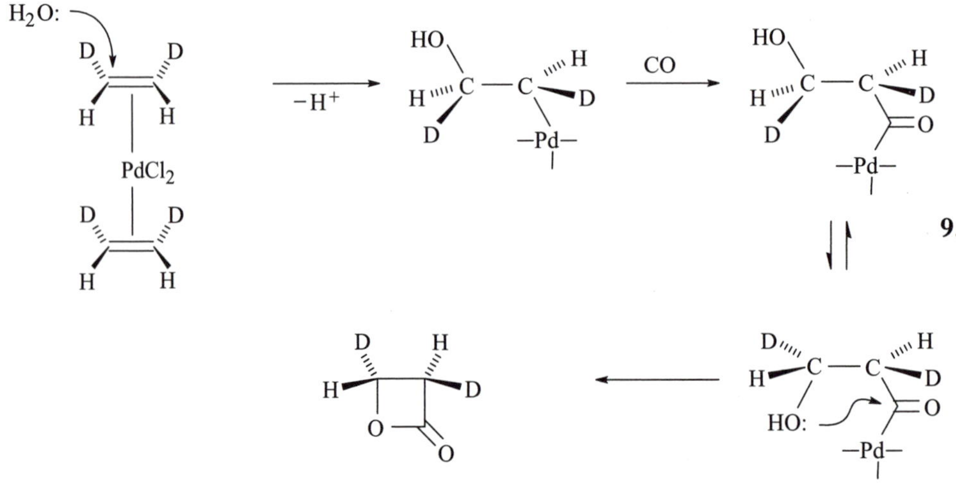

9.18

13

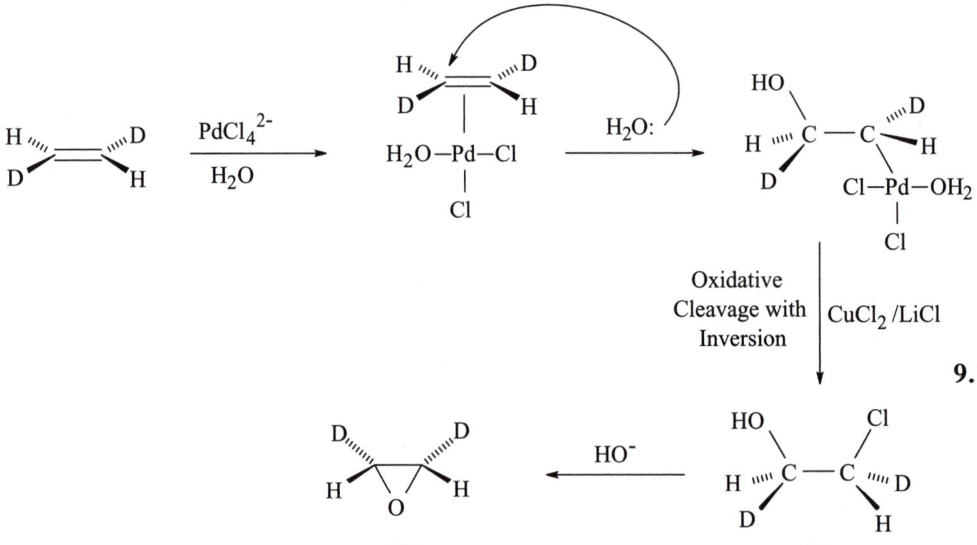

Oxidative
Cleavage with   CuCl₂ /LiCl
Inversion

9.19

15   14

³³J.E. Bäckvall, B. Åkermark, and S.O. Ljunggren, *op. cit.*

Exercise 9-2

Fill in the details of equation **9.19** by showing how the *trans* dideu-
terioalkene complex ends up as *threo*-**14** and then *cis*-dideuterioethyl-
ene oxide **15**. What must be the stereochemistry at the carbon originally attached to Pd after oxidative cleavage by CuCl$_2$ of the Pd–C
bond? What would the stereochemistry of **15** be if intramolecular 1,2-
insertion occurred instead?

In spite of these two experiments, evidence exists that the *cis*-intramole-
cular route occurs for hydroxypalladation. Recent reports by Henry and
Francis[34] describe experiments designed to probe both the kinetics and stereochemistry of the Wacker-Smidt process. Scheme **9.6** shows the two stereochemical outcomes possible starting with allylic alcohol **16**, designed so that
β-elimination—and thus acetaldehyde formation—are not possible. When
Cl⁻ concentrations were low and comparable to Wacker-Smidt conditions, **17**

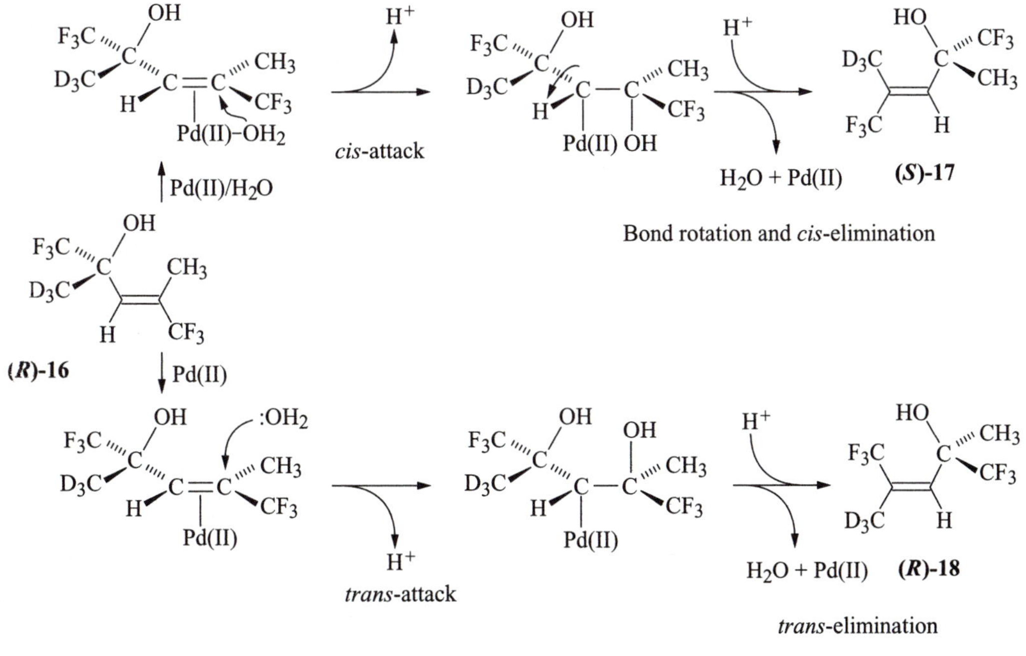

Scheme 9.6
Hydroxypalladation of 2-(Methyl-d$_3$)-4-methyl-1,1,1,5,5,5-hexafluoro-3-penten-2-ol

[34]J.M. Francis and P.M. Henry, *Organometallics*, **1991**, *10*, 3498; **1992**, *11*, 2832.

formed—a compound that indicated *cis* intramolecular attack by a coordinated OH ligand had occurred. At higher Cl⁻ concentrations (similar to those used by Bäckvall, *et al.*), **18**, a product that must result from *trans* intermolecular attack by H_2O, was produced instead.

Exercise 9-3

Verify that the stereochemical outcomes are those outlined in Scheme **9.6**.

All of the above-described results shed light upon the hydroxypalladation stage of the Wacker-Smidt process, but they fail to replicate the exact conditions under which the process is run. The question of whether the nucleophile is water from outside the coordination sphere that adds *trans* with respect to Pd or whether it is coordinated OH that adds *cis*, is still open to question. The final answer will require that investigations be run under conditions where rapid oxidation occurs to yield acetaldehyde and under circumstances that give a clear stereochemical picture of the mechanism. This will be challenging to chemists because stereochemical information is lost in the last step as the planar carbonyl group forms.

Steps c, d, and e

These steps represent β-elimination and 1,2-insertion reactions that result in Pd becoming attached to more substituted carbon. Slow, rate-determining loss of Cl⁻ from 16-electron intermediate **19** (step c) to give **20**, a 14-electron complex, probably precedes β-elimination (step d).

Once β-elimination has occurred, readers might imagine that the complexed enol simply leaves the coordination sphere and goes directly to acetaldehyde by an enol-keto tautomerism. When the Wacker-Smidt reaction is run in D_2O, however, there is no D-incorporation in the aldehyde, thus ruling out ligand dissociation-tautomerism.[35]

Exercise 9-4

Why would deuterium incorporation be expected into acetaldehyde if enol simply dissociates after step c? Explain.

Step f

Step f represents the last step in the formation of aldehyde—apparently as a result of β-elimination of proton from the α-OH group. At the same time, the oxidation state of palladium drops from $+2 \rightarrow 0$, probably by reductive elimination of HCl.

Step g

The last step in the cycle regenerates Pd(II) at the expense of Cu(II). The Cu(I) species that results is reoxidized to Cu(II) by either pure O_2 (single stage) or air (two-stage process).

Wacker-Smidt chemistry is also applicable to the laboratory-scale preparation of ketones. Equation **9.20** shows the conversion of 1-undecene to 2-undecanone. Internal double bonds are much less reactive than terminal, as exemplified in equation **9.21**, where a dieneester is oxidized to the corresponding ketone (a precursor in the synthesis of diplodialide A, a fungal metabolite) in good yield.[36]

85 %   **9.20**

Diplodialide A   **9.21**

Exercise 9-5

Explain why a ketone is formed and not an aldehyde when a monosubstituted alkene undergoes the Wacker-Smidt reaction. [Hint: See Chapter 8, Section **8-2-2**]

9-4   Hydrogenation

The addition of H_2 across a multiple bond, such as C=C, C≡C, or C=O, constitutes an important synthetic procedure—both at the laboratory and industrial scale. The reaction is usually run in the presence of a heterogeneous

[36]J. Tsuji and T. Mandai, *Tet. Letters*, **1978**, 1817.

catalyst. In this section, however, we will discuss the use of several homogeneous catalysts capable of promoting the saturation of multiple bonds (especially C=C) under mild conditions. We have already encountered an example of homogeneous catalytic hydrogenation, i.e., the modified Co catalyst used to promote formation of long-chain alcohols (Section **9-2**).

9-4-1   Hydrogenation Involving a Monohydride Intermediate

Two major pathways seem to occur with homogeneously catalyzed hydrogenation. One involves monohydrides (M-H), and the other involves dihydrides (MH$_2$). The M-H mechanism will be considered first in a rather limited manner, with much more detailed treatment reserved for the dihydride pathway.

RuCl$_2$(PPh$_3$)$_3$ serves usefully in the catalytic hydrogenation of alkenes, as shown in equations **9.22** and **9.23**[37] below, showing very high selectivity for terminal over internal double bonds.

$$9.22$$

$$9.23$$

The active catalytic species is probably the 16-electron complex, HRuCl(PPh$_3$)$_3$, **21**, formed according to equation **9.24**. The reaction probably begins with the addition of H$_2$ to RuCl$_2$(PPh$_3$)$_2$ to give a dihydrogen complex and not a dihydride (a dihydride would oblige Ru to take on a relatively unstable oxidation state of +4). The presence of Et$_3$N accelerates the formation of the active species by serving as a proton sponge to trap the HCl that forms by heterolytic cleavage of M-H and M-Cl bonds.

$$9.24$$

[37]P.S. Hallman, B.R. McGarvey, and G. Wilkinson, *J. Chem. Soc., A,* **1968**, 3143.

Exercise 9-6

Propose a mechanism for the conversion of the dihydrogen complex to HCl and **21** described in **9.24**.[38]

Monohydride **21**, with its vacant site now available, can bind to the alkene, giving **22**, which can then undergo 1,2-insertion to form alkylruthenium complex **23**. Finally, heterolytic cleavage with H_2 regenerates the catalyst and gives the alkane (equations **9.25** to **9.27** below).

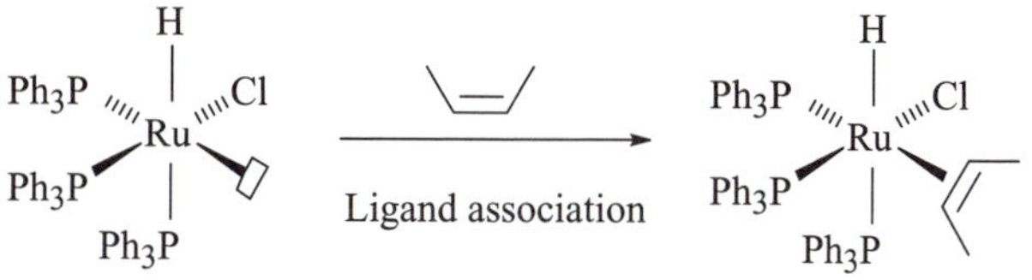

9.25

21   **22**

9.26

22   **23**

$+ CH_3CH_2CH_2CH_3$

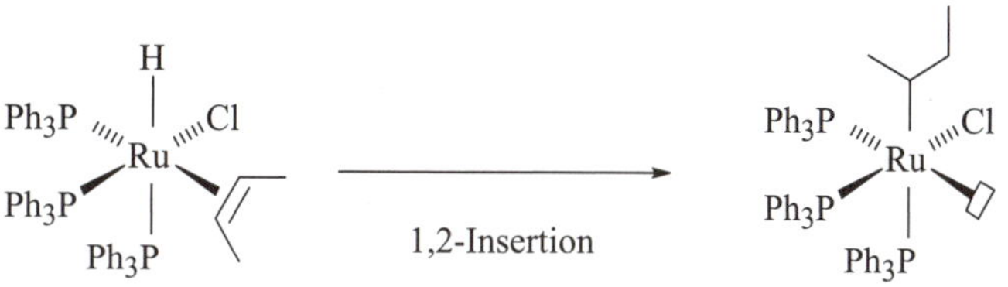

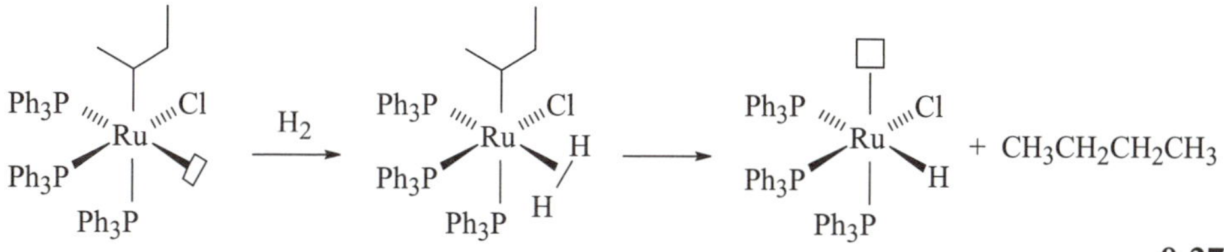

9.27

23

---

[38]See R.H. Crabtree and D.G. Hamilton, *J. Am. Chem. Soc.*, **1986**, *108*, 3124 for further insight into the mechanism.

Homogeneous hydrogenation also occurs with Co(II) in the form of $[Co(CN)_5]^{3-}$, known for over 50 years to promote hydrogenation of "activated" alkenes.[39] The catalytic species has 17 electrons, and it is not surprising that this complex, with its odd number of electrons, undergoes reaction through radical intermediates. Equations **9.28** to **9.30** provide a plausible mechanism for hydrogenation of an alkene. Equation **9.28** shows a bimolecular oxidative addition (Section **7-2**), which is then followed by **9.29** to give organic radical intermediate **24**. There must be at least one radical-stabilizing group attached to the double bond (such as R and/or R' = Ph, CO_2R, CN) in order for the reaction to proceed at a reasonable rate. Thus simple ("unactivated") alkenes (R and R' = alkyl) are unreactive to the cyanocobalt system. Homolytic cleavage of the Co–H bond (equation **9.30**) completes the cycle to regenerate catalyst and yield alkane.

$$2\,[(CN)_5\overset{\bullet}{Co}]^{3-} \quad + \quad H–H \quad \longrightarrow \quad 2\,[H–Co(CN)_5]^{3-} \qquad \textbf{9.28}$$

$$[H–Co(CN)_5]^{3-} \quad + \quad \underset{R}{\overset{}{\diagup}}\!\!=\!\!\underset{}{\overset{R'}{\diagdown}} \quad \longrightarrow \quad [\overset{\bullet}{Co}(CN)_5]^{3-} \quad + \quad R\!\diagup\!\underset{H}{\overset{\bullet}{\diagdown}}\!R'$$

R, R' = Ph, CO_2Me, CN   **24**

$$\textbf{9.29}$$

$$[H–Co(CN)_5]^{3-} \quad + \quad R\!\diagup\!\underset{H}{\overset{\bullet}{\diagdown}}\!R' \quad \longrightarrow \quad [\overset{\bullet}{Co}(CN)_5]^{3-} \quad + \quad R\!\diagup\!\underset{H}{\overset{\overset{H}{|}}{\diagdown}}\!R'$$

24

$$\textbf{9.30}$$

$$H_2 \quad + \quad \underset{R}{\overset{}{\diagup}}\!\!=\!\!\underset{}{\overset{R'}{\diagdown}} \quad \longrightarrow \quad R–CH_2–CH_2–R'$$

Equations **9.31**[40] and **9.32**[41] provide examples of Co(II)-catalyzed hydrogenation.

$$HO_2C\diagup\!\!=\!\!\diagdown\!CO_2H \quad \xrightarrow[{[Co(CN)_5]^{3-}/70°,\,1\ atm}]{H_2} \quad HO_2C–CH_2–CH_2–CO_2H$$

$$\textbf{9.31}$$

[39]M. Iguchi, *J. Chem. Soc. Jap.*, **1939**, *60*, 1287; R. Mason and D.W. Meek, *Angew. Chem. Int. Ed. Engl.*, **1978**, *17*, 183.

[40]M. Murakami, K. Suzuki, and J-W. Kang, *Nippon Kagaku Zasshi*, **1962**, *83*, 1226.

[41]J. Kwiatek, I.L. Mador, and J.K. Seyler, *J. Am. Chem. Soc.*, **1962**, *84*, 304.

$$9.32$$

9-4-2 Hydrogenation Involving a Dihydride Intermediate

By far the most well-studied and synthetically useful catalyst systems involving dihydride (MH_2) intermediates are those based upon Rh(I). The discovery by Wilkinson[42] (and independently and nearly simultaneously by Coffey[43]) in 1965 that $RhCl(PPh_3)_3$ could catalyze the hydrogenation of alkenes at atmospheric pressure (in the presence of a wide variety of unsaturated functional groups that are unaffected by the reaction conditions) was a momentous event in the history of organometallic chemistry. This breakthrough stimulated a great deal of research on the elucidation of catalytic mechanisms in general, and in particular on both the determination of the mechanism and the definition of the synthetic utility of homogeneous hydrogenation using Rh(I) complexes. So much so, that today $RhCl(PPh_3)_3$ is known as "Wilkinson's catalyst."

Equation **9.33** shows an example of hydrogenation using Wilkinson's catalyst. Employment of D_2 rather than H_2 clearly indicates the stereochemistry of *syn* addition typical of Rh-catalyzed hydrogenation.

$$9.33$$

The catalyst itself is prepared either from $RhCl_3 \cdot xH_2O$ through treatment with excess PPh_3 in hot ethanol (equation **9.34**) or by first making a chlorine-bridged dialkene complex and then allowing it to react with phosphine (equation **9.35**). The latter process allows phosphines other than PPh_3 to be coordinated to Rh.

$$RhCl_3(H_2O)_x \ + \ xs\ PPh_3 \ \xrightarrow[\text{EtOH/80°}]{} \ RhCl(PPh_3)_3 \ + \ PPh_3O$$

$$9.34$$

$$[Rh(cyclooctene)_2Cl]_2 \ + \ 6\ PPh_3 \ \longrightarrow \ 2\ RhCl(PPh_3)_3 \ + \ 2\ cyclooctane$$

$$9.35$$

[42]J.F. Young, J.A. Osborn, F.H. Jardine, and G. Wilkinson, *J. Chem. Soc., Chem. Commun.*, **1965**, 131.
[43]R.S. Coffey, Imperial Chemical Industries, Brit. Patent 1 121 642, **1965**.

Table 9-2 Relative Rates for Hydrogenation of C=C over Wilkinson's Catalyst at 25°C

Substrate	k x 10^2 (L/mol-sec)
Cyclohexene	31.6
1-Hexene	29.1
2-Methyl-1-pentene	26.6
Z-4-Methyl-2-pentene	9.9
E-4-Methyl-2-pentene	1.8
1-Methylcyclohexene	0.6
3,4-Dimethyl-3-hexene	<0.1

Steric hindrance about the C=C bond of the alkene seems to play a role in influencing the rate of hydrogenation according to Table **9-2**. Interestingly, the rate of H_2 addition to ethylene, a most sterically uncongested molecule, is quite slow relative to that for mono- and disubstituted alkenes. Ethylene apparently binds tightly to Rh early in the catalytic cycle, inhibiting subsequent addition of H_2 (step j of Scheme **9.7**).[44]

Scheme **9.7** depicts what most consider to be the mechanism for hydrogenation with Wilkinson's catalyst. The inner circle of reactions (surrounded by the diamond-shaped dotted-line box) represents the key catalytic steps based on the work of Halpern.[45] He showed that compounds **30** to **33**, while detectable and isolable, were not responsible for the actual catalytic process, and, moreover, the buildup of these intermediates during hydrogenation may even slow down the overall reaction. This work provided a valuable lesson for chemists trying to determine a catalytic mechanism—**compounds that are readily isolable are probably not true intermediates**. Through careful kinetic and spectroscopic studies, Halpern showed, by inference, that structures **26** to **29** were the true intermediates in the catalytic cycle. The following discussion will look at some of the key steps in the catalytic cycle to see how they correspond to the overall hydrogenation process.[46]

Steps a and b

Dissociation of **25** to **26**, a 14-electron intermediate (unless we count solvent coordination), followed by oxidative addition of H_2 seems at first glance implausible and unnecessary since the step is endothermic and a scheme involving 16- and 18-electron intermediates (**25** → **30** → **27**, steps f and g)

---

[44]For an excellent discussion of the scope and limitations of Rh(I)-catalyzed hydrogenation, see A.J. Birch and D.H. Williamson, *Org. React.*, **1976**, *24*, 1.

[45]J. Halpern, T. Okamoto, and A. Zakhariev, *J. Mol. Cat.*, **1976**, *2*, 65.

[46]See J.P. Collman, *et al.*, *op. cit.*, 531-535 for a good discussion of the mechanism of hydrogenation catalyzed by RhCl(PPh$_3$)$_3$.

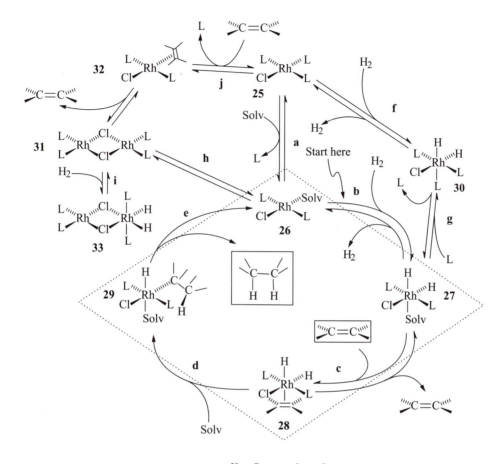

Key Steps: **a, b, c, d, e**

Scheme 9.7

Mechanism of Hydrogenation
with Wilkinson's Catalyst

L = PPh₃; Solv: EtOH, THF

could also occur readily. Halpern estimated, however, that the rate of H₂ addition to **26** was 10,000 times greater than to **25**, even though he had not actually isolated **26**. Several years later Wink and Ford[47] generated **26** as a transient species and observed it spectroscopically. They also measured the rate of H₂ addition and found it to be close to Halpern's value.

Dimerization of **26** to **31** followed by hydrogen addition to give **33** seems to occur (steps h and i). Complex **33**, however, does not take part in the catalytic cycle and is a "dead-end" pathway. Under normal hydrogenation conditions, **26** reacts rapidly to give **27**.

[47]D. A. Wink and P. C. Ford, *J. Am. Chem. Soc.*, **1985**, *107*, 1794 and **1986**, *108*, 4838.

Step c

Step c is a straightforward coordination of alkene ligand onto dihydride **27**. Assuming that solvent only very weakly coordinates to Rh in **27**, addition of the alkene to give 18-electron **28** should be facile. It should be pointed out that strongly coordinating alkenes such as ethylene, however, bind preferentially to **25** to give **32** (step j), a stable complex that does participate in the catalytic cycle.

Step d

This step is rate determining and actually is a combination of two fundamental steps. First, 1,2-insertion of the alkene ligand gives **34** (see equation **9.36** below; structures **34** to **37** may be considered distorted square pyramids), where the alkyl and hydride ligands are *trans*. In order for step e (reductive elimination) to occur, these two ligands must be situated *cis* with respect to each other. This requires that an isomerization occur. While it is not known just how this happens, calculations[48] on a model system where PH_3 is used instead of PPh_3 indicate that a reasonable scenario for isomerization might take place according to equation **9.36**. Here, **35** (**34** if L = PPh_3) undergoes sequential hydride and chloride migration (both occur with low activation energy) to yield **36** before Rh–C bond rotation in the ethyl group occurs to give **37** (**29** if L = PPh_3). The observation of a primary deuterium isotope effect[49] ($k_{obs}^{Rh-H}/k_{obs}^{Rh-D} = 1.15$) provides additional evidence that the rate-determining step is one that involves M–H bond breaking.

34: R = Ph
35: R = H

36

37

(**29** if R = Ph)

9.36

[48]C. Daniel, N. Koga, J. Han, X.Y. Fu, and K. Morokuma, *J. Am. Chem. Soc.*, **1988**, *110*, 3773.
[49]Primary kinetic deuterium isotope measurements are useful in determining whether bond breaking between a hydrogen and some other atom has occurred before or during the rate-determining step in a reaction mechanism. If so, the rate of reaction involving X–H bond breaking (X = a heavy atom) will be slower when deuterium is present than the case for hydrogen, and $k_{obs}^{X-H}/k_{obs}^{X-D} > 1$. If $k_{obs}^{X-H}/k_{obs}^{X-D} = 1$, then there is no isotope effect and X–H bond breaking likely occurs after the rate-determining step. See T.H. Lowry and K.S. Richardson, *Mechanism and Theory in Organic Chemistry*, 3rd ed. Harper and Row, New York, 1987, 232-244 for a good discussion of kinetic isotope effects.

Exercise 9-7

Based on your knowledge of the *trans* and *cis* effects (Chapter 7), why should the rearrangement of **35** to **37** be energetically reasonable?

Step e

Now that the hydride and alkyl groups are *cis* with respect to each other, reductive elimination readily occurs to yield alkane and regenerate **26**. We know that RE takes place with retention of configuration. This result, combined with effective *syn* addition of Rh and hydrogen during the 1,2-insertion, means that both hydrogens ultimately add to the same face of the alkene C=C bond.

Molecular orbital calculations,[50] performed for each step of the catalytic cycle (solvent effects were neglected and L = PH_3), verified that the mechanism postulated by Halpern is reasonable. The caveat to bear in mind with the Halpern mechanism or the mechanism of any catalytic process is that variations in alkene, solvent, and phosphine ligands may change the pathway or the rate–determining step.

Cationic Hydrogenation Catalysts

The discovery of Wilkinson's catalyst led to the development of a new class of complexes capable of promoting hydrogenation; these have the general formula L_nM^+ (M = Rh or Ir). The Rh series was reported by Schrock and Osborn.[51] Equation **9.37** demonstrates how such a complex may be prepared. The cationic Rh(I) complex (**38**) interacts with a solvent such as THF or acetone to give a 12-electron, "unsaturated" diphosphine intermediate (**39**), which is considered to be the active catalyst. The catalytic cycle begins this time with alkene binding, followed by oxidative addition of H_2.

$$\text{38} \qquad\qquad\qquad\qquad \text{39} \qquad\qquad \textbf{9.37}$$

[50]C. Daniel, et al., *op. cit.*; N. Koga and K. Morokuma, *Chem. Rev.*, **1991**, *91*, 823.
[51]R.R. Schrock and J.A. Osborn, *J. Am. Chem. Soc.*, **1976**, *98*, 2134 and references therein.

Although cationic Rh complexes have found some use in reducing alkenes to alkanes, converting alkynes to *cis*-alkenes, and transforming ketones to alcohols, the most important function of these catalysts has been to promote asymmetric hydrogenation of a C=C bond. Equation **9.38** shows a key step in the industrial-scale synthesis of the drug, L–Dopa (**40**), used to treat Parkinson's Disease.[52] The reduction proceeds with remarkable stereoselectivity producing the *S* enantiomer in 94% ee.[53] Note that the diphosphine ligand DIPAMP (**41**) possesses two stereogenic centers and provides a chiral environment (when complexed with Rh) during the course of the catalytic cycle. A closer examination of the mechanism of asymmetric hydrogenation will appear in Chapter 11 when it is considered along with other methods for inducing chirality into molecules.

$$9.38$$

*: Stereocenter

41

40

Since Ir resides below Rh in the Periodic Table, we would expect it to parallel rhodium's behavior as a hydrogenation catalyst. Unfortunately, the Ir analog of Wilkinson's catalyst, $IrCl(PPh_3)_3$, is not effective in saturating C=C bonds because phosphine is more tightly bound to Ir (typical of third–row transition elements) than with Rh, and thus the catalytically-active species, $IrCl(PPh_3)_2$, does not readily form. Cationic Ir complexes, on the other hand, are even more active in promoting hydrogenation than Wilkinson's catalyst. Developed mainly by Crabtree,[54] cationic Ir complexes of the general form

[52]W.S. Knowles, *J. Chem. Ed.*, **1986**, *63*, 222.

[53]The term "ee" stands for *enantiomeric excess*, that is the excess of one enantiomer over the other (%R - %S). For example, if a synthesis produced 90% S and 10% R, the ee = 90% - 10% = 80%. This means that the mixture consists of 80% S and 20% racemic modification (R + S). In the example used above, *ee* = *94%* indicates the mixture consists of 97% S and 3% R.

[54]R. Crabtree, *Accounts Chem. Res.*, **1979**, *12*, 331.

(COD)Ir(L)(L′) catalyze saturation of even tetrasubstituted double bonds almost as fast as that for the mono- and disubstituted variety. Apparently, in the presence of H_2 and a noncoordinating solvent such as CH_2Cl_2, an active, transient, 12-electron species—"Ir(L)(L′)$^{+}$"—forms along with the conversion of the COD ligand to cyclooctane (equation **9.39** below).

$$\text{(COD)} \quad \xrightarrow[\text{Solvent}]{H_2} \quad (\text{Solv})_2\overset{+}{\text{Ir}}(\text{Py})(\text{PCy}_3) \quad + \quad \bigcirc$$

(COD)   **9.39**

This highly unsaturated complex with its positive charge is a relatively hard Lewis acid, binding not only alkene and hydrogen but also polar ligands such as alcohol and carbonyl oxygen. Equation **9.40** shows[55] the remarkable stereoselectivity obtained when enone **42** is hydrogenated in the presence of [Ir(COD)(Py)(PCy₃)]PF₆. The high preference for the isomer in which the methyl group is *trans* to H at the junction of the two rings is probably due to the formation of intermediate **43** in which hydrogen, alkene, and alcohol groups bind simultaneously to Ir. Such high stereoselectivity is typical in cyclic systems where a polar group resides near a C=C bond.

$$\textbf{42} \quad \xrightarrow[\text{CH}_2\text{Cl}_2]{\text{Ir(COD)(Py)(PCy}_3)/\text{H}_2} \quad \textbf{43}$$

42   **43**

9.40

9-5   Monsanto Acetic Acid Process

Acetic acid is produced industrially on a large scale. Although useful in some commercial processes as is, most of the acetic acid produced is subsequently converted to acetic anhydride—a valuable acetylating agent employed in synthesizing cellulose acetate films and aspirin. For many years, the major production method of acetic acid was through the Wacker Chemie process. The ready availability of CH_3OH from synthesis gas using a heterogeneously catalyzed reaction (equation **9.41**), however, sparked interest in trying to discover a

[55] G. Stork and D. E. Kahne, *J. Am. Chem. Soc.*, **1983**, *105*, 1072.

procedure to insert CO directly into the C–O bond of methanol (**9.42**). Such a process would then be based entirely on synthesis gas if a suitable catalyst could be found.

$$2 \ H_2 + CO \rightarrow CH_3OH \hspace{4cm} \textbf{9.41}$$

$$CH_3-OH + CO \rightarrow CH_3-\overset{\overset{\displaystyle O}{\|}}{C}-OH \hspace{3cm} \textbf{9.42}$$

In the mid 1960s, the German-based chemical company BASF, developed a methanol carbonylation process (equation **9.42**) using a mixture of $Co_2(CO)_8$ and HI as catalyst. Unfortunately, severe conditions (210° and 700 atm) were required to produce CH_3OH rapidly enough and in sufficient amount to be commercially acceptable. A few years later Monsanto announced a significant breakthrough in the quest to produce acetic acid directly from methanol. Instead of Co, an Rh–HI system provided the necessary catalytic activity to form acetic acid as virtually the sole product at 180° and 30 to 40 atm. The molar concentration of catalyst required, moreover, was about 100 times less than for the Co-catalyzed process. Once Monsanto perfected its new procedure on a tonnage scale, all previous operations became rapidly obsolete. Again we see an example of how rhodium chemistry is superior to that of cobalt.

Scheme **9.8** diagrams catalytic cycles involved in methanol carbonylation. Note that there are actually two major interlocking cycles—one involving the Rh complex and the other designed to generate CH_3I from CH_3OH. Almost any Rh(III) salt and source of I^- are suitable precursors for generating the active catalytic complex, $[Rh(CO)_2I_2]^-$ (**44**). The overall rate of carbonylation is independent of the concentration of CO but dependent upon Rh and iodide concentrations according to the following:

$$\text{Rate} = k[\textbf{44}][CH_3I]$$

The rate law is consistent with other observations that indicate oxidative addition of CH_3I to **44** (step a) is the rate-determining step, consisting of S_N2 displacement (see Chapter 7, Section **7-2-2**) of I^- on CH_3I by the negatively charged and highly nucleophilic **44**.[56] After OA of CH_3I, facile methyl migration (step b), addition of one more CO ligand (step c), and reductive elimination occur to regenerate **44** and produce acetyl iodide (step d)—which

---

[56]Consistent with the S_N2 mechanism are the observations that the rate of reaction is 10 times slower when Br^- is the co-catalyst instead of I^- and that the rates of carbonylation of methyl, ethyl, and n-propyl alcohols parallel the rates of ordinary S_N2 substitution of the corresponding alkyl halides with methyl > ethyl > n-propyl.

subsequently undergoes rapid hydrolysis to acetic acid (step e). The by-product of the last step, HI, then reacts with additional CH_3OH to yield CH_3I (step f), allowing a new carbonylation cycle to occur.[57]

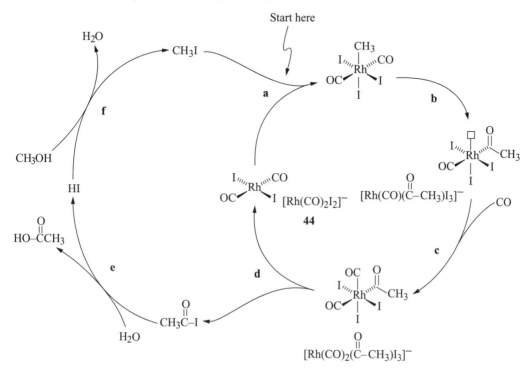

Scheme 9.8
Mechanism of Rh-Catalyzed
Carbonylation of Methanol

Exercise 9-8

The rate of carbonylation of 2-propanol to give a mixture of butanoic and 2-methylpropanoic acids is actually a little faster than that for carbonylation of ethanol and up to seven times more rapid than that for 1-propanol. If OA is still the rate-determining step in the pathway, what can be said about the mechanism of that step when 2-propanol is the starting material? [Hint: See Chapter 7, Section **7-2-3.**]

[57]For a good discussion of the mechanism of Rh-catalyzed carbonylation, see a paper written by chemists who were instrumental in developing the process at Monsanto: D. Forster and T. W. Dekleva, *J. Chem. Ed.*, **1986**, *63*, 204 and references therein.

Since acetic anhydride is more useful to the chemical industry than acetic acid, there was economic incentive to develop a process that would yield the anhydride directly without first producing the acid as a separate operation. By the early 1980s, Eastman Chemicals in conjunction with Halcon Chemical Company developed a procedure that provided acetic anhydride using technology similar to the Monsanto process, and since 1983 a plant run by Eastman has produced anhydride in excess of 225,000 metric tons per year. The Eastman-Halcon (E-H) operation amounts formally to inserting CO into the C-O bond of methyl acetate according to equation **9.43**.[58]

$$
CH_3-O-\overset{\overset{\displaystyle O}{\|}}{C}CH_3 + CO \rightarrow CH_3-\overset{\overset{\displaystyle O}{\|}}{C}-O-\overset{\overset{\displaystyle O}{\|}}{C}-CH_3 \qquad\qquad \textbf{9.43}
$$

We can best understand the process by looking at Scheme **9.9**, which is largely a revisit to Scheme **9.8**. Note that the reagents in brackets are those for the Monsanto methanol carbonylation. The E-H process uses methyl acetate instead of methanol as one of the starting materials and acetic acid rather than H_2O to react with acetyl iodide at the end.

In spite of the similarities between the two carbonylations, there are some significant differences. The E-H scheme requires a reducing atmosphere of H_2 to be present in order to keep sufficient Rh in the form of **44**.[59] A second difference is that the E-H process uses a cationic promoter.[60] Exhaustive studies indicate that Li^+ is the best cation for the job, and, at low concentrations, it plays a role in the overall kinetics of the carbonylation. A third cycle in Scheme **9.9** attempts to rationalize the role of Li^+ in the overall reaction, based on work by researchers at Eastman.

The Monsanto and E-H processes represent triumphs in the application of organotransition metal chemistry to catalysis, using homogeneous transition metal compounds to promote production of valuable materials cheaply, efficiently, and selectively. Without question, fundamental work accomplished previously on the understanding of basic organometallic reaction types helped tremendously the efforts required to unravel the intricacies of these and other cycles we have already considered. These catalytic processes and those to be considered in later chapters provide beautiful illustrations of not only the basic reaction types in action, but also how scientific investigations can lead to practical applications of major economic significance.

---

[58]S. W. Polichnowski, *J. Chem. Ed.*, **1986**, *63*, 206 and J.R. Zoeller, J.D. Cloyd, J.L. Lafferty, V.A. Nicely, S.W. Polichnowski, and S.L. Cook, *Rhodium-Catalyzed Carbonylation of Methyl Acetate* in *Homogeneous Transition Metal Catalysis*, W.R. Moses and D.W. Slocum, Eds., American Chemical Society, Washington, D.C., 1992, 395-418.

[59]Unless H_2 is present in low concentration, the anhydrous conditions of the E-H process produce $[Rh(CO)_2I_4]^-$, an inactive catalyst.

[60]The presence of Li^+ apparently serves to maximize the amount of Rh present in the form of **44**.

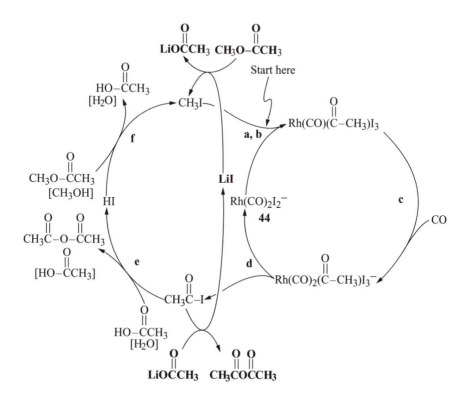

Scheme 9.9
Eastman-Halcon Carbonylation
of Methyl Acetate

9-6   Future Directions for Transition Metal Catalysis

Although vigorous efforts to develop and commercialize novel transition metal-catalyzed processes continue, relatively few new ones have come on stream over the last decade. The causes of this phenomenon are probably more economic than scientific in origin. Worldwide, the chemical industry is already one of most efficient manufacturing enterprises extant. Hydroformylation and the Monsanto acetic acid synthesis, for example, both provide product with high selectivity and efficiency. The proven effectiveness of current processes and the sluggish economy, which has persisted over the last few years, make companies reluctant to invest in new technology unless it clearly is superior to the old. Yet as the economy improves, catalytic processes just now being developed will become more attractive commercially. The impact of current operations on the environment, more-over, will undoubtedly provide impetus either to use recently discovered and more ecologically sound processes or to develop totally new approaches that minimize pollution with the least cost to the manufacturer. In this sec-tion we will examine two areas where the use of homogeneous transition metal catalysts holds promise for the future of the chemical industry.

9-6-1   Processes Based upon Synthesis Gas

Fischer-Tropsch Synthesis of Hydrocarbons

We have already seen how synthesis gas (or more simply, "syngas") is used as a source of carbon, hydrogen, and oxygen in hydroformylation or the synthesis of methanol and acetic acid. Syngas is attractive as a feedstock because it derives from not only petroleum, natural gas, or coal, but also biomass. Because of the world's seemingly insatiable demand for petroleum-based fuels, using syngas to produce gasoline or diesel fuel seems highly desirable. Based on the work of Fischer and Tropsch[61] in Germany about 70 years ago, there is indeed a process available that yields liquid hydrocarbons suitable for use as fuels, according to equation **9.44** below:

$$CO + H_2 \xrightarrow[\substack{20\ atm,\ Fe}]{220\text{-}240°} \begin{array}{l} CH_3(CH_2)_nCH_2OH \quad + \quad CH_3(CH_2)_nCH=CH_2 \\ \quad\quad + \quad CH_3(CH_2)_nCH_3 \quad + \quad H_2O \end{array} \qquad \textbf{9.44}$$

The Fischer-Tropsch (F-T) synthesis was commercialized in 1936 and supple-mented Germany's supply of gasoline, diesel fuel, and other petrochemicals throughout World War II. Since 1945, only South Africa has used the F-T reaction on a large scale to produce fuels and petrochemicals—not only because it possesses large reserves of coal (the source of syngas), but also because it wishes to remain independent of external oil supplies. Even under the most favorable conditions using the best heterogeneous catalysts, however, application of the F-T reaction is not yet economically feasible elsewhere.

Although usually promoted by transition metals (e.g., Co or Fe), the F-T reaction is an example of heterogeneous catalysis when employed commer-cially. Homogeneous catalysts using these same metals have been developed that mimic the results of the heterogeneous process, yet these are not satisfac-tory on an industrial scale.[62] Studies using homogeneous catalysts have, how-ever, helped to shed light upon the mechanism of the F-T reaction, but these investigations are beyond the scope of our discussion.

[61]F. Fischer and H. Tropsch, *Brennst. Chem.*, **1923**, *4*, 276 and Ger. Patent, 484 337, 1925.
[62]For a good discussion of the F-T reaction and studies using homogeneous catalysts, see C. Masters, *Adv. Organomet. Chem.*, **1979**, *17*, 61 and J.P. Collman, *et al.*, *op. cit.*, 653-660.

Synthesis of Ethylene Glycol

Because homogeneous catalysts often afford higher selectivity in the production of specific products, chemists have directed a great deal of research effort toward using these catalysts to promote conversion of syngas to certain homologated[63] compounds—more specialized than long-chain hydrocarbons. One such compound is ethylene glycol ($HO\text{-}CH_2\text{-}CH_2\text{-}OH$), used in making polyester fibers and films and serving as the main ingredient for antifreeze in automobiles. According to equation **9.45**, ethylene glycol could result directly from the reaction of two equivalents of CO with three of H_2.

$$2\ CO + 3\ H_2 \rightarrow HO\text{-}CH_2\text{-}CH_2\text{-}OH \qquad\qquad\qquad \textbf{9.45}$$

Researchers at Union Carbide successfully demonstrated that a soluble Rh-carbonyl complex could catalyze formation of ethylene glycol from syngas mixtures.[64] Due to the severe reaction conditions (>200° and >500 atm), however, the process was not commercially feasible. The mechanism of the reaction, which must account for the homologation, is not known. Zirconium complexes, such as Cp_2ZrH_2, can promote the formation of ethylene glycol when present in stoichiometric amounts.[65] Evidence indicates that the reaction occurs according to Scheme **9.10** where, due to the oxophilic[66] nature of early

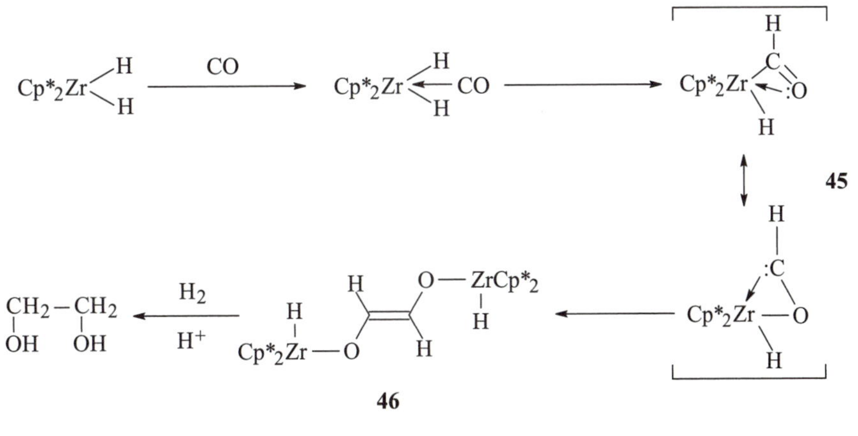

Scheme 9.10
Zr-Promoted Formation of
Ethylene Glycol from Synthesis Gas

[63]Homologation is the process of adding carbon atoms to a moiety or group already present, e.g., building up a carbon chain using CO as the carbon source.
[64]R.E. Pruett and W.E. Walker, U.S. Patent 3 957 857, 1976.
[65]J.M. Manriquez, D.R. McAlister, R.D. Sanner, and J.E. Bercaw, *J. Am. Chem. Soc.*, **1978**, *100*, 2716.
[66]Early transition metals, especially in a highly oxidized state, have a high affinity for oxygen in any form and combine with that element to produce strong M-O bonds.

transition metals such as Zr, an unstable carbene-like intermediate (**45**) is stabilized by carbonyl oxygen through resonance. Dimerization of **45** gives **46**; subsequent hydrogenation and protonation yields the glycol. It is unclear whether the mechanism involving Zr complexes is similar to that with Rh.

Water Gas Shift Reaction

For some processes, H_2 is the more valuable of the two components of syngas. Certainly procedures to yield methanol or ethylene glycol require an excess of H_2 over CO. The Haber-Bosch process for producing NH_3 from H_2 and N_2 requires the absence of CO entirely. Thus it is clear that chemistry designed to obtain an excess of H_2 over CO would have great economic value. Equation **9.46** below (also **9.1**) describes the Water Gas Shift Reaction (WGSR).[67]

$$CO + H_2O \rightarrow H_2 + CO_2 \hspace{4cm} \textbf{9.46}$$

The reaction is quite slow at room temperature, but the equilibrium favors products (ΔG = -6.7 kcal/mole). The industrial version of the WGSR is run using a heterogeneous catalyst such as chromium-activated iron oxide or Co-Mo oxides and requires rather severe conditions (>400° and >200 atm). The high temperature required coupled with the negative entropy change (ΔS_{298} = -10.0 eu, making the $T\Delta S$ term large and positive at high temperatures) associated with the WGSR means that the position of equilibrium becomes much less favorable toward the product side under such reaction conditions. In recent years chemists have discovered several soluble transition metal catalysts, such as Ru(II)-carbonyl[68] or Pt[P(i-Pr)$_3$]$_3$,[69] that allow the WGSR to occur under milder conditions than those used on an industrial scale. Investigations on the behavior of these catalysts have provided useful information about the mechanism of the WGSR under both homogeneous and heterogeneous conditions. None of these catalysts is used commercially, but the mechanistic studies that followed their discovery are worth comment because important organometallic reaction types are involved.

Scheme **9.11** depicts a catalytic cycle proposed for an Ir-catalyzed WGSR reported by Ziessel.[70] The use of visible light in addition to the transition metal catalyst allowed the reaction to proceed at room temperature, atmospheric pressure, and neutral pH—mild conditions indeed! Once a CO ligand is attached to the metal (step a), water (or OH⁻) nucleophilically adds to the ligand (step b). The rate-determining step in this case is the loss of CO_2 (step c).

[67]The by-product from the WGSR is CO_2, which, due to its acidity, may be removed from the gaseous mixture by "scrubbing" with a base such as lime.
[68]M.M. Taqui Khan, S.B. Halligudi, and S. Shukla, *Angew. Chem. Int. Ed. Engl.*, **1988**, *27*, 1735.
[69]T. Yoshida, Y. Ueda, and S. Otsuka, *J. Am. Chem. Soc.*, **1978**, *100*, 3941.
[70]R. Ziessel, *Angew. Chem. Int. Ed. Engl.*, **1992**, *30*, 844.

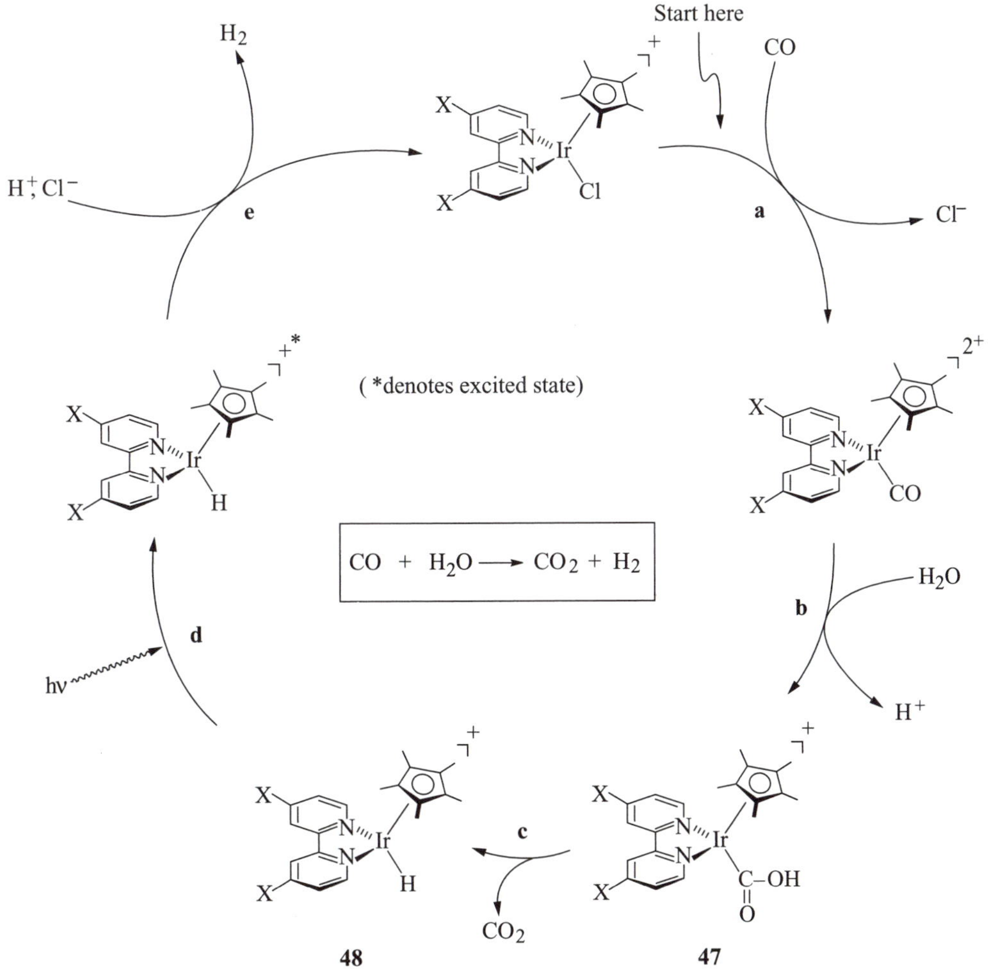

Scheme 9.11
Ir-Catalyzed Photochemical
Water Gas Shift Reaction

It is tempting to ascribe the loss of CO_2 to a β–elimination process, but this would require a loss of ligand in order to create an active site. This is not likely since the ligands already present are Cp^* and bipyridyl (bipy).

Exercise 9-9

Why is the loss of ligand in order to create a vacant site in complex **47** unlikely?

Equation **9.47** below describes a more reasonable pathway for loss of CO_2 in which the η^1-CO_2 ligand is replaced by H^+ supplied by the reaction medium.

$$9.47$$

Support for this pathway rests upon the observation that, when the X group on the bipy ligand in **47** is electron withdrawing, the rate of CO_2 loss increases. Upon photoactivation (step d), the Ir complex gains a proton and then loses H_2 by reductive elimination (step e).

Exercise 9-10

Why should an electron withdrawing X group in **47** increase the rate of CO_2 loss?

Scheme **9.12** describes an alternate pathway for Pt complexes[71] in which water adds oxidatively to the metal (step b). Because a CO ligand is also attached, a carbonyl insertion may occur to give intermediate **49** (step c). Decarboxylation and protonation (step d) then yield dihydride **50**, which loses H_2 by RE (step e).

9-6-2   Specialty Chemicals

It should be clear by now that, except for a few prominent examples, most large scale industrial chemical processes are catalyzed by heterogeneous catalysts, due to their advantages of lower cost and greater separability from products. Heterogenous catalysts, because they often lack the selectivity of soluble transition metal complexes, promote reactions that yield product mixtures. These mixtures, which would be unacceptable in a laboratory setting, are often tolerable and even desirable in an industrial venue—provided the product properties are suitable for the application. As the science of organometallic

---

[71]T. Yoshida, *et al.*, *J. Am. Chem. Soc.*, **1978**, *op. cit.*

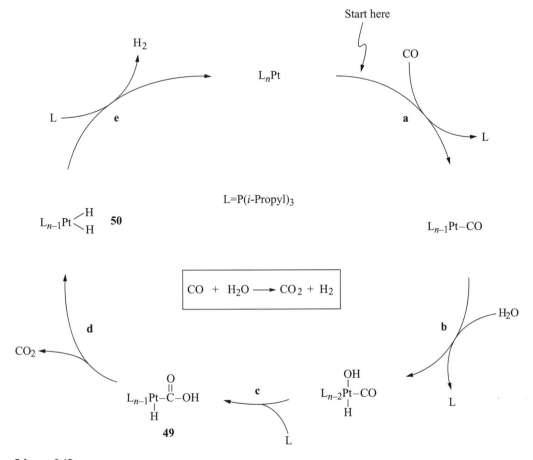

Scheme 9.12
An Alternate Mechanism for
the Water Gas Shift Reaction

chemistry has developed over the last two decades, it has become apparent that transition metal complexes can serve as extremely efficient and highly selective catalysts. Not only can these catalysts control the chemoselectivity of the reaction but also its regioselectivity and—more importantly—stereoselectivity.

The demand for new compounds (specialty chemicals) that can alleviate human ailments and safely control agricultural pests continues to increase. Many of these compounds are chiral, sometimes possessing several stereogenic centers. Usually, only one enantiomer or diastereomer is active. Racemic modifications are unacceptable because 50% of the drug or pesticide is at best useless and often even toxic to the patient or the environment. The chemical industry and particularly manufacturers of pharmaceuticals and agricultural chemicals have focused their efforts recently on developing homogeneous catalysts that can control the selectivity of key steps in the synthesis of these specialty compounds. The expense of catalyst material and the cost of developmental research are both high in designing these selective catalysts. In these endeavors, however, selectivity becomes more important than the economies derived from use of heterogeneous catalysts. The rest of this section will briefly explore two examples where homogeneous catalysis played a major role in the successful synthesis of biologically important molecules of commercial significance.

Naproxen

We have already seen (equation **9.38**) how asymmetric hydrogenation was used in one step of the synthesis of L-Dopa. Equation **9.48** shows the last step of a synthesis developed by Monsanto[72] where asymmetric hydrogenation yields naproxen, a widely-prescribed (now available over the counter), non-steroidal, anti-inflammatory drug.[73] (-)-BINAP[74] (**51**) serves as the chiral auxiliary ligand for the Ru(II) complex (**52**), which catalyzes the formation of the S enantiomer[75] in 96 to 98% ee.

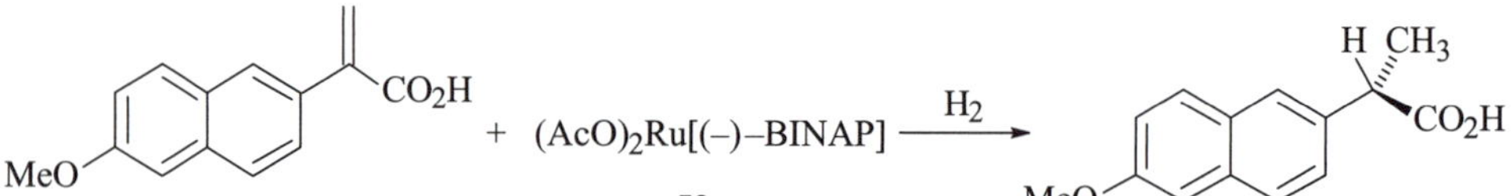

9.48

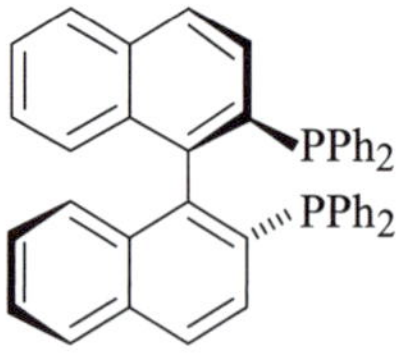

---

[72]Monsanto, *World Pat.*, WO9,015,790A, **1991**.

[73]Naproxen is now available as the sodium salt in an over-the-counter form, which is called naproxen sodium.

[74]T. Ohta, H. Takaya, M. Kitamura, K. Nagai, and R. Noyori, *J. Org. Chem.*, **1987**, *52*, 3174. Note that BINAP lacks stereogenic centers but is chiral nonetheless.

[75]The R enantiomer is a liver toxin.

(–)–Menthol

(-)-Menthol (**53**) is an important fragrance ingredient that finds use in food and cold remedies. Although it is obtainable from natural sources, workers at the Takasago Perfumery in Japan developed a route to **53** outlined in Scheme **9.13**.[76] Run on tonnage scale, the process now accounts for about one-third of the world's supply of (-)-menthol. The key step in the synthesis is the stereoselective isomerization of (*E*)-vinylamine **54** to enamine **55**. This step induces chirality at the three-position and is catalyzed by [Rh(-)-BINAP(COD)]ClO$_4$.

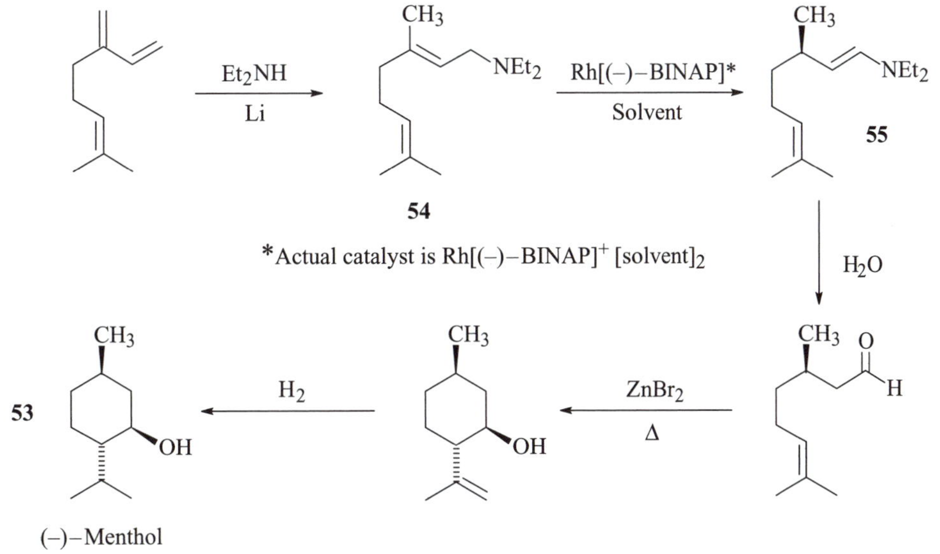

Scheme 9.13
Chiral Synthesis of
(-)-Methanol

Double bond migration such as that involved in **54** going to **55** is an important reaction, useful on both an industrial and laboratory scale; its mechanism has been well studied. Equations **9.49** and **9.50** describe two major pathways for migration—the former involving metal hydride insertion and β-elimination steps and the latter a 1,3-hydride shift via a η^3 π-allyl intermediate. The researchers at Takasago claimed a third mechanism for bond migration, outlined in equation **9.51**, which they termed "nitrogen triggered."

Scheme **9.14** describes the nitrogen-triggered pathway. Steps a and b provide an activated Rh-BINAP complex (**56**) by associative substitution of

[76]S. Inoue, H. Takaya, K. Tani, S. Otsuka, T. Sato, and R. Noyori, *J. Am. Chem. Soc.*, **1990**, *112*, 4897 and R. Noyori and H. Takaya, *Acc. Chem. Res.*, **1990**, *23*, 345.

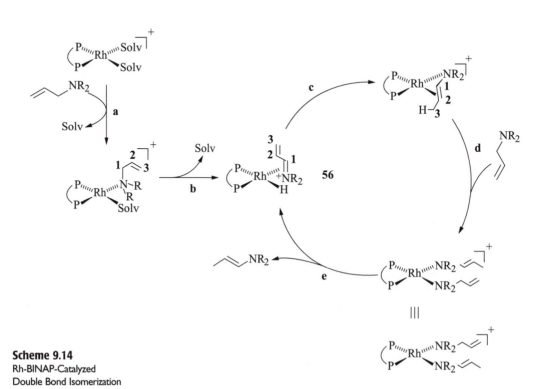

Scheme 9.14
Rh-BINAP-Catalyzed
Double Bond Isomerization

solvent (EtOH or THF) and substrate followed by β-elimination to form an iminium complex. The stereochemically key step, c, adds a proton to the three-position and gives the complexed enamine. In step d the η^3-enamine ligand shifts to η^1 as a new molecule of allyl amine adds to the complex. The nitrogen plays a key role in activating the Rh, particularly in step d, the rate-determining step of the overall cycle. Step e liberates enamine product and regenerates **56**.

Exercise 9-11

Predict the product of the following transformation. Indicate the expected stereochemistry of the product.

$$^+Rh[(+)-BINAP]$$

(E)-1-Diethylamino-3-methyl-2-hexene $\xrightarrow{\hspace{3cm}}$

The isomerization of the double bond in the synthesis of (-)-menthol is completely stereospecific. Evidence obtained using vinyl amine specifically deuterated at C-1 indicates that the transfer of a proton from C-1 to C-3 is suprafacial,[77] as indicated by equation **9.52** below:

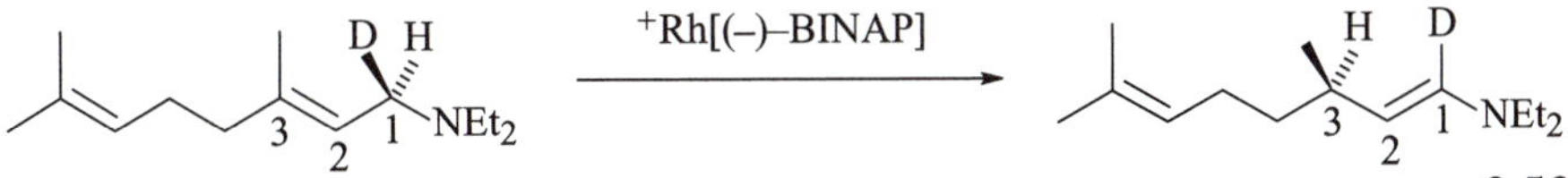

9.52

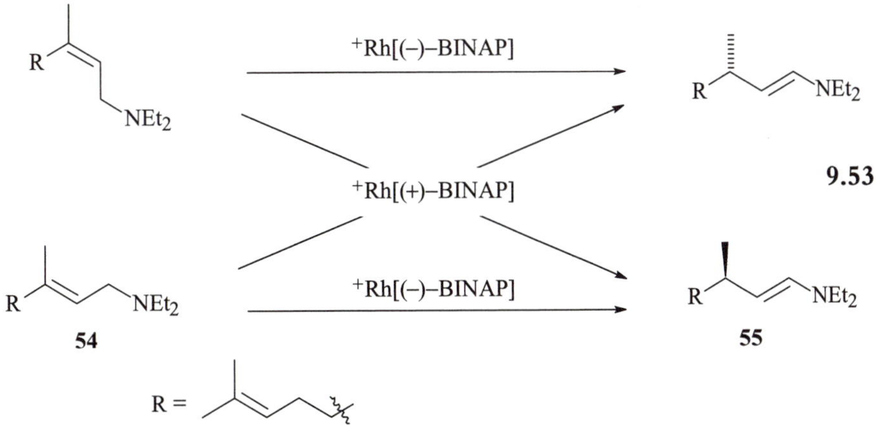

9.53

54   55

R =

---

[77]With regard to sigmatropic shifts such as a 1,3–hydride shift, the word "suprafacial" means that the migrating group moves along the same face of the π system from one end to the other. An "antarafacial" migration results in the movement of the group to the opposite face of the π system. See J.E. McMurry, *Organic Chemistry*, 4th. ed. Brooks/Cole, Pacific Grove, CA, 1996, Chapter 31, especially 1230-1236, and Chapter 10 of T.H. Lowry and K.S. Richardson, *Mechanism and Theory in Organic Chemistry, op. cit.*, for a more detailed definition with other examples.

Equation **9.53** shows how the chirality of the BINAP ligand and the stereochemistry of the trisubstituted double bond in **54** control the stereochemistry of the methyl group at C-3 in **55** upon double bond migration.

Biological Double Bond Migration

It is interesting to note that this Rh-catalyzed isomerization mimics a similar double bond migration found in nature in the biosynthesis of cholesterol and terpenes in which isopentenyl pyrophosphate (**57**) is converted reversibly to dimethylallyl pyrophosphate (**58**), shown in equation **9.54**.[78] The stereochemistry of this reaction is also suprafacial.

$$CH_3 \quad H_E \quad H_Z \quad H_S \quad H_r \quad OPP \xrightarrow[\text{Isomerase}]{} \quad H_E \quad CH_3 \quad H_Z\!-\!C \quad H_r \quad H_S \quad OPP$$

57   OPP = Pyrophosphate   **58**   **9.54**

The syntheses of Naproxen and (-)-menthol represent but two of the numerous uses of soluble transition metal complexes to catalyze, often stereoselectively, key steps in the production of biologically important compounds in the laboratory or on an industrial scale.[79] Discussions in Chapter 11 explore more fully the use of these catalysts when they describe how organometallic chemistry may be applied to the synthesis of other organic compounds.

---

[78]For a good discussion of the role of double bond isomerization in steroid and terpene biosynthesis, see C.D. Poulter and H.C. Rilling, *Prenyl Transferases and Isomerases*, in *Biosynthesis of Isoprenoid Compounds*, vol. 1. J.W. Porter and S.L. Spurgeon, Eds., Wiley, New York, 1981, 161-224.
[79]See J.W. Parshall and S.D. Ittel, *op. cit.* for a good survey of homogeneous transition metal catalysis applied to the synthesis of specialty chemicals.

Problems

9-1   The complex $Rh(H)(CO)_2(PPh_3)_2$ can be used in the catalytic synthesis of pentanal from an alkene having one less carbon atom. Propose a mechanism for this synthesis. In your mechanism indicate the reaction type of each step. Identify the catalytic species.

9-2   Increasing the concentration of CO during Co-catalyzed hydroformylation not only retards the reaction rate, but also suppresses the tendency of the reaction to add a formyl group (CHO) to a carbon other than those contained in the C=C bond of the starting alkene (i.e., an isomerization of the double bond does not occur during the catalytic cycle). Explain.

9-3   Draw a plausible catalytic cycle for the hydrogenation of (Z)-2-butene using the cationic complex $[Rh(I)\text{-}DIPHOS]^+$ that was mentioned in Section **9-4-2**.

9-4   Propose a mechanism for step e of Scheme **9.14**. This transformation requires more than one step. Be sure to count electrons for each species in your pathway.

9-5   Below is the catalytic cycle associated with the Heck olefination reaction. For each step marked with a letter, attach the name of one of the fundamental types of organometallic reactions (e.g., nucleophilic abstraction or ligand substitution).

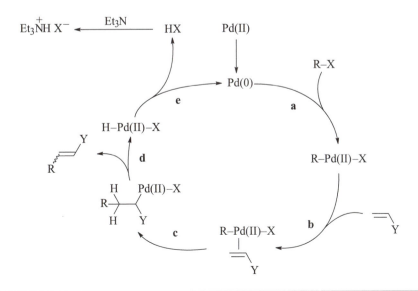

9-6   Consider the transformation of an alkyl bromide to an aldehyde shown below:

1-bromobutane $\longrightarrow$ pentanal + HBr

Assume that the reaction is catalyzed by $HFe(CO)_4^-$ in an atmosphere rich in CO and H_2. Propose a catalytic cycle that would show this catalysis and account for the formation of the two products. Label each step in your cycle with terms corresponding to the fundamental organometallic reaction types.

9-7   The transformation below is reminiscent of Wacker chemistry described in this chapter.

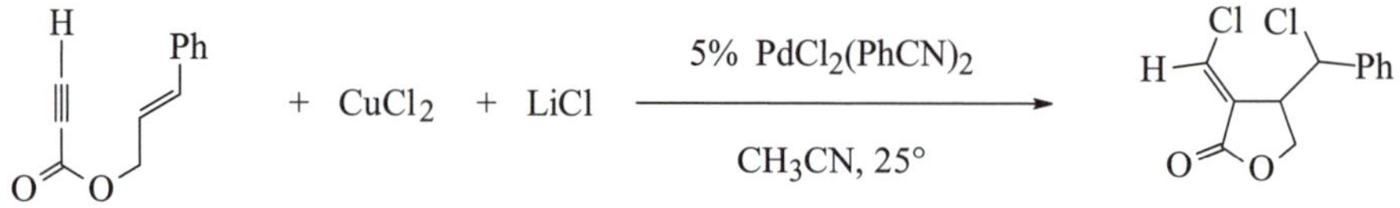

A catalytic cycle has been proposed and is shown below.[80] One of the steps is an example of a relatively rare type of reaction discussed in Chapter 8, Section **8-1-2**. Describe in detail what is happening in each of the four major transformations in the cycle. What is the function of $CuCl_2$?

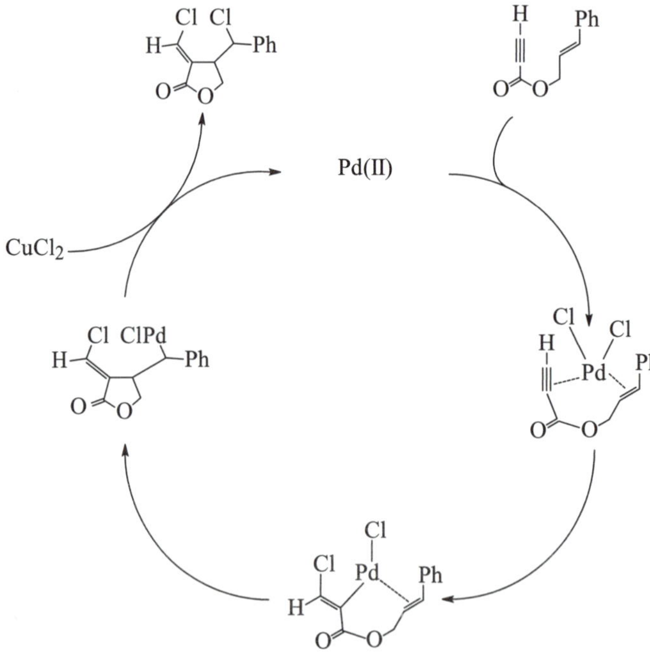

---

[80]S. Ma and X. Lu, *J. Org. Chem.*, **1993**, *58*, 1245.

9-8   Catalytic, selective functionalization of hydrocarbons is one of the most important goals of organometallic chemistry. A recent report in the literature describes the Rh-catalyzed conversion of benzene to benzaldehyde under photochemical conditions.[81]

$$\text{Ph-H} + \text{CO} \xrightarrow{\text{Rh(PMe}_3)_2(\text{CO})\text{Cl}} \text{PhCHO}$$

Assuming that the first step in the catalytic cycle is photo-activation of the Rh complex, propose a possible mechanism for the transformation above.

[81]G.P. Rosini, W.T. Boese, and A.S. Goldman, *J. Am. Chem. Soc.*, **1994**, *116*, 9498.

10

Transition Metal-Carbene and Transition Metal-Carbyne Complexes

Structure, Preparation, and Chemistry; Metathesis and Polymerization Reactions

Disubstituted carbon atoms may bind directly to a transition metal producing a formal double bond between the metal and the carbon. The divalent carbon ligand possesses distinctive properties in comparison to ligands discussed in Chapters 4 to 6, not only with regard to its structure and bonding characteristics, but also in terms of its chemistry. Complexes containing these ligands are called *metal-carbene complexes*. Metal-carbene complexes have the general structure shown in **1**, where X and Y may be alkyl, aryl, H, or heteroatoms (O, N, S, halogens). The first carbene complex (**2**) was reported by Fischer and Maasböl in 1964,[1] and since then chemists have learned much about these interesting compounds (see Chapter 6, Section **6-1-2** for an introductory discussion of metal-carbene complexes).

$$L_nM = C \begin{smallmatrix} X \\ Y \end{smallmatrix} \qquad\qquad (CO)_5W = C \begin{smallmatrix} OMe \\ Ph \end{smallmatrix}$$

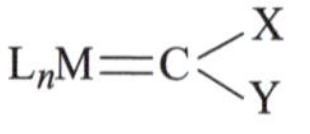

1 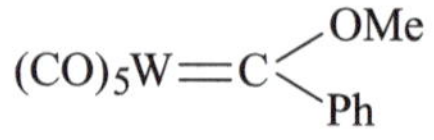   **2**

Metal-carbene complexes undergo numerous reactions, several of which are useful in the synthesis of complex organic molecules. These complexes are also

[1]E.O. Fischer and A. Maasböl, *Angew. Chem. Int. Ed. Eng.*, **1964**, *3*, 580.

intermediates in processes such as *metathesis* (Section **10-5**) and *ring opening metathesis polymerization* (Sections **10-5-3**). In this chapter we will discuss the structure, synthesis, and chemistry of carbene complexes, also emphasizing metathesis and polymerization. Metal-carbene complexes will appear again in Chapter 11, where some of the applications of their chemistry to organic synthesis will be discussed.

10-1   Structure of Metal-Carbene Complexes

The word "carbene" stems from the name given to free, disubstituted carbon compounds with general structure **3** (X and Y same as in **1**).[2] Because the central carbon atom does not possess an octet of electrons, free carbenes are electron deficient and extremely reactive. They are so reactive that some carbenes insert themselves into normally inert alkane C-H bonds (equation **10.1**) or react with alkenes to form cyclopropanes (equation **10.2**), a synthetically useful transformation.

$$:C\!\!\begin{array}{c} \diagup X \\ \diagdown Y \end{array}$$

3

$$CH_3\text{--}CH_2\text{--}\overset{\overset{\displaystyle H}{\big|}}{C}H\text{--}CH_2\text{--}H \;+\; :CH_2 \longrightarrow CH_3\text{--}CH_2\text{--}\overset{\overset{\displaystyle H}{\overset{\big|}{\underset{\big|}{C}H_2}}}{C}H\text{--}CH_3 \qquad \textbf{10.1}$$

$$+ \; CH_3\text{--}CH_2\text{--}CH_2\text{--}CH_2\text{--}CH_2\text{--}H$$

$$\overset{H}{\underset{Me}{\diagdown}}C\!\!=\!\!C\overset{H}{\underset{Me}{\diagup}} \;+\; :CCl_2 \longrightarrow \begin{array}{c} Cl \diagdown \underset{}{C} \diagup Cl \\ H \diagdown C\!\!-\!\!C \diagup H \\ Me \diagup \qquad \diagdown Me \end{array} \qquad \textbf{10.2}$$

Transition metal-carbene complexes are not produced from free carbenes; moreover, they do not produce free carbenes. It is, however, useful to think of metal-carbene complexes as a construct of a free carbene and a metal fragment. Before we actually discuss the structure of carbene complexes, we will review some information regarding free carbenes.

---

[2]We will call such compounds *free carbenes* to distinguish them from metal complexes possessing divalent carbon ligands.

Free carbenes exist in two different electronic states: singlet and triplet,[3] which are represented in structures **4** and **5**. The singlet state has one lone electron pair, while the triplet state has two unpaired electrons.

singlet	triplet
4	**5**

We can think of the singlet state of a free carbene, in terms of localized hybrid orbitals, as a bent molecule (**6**) harboring three sp^2 orbitals (one of which is doubly occupied and nonbonding) and an empty, perpendicularly situated $2p$ orbital. A triplet carbene (really a diradical) would be a less-bent species (**7**) with two singly occupied, orthogonal $2p$ orbitals and two sp orbitals that are involved in bonding to substituents.

6	**7**

Figure **10-1** shows the molecular orbital picture of a free carbene. Note that for the singlet state (shown on the right side of the figure) there are two bonding orbitals, MO-1 and MO-2 (the HOMO resembles an sp^2 orbital), and a relatively low-lying empty orbital, MO-3, (the LUMO resembles a $2p$ orbital). If MO-2 and MO-3 are close in energy, the two orbitals tend to be singly occupied and a triplet state exists (shown on the left side of the figure). If the energy gap widens, then the ground state of the carbene consists of a doubly-occupied MO-2 and an empty MO-3, a singlet state.

Whether a free carbene exists in the ground state as a singlet or as a triplet depends upon the nature of the substituents, X and Y, that are attached to carbon. When X,Y = alkyl or H, the triplet state is usually the ground state. A singlet ground state occurs, on the other hand, when X and Y are heteroatoms such as N, O, S, or the halogens. Figure **10-2** describes the influence of a heteroatomic substituent such as Cl (the MOs shown on the left side of

---

[3]The designations singlet and triplet refer to the spin multiplicity of the electronic state. Spin multiplicity is calculated by 2S + 1 = multiplicity, where S is the total spin for all the electrons. The spin of the first electron in a pair is arbitrarily designated + ½, while the second is −½. For every electron pair, S = ½+ (-½) = 0. If all electrons in a species are paired, S = 0, and the multiplicity is 2(0) + 1 = 1, which signifies a singlet. On the other hand, if two electrons are unpaired, S = ½ + ½ = 1. The multiplicity is 2(1) + 1 = 3, which designates a triplet.

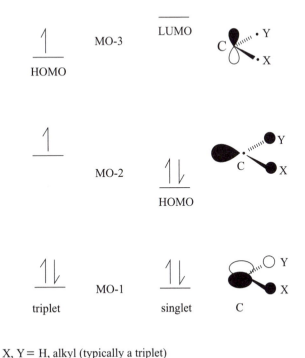

X, Y = H, alkyl (typically a triplet)

X, Y = Cl, O, N, S (typically a singlet)

Figure 10-1
Molecular Orbital Diagram for
Triplet and Singlet Free Carbene

Figure **10-2** are the same as those in Figure **10-1**). Imagine that a Cl atom mixes with a free carbene (:CH$_2$ in this case) to give :CHCl. This mixing results in drastic lowering in energy of MO-3 to give a new π MO called MO-3′. MO-2 is lowered slightly due to the electronegativity of Cl to give a new σ MO designated MO-2′. The result is a considerable energy gap between MO-2′ (the HOMO) and MO-3′* (the LUMO). The gap is large enough that a singlet (no unpaired electrons) now becomes the ground state when Cl is attached to carbon. This effect is general for all heteroatoms such as N, O, S, and the halogens. Substituents such as a hydrogen atom or an alkyl group have approximately the same electronegativity as carbon and also lack filled π orbitals, so the interactions to lower the σ and raise the π MOs do not occur; a triplet state results.

In Chapter 6, Section **6-1-2** it was mentioned that, just as there are two types of free carbenes, there are also two kinds of metal-carbene complexes. As with free carbenes, the two varieties of metal-carbene complexes depend upon the nature of the substituents X and Y. When either or both substituents bound

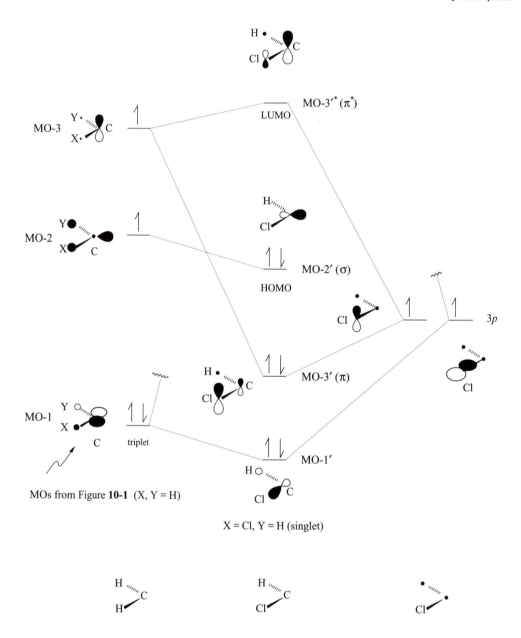

Figure 10-2
Influence of a Heteroatom Substituent
on the Electronic State
of a Free Carbene

to $C_{carbene}$[4] are heteroatomic, the resulting complex is called a *Fischer-type* carbene complex (structure **2** below). Several years after the first Fischer-type carbene was reported, Schrock and co-workers[5] discovered species where the substituents X and Y attached to $C_{carbene}$ were H or alkyl. Such metal-carbene complexes have since become known as *Schrock-type* carbene complexes or *alkylidenes* (structure **8** shows an example of a Schrock-type carbene com-

$$(CO)_5\,W = C \begin{cases} OMe \\ Ph \end{cases}$$

2

8

Fischer-type   Schrock-type

plex). The two kinds of carbene complexes differ in several ways. As will be discussed in Section **10-3**, Fischer-type metal-carbene complexes tend to undergo attack at $C_{carbene}$ by nucleophiles, and thus are termed *electrophilic* (equation **10.3**). Schrock-type carbene complexes, on the other hand, undergo attack by electrophiles at $C_{carbene}$, and are considered to be *nucleophilic* species (equation **10.4**). Because of their chemical behavior, Fischer-type carbene complexes are more properly called *electrophilic metal-carbene complexes* and Schrock-type carbenes *nucleophilic metal-carbene complexes*. We will use the two names for each type of carbene complex interchangeably. Table **10–1** summarizes some other differences between the two carbene complex types.

$$(CO)_5W = C \begin{cases} OMe \\ Ph \end{cases} \quad + \quad :Nuc^- \longrightarrow (CO)_5\overset{-}{W}-\underset{Ph}{\overset{OMe}{\underset{|}{\overset{|}{C}}}}-Nuc$$

10.3

$$Cp_2(CH_3)Ta = CH_2 \;+\; E^+ \longrightarrow Cp_2(CH_3)\overset{+}{Ta}-CH_2-E$$

10.4

[4]The carbon atom directly attached to the metal in a metal-carbene complex is called $C_{carbene}$.
[5]R.R. Schrock, *J. Am. Chem. Soc.*, **1974**, *96*, 6796.

Table 10-1 Fischer- and Schrock-Type Carbene Complexes

Characteristic	Fischer-Type	Schrock-Type (Alkylidenes)
Typical metal [ox. state]	Middle to late transition metal [Fe(0), Mo(0), Cr(0)]	Early transition metal [Ti(IV), Ta(V)]
Substituents attached to $C_{carbene}$	At least one electronegative heteroatom, e.g., O or N	H or alkyl
Typical ligands also attached to metal	Good π acceptor, e.g., CO	Good σ or π donor, e.g., Cp, Cl, alkyl
Electron count	18 e$^-$	10-18 e$^-$
Typical chemical behavior	Nucleophile attacks at $C_{carbene}$	Electrophile attacks at $C_{carbene}$
Ligand type	L	X_2

There are several approaches to understanding the structure of metal carbene complexes. Perhaps the simplest and most familiar is resonance theory. Structures **9** to **12** represent several possible contributing resonance structures of metal carbene complexes. Structures **10** and **11** seem to be important contributors for Fischer-type carbene complexes as indicated by experiments and calculations.

For example, calculations indicate that the barrier to rotation about the

M–C bond of $(CO)_5Cr=C(OH)(H)$ is less than 1 kcal/mole.[6] In carbene **13a** the barrier to rotation about the C–N bond is considerably higher at 25 kcal/mole. The C–N bond distance, moreover, is significantly less than that expected for nitrogen attached to an sp^2 hybridized carbon by a single bond (suggesting a C=N bond, as shown in structure **11** above or **13b** on the next page).

[6]H. Nakatsuji, J. Ushio, S. Han, and T. Yonezawa, *J. Am. Chem. Soc.*, **1983**, *105*, 426.

$$(CO)_5Cr=C\overset{NMe_2}{\underset{Me}{\diagdown}} \longleftrightarrow (CO)_5\overset{-}{Cr}-C\overset{\overset{+}{NMe_2}}{\underset{Me}{\diagdown}}$$

13a   **13b**

It should be pointed out, however, that resonance theory is misleading regarding the true nature of the M–C bond in a Fischer-type carbene complex. It turns out that the barrier to rotation about the M–C bond is low because a π bond exists regardless of the degree of rotation. Molecular orbital theory (described later in this chapter) is much better at explaining the apparent dichotomy between the presence of a formal M=C bond and a low rotational energy barrier.

The picture is quite different for nucleophilic metal-carbene complexes. Here contributing structures **9** and **12** seem to make the most contribution to the overall structure. Support for this observation comes from temperature dependent NMR measurements[7] of the M–C rotational barriers of various Ta carbenes. The values obtained range from 12 to 21 kcal/mole, and seem to indicate considerable double bond character (structure **9**).

Another approach toward understanding the bonding and structure of metal carbenes involves interaction of localized orbitals of the metal and carbon atom. We may envision electrophilic carbene complexes to be the result of bonding of a singlet free carbene to the metal (Figure **10-3a**). The free carbene acts as a σ donor through its filled sp^2 orbital (MO-2, Figure **10-1**) and a π acceptor via back donation of electrons from a filled metal d orbital to the empty $2p$ orbital (MO-3, Figure **10-1**) of the free carbene. The picture is exactly analogous to bonding of a CO ligand, except that carbene is a better σ donor and poorer π acceptor than CO (which makes the M–$C_{carbene}$ bond weaker than an M–CO bond). The carbon in a Fischer-type carbene complex is, therefore, formally an **L-type ligand** donating two electrons to the metal. Upon addition of the free carbene to the metal, the metal's oxidation state remains unchanged.

Figure **10-3b** pictures the formation of a Schrock-type carbene complex. Here a triplet free carbene interacts with the metal, and the double bond forms because two singly occupied $2p$ orbitals from carbon and two singly occupied metal orbitals contribute one electron to form both a σ and π bond. Because highly oxidized, early-transition metals typically are involved, we can think of the carbon fragment as both a **σ and** a π donor, contributing a total of four electrons to the metal (still two electrons according to the neutral ligand scheme). The carbon is thus an **X₂-type ligand** that causes an oxidation state change of +2 on the metal.

7J.D. Fellmann, G.A. Rupprecht, C.D. Wood, and R.R. Schrock, *J. Am. Chem. Soc.*, **1978**, *100*, 5964.

Figure 10-3
(a) Interaction between a Singlet Free Carbene and a Metal
(b) Interaction between a Triplet Free Carbene and a Metal

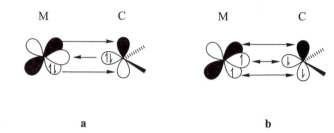

a   b

Molecular orbital calculations have been reported on complexes **14** and **15**, the former a good example of a Fischer-type carbene complex and the latter a Schrock-type.[8]

14   15

Figure **10-4a** shows the interaction of the carbene fragment with appropriate metal orbitals. For the sake of simplification, the free carbene orbitals are shown without interaction of the heteroatom (see Figure **10-1**). The key features of this interaction include σ donation from the carbon (MO-2) to the d_{z^2} metal orbital, and back donation of π electrons from the filled d_{yz} metal orbital to the ligand (MO-3). There is significant energy separation of the frontier orbitals[9] in both metal and ligand, which leads to all electrons being paired.

When the metal is niobium, the two metal frontier orbitals, d_{yz} and d_{z^2}, are closer in energy and match the corresponding orbitals of triplet free carbene. Figure **10-4b** provides a picture of the interactions involved (only the important orbital interactions are shown). The electrons in the M–C bond are much more equally distributed between the metal and $C_{carbene}$ (because the energy levels of the metal and carbon orbitals are comparable) than with an electrophilic carbene complex. In resonance terminology this means that structure **9** becomes an important contributor to the overall structure of this Schrock-type carbene complex.[10]

[8]T.E. Taylor and M.B. Hall, *J. Am. Chem. Soc.*, **1984**, *106*, 1575.
[9]The HOMO and the LUMO comprise the frontier orbitals.
[10]There exists a relevant analogy between Schrock-type carbene complexes and ylides. The term *ylide* typically refers to a dipolar compound where carbon is attached to a nonmetal such as N, P, or S. An ylide shows nucleophilic character at the carbon with two resonance structures providing significant contribution to the overall structure of the molecule, i.e., $R_xY=CR_2 \leftrightarrow R_xY^+{-}^-CR_2$. For an introduction to the chemistry of ylides, see J. March, *Advanced Organic Chemistry*, 4th ed. Wiley-Interscience, New York, 1992, 38–40 and 956–963 and references therein.

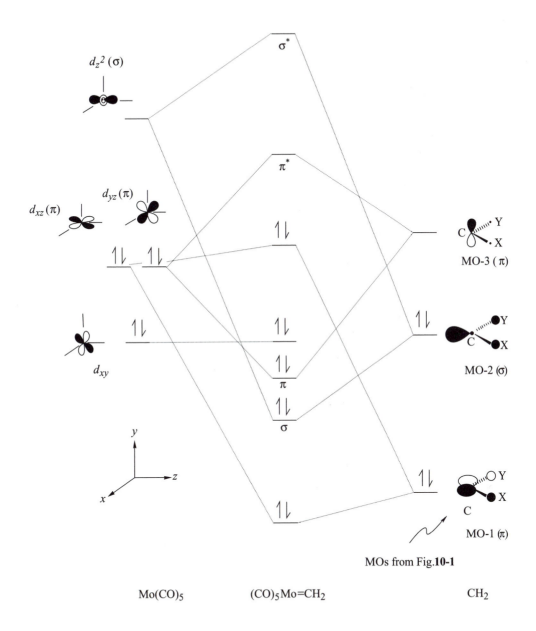

$d_z{}^2 (\sigma)$

$d_{yz} (\pi)$

$d_{xz} (\pi)$

d_{xy}

σ^*

π^*

π

σ

y

z

x

C ⋯Y
⋯X

MO-3 (π)

Y
C
X

MO-2 (σ)

O Y
X
C

MO-1 (π)

MOs from Fig.**10-1**

Mo(CO)$_5$ (CO)$_5$Mo=CH$_2$ CH$_2$

Figure 10-4a
Molecular Orbital Picture
of a Fischer-Type
Carbene Complex

Figure 10-4b
Molecular Orbital
Picture of a
Schrock-Type
Carbene Complex

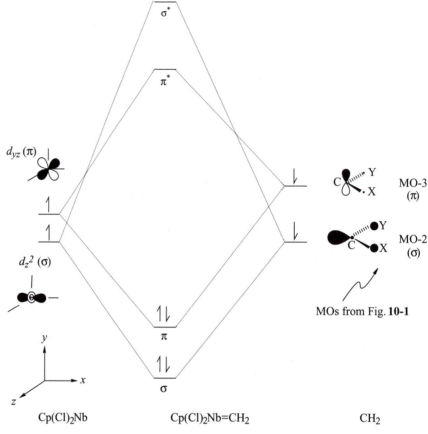

$$Cp(Cl)_2Nb \qquad Cp(Cl)_2Nb{=}CH_2 \qquad CH_2$$

10-2   Synthesis of Metal-Carbene Complexes

10-2-1 Fischer-Type (Electrophilic) Carbene Complexes

There are numerous synthetic routes to Fischer-type carbene complexes.[11]
These routes may be fundamentally classified as one of the following:

    A. Replacement or modification of an existing noncarbene ligand
    B. Modification of an existing carbene ligand

In this section, our discussion focuses on the type A pathway. Since modification of an existing carbene ligand to produce a new carbene complex is an

---

[11]Several good reviews exist that describe synthetic routes to electrophilic carbene complexes.
These include F.J. Brown, *Prog. Inorg. Chem.*, **1980**, *27*, 1; K.H. Dötz, *et al.*, *Transition Metal Carbene
Complexes,* Verlag Chemie, Deerfield Beach, FL, 1983, pp. 1-68; and M.A. Gallop and W.R.
Roper, *Adv. Organomet. Chem.*, **1986**, *25*, 121.

example of a type of reaction carbene complexes may undergo, we will post-
pone discussion of type B reactions until Section **10-3**.

The following are a few general methods for the preparation of Fischer-
type carbene complexes:

Formation of Carbene Complexes by Nucleophilic Attack at a Carbonyl Ligand

Equation **10.5** shows the original work of Fischer and Maasböl in preparing
the first reported metal-carbene complex. Attack by the carbanion (see equation **8.24** in Chapter 8, Section **8-2-1**) at the carbon atom of a carbonyl lig-
and, followed by methylation, yields the corresponding carbene.

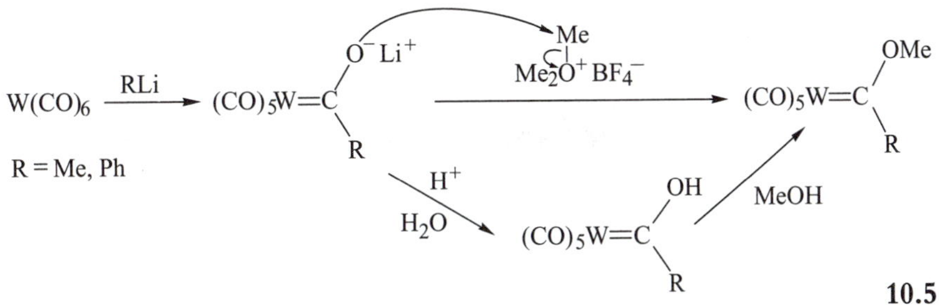

$$W(CO)_6 \xrightarrow{RLi} (CO)_5W{=}C\overset{O^-Li^+}{\underset{R}{\Big\langle}} \qquad \xrightarrow[Me_2O^+ \, BF_4^-]{Me} \qquad (CO)_5W{=}C\overset{OMe}{\underset{R}{\Big\langle}}$$

R = Me, Ph

$$\xrightarrow[H_2O]{H^+} (CO)_5W{=}C\overset{OH}{\underset{R}{\Big\langle}} \xrightarrow{MeOH}$$

10.5

This is a versatile reaction since one may use different attacking carbanionic
groups (RLi, R = alkyl, aryl, or silyl) and different O-alkylating agents.
Carbonyl complexes of several different metals, such as Mo, Mn, Rh, Fe, and
Ni undergo this transformation.

Exercise 10-1

Propose a synthetic route to **16**, a silicon-containing carbene complex.

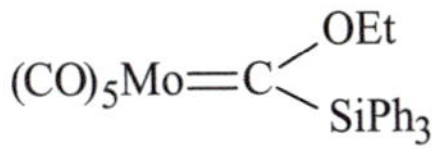

$$(CO)_5Mo{=}C\overset{OEt}{\underset{SiPh_3}{\Big\langle}}$$

16

Formation of Cyclic Carbene Complexes

The cyclic carbene complex (**17**) formed in equation **10.6**[12] is analogous to the original Fischer-type carbene complex. The ring, however, offers an added complexity that is useful in the realm of organic synthesis.

$$M = Mo, n = 2$$
$$M = Fe, n = 1$$

10.6

The first step forms an anionic acyl complex that undergoes intramolecular S_N2 displacement of bromide to yield **17**.

Exercise 10-2

What is the mechanism of the first step in equation **10.6**?

Synthesis of N-Substituted Carbene Complexes

Nitrogen often appears attached directly to carbon in Fischer-type carbene complexes. Equations **10.7**[13] and **10.8**[14] provide two examples of N-substituted carbene complex synthesis. The first procedure involves attack by the amide on one of the carbonyls (analogous to equation **10.5**), followed by alkylation.

$$Cr(CO)_6 \xrightarrow[\text{2) } Et_3O^+ BF_4^-]{\text{1) } Et_2NLi} (CO)_5Cr = C\begin{array}{l} OEt \\ NEt_2 \end{array}$$

10.7

Isonitrile form

$$Et_3P(Br)_2\overset{-}{Pt}-C\equiv\overset{+}{N}-Me$$

$$Et_3P(Br)_2Pt = \overset{..}{C} = N-Me$$

$$Et_3P(Br)_2 Pt = C\begin{array}{l} NMe \\ \overset{+}{Y}-H \\ R \end{array}$$

$$Et_3P(Br)_2Pt = C\begin{array}{l} H \\ NMe \\ Y-R \end{array}$$

$$R-\overset{..}{Y}-H \qquad Y = O, S, N$$

10.8

[12]H. Adams, *et al.*, *Organometallics*, **1990**, *9*, 2621.
[13]E.O. Fischer and H-J. Kollmeier, *Angew. Chem. Int. Ed. Eng.*, **1970**, *9*, 309.
[14]E.M. Badley, J. Chatt, and R.L. Richards, *J. Chem. Soc., A*, **1971**, 21.

The second method involves attack of R–Y-H on the electron-deficient carbon of the isonitrile ligand, followed by proton transfer. It is adjustable with regard to nucleophile since amines (Y = N), unhindered alcohols (Y = O), and thiols (Y = S) add to the isonitrile carbon to form the corresponding diamino, aminoalkoxy, and aminothio carbenes. Isonitrile complexes of several different metals undergo this reaction, although Pd and Pt complexes have been the most thoroughly investigated.[15]

Preparation of Carbene Complexes by Scission of the C=C Bond of Electron-Rich Alkenes
Producing Fischer-type carbene complexes by C=C scission is the least general, but arguably the most mechanistically interesting, of the methods presented. Equation **10.9** indicates the overall pathway to carbene **18**.[16]

$$R = Me, Et \qquad\qquad 18 \qquad\qquad 10.9$$

The mechanism of this reaction is not clear, but only alkenes tetrasubstituted with nitrogen seem to react efficiently. Scheme **10.1** indicates a possible pathway to carbene starting from $Mo(CO)_6$. Carbonyl complexes of other metals, such Ru, Os, and Ir, also undergo this reaction.

10-2-2 Schrock-Type (Nucleophilic) Carbene Complexes
Compared with electrophilic carbene complexes, relatively few procedures exist for preparing alkylidenes. Generally these complexes are more labile than corresponding Fischer-types due to the lack of stabilizing heteroatom substituents. Obvious precursors for alkylidenes ought to be alkyl ligands; loss of an α-H would give the corresponding carbene complex as shown in equation **10.10**.

$$10.10$$

[15]H. Fischer in *The Chemistry of the Metal-Carbon Bond*, Vol. 1, F.R. Hartley and S. Patai, Eds., Wiley-Interscience, New York, 185.
[16]M.F. Lappert and P.L. Pye, *J. Less-common Met.*, **1977**, *54*, 191.

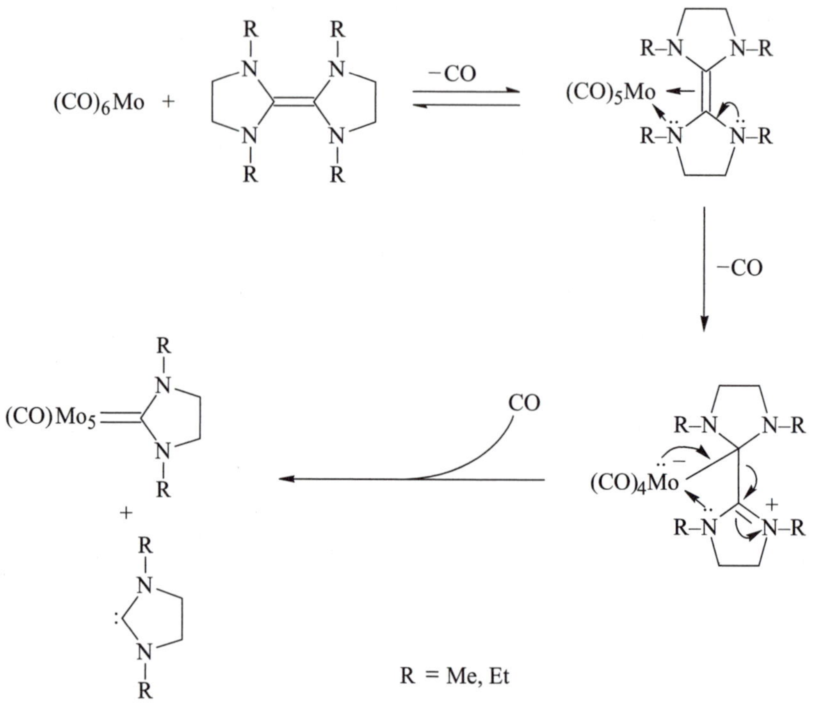

Scheme 10.1
A Possible Mechanistic Pathway
for Alkene Scission

One problem with this approach is the ever-present, rapid, competing side reaction of β-elimination (Chapter 8, Section **8-1-2**). As we shall see, α-elimination is indeed used to produce alkylidenes, but the circumstances of this reaction are unique. Other approaches that we will discuss include decomposition of bridged alkylidenes and direct attack of a free carbene precursor on a metal.

Loss of the α Hydrogen

The early work of Schrock and co-workers[17] involved synthesis and characterization of Group 5 alkylidenes involving Ta and Nb. Scheme **10.2** describes some of the first preparations of alkylidenes. The key reaction here is loss of a proton attached to C_α to form an M=C bond (α-elimination). In all cases β-elimination is blocked due to lack of a β-hydrogen.

---

[17]For a good summary of this work, see R.R. Schrock, *Acc. Chem. Res.*, **1979**, *12*, 98 and references therein.

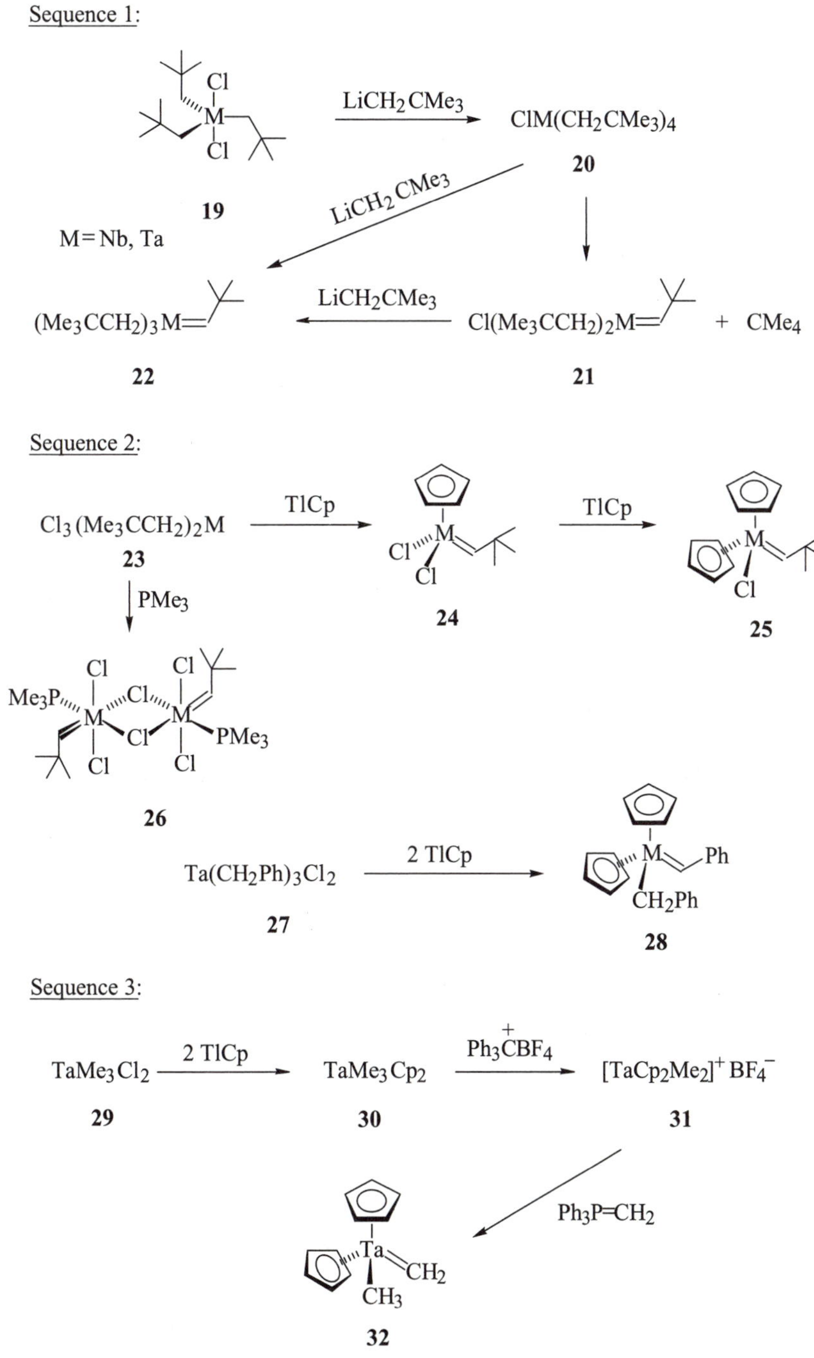

Sequence 1:

Sequence 2:

Sequence 3:

Scheme 10.2
Preparation of Group 5
Schrock-Type Carbene Complexes

Sequence 1. The first sequence of reactions in the scheme begins when 10-electron **19** reacts with neopentyllithium to give the tetraalkylchloro complex **20** via a ligand substitution reaction. Intermediate **20** either reacts directly with additional neopentyllithium to give **22** [via (neopentyl)$_5$M], or it decomposes to the chloroalkylidene complex **21**, which then reacts with another equivalent of neopentyllithium, replacing the chloro group with a neopentyl ligand. Regardless of the pathway, the reaction is remarkable because somehow an alkane (2,2-dimethylpropane) and an alkylidene form as products. Equation **10.11** shows how this may occur.

$$\text{10.11}$$

The driving force for the reaction clearly must be the reduction of steric hindrance that occurs upon expulsion of one of the sterically demanding neopentyl groups. Protons attached at the α-position of the neopentyl groups also may be somewhat agostic (**33**) (see Chapter 6, Section **6-2-3**), since the Group 5 metal is electron deficient [M(V) and d^0]. If so, the α C-H bond would already be weakened, making hydrogen transfer to alkyl more favorable energetically.

Sequence 2. Treatment of the dineopentyl complex (**23**) with cyclopentadienylide ion displaces one of the chloro groups with a Cp ligand, with simultaneous expulsion of alkane to give **24**. Again, the steric bulkiness of the Cp group actuates the loss of alkane by α-elimination. A second Cp-Cl ligand exchange results in formation of **25** (an 18-e$^-$ complex). Conversion of **27** to **28** is analogous and demonstrates that the reaction also occurs with a benzyl ligand.

    Conversion of **23** to **26** is possible due to the steric and electronic properties of PMe$_3$. The dimeric **26** contains 14 electrons/metal atom, and its electron deficiency is stabilized by the presence of phosphine, a strong σ donor. The phosphine also has enough steric bulk so that, as it attaches to the metal, steric hindrance increases sufficiently to allow *intra*molecular α-elimination to occur. An X-ray structure of **26** (shown partially as structure **34**, M = Ta) is interesting.[18] It shows that the bond angle involving the metal, α-, and β-carbons is 161.2°, much higher than the expected 120° for the sp^2 alkylidene α-carbon. The bond angle involving metal, $C_{carbene}$, and attached hydrogen is 84.8° significantly less than 120°! The bond distance between Ta and C is only 189.8 pm, considerably shorter than Ta=C bond distances in other Ta-carbene complexes (average distance = 204 pm). Structure **35** shows a possible explanation for these observations. Steric hindrance apparently forces the M-$C_{carbene}$-C_β bond angle to increase beyond that expected for an sp^2-hybridized carbon. The electron deficiency at Ta (14 e⁻s) provides an opportunity for extra electron density to be supplied through an agostic interaction between the α-hydrogen and an empty d orbital on the metal. In effect, the bond between Ta and carbon is a three-center, six-electron bond.

34   **35**

Sequence 3. The last reaction sequence in Scheme **10.2** details the preparation of a methylidene complex (**32**) starting with **29**. Attempts to observe α-elimination analogous to the reactions described above in **30** were unsuccessful; only decomposition was observed. Electrophilic abstraction of a methyl group from **30** using trityl cation (Ph_3C^+), however, produced tantalonium salt **31**. Schrock realized here the analogy between the tantalonium salt and phosphonium salts, the precursors to ylides in the Wittig reaction. Protons attached to the α-carbon in phosphonium salts are relatively acidic and may be abstracted with bases of varying strength depending upon the nature of other substituents attached to the α-carbon. Schrock reasoned that a methyl proton (the α proton) in **31** could be removed by a strong base. Use of a phosphorus ylide,[19] $(CH_3)_3P=CH_2$, as well as other bases, successfully confirmed his prediction, yielding methylidene **32** as a pale green crystalline solid.

---

[18]A.J. Schultz, J.M. Williams, R.R. Schrock, G.A. Rupprecht, and J.D. Fellmann, *J. Am. Chem. Soc.*, **1979**, *101*, 1593.
[19]See Footnote 10.

Decomposition of Bridged Alkylidene Complexes

Another approach to alkylidenes involves Group 4 metals, particularly Zr. Treatment of a vinyl Cp_2Zr complex (**36**) with diisobutylaluminum hydride gives bridged alkylidene **37**, where the alkylidene is shared (μ bonding) between Zr and Al (equation **10.12**).[20] The dialkylaluminum hydride's Al-H bond adds across the vinyl C=C in a manner analogous to hydroboration such that the chloro ligand (attached to Zr) can act as a Lewis base complementary to electron-deficient Al.

$$\text{10.12}$$

Once **37** forms, treatment—first with Ph_3P, and then with hexamethylphosphoramide (HMPA)—results in formation of alkylidene **38** (Electron-rich phosphines bind readily to electron-deficient Zr(IV) complexes such as **37**). Once the phosphine binds to Zr, HMPA (a highly polar molecule and a Lewis base) attacks Al (equation **10.13**).[21] Fragmentation results in formation of the alkylidene and an HMPA-Al complex.

$$\text{10.13}$$

Titanium carbene complexes also exist; perhaps the most famous of these is methylidene **39**. Although **39** probably occurs fleetingly as an intermediate in Schrock-type carbene complex reactions, it is too reactive to be isolated or observed spectroscopically. Two related compounds, however, have been syn-

---

[20]F. W. Hartner, Jr. and J. Schwartz, *J. Am. Chem. Soc.*, **1981**, *103*, 4979.

[21]F. W. Hartner, Jr., J. Schwartz, and S.M. Clift, *J. Am. Chem. Soc.*, **1983**, *105*, 640.

thesized:[22] $Cp_2TiCH_2(PEt_3)$, in which the phosphine donates electrons to Ti to stabilize the complex; and the immediate precursor to **39**, known as Tebbe's reagent (**40**, equation **10.14**[23]).

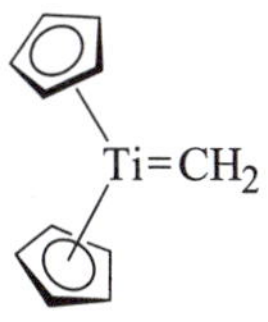

39

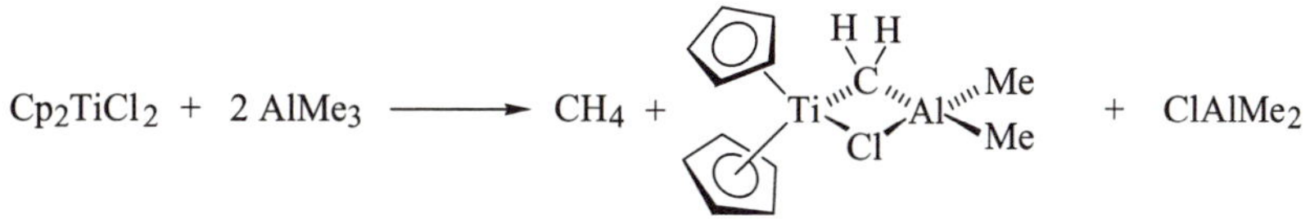

$$Cp_2TiCl_2 + 2 AlMe_3 \longrightarrow CH_4 + \underset{\textbf{40} \quad \text{(Tebbe's Reagent)}}{\text{[Ti–CH}_2\text{–AlMe}_2\text{Cl complex]}} + ClAlMe_2$$

10.14

Equation **10.14** provides another example of the use of Al compounds to form bridging alkylidenes, providing stabilization for the very reactive Group 4 carbenes. Later in this chapter (Section **10-3-2**) we shall encounter some of the synthetically-useful chemistry of Tebbe's reagent.

Direct Attack of a Free Carbene Precursor onto a Metal

Diazoalkanes, $RR'C=N=N$ (R,R' = H, alkyl, aryl) serve as precursors to free carbenes through thermal- or photochemical-induced loss of N_2 according to equation **10.15**.

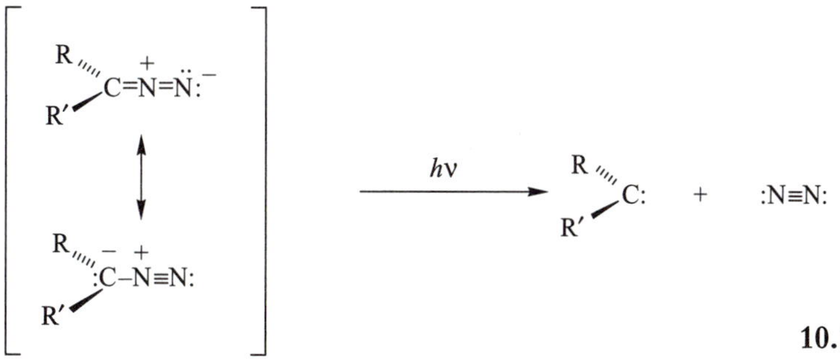

10.15

[22]S.H. Pine, *Org. Reactions*, **1993**, *43*, 1.
[23]F.N. Tebbe, G.W. Parshall, and G.S. Reddy, *J. Am. Chem. Soc.*, **1978**, *100*, 3611.

Although transition metal complexes do not react directly with free carbenes, low valent Group 7 to 9 metals in particular, react with diazoalkanes to produce Schrock-type metal carbenes. Equation **10.16** shows an example of this reaction. The complex must either be unsaturated or possess a labile L-type ligand so that the reaction can occur. The intermediate in this reaction has never been isolated, but it is unlikely to be a free carbene.

$$L_nM \ + \ RCH_2N \xrightarrow{\quad \Delta \quad} L_{n-1}M{=}C{\overset{R}{\underset{H}{\diagup}}} \ + \ N_2(g) \qquad \textbf{10.16}$$

A good example of this type of carbene synthesis appears in equation **10.17**, which shows that a variety of diazoalkanes react with the Os complex.[24]

$$OsCl(NO)(PPh_3)_3 \xrightarrow{\ RCHN_2\ } Cl(NO)(PPh_3)_2Os{=}C{\overset{R}{\underset{H}{\diagup}}} \ + \ N_2(g) \ + \ PPh_3$$

R = H, Me, p-Tolyl   **10.17**

Exercise 10-3

Propose a synthesis of $(NO)(PMe_3)_3Ir{=}C(H)(Me)$ starting with an appropriate Ir complex.

10-3   Reactions of Metal-Carbene Complexes

Transition metal-carbene complexes possess several sites where nucleophiles, electrophiles, oxidizing agents, and protic acids might attack; these are depicted in Figure **10-5**.

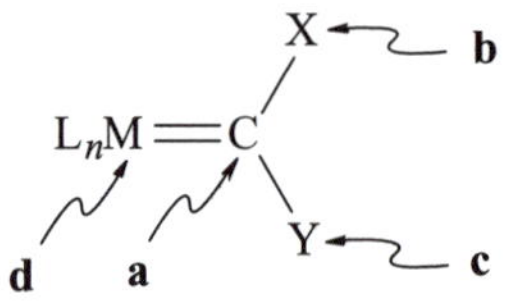

Figure 10-5
Reactive Sites of Transition
Metal-Carbene Complexes

In the following discussion we will focus primarily on the reaction of nucleophiles at $C_{carbene}$, (site **a**), at the substituents attached to $C_{carbene}$ (sites **b** and **c**), and at the metal (site **d**). We also discuss reactions of electrophiles at sites **a** and **d**.

[24]M.A. Gallop and W.R. Roper, *Adv. Organomet. Chem.*, **1986**, *25*, 157.

Caveat Regarding the Reactivity of Metal Carbenes

As mentioned in Section **10-1**, it is convenient to classify carbene complexes on the basis of their reactivity at $C_{carbene}$ (site **a**). Fischer-type carbene complexes tend to undergo nucleophilic attack at this position, while Schrock-type alkylidenes undergo attack by electrophiles at site **a**. While this is a useful generalization, applicable in most cases, there are several exceptions to this pattern of reactivity. For example, we shall encounter alkylidenes that undergo attack by nucleophiles at $C_{carbene}$, indicating that there is a spectrum of reactivities possessed by metal-carbene complexes. Metal oxidation state, overall charge on the complex, position of the metal in the Periodic Table, and electronic properties of ligands influence the reactivity of the metal carbene such that the line of demarcation between the reactivity patterns of Fischer- and Schrock-type carbene complexes is indistinct at times.

10-3-1 Nucleophilic Reactions

Site a

In spite of the caveat expressed above, the most common reaction that Fischer-type carbene complexes undergo is attack by a nucleophile at $C_{carbene}$. It is interesting that such reactivity should occur, since partial charge calculations indicate typically a higher positive charge at the carbon in CO ligands than at $C_{carbene}$.[25] The key to understanding the electrophilicity of these complexes, however, is the application of frontier molecular orbital theory. Electrophilic carbene complexes usually have a relatively low energy LUMO, with a large lobe on $C_{carbene}$, and a much smaller lobe on M. Thus, as the nucleophile approaches the carbene complex, there is a much better bonding overlap between the HOMO of the nucleophile and the carbene LUMO at $C_{carbene}$ than at the metal. Figure **10-6** illustrates this concept for attack on $(CO)_5Cr=C(H)(OH)$.

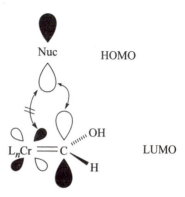

Figure 10-6

Frontier Orbital Interactions in Nucleophilic Attack on Fischer-Type Carbene Complexes

[25]T.F. Block, R.F. Fenske, and C.P. Casey, *J. Am Chem. Soc.*, **1976**, *98*, 441 and N.M. Kostić and R.F. Fenske, *Organometallics*, **1982**, *1*, 974.

It is noteworthy that the LUMO of a Fischer-type carbene resembles the LUMO of a typical ester such as methyl acetate (Figure **10-7a**), or a ketone such as acetone (Figure **10-7b**). In fact, the analogy between Fischer carbenes and carbonyl compounds should be useful to readers already familiar with organic chemistry. Indeed, the chemistry of these two types of compounds, seemingly so different in structure, is similar in several ways.

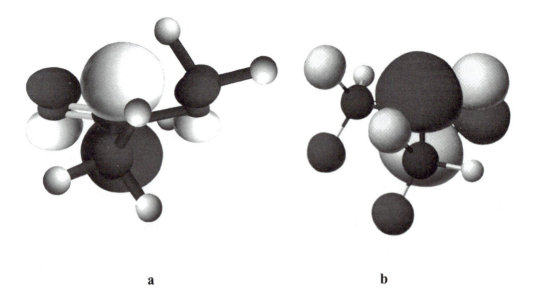

a   b

Figure 10-7
LUMOs of Methyl Acetate (a)
and Acetone (b)

Equation **8.25** (Section **8-2-1** in Chapter 8, and reproduced below) already has provided an illustration of nucleophilic attack on a Fischer-type carbene, and the analogy with aminolysis.

$$(CO)_5\,Cr{=}C\big\langle^{OMe}_{Ph} \ + \ RNH_2 \ \longrightarrow \ (CO)_5Cr{=}C\big\langle^{NHR}_{Ph} \ \longleftrightarrow \ (CO)_5\overset{-}{Cr}{-}C\big\langle^{\overset{+}{N}HR}_{Ph}$$

$$41$$

$$O{=}C\big\langle^{OR'}_{R'} \ + \ RNH_2 \ \longrightarrow \ O{=}C\big\langle^{NHR}_{R'} \ + \ R'\,OH \qquad \textbf{8.25}$$

The driving force for the exchange of groups at $C_{carbene}$ is formation of a C–N bond that is stronger than C–O. Structure **41** (comparable to resonance contributor **11** in Section **10-1**), which shows considerable double bond character between $C_{carbene}$ and the heteroatom, takes on increasing importance when nitrogen is present instead of oxygen because a positively charged nitrogen atom is more stable than the correspondingly charged oxygen.

Fischer measured the kinetics of carbene aminolysis (specifically for the reaction shown in equation **8.25**) and derived a rate law according to the following:[26]

$$\text{Rate} = k[\text{MeNH}_2][\text{HX}][\text{Y}][\text{carbene}]$$

where   HX = a proton donor, e.g., the amine or a protic solvent
                     Y = a proton acceptor, e.g., the amine or solvent

The rate law is consistent with the mechanism shown in Scheme **10.3**.

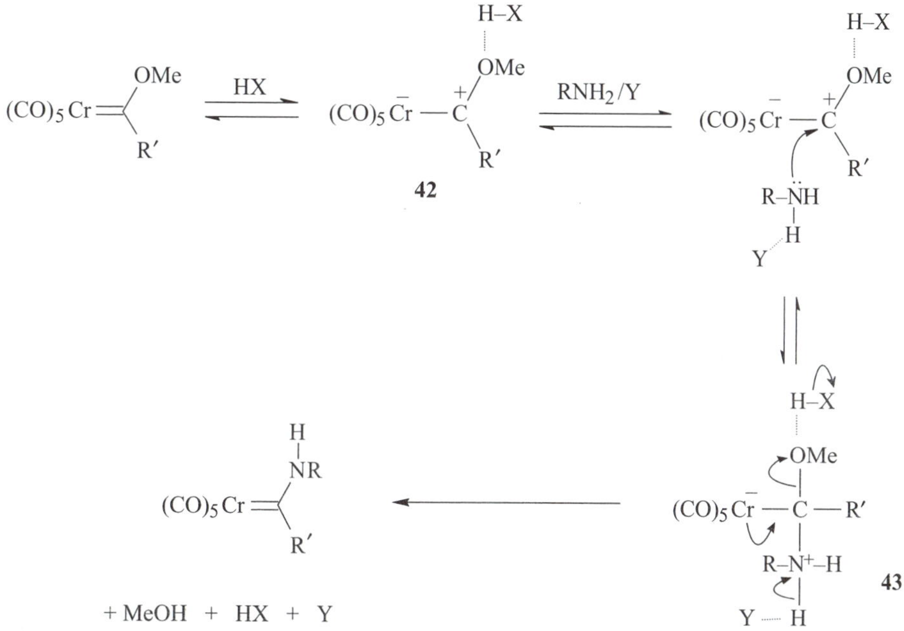

Scheme 10.3
Aminolysis of Fischer-Type
Carbene Complexes

[26]B. Heckl, H. Werner, and E.O. Fischer, *Angew. Chem. Int. Ed. Eng.*, **1968**, 7, 817 and H. Werner, E.O. Fischer, B. Heckl, and C.G. Kreiter, *J. Organomet. Chem.*, **1971**, 28, 367.

The first step is the formation of H-bonded intermediate **42** in which $C_{carbene}$ takes on substantial cationic character. Next, intermolecular attack by the amine in the presence of Y provides tetrahedral intermediate **43**, which then breaks down into products. The reaction is sensitive to steric hindrance, with ammonia and primary amines reacting rapidly (several orders of magnitude faster than aminolysis of carboxylic acid esters) and secondary amines reacting much more sluggishly.

Thiols and thiolates react with O–substituted Fischer carbenes in an analogous manner as shown in equation **10.18**.[27]

$$(CO)_5M=C\underset{R}{\overset{OMe}{<}} \xrightarrow{R'SH} (CO)_5M=C\underset{R}{\overset{SR'}{<}} + MeOH$$

M = Cr, W; R = Me, Ph; R′ = Me, Et, Ph   **10.18**

Organolithium compounds can serve as sources of carbon nucleophiles to drive substituent group exchange at $C_{carbene}$— thus providing a route to alkylidenes according to equation **10.19**.[28] This reaction is limited to lithium reagents that do not possess H atoms attached to the carbon bonded to Li; otherwise, rearrangement (equation **10.20**) occurs to give an alkene.[29]

$$(CO)_5W=C\underset{Ph}{\overset{OMe}{<}} \xrightarrow[-78°]{PhLi} (CO)_5\overset{-}{W}-\underset{Ph}{\overset{\overset{OMe}{|}}{C}}-Ph \xrightarrow{HCl/-78°} (CO)_5W=C\underset{Ph}{\overset{Ph}{<}} + MeOH$$

10.19

$$(CO)_5W=C\underset{Ph}{\overset{OMe}{<}} \xrightarrow[\text{2) HCl/}-78°]{\text{1) MeLi}} \left[(CO)_5W=C\underset{Ph}{\overset{\overset{H}{\curvearrowright}CH_2}{}} \right] \xrightarrow{25°} (CO)_5W-\underset{Ph}{||}$$

+ MeOH   **10.20**

Sites b and c

In organic chemistry, carbonyl compounds can undergo reaction at the carbon α to a C=O group, as well as nucleophilic attack on the carbonyl carbon. The electron-withdrawing nature of the carbonyl group renders protons at the α position acidic. In Fischer-type carbene complexes, hydrogen atoms α to $C_{carbene}$ are analogously acidic. Such reactivity is synthetically useful. Loss of an α hydrogen generates a carbanion analogous to an enolate ion that then undergoes reaction with electrophiles such as D^+, R^+, or $R-C=O^+$, the last two

[27]For examples of several nucleophilic substitutions, see F.J. Brown, *op. cit.*, 31 ff.

[28]C.P. Casey and T.J. Burkhardt, *J. Am. Chem. Soc.*, **1973**, *95*, 5833.

[29]C.P. Casey, L.D. Albin, and T.J. Burkhardt, *J. Am. Chem. Soc.*, **1977**, *99*, 2533.

affording routes to new carbenes. Equation **10.21**[30] shows specific deuteration; **10.22**,[31] alkylation using allyl bromide; and **10.23**,[32] an aldol condensation. More examples that demonstrate the usefulness of this chemistry to generate complex organic molecules appear in Chapter 11.

$$(CO)_5Cr=C\underset{Me}{\overset{OMe}{<}} \quad \xrightarrow[\text{2) DCl}]{\text{1) BuLi}} \quad (CO)_5Cr=C\underset{CH_2D}{\overset{OMe}{<}} \qquad \textbf{10.21}$$

$$(CO)_5Cr= \quad \xrightarrow[\text{2)} \overset{}{\underset{Br}{\diagdown}}]{\text{1) BuLi}} \quad (CO)_5Cr= \qquad \textbf{10.22}$$

$$(CO)_5Cr=C\underset{Me}{\overset{OMe}{<}} \quad \xrightarrow[\text{2) PhCHO}]{\text{1) BuLi}} \quad \left[(CO)_5Cr= \overset{OMe}{\underset{Ph}{\diagup}} \right] \quad \longrightarrow \quad (CO)_5Cr= \overset{OMe}{\underset{Ph}{\diagup}} \qquad \textbf{10.23}$$

Protonation followed
by elimination of OH⁻ **10.23**

Exercise 10-4

Propose a synthesis of spirocyclic carbene complex **B** starting with **A**.

$$(CO)_5W= \qquad \qquad \longrightarrow \qquad \qquad (CO)_5W=$$

A   **B**

Site d

Carbonyls are the most common ligand found in electrophilic carbene complexes. They may be exchanged readily for other ligands, usually by a dissociative substitution pathway (see Chapter 7, Section **7-1-3**). Equation **10.24** describes studies involving ligand substitution with phosphines, perhaps the most typical nucleophile used.

$$(CO)_5M=C\underset{R'}{\overset{OMe}{<}} \quad \xrightarrow{PR_3} \quad cis\text{-} + trans\text{-}(CO)_4(R_3P)M=C\underset{R'}{\overset{OMe}{<}}$$

M= Cr, Mo, W; R′ = Ph, Me, *i*-Pr; R = Et, *n*-Bu, Ph, Me   **10.24**

[30]C.P. Casey, R.A. Boggs, and R.L. Anderson, *J. Am. Chem. Soc.*, **1972**, *94*, 8947.
[31]C.P. Casey, in *Transition Metal Organometallics in Organic Synthesis*, vol. 1, H. Alper, Ed., Academic Press, New York, 1976.
[32]C.P. Casey and W.R. Brunsvold, *J. Organomet. Chem.*, **1974**, *77*, 345.

Table 10-2 Rate Parameters for Ligand Dissociation of Cr Complexes

L	k_{rel}	$\Delta H^{\ddagger}$ (kcal/mole)
=C(OMe)(Me)	240,000	27.6
–CO	11	38.7
–P(Cy)$_3$[a]	1	40.4

[a]Cy = cyclohexyl

It is worth noting that the presence of a carbene ligand in Cr carbonyl complexes provides a kinetic stimulus toward CO dissociation, especially compared to non-carbene analogs. Table **10–2**[33] shows the effect of the carbene ligand on the tendency for CO to dissociate in Cr complexes according to equation **10.25**.

$$(CO)_5CrL \ + \ PCy_3 \ \xrightarrow[\text{Decane/58.8}^\circ]{} \ (CO)_4CrL(PCy_3) \ + \ CO$$

10.25

The driving force for such high CO ligand lability probably rests with the ability of the heteroatom attached to $C_{carbene}$ to donate electrons by resonance to the metal, which becomes electron deficient upon loss of CO.

10-3-2 Electrophilic Reactions

Certain metal-carbene complexes, especially alkylidenes involving the early transition metals, behave as nucleophiles, reacting with electron-deficient species at $C_{carbene}$. Figure **10–8** shows the HOMO of $H_3Nb=CH_2$, a hypothetical Schrock-type carbene complex. Note the high electron density at $C_{carbene}$, indicating a position readily susceptible to attack by electrophiles. Resonance theory also indicates that Schrock-type metal carbene complexes possess substantial negative charge density at $C_{carbene}$ with structures **9** and **12** (Section **10–1**) having greatest contribution to the overall structure. We again emphasize that it is useful to think of alkylidenes as behaving like phosphorus ylides (see footnote 10).

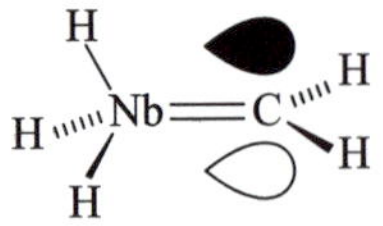

Figure 10-8
The HOMO of $H_3Nb=CH_2$

In our discussion we will focus on electrophilic reactions that occur at $C_{carbene}$, site **a**. Equations **10.26, 10.27,** and **10.28** provide some examples of alkylidene chemistry.

[33]F.J. Brown, *op. cit.*, 45.

$$Cp_2(CH_3)Ta{=}CH_2 \ + \ AlMe_3 \ \longrightarrow \ Cp_2(CH_3)\overset{+}{Ta}{-}CH_2{-}\overset{-}{Al}Me_3$$

$$10.26$$

$$Cl(NO)(PPh_3)_2Os{=}CH_2 \ + \ AlMe_3 \ \xrightarrow{\ HCl\ } \ Cl_2(NO)(PPh_3)_2Os{-}CH_2{-}H$$

$$10.27$$

$$Cp_2(CH_3)Ta{=}CH_2 \ + \ CD_3I \ \longrightarrow \ \left[I(CH_3)Cp_2Ta{-}CH_2{-}CD_3 \right] \ \longrightarrow \ ICp_2Ta - \overset{CD_2}{\underset{CH_2}{\|}}$$

$$+$$

$$CH_3D$$

$$10.28$$

In equation **10.26** the Ta carbene complex is a Lewis base (at $C_{carbene}$), complexing strongly with the acidic $AlMe_3$. The metal in the Os complex (**10.27**) is relatively electron rich and bonded to a carbon lacking heteroatom substituent. $AlMe_3$ either complexes directly with Os, followed by protonation with HCl, or it reacts with HCl to give $AlMe_3Cl^- \ H^+$, which then protonates $C_{carbene}$. The alkylidene behaves, therefore, as a nucleophilic Schrock-type carbene complex. In **10.28** a Ta complex again reacts with an electrophile initially via S_N2 substitution at $C_{carbene}$.

Exercise 10-5

In equation **10.28** some steps are missing. Provide these steps, emphasizing the transformation from the intermediate to the final products.

Schrock-type carbene complexes undergo reaction with multiple bonds via four-center metallacyclic intermediates (**44**). Section **10-5** considers what occurs when alkylidenes react with alkenes, a reaction known as metathesis. Below are examples of Schrock-type carbene complexes reacting with polar multiple bonds such as $C{\equiv}N$ and $C{=}O$.

$$L_nM{=}CR_2 \ + \ Y{=}CR'_2 \ \longrightarrow \ \begin{matrix} L_nM{-}CR_2 \\ | \quad \ | \\ Y{-}CR'_2 \end{matrix}$$

$$\textbf{44}$$

M = Early transition metal

Y = N, O, C

R, R' = H, Alkyl

Exercise 10-6

Assuming that alkylidenes act as nucleophiles, show how metallacycles such as **44** could form when a compound containing a C=O bond reacts with $L_nM=CR_2$.

Equation **10.29** shows a Ta alkylidene reacting with benzonitrile to give **46** as a mixture of E and Z isomers. Presumably, the reaction goes through metallacycle **45** as an intermediate. Ring opening of **45** gives **46**. The driving force behind this last step is probably because Ta, as an early transition metal, prefers to bond to the more electronegative nitrogen rather than the less electronegative carbon.

$$Cl_2(Cp)Ta=CH-CMe_3 + PhC\equiv N \longrightarrow Cl_2(Cp)Ta-\!\!\!\!\begin{array}{c} H \\ | \\ C-CMe_3 \\ | \\ N=C-Ph \end{array}$$

45

$$Cl_2(Cp)Ta=N \diagup \diagdown ^{H}_{Ph \quad CMe_3}$$

46

$$Cl_2(Cp)Ta=N \diagup \diagdown ^{CMe_3}_{Ph \quad H}$$

10.29

In Section **10-2-2** we discussed the synthesis of Tebbe's reagent, a bridged Ti-methylidene complex. This reagent converts a C=O group to an alkene in a manner analogous to a phosphorus ylide (Wittig reagent). Although other metal-alkylidene complexes besides Ti-methylidene can be used to generate alkenes, the most useful alkylidene for this task is Tebbe's reagent. Conversion of a carbonyl to a terminal alkene may seem to be of limited utility until one realizes that, unlike phosphorus ylides that react only with aldehydes and ketones, Tebbe's reagent will react with a variety of carbonyl compounds such as esters, thioesters, and amides, in addition to aldehydes and ketones. Equation **10.30** provides an example of alkene formation from an ester reacting with Tebbe's reagent:[34]

$$\begin{array}{c} \text{(benzofuranone)} \end{array} + Cp_2Ti\diagup \diagdown ^{CH_2}_{Cl} AlMe_2 \longrightarrow \text{(benzofuran)}=CH_2$$

Tebbe's reagent   85 %   **10.30**

<u>Wittig reaction:</u>

$$\text{(cyclohexanone)}=O + CH_2=PPh_3 \longrightarrow \text{(methylenecyclohexane)}=CH_2 + O=PPh_3$$

[34] For a recent review that details the scope and utility of methylenation of carbonyl compounds using Tebbe's reagent, see S.H. Pine, *Org. React.*, **1993**, *43*, 1.

10-3-3 Electrophilic Alkylidenes

At the beginning of Section **10-3** we commented that metal-carbene complexes exhibit a spectrum of reactivities with nucleophiles and electrophiles, especially at $C_{carbene}$. Carbene complexes of middle transition metals (Groups 7 to 9) without heteroatomic substituents at $C_{carbene}$ may show electrophilic behavior depending upon: the nature of other ligands, oxidation state of the metal, and overall charge on the complex. From some observations listed below we may be able to discern a pattern of reactivity.[35]

1. $Cl(NO)(PPh_3)_2Os=CH_2$ reacts with the electrophile H^+ (equation **10.27**) but not with CH_3I, indicating a relatively weak reactivity toward electrophiles. $[I(CO)_2(PPh_3)_2Os=CH_2]^+$ readily reacts with nucleophiles.

2. $(CO)_2(PPh_3)_2Ru=CF_2$ reacts with electrophiles, but $Cl_2(CO)(PPh_3)_2Ru=CF_2$ reacts with nucleophiles and not at all with electrophiles.

3. $[Cp(NO)(PPh_3)Re=CH_2]^+$ reacts with both electrophiles and nucleophiles.

4. $[Cp(CO)_3M=CH_2]^+$ (M = Cr, Mo, W) reacts with nucleophiles but not electrophiles, even though neutral methylene complexes involving these metals are nucleophilic.

Observation 1 indicates that the overall charge is important in determining reactivity of Group 7 to 9 carbene complexes; adding a positive charge makes the Os complex electrophilic. The Ru complexes compared in observation 2 differ in oxidation state of Ru (assuming that the carbene is an L-type ligand), with the latter an electrophilic Ru(II) complex and the former nucleophilic and Ru(0). The Re complex, described in observation 3, is transitional between nucleophilic and electrophilic reactivity. There apparently is a balance between the election donating properties of the Cp and phosphine ligands, and the overall charge on the complex, giving the carbene borderline reactivity. Simply increasing the overall charge to +1 in normally nucleophilic Group 6 metal carbenes causes these complexes to be electrophilic according to observation 4. On the basis of these observations, the largest influence upon reactivity at $C_{carbene}$ is the overall charge on the complex; the more positive the charge, the more electrophilic the species.

Cationic Fe alkylidenes, compounds that have generated much research interest, typically show electrophilic behavior, reacting with a variety of nucleophiles at $C_{carbene}$. Several routes are now available to synthesize these carbenes; equations **10.31**[36] and **10.32**[37] demonstrate two of these pathways.

[35]M.A. Gallop and W.R. Roper, *op. cit.*, 127–129.
[36]P.W. Jolly and R. Pettit, *J. Am. Chem. Soc.*, **1955**, *88*, 5044.
[37]C.P. Casey, W.H. Miles, and H. Tukada, *J. Am. Chem. Soc.*, **1985**, *107*, 2924.

$$Cp(CO)_2Fe-CH_2-OCH_3 \xrightarrow{\text{H}^+} [Cp(CO)_2Fe=CH_2]^+ + CH_3OH \qquad \textbf{10.31}$$

$$Cp(CO)_2Fe-\!\!\!/\!\!\!< \xrightarrow[-78°]{\text{HBF}_4} [Cp(CO)_2Fe=\!\!\!<\,]^+BF_4^- \qquad \textbf{10.32}$$

The first procedure involves ionization of a leaving group attached to $C_{carbene}$ (perhaps more accurately described as an electrophilic abstraction: see Chapter 8, Section **8-4-2**). The second procedure occurs when an electrophile (usually H$^+$) undergoes electrophilic addition (Section **8-4-2**) to a η^1-vinyl complex. The cationic iron complexes produced are usually thermally unstable and may either react with a nucleophile or rearrange at low temperature to an alkene complex via a 1,2-H-shift (Scheme **10.4**):

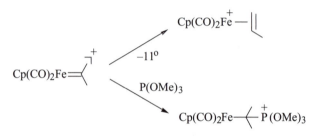

Scheme 10.4

Reactions of Electrophilic
Fe Alkylidenes

Cyclopropane Formation

The cycloaddition of free carbenes with alkenes to give cyclopropanes is well known.[38] Electrophilic metal-carbene complexes can also serve as carbene transfer agents yielding the corresponding cyclopropane according to equation **10.33**.[39]

$$(CO)_5Cr=\!\!\!<^{OMe}_{Ph} + {}^{EtO_2C}_{H}\!\!>=<^{CO_2Et}_{H} \xrightarrow{110°} \text{cyclopropane products} \qquad \textbf{10.33}$$

The temperature required for cyclopropanation using Fischer carbene complexes is sufficiently high that dissociation of a CO ligand can occur. An alkene to which at least one electron withdrawing group is attached is required for the reaction; simple alkenes do not react. Cycloaddition is usually stereospecific with regard to maintaining the stereochemistry about the C=C bond. Scheme **10.5** shows the generally accepted mechanism for cyclopropanation.[40]

---

[38]J. March, *op. cit.*, 866-873 and references therein.
[39]K.H. Dötz and E.O. Fischer, *Chem. Ber.*, **1972**, *105*, 1356.
[40]F.J. Brown, *op. cit.*, 93.

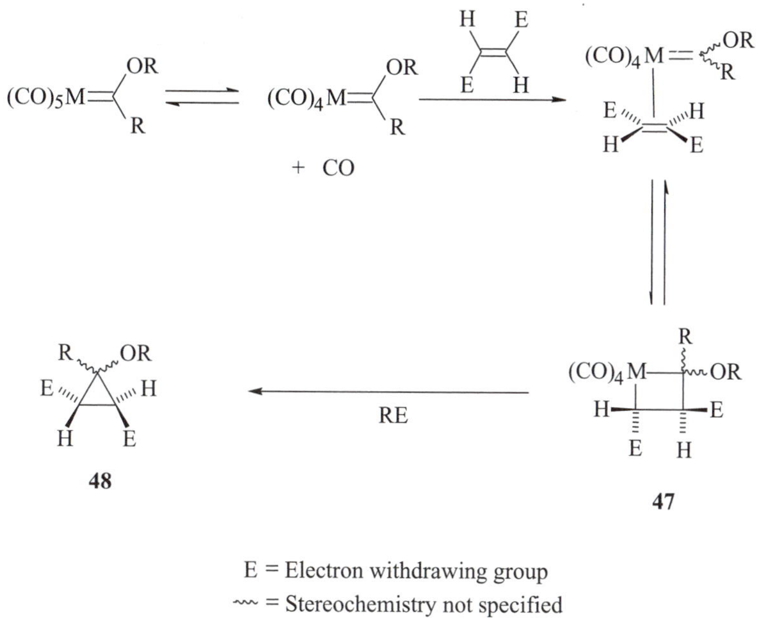

E = Electron withdrawing group

∿ = Stereochemistry not specified

Scheme 10.5
A Mechanistic Pathway for Cyclopropanation
Using Fischer-Type Carbene Complexes

The pathway in Scheme **10.5** shows loss of a CO ligand followed by complexation of the alkene to the metal. Rearrangement to a four-membered ring metallacycle (**47**) is followed by reductive elimination to provide the cyclopropane (**48**).

Cyclopropanation using Fischer-type carbene complexes is not a good general synthetic route to three-membered rings because of the limited number of reactive alkenes, relatively high reaction temperatures required, and variable product yields. Offering more synthetic utility is the use of cationic alkylidene complexes, with Fe complexes being the most thoroughly studied.[41] These react at low temperatures with a variety of alkenes bearing alkyl and aryl substituents; equation **10.34**[42] shows a common example with the cationic alkylidene precursor usually generated *in situ*.

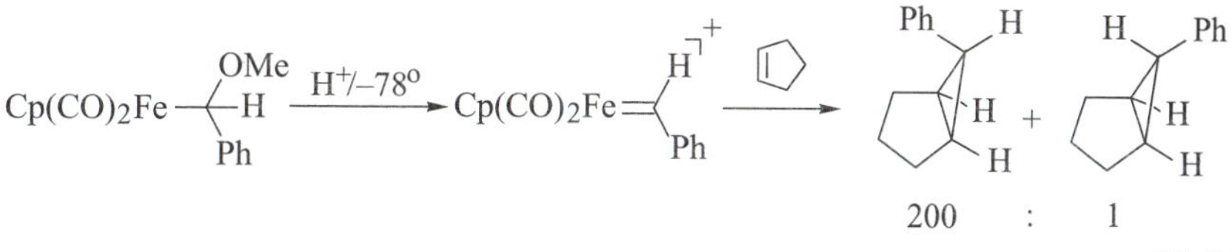

10.34

[41]For information on the scope and synthetic utility of cyclopropanation using electrophilic alkylidenes, see M. Brookhart and W.B. Studabaker, *Chem. Rev.*, **1987**, *87*, 411.
[42]M. Brookhart, M.B. Humphrey, H.J. Kratzner, and G.O. Nelson, *J. Am. Chem. Soc.*, **1980**, *102*, 7802.

The mechanism of cyclopropanation using cationic alkylidenes was not clear for several years. Brookhart and Casey have independently accumulated a great deal of evidence that indicates a different pathway than that for Fischer carbene complexes. Scheme **10.6** shows several possible pathways to cyclopropanes that have different stereochemical consequences.

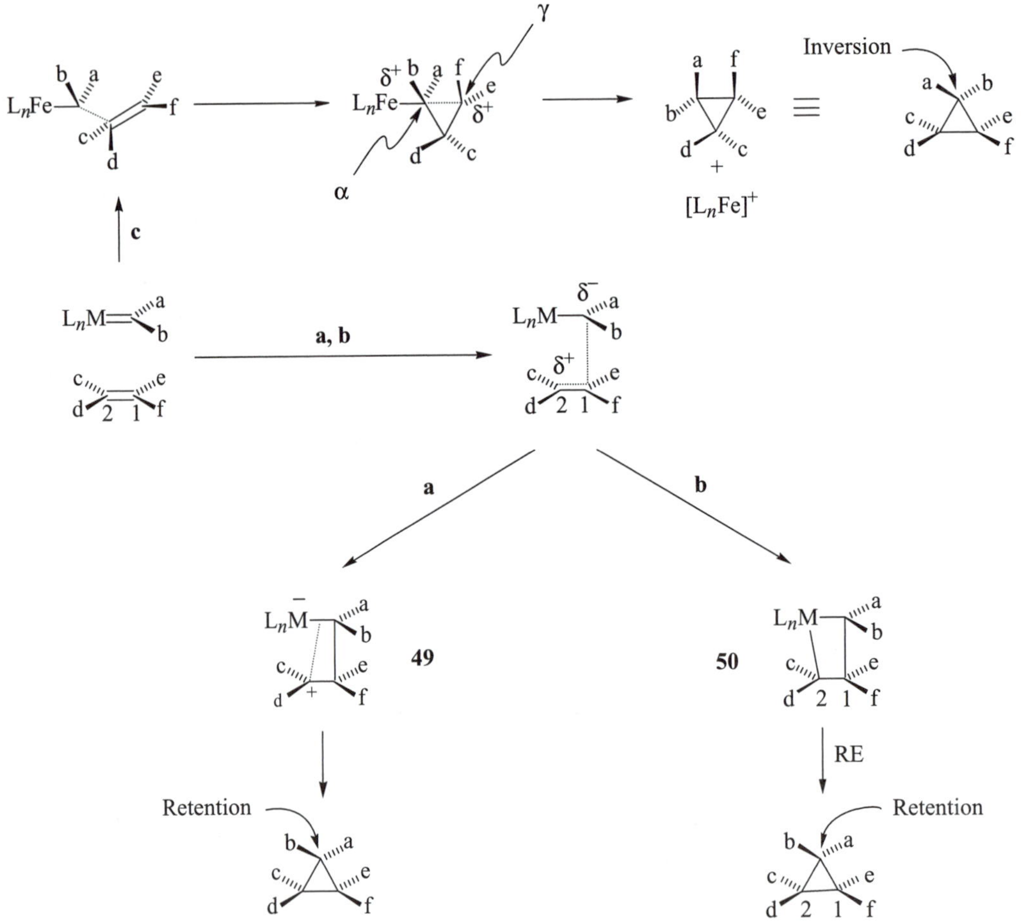

Scheme 10.6
Mechanistic Pathways for Cyclopropanation
Using Electrophilic Alkylidenes

Path **a** involves attack by electrophilic $C_{carbene}$ onto the alkene, typical of the first step of electrophilic addition to alkenes in organic chemistry. A transition state (**49**) forms that allows ring closure to occur at $C_{carbene}$ with retention of configuration. Path **b** occurs by the same initial electrophilic attack, but subsequently a σ bond forms between the metal and the other original alkene carbon (C-2) to give a metallacyclic intermediate (**50**). Cyclopropane formation occurs with retention of configuration by reductive elimination. Path **c** shows backside closure with electrophilic cleavage at the Fe-C_α bond induced by attack of the carbon at C_γ. Inversion of configuration at C_α occurs.

Both Brookhart's and Casey's groups conducted experiments to probe the stereochemical nature of cyclopropanation. Scheme **10.7** shows the result of Casey's[43] work using a model reaction that mimics pathway **b** or **c** in Scheme **10.6**. The topmost equation of Scheme **10.7** indicates the scope of the Ag^+-promoted cyclopropanation. Starting with the *threo* isomer of Fe-alkyl complex **51**, loss of X by electrophilic abstraction using Ag^+ sets up pathway **b** with retention of configuration, or **c** involving inversion of configuration. Only *cis*-1,2-dideuteriocyclopropane was observed. This result, and those of similar experiments performed by the Brookhart group[44] also showing inversion of configuration, provide strong support for path **c** as the mechanistic course of cyclopropanation using electrophilic alkylidenes.[45]

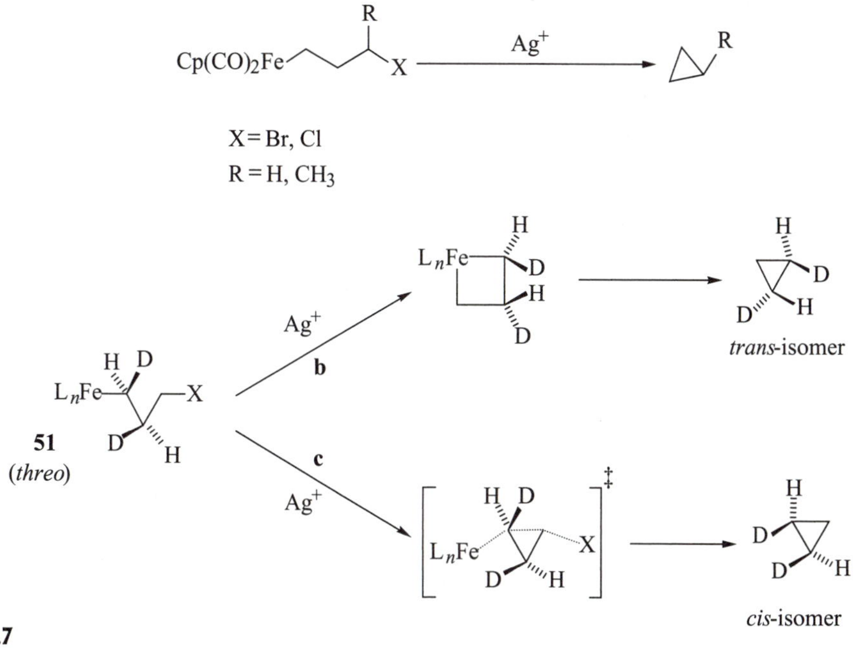

Scheme 10.7
Stereochemical Examination of
Electrophilic Alkylidene Cyclopropanation

[43]C.P. Casey and L.J. Smith Vosejpka, *Organometallics*, **1992**, *11*, 738.
[44]M. Brookhart, et al., *J. Am. Chem. Soc.*, **1991**, *113*, 927 and M. Brookhart and Y. Liu, *J. Am. Chem. Soc.*, **1991**, *113*, 939.
[45]Another possible route to cyclopropanes involves reaction of $R(CO)CHN_2$ (a free carbene precursor) and alkenes in the presence of catalytic amounts of dirhodium complexes (Rh_2L_4). Rh-carbene intermediates are probably present in the catalytic cycle. See A. Padwa, M.P. Doyle, et al., *J. Am. Chem. Soc.*, **1992**, *114*, 1874 and references therein for more information about this method.

Exercise 10-7

Start with the *erythro* isomer of **51** and follow the course of pathways **b** and **c**. What should the stereochemistry of the dideuteriocyclopropane products be as the result of the two pathways?

10-4   Metal-Carbyne Complexes

10-4-1   Structure

The pioneering work of E.O. Fischer and co-workers led to the preparation of the first stable metal carbenes. It is perhaps fitting that Fischer[46] reported the first syntheses of complexes containing an M≡C bond nine years later. Complexes **52**, called *metal-carbyne complexes* or *alkylidynes* (if R = alkyl), were discovered serendipitously in the course of attempts to develop new methods for preparing carbene complexes.

52

M = Cr, Mo, W   X = Cl, Br, I   R = Me, Et, Ph

Since Fischer's breakthrough, hundreds of new carbyne complexes involving mainly Group 6, 7, 8, and 9 metals have been prepared, their structures determined, and their chemistry explored. Some examples of metal carbynes appear as structures **53, 54**, and **55**.

53   **54**   **55**

[46]E.O. Fischer, *et al.*, *Angew. Chem. Int. Ed. Eng.*, **1973**, *12*, 564.

Tungsten complex **53** (also known as an *alkylidyne*) was first reported by Schrock[47] in 1978, and represents a very different type of carbyne complex than that first synthesized by Fischer. Structure **54** portrays a dirhenium–dicarbyne complex, also prepared by Schrock,[48] and rather unusual in that it lacks bridging ligands that support the metal-metal bonding. The μ_3-bridging carbyne complex[49] (**55**) demonstrates that the carbyne ligand may bond to more than one metal atom. While **55** is interesting[50] structurally, we will confine our discussion to *terminal* carbyne complexes (nonbridging) as represented by structures **52** to **54**.

There are several parallels between the chemistry of the carbene ligand and that of the carbyne. The classification of carbyne complexes into two major structural types—Fischer and Schrock—is perhaps the most obvious of these parallels. Complex **52** represents the prototype for a Fischer-type metal carbyne complex: a low-valent metal with π-accepting CO ligands attached. Structure **53**, on the other hand, is a classic example of a Schrock-type metal carbyne complex because a high-valent metal is present with electron-donating ligands attached. Atoms attached to the carbon directly bound to the metal ($C_{carbyne}$ in the case of carbyne complexes), helpful in distinguishing between Fischer-type and Schrock-type carbene complexes (i.e., heteroatoms for the former, and H and C for the latter), are less important in the case of carbyne complexes. It is convenient to classify carbyne complexes according to these two types, but some caution must be exercised since the scheme breaks down for some of these complexes. For instance, the metal in tungsten-benzylidyne complex **56** is low valent and relatively electron-rich, yet the ligands attached

$$Me_2P \quad \underset{Me_2P}{\overset{Br}{\rule{0pt}{0pt}}} \quad W \equiv C - Ph$$

56

[47]D.N. Clark and R.R. Schrock, *J. Am. Chem. Soc.*, **1978**, *100*, 6774.

[48]R. Toreki, R.R. Schrock, and M.G. Vale, *J. Am. Chem. Soc.*, **1991**, *113*, 3610.

[49]M. Green, S.J. Porter, and F.G.A. Stone, *J. Chem. Soc., Dalton Trans.*, **1983**, 513.

[50]Bridging carbynes are important ligands in metal cluster compounds (Chapter 12) and may play a role in surface-catalyzed reactions. Methine (C–H), bound to more than one metal atom, may also be involved in intermediate stages of the Fischer-Tropsch reaction (Chapter 9, Section **9-6-1**) (A.T. Bell, *Cat. Rev. Sci. Eng.*, **1981**, *23*, 203).

[51]J. Manna, T.M. Gilbert, R.E. Dallinger, S.J. Geib, and M.D. Hopkins, *J. Am. Chem. Soc.*, **1992**, *114*, 5870.

are neither strong π donors nor π acceptors.[51] Another example of a carbyne complex intermediate between the Fischer and Schrock structural types would be $Cl(CO)(PPh_3)_2Os\equiv C\text{-}Ph$, analogous to the Group 8 carbene complexes mentioned in Section **10-3-3**.

Figure **10-9a** provides a rationalization for the bonding of a univalent carbon fragment to a metal in Fischer-type carbyne complexes. The carbon fragment possesses three electrons,[52] two in an sp orbital and one distributed between two degenerate $2p$ orbitals. The metal fragment has filled and unfilled

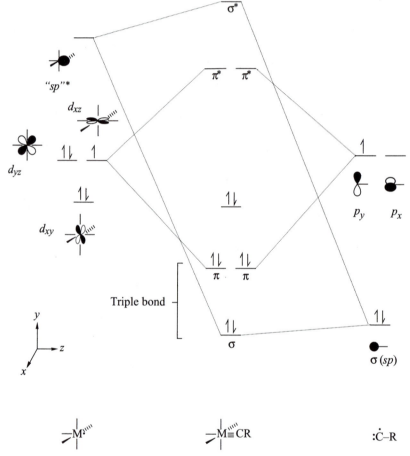

Figure 10-9a
Molecular Orbital Bonding Scheme
for a Fischer-Type Carbyne Complex
*The sp orbital is similiar to d_z^2

[52]Some models for carbyne bonding use a two-electron $^+C\text{-}R$ ligand bonded to a L_nM^- fragment. Use of a three-electron carbon ligand is more consistent with the electron count (a 3-e$^-$ LX ligand according to the neutral ligand method) associated with carbyne ligands.

orbitals of comparable symmetry such that one σ and two π bonds form. Under the neutral ligand method of electron counting, the carbyne ligand is considered an LX ligand.

The same type of bonding scheme could accommodate Schrock-type carbyne complexes involving high-oxidation-state metals. In this case we consider the carbyne ligand to be $R–C^{3-}$ or an X_3 ligand (still a 3-e^- ligand according to the neutral ligand model). The carbyne fragment forms three bonds to the electron-deficient metal acting as both a σ and π donor (Figure **10-9b**).[53]

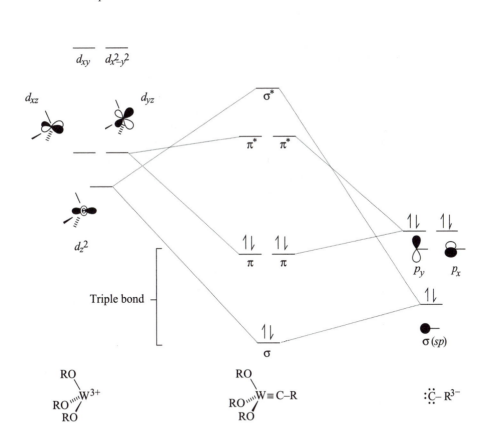

Figure 10-9b
Molecular Orbital Bonding Scheme
for a Schrock-Type Carbyne Complex

[53]For a summary of the use of MO calculations on carbyne complexes, see H. Fischer, *et al.*, *Carbyne Complexes*, VCH Publishers, New York, **1988**, 60–98.

The triple bond in metal carbynes is shorter than the comparable double bond in metal carbenes (*ca.* 165 to 185 pm vs. >200 pm). The M-C-R bond angle, while not exactly 180°, is usually larger than 170°.

Exercise 10-8

The $M \equiv C$ bond distance for $Br(CO)_4Cr \equiv C$-Ph is 168 pm, while that for $Br(CO)_4Cr \equiv C$-NEt_2 is 172 pm. Consider the nature of the two substituents attached to $C_{carbyne}$ and explain the difference in bond lengths in terms of resonance theory.

10-4-2 Synthesis

Equation **10.35** shows the chemistry used to produce **52** and similar complexes; this is not only the original procedure employed by Fischer, but also constitutes perhaps the most versatile method for the preparation of Fischer-type carbyne complexes. The reaction begins with an electrophilic abstraction of the alkoxy group from starting carbene **57** to give the cationic carbyne complex **58**. Carbyne ligands are such powerful π acceptors that they exert a strong *trans* effect. If the group *trans* to $C_{carbyne}$ is a π acceptor such as CO or PR_3, the M-CO or M-P bond is relatively weak. A halide displaces the phosphine ligand to give the final neutral carbyne complex **59**.

R = Alkyl, Aryl; M = Mo, W; X = Cl, Br

10.35

Exercise 10-9

If R in **58** is NR'_2 (R$'$ = alkyl, aryl), show that the molecule is resonance stabilized.

Scheme **10.8** shows routes that Schrock[54] used to produce Ta carbyne complex **62** involving either α-hydrogen abstraction or α-elimination, two widely-used methods for producing alkylidynes. Path **a** involves loss of Cl⁻ and addition of two equivalents of PMe₃ to give **60**. Deprotonation of **60** with an ylide in the presence of excess PMe₃ yields **62**. Path **b** involves displacement of one Cl ligand with neopentyl to give **61**. Attachment of PMe₃ to Ta creates sufficient steric hindrance for an intramolecular α -elimination to occur yielding **62** and one equivalent of 2,2-dimethylpropane.

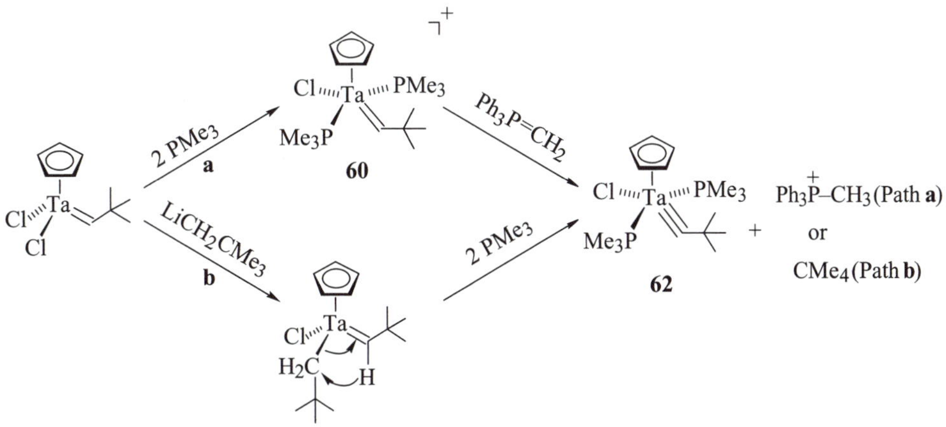

Scheme 10.8
Two Routes to Schrock-Type
Carbyne Complexes

10-4-3 Reactions

As with carbene complexes, alkylidynes display a range of reactivity with electrophiles and nucleophiles. Molecular orbital calculations show that even cationic Fischer-type carbyne complexes are polarized as $M^{\delta+} \equiv C^{\delta-}$; neutral Fischer- and Schrock-type carbyne complexes have an even greater negative charge on $C_{carbyne}$.[55] If all reactions between carbyne complexes and other species were charge-controlled, we would predict that nucleophiles would always attack at the metal and electrophiles at $C_{carbyne}$. As we should expect by now, the picture is more complicated in practice.

Cationic Fischer-type carbyne complexes react with nucleophiles exclusively at $C_{carbyne}$. Calculations show that the LUMO of these complexes possess a large lobe centered about $C_{carbyne}$ and a much smaller lobe at the metal

[54]R.R. Schrock, et al., J. Am. Chem. Soc., **1978**, 100, 5962.
[55]J. Ushio, H. Nakatsuji, and T.Yonezawa, J. Am. Chem. Soc., **1984**, 106, 5892.

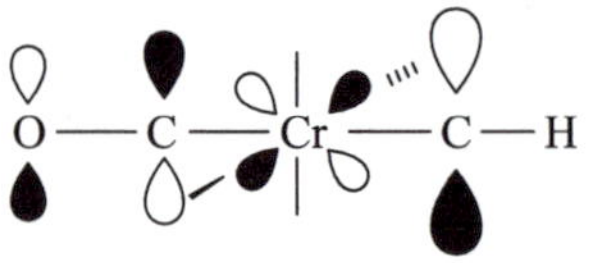

Figure 10-10a
LUMO of a Cationic Fischer-Type
Carbene Complex

$$[\,(CO)_5\,Cr\equiv CH\,]^+$$

(Figure **10-10a**). Instead of being controlled by charge density, such reactions are frontier orbital controlled. Equation **10.36**[56] shows an example of cationic Group 7 carbyne complexes reacting with a variety of nucleophiles, a transformation that is also an excellent method for converting carbynes to carbenes.

$$[\,Cp(CO)_2\,M\equiv C\text{–}Ph]^+ \;+\; Nuc\!:^- \;\longrightarrow\; Cp(CO)_2M\!=\!\!\!<\!\!\!\begin{array}{c} Nuc \\ Ph \end{array}$$

$$Nuc\!:^- = F^-,\,Cl^-,\,Br^-,\,I^- \quad M = Mn,\,Re \qquad\qquad\qquad \textbf{10.36}$$

Nucleophiles react differently with neutral Fischer-type carbyne complexes because now the LUMO and next higher unoccupied orbital (close to the LUMO in energy) of these complexes do not show a distinct region of electron deficiency at $C_{carbyne}$. In addition to $C_{carbyne}$, the metal and the ligands (such as the carbon of a CO ligand) offer sites suitable for nucleophilic attack (Figure **10-10b**, which shows only the lobes on $C_{carbyne}$ and Cr). Equation **10.37** shows the attack of a phosphine or phosphite on the metal of neutral Group 6 carbyne complexes.[57]

$$trans\text{-}Br(CO)_4M\equiv C\text{–}Ph \;+\; L \longrightarrow mer\text{-}Br(CO)_3(L)M\equiv C\text{–}Ph \;+\; CO$$

$$L = PPh_3,\, P(OPh)_3 \quad M = Cr,\, W \qquad\qquad\qquad \textbf{10.37}$$

[56]E.O. Fischer, C. Jiabi, and S. Kurt, *J. Organomet. Chem.*, **1983**, *253*, 231 and E.O. Fischer and J. Chen, *Huaxue Xeubao*, **1985**, *43*, 188.
[57]M.H. Chisholm, K. Folting, D.M. Hoffman, and J.C. Huffman, *J. Am. Chem. Soc.*, **1984**, *106*, 6794.

LUMO   LUMO + 1

Figure 10-10b $Cl(CO)_4 Cr \equiv C-H$
LUMOS of a Neutral Fischer-Type
Carbyne Complex Showing
Only the M–C$_{carbyne}$ Interactions

The reaction is first-order in the presence of excess phosphine with a positive entropy of activation, suggesting that ligand substitution occurs via a dissociative pathway.

Schrock-type carbynes and Group 8 (M = Os, Ru) alkylidynes react with electrophiles, typically at C$_{carbyne}$. Equation **10.38** shows electrophilic addition of HCl across the M≡C of an Os carbyne complex in a manner reminiscent of Markovnikov addition of HCl across an unsymmetrical C≡C. The reaction presumably begins by attack of H$^+$ at C$_{carbyne}$ followed by ligand substitution by Cl$^-$.[58]

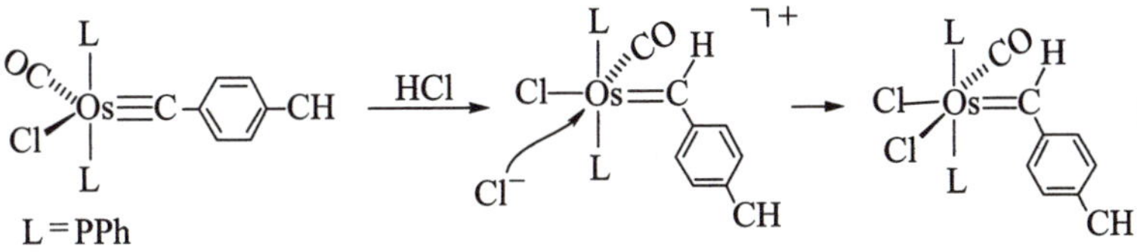

Reaction of the Mo Schrock-type carbyne complex in equation **10.39** with HBF$_4$ results in protonation at C$_{carbyne}$, followed by rearrangement to the thermodynamically more stable Mo–H. The BF$_4^-$ ion is such a weakly coordinating ligand that substitution at the metal does not occur.[59]

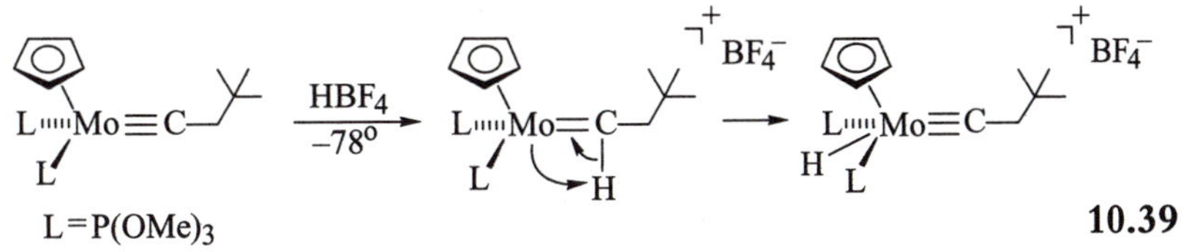

[58] G.R. Clark, K. Marsden, W.R. Roper, and L.J. Wright, *J. Am. Chem. Soc.*, **1980**, *102*, 6570.
[59] M. Bottrill, M. Green, A.G. Orpen, D.R. Saunders, and I.D. Williams, *J. Chem. Soc., Dalton Trans.*, **1989**, 511.

Other electrophiles such as Lewis acids react with carbynes. Although initial attack often occurs at $C_{carbyne}$, the final product may be the result of subsequent reactions that are difficult to rationalize mechanistically.[60]

10-5   Metathesis

Equation **10.40** shows a formal interchange of substituent groups (typically alkyl groups or H) attached to the C=C bond of either the same, or two different, alkenes. The name of such reactions is *olefin*[61] *metathesis* or simply *metathesis*. Discovered independently in the mid 1950s by workers at DuPont, Standard Oil of Indiana, and Phillips Petroleum, metathesis is a reaction catalyzed either homo- or heterogeneously, primarily by complexes of Mo, W, Re, and some Group 4 and 5 metals. The reaction often occurs under very mild conditions (room temperature and 1 atm pressure).

10.40

The original metathesis reactions involved the transformation of low molecular weight alkenes to other simple alkenes (equation **10.41**). Conversion

10.41

[60]The synthesis and reactivity of metal carbynes have been reviewed. See M.A. Gallop and W.R. Roper, *Adv. Organomet. Chem.*, **1986**, *25*, 121; H.P. Kim and R.J. Angelici, *Adv. Organomet. Chem.*, **1987**, *27*, 51; and A. Mayr and H. Hoffmeister, *Adv. Organomet. Chem.*, **1991**, *32*, 227.
[61]Olefin is the older term for alkene. The name derives from two Latin words—*oleum* (oil) and *ficare* (to make)—that were combined to describe the reaction of ethene (a gas) with Cl_2 to give $ClCH_2\text{-}CH_2Cl$ (a liquid or "oil"). Chemists, especially those employed in industrial settings, often use the word olefin interchangeably with alkene.

of propene to ethene and butenes (the Triolefin Process) was commercialized in 1966 but later discontinued as ethylene and butenes became available less expensively from other processes.[62]

Metathesis continues to be used for production of specialty chemicals, and interest in using it as a means to generate new materials remains strong. The economic importance of olefin metathesis to the chemical industry has been the stimulus for research on the mechanism of this reaction. In addition to providing obvious monetary benefits, basic study on the metathesis reaction promoted a synergistic melding of the disciplines of catalysis and organometallic chemistry, the result of which was a significant advancement of our knowledge of carbene complexes and their role in alkene interchange and polymerization.

10-5-1 Mechanism of Metathesis

Equation **10.41** shows only two possibilities for exchange of alkyl substituents during metathesis. Several other product combinations are possible and, given enough time for the reaction to reach equilibrium, all possible products form in amounts based upon their thermodynamic stabilities (see equation **10.42** below and exercise **10–10** for examples). The reaction may be manipulated to achieve desired product combinations. For example, a self-metathesis of a terminal alkene will produce ethene as one of the products. The reaction may be driven toward formation of ethene and an internal alkene by collecting volatile ethene as it forms. The presence of high concentrations of ethene during metathesis, on the other hand, converts an internal alkene to a terminal one.

$$\diagup\!\!\!\diagup + \diagup\!\!\!\diagup\quad \xrightleftharpoons{\text{L}_n\text{M}}\quad = + \diagdown\!\!\diagup + \diagdown\!\!\!\diagup + \diagdown\!\!\diagup$$

$$(+E\text{-isomers})\qquad\qquad \textbf{10.42}$$

Exercise 10-10

Assume 2-pentene and 2-hexene undergo metathesis. At equilibrium, what are all the possible alkenes that could be present, neglecting stereochemistry about the double bond? Remember to consider self-metathesis reactions.

[62]For a summary of the history of the metathesis reaction written by two of its co-discoverers, see R.L. Banks, *Chemtech*, **1986**, *16*, 112 and H. Eleuterio, *Chemtech*, **1991**, *21*, 92.

Elucidation of the mechanism of metathesis, as with any mechanistic pathway, requires knowing which bonds are broken and which are formed. Over the years investigators of this mechanism have considered three major pathways:

1.   Transfer of alkyl groups from one C=C bond to another, as shown in equation **10.43**.

$$R-CH=CH-R$$
$$+$$
$$R'-CH=CH-R'$$
$$\xrightarrow{\quad L_nM \quad}$$
$$R-CH=CH-R'$$
$$+$$
$$R'-CH=CH-R$$

(One possiblility)

10.43

2.   A *pairwise* breakage of C=C bonds followed by the construction of new C=C bonds, as shown in equation **10.44**.

$$R-CH=CH-R$$
$$L_nM$$
$$R'-CH=CH-R'$$

63

$$\begin{array}{c} R \\ | \\ CH \\ || \\ CH \\ | \\ R' \end{array} -L_nM- \begin{array}{c} R \\ | \\ CH \\ || \\ CH \\ | \\ R' \end{array}$$

10.44

3.   A *non-pairwise* breakage of C=C bonds followed by formation of new C=C bonds (Scheme **10.10**).

A single experiment (Scheme **10.9**), involving self-metathesis of a symmetrical alkene, clearly indicated that Mechanism **1** could not occur.[63] A mechanism involving alkyl group transfer by C-C bond cleavage should give several different labeled 2-butene isomers, while cleavage of the C=C bond would give only one structure (neglecting stereoisomers). Since the only product isolated upon equilibrium contained four deuterium atoms (d_4-2-butene), the results are consistent with cleavage of the double bond (Mechanisms **2** or **3**).

Study of organic chemistry tells us that C=C bonds are rather strong (ca. 160 kcal/mole), and yet metathesis, remarkably, is a reaction capable of breaking these bonds and forming new ones. The question remains: **how do C=C bonds break and then reform?**

In 1967 Bradshaw[64] proposed (equation **10.44**) that metathesis occurs first by coordination of the two alkenes to the metal, followed by cycloaddition

[63]N. Calderon, E.A. Ofstead, J.P. Ward, W.A. Judy, and K.W. Scott, *J. Am. Chem. Soc.*, **1968**, *90*, 4133.

[64]C.P.C. Bradshaw, E.J. Howman, and L. Turner, *J. Catal.*, **1967**, *7*, 269.

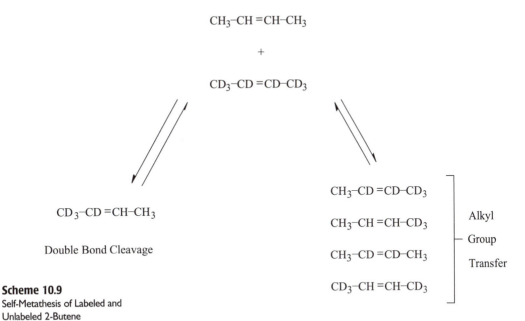

Scheme 10.9
Self-Metathesis of Labeled and
Unlabeled 2-Butene

of the alkenes to form a metal-coordinated cyclobutane (**63**). Finally, the four-membered ring breaks apart in a retrocycloaddition to give two new alkenes. Chemists soon labeled Bradshaw's pathway the "pairwise" or "diolefin" mechanism (Mechanism **2**).

In 1970, Hérisson and Chauvin[65] performed an experiment described in equation **10.45**. They isolated three major products, **64** (C_{10}), **65** (C_9), and **66** (C_{11}), upon metathesis of cyclopentene and 2-pentene. Product **64** is expected as the direct result of pairwise metathesis of the two starting materials; the other two products could result from subsequent reactions of **64** with 2-pentene. At equilibrium all products would be present in near statistical distribution, according to the pairwise mechanism. What troubled Hérisson and Chauvin, however, was the observation that, **upon quenching the reaction well before equilibrium could be achieved, a statistical distribution of the three products was already present.**

Exercise 10-11

Show how **65** and **66** (equation **10.45**) could form using the pairwise mechanism for metathesis.

[65]J.L. Hérisson and Y. Chauvin, *Makromol. Chem.*, **1970**, *141*, 161.

The pathway (Mechanism **3**) that Hérisson and Chauvin proposed to explain their results is outlined in Scheme **10.10**. Two key intermediates in this pathway are an alkene-metal carbene complex (**67**) and a metallacyclobutane (**68**), formed through concerted cycloaddition of the M=C and C=C bonds. A key feature of the mechanism, due to the unsymmetrical structure of **68**, is its explanation of randomization early in the course of reaction. **The Hérisson-Chauvin mechanism does not require a specific pair of alkenes to interact directly for metathesis to occur, hence the name "non-pairwise" mechanism.**

$$L_nM-CH_2R \longrightarrow L_nM=CHR \qquad \text{(Initiation)}$$

$$L_nM=CHR \;+\; R'CH=CHR' \;\rightleftharpoons\; \begin{array}{c} L_nM=CHR \\ | \\ R'CH=CHR' \end{array} \quad \mathbf{67}$$

(Propagation)

$$\begin{array}{c} L_nM \\ \| \\ R'CH \end{array} + \begin{array}{c} HCR \\ \| \\ HCR' \end{array} \;\rightleftharpoons\; \begin{array}{c} L_nM\!-\!CHR \\ | \quad\quad | \\ R'C\!-\!CHR' \\ H \end{array} \quad \mathbf{68}$$

$$L_nM=CHR' \;+\; R'CH=CHR' \xrightarrow{\text{etc.}}$$

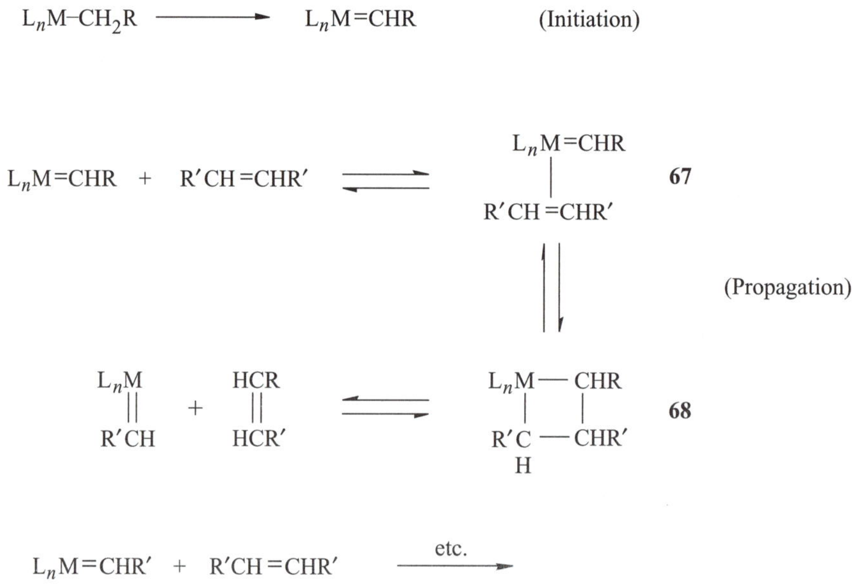

Scheme 10.10
Non-Pairwise Mechanism
for Metathesis

Exercise 10-12

Starting with L_nM=CH-Me or L_nM=CH-Et, cyclopentene, and 2-pentene, show how the non-pairwise mechanism could account for initial formation of **65** and **66** as well as **64** (equation **10.45**).

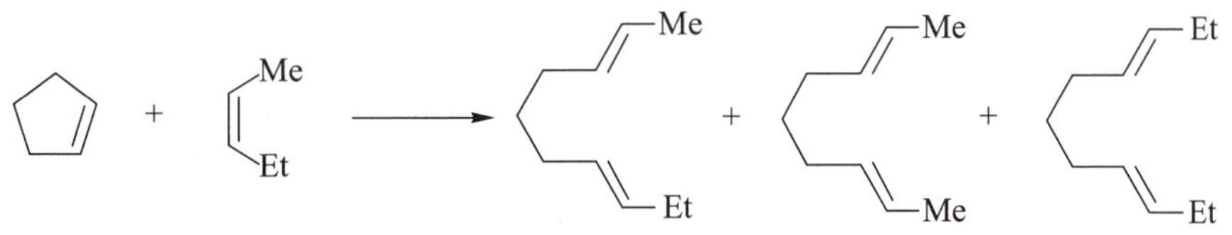

10.45

The proposal of Mechanism **3** was bold for its day because Fischer-type carbene complexes had just been discovered a few years earlier, and those known
were not alkylidenes. The carbene complexes prepared before 1970 also did
not catalyze olefin metathesis. With the discovery of Schrock-type carbene
complexes and the demonstration that some alkylidenes could promote
metathesis, the non-pairwise mechanism became more plausible. It was, however, the elegant work of Katz and co-workers that provided the most substantial support for the Hérisson-Chauvin mechanism.

First, Katz[66] conducted an experiment similar to that of Hérisson and
Chauvin (equation **10.46**), which he termed the "double cross" metathesis. If
Mechanism **2** were operative, the product ratios [**70**]/[**69**] and [**70**]/[**71**]
should be zero when concentrations were extrapolated back to the very beginning of the reaction (t_0), since **70**, the double cross product, would have to
form after the symmetrical products **69** and **71**.

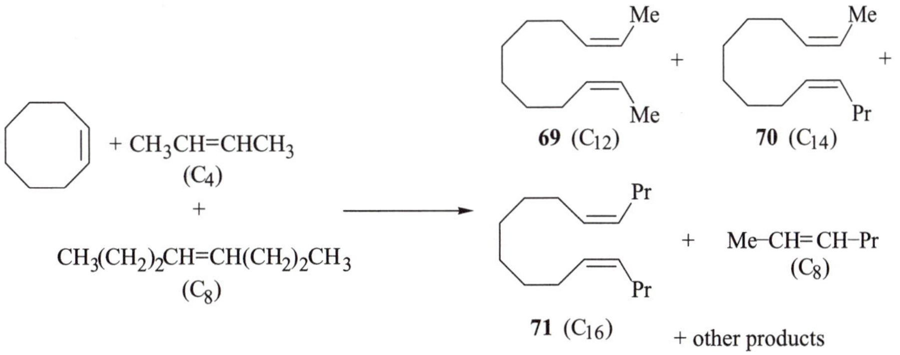

10.46

---

[66]T.J. Katz and J. McGinness, *J. Am. Chem. Soc.*, **1975**, *97*, 1592 and **1977**, *99*, 1903.

The non-pairwise mechanism (Mechanism **3**) should provide some of the unsymmetrical **70** quickly, thus making the ratios [**70**]/[**69**] and [**70**]/[**71**] non-zero at t_0. When the experiment was run, Katz found values of [**70**]/[**69**] = 0.4 and [**70**]/[**71**] = 11.1 at t_0.

In spite of this strong evidence supporting the non-pairwise mechanism, others objected that the pairwise mechanism could explain the results of Katz's experiment if another step in the pairwise mechanism were rate determining. Consider Scheme **10.11**, in which initial "single cross" metathesis is rapid to form the C_{12} alkene (**69**). If **69** sticks to the metal, and 4-octene attacks in a rate-determining step (step b) to displace one of the alkene groups to give **72**, then metathesis would give "double cross" product **70**. If **69** sticks to the metal long enough, several displacements and subsequent metatheses could occur, providing a statistical distribution of products, even at the beginning of the reaction. If, on the other hand, pairwise metathesis (step a) is rate determining, even this modified pairwise mechanism fails to explain the results of Katz, and the earlier results of Hérisson and Chauvin, since the rate-determining step would produce only the single cross product. The proposition that an alkene could stick to the metal and subsequently be displaced by another alkene was called the "sticky olefin" hypothesis, and it constitutes a modification of Mechanism **2**.[67]

At this point, the non-pairwise (Mechanism **3**) and sticky-olefin-modified pairwise processes (Mechanism **2**) both could explain the double cross experiment. More definitive proof was required, however, and this was provided independently by Grubbs and Katz through some clever experiments. In one experiment,[68] metathesis of a 1 : 1 mixture of dialkene **73** and deuterated partner **74**, gives cyclohexene and three ethenes (equation **10.47**). The design of the experiment was astute in at least two respects. First, cyclohexene, though a simple disubstituted alkene, is a notable exception to the rule that sterically unhindered alkenes undergo metathesis; in this case it does not react with starting material or other products in the presence of the Mo catalyst. Second, the ethenes are sufficiently volatile that they could be removed and collected as formed, thus preventing their reaction with starting material.

$$CH_2=CH_2$$
$$+$$
$$CH_2=CD_2$$
$$+$$
$$CD_2=CD_2$$

73   **74**

10.47

[67]N. Calderon, *Acc. Chem. Res.*, **1972**, *5*, 127.
[68]R.H. Grubbs, P.L. Burk, and D.D. Carr, *J. Am. Chem. Soc.*, **1975**, *97*, 3265.

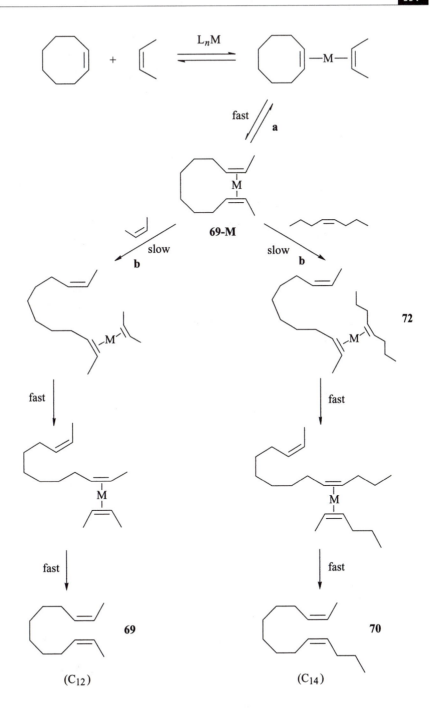

Scheme 10.11
"Sticky Olefin" Hypothesis

Because of these constraints, the initial products observed would truly be those formed during the first metathesis. For example, secondary reactions that could occur via the sticky-olefin-modified pairwise pathway (Mechanism **2**) and lead to scrambled products, would not take place at the initial stages of the reaction. A non-pairwise mechanism predicts that the ratio of $d_0 : d_2 : d_4$-ethenes should be $1 : 2 : 1$, while the sticky-olefin-metathesis mechanism does not; the experiment showed the $1 : 2 : 1$ ratio.

Addition of excess d_0-ethene to the reaction mixture did not affect the ethene-d_2/d_4 product ratio, which means that the ethene products do not react with each other once formed. Moreover, mass spectral analysis of **73** and **74**, both before the reaction started, and after a short reaction time, showed no scrambling of label between the two starting materials. This suggests that **73** and **74** must interact separately with the metal catalyst during metathesis and do not interact with each other before metathesis. Mechanism **3** best explains all of these results.

Equation **10.48** describes another experiment, performed by Katz, and similar to Grubbs' earlier work.[69] A $1 : 1$ ratio of d_0 and d_4-divinylbiphenyls gave a $1 : 2 : 1$ ratio of ethenes, as above. Again, the products are such that secondary metatheses do not occur.

10.48

Since the definitive experiments of Katz and Grubbs, the non-pairwise mechanism is the accepted pathway for metathesis. Subsequent investigations have supported this pathway.[70] Key intermediates in the non-pairwise

[69]T.J. Katz and R. Rothchild, *J. Am. Chem. Soc.*, **1976**, *98*, 2519.

[70]For general reviews covering the mechanism of metathesis, see R.H. Grubbs, *Prog. Inorg. Chem.*, **1978**, *24*, 1 and T.J. Katz, *Adv. Organomet. Chem.*, **1977**, *16*, 283.

mechanism are metal-carbene and metallacyclobutane complexes. Both have been prepared and shown to catalyze metathesis, and more recently both species have actually been observed in the same reaction mixture and shown to interconvert during it, thus offering additional support for Mechanism **3**.[71]

Exercise 10-13

Based upon your knowledge of alkene metathesis, what is the mechanism of the following transformation?

$$2\ Ph-C \equiv C-H \quad \xrightarrow[500^{\circ}]{MoO_3/SiO_2} \quad Ph-C \equiv C-Ph \quad + \quad H-C \equiv C-H$$

10-5-2 Metathesis Catalysts

Metathesis catalysts may be either homo- or heterogeneous. While complexes of Mo, W, and Re seem to show the most activity, metathesis may be catalyzed, in some instances, by Ti and Ta species. Most homogeneous catalysts consist of mixtures of metal halides and some main group metal alkylating agent, such as $WCl_6/EtOH/EtAlCl_2$, $Me_4Sn/WOCl_4$, or $MoCl_2(NO)_2(PR_3)_2/Al_2Me_3Cl_3$. An example of a heterogeneous, supported catalyst system would be Re_2O_7/Al_2O_3.[72]

A key question in the discussion of metathesis is this: **if metal-carbene complexes are true catalysts, produced during the initiation step(s) of the mechanism, how are they formed?** As we see from above, neither the homogeneous nor heterogeneous catalyst systems are metal carbene complexes. Moreover, several carbene complexes of Mo and W have been prepared, and these typically either catalyze metathesis poorly, or not at all. Clearly, carbene complexes must form from the components of the system, but how? This question remains without a satisfactory answer in spite of great efforts to seek an explanation. Equation **10.49** shows how a carbene complex might form from a homogeneous system. In this case the alkylating agent reacts with the M–Cl bond to give an alkylated metal complex. Alpha-elimination occurs, followed by reductive elimination to give an active catalyst. In limited cases this sequence has been observed.[73]

[71]J. Kress, J.A. Osborn, R.M.E. Greene, K.J. Ivin and J.J. Rooney, *J. Am. Chem. Soc.*, **1987**, *109*, 899.

[72]For a good discussion of metathesis catalysts, see R.H. Grubbs in *Comprehensive Organometallic Chemistry: The Synthesis, Reactions, and Structures of Organometallic Compounds*, vol. 8, G. Wilkinson, Ed., Pergamon Press, New York, 1982, 513-525.

[73]R.H. Grubbs and C.R. Hoppin, *J. Chem. Soc., Chem. Commun.*, **1977**, 634.

$$L_n(Cl)W-CH_3 + (CH_3)_4Sn \longrightarrow \left[L_nW \overset{CH_3}{\underset{CH_3}{\diagdown}} \right] \longrightarrow L_{n-1}\overset{H}{\underset{CH_3}{W}}=CH_2 \longrightarrow \begin{array}{c} L_nW=CH_2 \\ + \\ CH_4 \end{array}$$

10.49

While equation **10.49** is perhaps apropos for describing catalyst generation from mixtures of metal halide and alkylating agent, another question presents itself: **how do metal–carbene complexes form when no alkylating agent is present?** Equation **10.50** provides a possible scenario for metal-carbene formation. The sequence begins with 1,2–insertion of the alkene followed by α–elimination. While α–elimination may not be favorable, the presence of just a few metal–carbene sites could be sufficient for catalytic activity. In support of this hypothesis is the observation of M-H bonds in some supported catalyst systems:[74]

$$\begin{array}{c} R-CH=CH-R \\ | \\ M-H \end{array} \longrightarrow \begin{array}{c} R-CH-CH_2-R \\ | \\ M \end{array} \longrightarrow \begin{array}{c} R-C-CH_2-R \\ || \\ M-H \end{array} \quad \textbf{10.50}$$

A definitive answer to how catalytically active metal carbenes form awaits further experimentation.

10-5-3 Industrial Uses for Metathesis

Most cyclic alkenes (cyclohexene being the notable exception) undergo metathesis, but instead of dimerizing to form cyclodienes (equation **10.51**), they polymerize instead to form polyalkenamers (equation **10.52**). Because metathesis involves rupturing the C=C and opening up the ring, the process is called *ring opening metathesis polymerization* or simply, ROMP.

10.51

10.52

cis or *trans* double bonds

[74]D. T. Laverty, J.J. Rooney, and A. Stewart, *J. Catal.*, **1976**, *45*, 110.

This remarkable reaction is used today in the chemical industry, and promises to be an important means toward the production of interesting new materials.

The first good example of ROMP involved cyclopentene.[75] Depending upon the catalyst, good stereoselectivity was possible, producing either all *cis* or all *trans* polymer (equation **10.53**). The reasons for this selectivity are not well understood, although several research groups have proposed models that explain how *cis* or *trans* double bonds may form.[76]

$$10.53$$

Equation **10.54** shows another case where 1-methylcyclobutene polymerizes to form polyisoprene, with *cis* stereochemistry about the C=C. The properties of this polymer are similar to natural rubber, which is also *cis*-polyisoprene. There has been general interest in ROMP because cycloalkenes often polymerize to give materials with elastomeric (rubber-like) properties.

$$10.54$$

One of the most interesting examples of ROMP involves polymerization of norbornene[77] (**75**, Scheme **10.12**). Using a variation of Tebbe's reagent as a catalyst, Grubbs[78] was able to prepare titanobicyclopentane **76**. Complex **76** is a catalyst for polymerization of **75**, but there is an interesting twist to this polymerization. Grubbs termed the process an example of a "living" polymerization because chain growth continued until the supply of monomer was exhausted. Chain growth was also linear as a function of the number of

[75]N. Calderon and R.L. Hinrichs, *Chemtech*, **1974**, *4*, 627.

[76]N. Calderon, J.P. Lawrence, and E.A. Ofstead, *Adv. Organomet. Chem.*, **1979**, *17*, 449; see especially 481–482.

[77]Norbornene possesses a significant amount of ring strain that provides the driving force for essentially irreversible ROMP.

[78]R.H. Grubbs and W. Tumas, *Science*, **1989**, *243*, 907 and references therein.

equivalents of **75**, providing polymer chains with very narrow molecular weight range.[79] Chain termination, in this case, was slow compared to initiation and chain propagation. Therefore, lacking any factors that could interfere with chain growth (presence of oxygen or moisture), the polymer chain awaited "feeding" with additional monomer. This process has possibilities for creating polymers that have homogeneous blocks. For instance, after a certain number of equivalents of **75** was added, another cycloolefin (**77**) could be introduced to give a polymer consisting of two different blocks. Block copolymers often have properties entirely different from homopolymers consisting of only one type of monomer. Scheme **10.12** shows a block polymerization process starting with **75**.

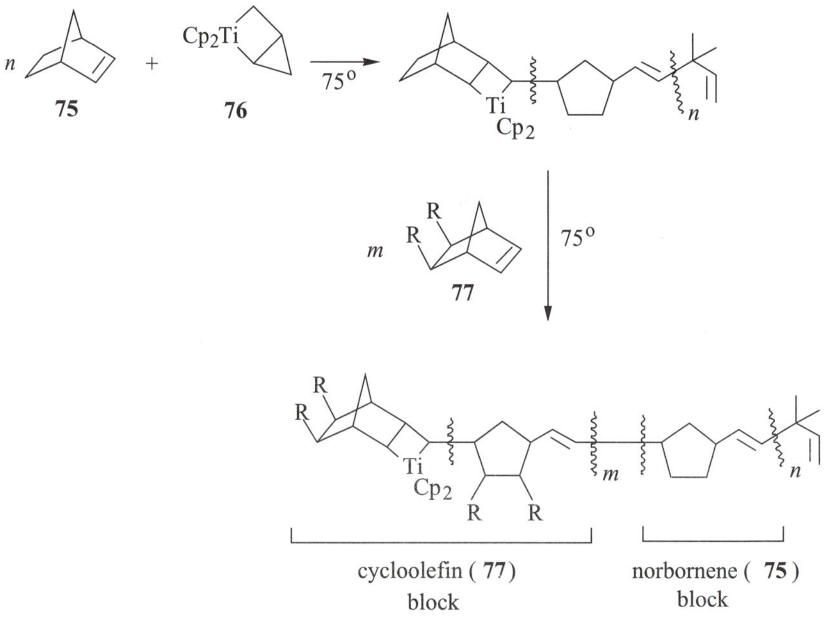

Scheme 10.12
"Living" Ring Opening Metathesis
Polymerization of Norbornenes

[79]Schrock has prepared alkylidene complexes of Mo and W whose activity compared to most Group 6 metathesis catalysts is "toned down" so that they are unreactive to ordinary internal alkenes but actively catalyze polymerization of strained cyclic alkenes, such as norbornene. See R.R. Schrock, *Acc. Chem. Res.*, **1990**, *23*, 158 and references therein.

Recent years have seen rapid advancements in the development of ROMP catalysts, some of the most promising of which are Ru complexes that allow ROMP to occur in an aqueous medium. Water is an attractive solvent for ROMP, not only because its use allows reaction conditions to be less stringent, but also because it offers greater environmental acceptability compared to hydrocarbon-based liquids. Some of the newer catalysts permit the presence of polar substituents in the molecules undergoing metathesis, a condition impossible with older catalysts due to their strong Lewis acidity.[80]

The Shell Higher Olefin Process (SHOP) is another example of the industrial importance of olefin metathesis. We discussed earlier in Chapter 9 (Section **9-2-2**) how phosphine-modified hydroformylation could be used to convert C_{10}-C_{16}-alkenes to the corresponding primary alcohols, compounds used in making detergents and plasticizers. The SHOP process provides a means of producing the linear alkenes needed for hydroformylation, and was developed several years ago by the Shell Oil Company. Scheme **10.13** diagrams this process, which begins with a Ni-catalyzed oligomerization[81] to give alkenes of various chain length. These are distilled into three different fractions corresponding to short-, medium-, and long-chain alkenes. The medium chains are used in the later stages of the process, but the short- and long-chain olefins are isomerized from terminal to internal alkenes. These are metathesized over a heterogeneous catalyst of MoO_3/Al_2O_3, or one that is Re-based. The resulting medium-weight alkenes (combined with a medium-length fraction from above) then undergo hydroformylation, the catalyst for which isomerizes internal double bonds to terminal before hydroformylation and subsequent hydrogenation occur. The olefin metathesis step in the SHOP process is by far the largest application of that reaction, with hundreds of thousands of tons of alkenes processed each year.[82]

10-6   Ziegler-Natta Polymerization of Alkenes

Consumers encounter polyethylene (PE) every day in the form of packaging material. The chemical industry worldwide produces literally billions of pounds of the polymer each year, not only because ethene is cheap and readily available from the petrochemical industry, but also because the resulting polymer has so many useful applications. Around 40 years ago interest was strong among chemists in finding ways of polymerizing ethene and other alkenes under mild conditions. Ziegler in Germany reported in 1952 a method of producing

[80]R.H. Grubbs and W. Tumas, *op. cit.* and *Chemical and Engineering News*, **1993**, 19.

[81]Oligomers are low molecular weight polymers. They are analogous to peptides in the realm of biochemistry whereas proteins would be analogous to high molecular weight polymers.

[82]A.J. Berger, U.S. Patent 3 726 938, 1973; P.A. Verbrugge and G.J. Heiszwolf, U.S. Patent, 3 776 975, 1973; E.R. Freitas and C.R. Gum, *Chem. Eng. Progr.*, **1979**, 75, 73; and W. Keim, F.H. Kowaldt, R. Goddard, and C. Krüger, *Angew. Chem. Int. Ed. Eng.*, **1978**, 17, 466.

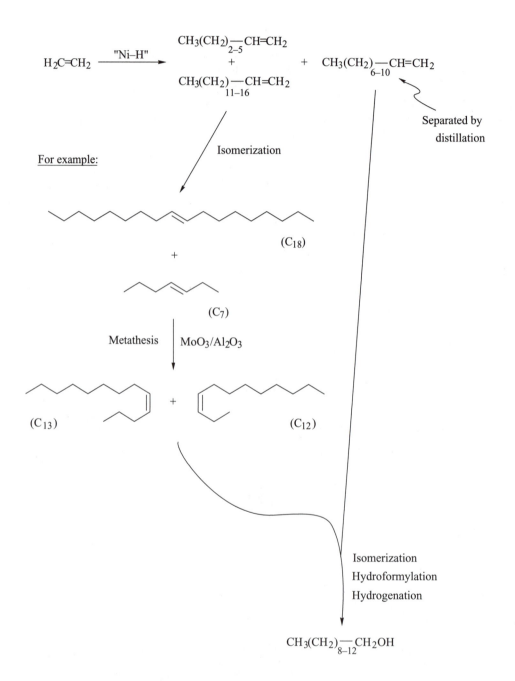

Scheme 10.13
The Shell Higher Olefin Process

oligomers of ethene under conditions of low temperature and pressure in the presence of trialkylaluminum compounds. A year later he discovered that, by adding small amounts of TiCl$_4$ to the system, he could obtain samples of high molecular weight PE with useful mechanical properties. Italian chemist Natta seized upon Ziegler's discovery by using a Ti-Al system to catalyze polymerization of propene, yielding a high molecular weight polymer with good mechanical strength.[83] Up to this time, attempts to polymerize propene under the same high-temperature, radical conditions used to make PE, always produced a low molecular weight, gooey material (often called "road tar" by frustrated chemists) with no commercial value. The years that followed Ziegler and Natta's original discoveries brought the commercialization of processes to manufacture both "high density"[84] PE and polypropylene (PP) as well as other polyolefins, using early transition metals as co-catalysts mixed with alkylaluminum compounds. The scientific and commercial importance of their work was so significant that in 1963 Ziegler and Natta shared the Nobel Prize in Chemistry.

10-6-1 Mechanism of Ziegler-Natta Polymerization

The mechanism of Ziegler-Natta (Z-N) catalysis was unclear for decades after the original reports by Ziegler and Natta. The active catalyst forms from a mixture of a halide of high-valent Ti or V, and an alkyl or chloroalkylaluminum compound, with either component being an ineffective catalyst when present alone. Catalysis may either be homogeneous[85] or heterogeneous, with the latter preferred on an industrial scale. Cossee proposed a mechanism in 1964[86] that is outlined in Scheme **10.14a**. The essential features of Cossee's mechanism require that a coordinatively unsaturated Ti- or V-alkyl (the alkyl group arising from the aluminum alkyl co-catalyst) complex forms, followed by 1,2-insertion of an alkene into the Ti-C bond. Eventually, chain termination occurs, most likely by β-elimination or chain transfer. Chain transfer occurs by complexation of another chain, 1,2-insertion, and β-elimination (see Scheme **10.14a**).

[83]See P. Pino and R. Mülhaupt, *Angew. Chem. Int. Ed. Eng.*, **1980**, *19*, 857 for a brief history of early work in transition metal-catalyzed polymerization.

[84]PE produced under high-temperature, radical conditions (called "low density" PE) has lower density, poorer mechanical strength, and a lower temperature range for use than the PE resulting from Ziegler-Natta catalysts. Low density PE has a great deal of chain branching along the polymer backbone, whereas the high density variety is much more linear. The lack of chain branching in high density PE means that carbon chains can fit together into regular, repeating "crystalline" regions similar to the regular packing patterns of monomeric crystals. The "crystalline" regions of high density PE provide high mechanical strength and higher density due to close packing of atoms in the aligned chains. For more information about the relationship between the molecular structure and physical properties of polymers, see H.R. Alcock and F.W. Lampe, *Contemporary Polymer Chemistry*, 2nd ed., Prentice-Hall, Englewood Cliffs, NJ, 1990, chap. 17.

[85]There are indications that even catalysts that appear to be soluble are actually present as small aggregate particles.

[86]P. Cossee, *J. Catal.*, **1964**, *3*, 80; E.J. Arlman and P. Cossee, *J. Catal.*, **1964**, *3*, 99.

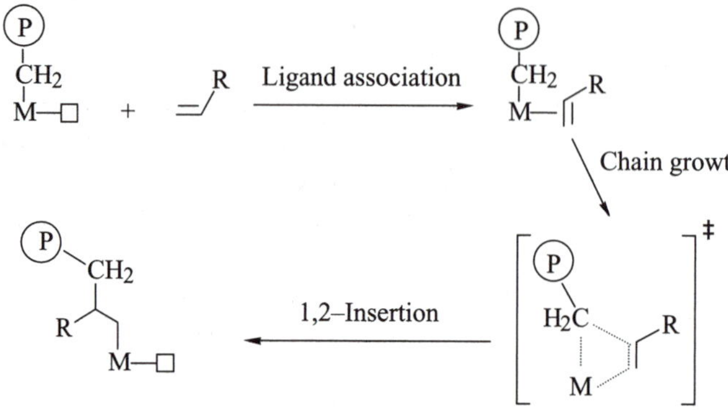

After many insertions, chain termination:

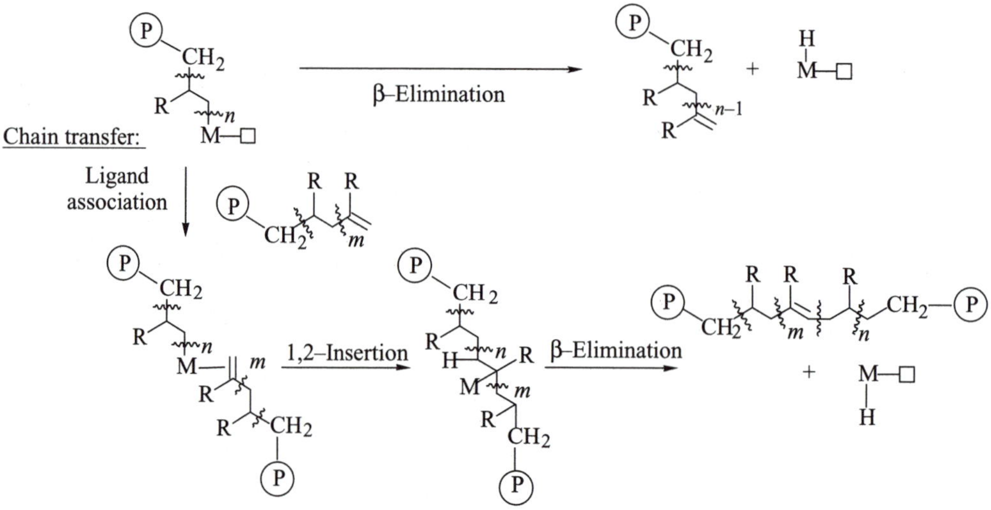

Chain transfer:

(P) : Polymer chain

□ : Empty coordination site

M : Ti or V

Scheme 10.14a

The Cossee Mechanism
for Z-N Polymerization

Although the Cossee mechanism accounted for much of the experimental data related to Z-N polymerization, it was a bold proposal because 1,2-insertions into M–C bonds of early transition metal complexes were unknown at the time. Metal alkyls, should they form by 1,2-insertion into an M–C bond, have a high propensity to undergo loss of a β-hydrogen by 1,2-elimination. Thus chemists asked, even if insertion could occur into the M–C bond, how could the polymer grow in light of facile β-elimination? The answer, we now know, probably rests in the reluctance of some high-valent early transition metal alkyls to undergo β-elimination due to the lack of *d* electrons (see Chapter 8, Section **8-1–2**).

Nevertheless, the lack of evidence for direct M–C alkene insertion was a nagging problem associated with the Cossee mechanism. Several years later, Green and Rooney[87] proposed an alternative mechanism (Scheme **10.14b**) that also accounted for Z-N catalysis. The mechanism resembles a metathesis-like pathway, starting with an α-elimination to give a metal-carbene hydride followed by cycloaddition with the alkene monomer to form a metallacyclobutane. Reductive elimination finally yields a new metal alkyl with two more carbon atoms in the growing chain. The Green-Rooney mechanism, while plausible overall, requires an α-elimination, a process that is difficult to demonstrate.

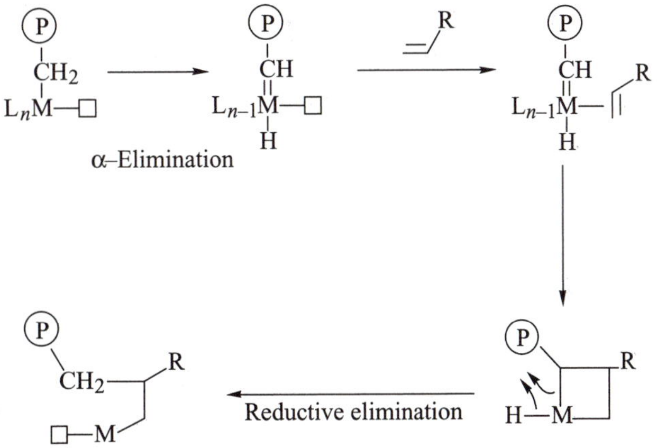

Scheme 10.14b
The Green-Rooney Mechanism
for Z-N Polymerization

[87]K.J. Ivin, J.J. Rooney, C.D. Stewart, M.L.H. Green, and J.R. Mahtab, *J. Chem. Soc., Chem. Commun.*, **1978**, *604*; M.L.H. Green, *Pure Appl. Chem.*, **1978**, *100*, 2079.

Elucidating mechanistic pathways for catalytic processes is always demanding work, but trying to make sense of Z–N catalysis was especially difficult because the structures of neither the active catalysts nor the intermediates and products were completely determined when Cossee, and later Green and Rooney, proposed their routes. In recent years, the efforts of several research groups strongly support the Cossee mechanism as the correct pathway. For instance, Watson[88] reported that an alkyl Lu(III) complex[89] could undergo a 1,2-insertion of propene. The new Lu–alkyl could continue to grow in length, or undergo reversible reactions of both β-hydride and β-alkyl elimination (Scheme **10.15**).

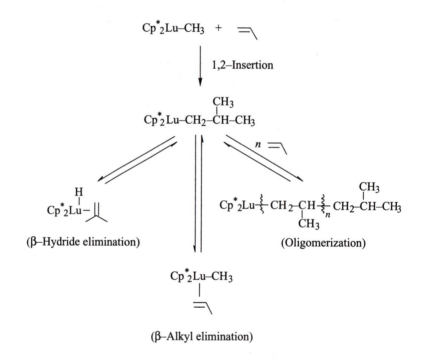

Scheme 10.15
Insertion of Propene
into a Lu-C Bond

Eisch[90] showed that an alkyne could insert into a Ti–C bond, the first such demonstration of this phenomenon (equation **10.55**).

[88]P.L. Watson, *J. Am. Chem. Soc.*, **1982**, *104*, 337.
[89]Lanthanides resemble early transition metals in their chemistry.
[90]J.J. Eisch, A.M. Piotrowski, S.K. Brownstein, E.J. Gabe, and F.L. Lee, *J. Am. Chem. Soc.*, **1985**, *107*, 7219.

$$\text{Ph–C} \equiv \text{C–SiMe}_3 \; + \; \text{Cp}_2\text{TiCl}_2 \bullet \text{MeAlCl}_2 \longrightarrow$$

Ph   SiMe$_3$

H$_3$C   TiCp$_2$ (+)

10.55

The experiments just described point out the feasibility of the 1,2-M-C insertion described by the Cossee mechanism, but they fail to distinguish between it and the Green-Rooney pathway. Grubbs[91] reported definitive evidence in support of the Cossee mechanism when he measured the rate of polymerization (in the presence of catalyst **78**) of a 1:1 mixture of $H_2C=CH_2$ and $D_2C=CD_2$ (equation **10.56**). There was no kinetic isotope effect, thus supporting the Cossee mechanism.

$$
\begin{array}{c}
D_2C=CD_2 \\
+ \\
H_2C=CH_2 \\
(1:1)
\end{array}
\xrightarrow{\text{Cp}_2(\text{Et})(\text{Cl})\text{Ti/EtAlCl}_2 \, (\textbf{78})}
CH_3CH_2(CH_2CH_2)_m\,(CD_2CD_2)_n\,H
$$

10.56

$m:n = 1:1$

(Random order of deuterated and nondeuterated C_2 units)

Exercise 10-14

Explain how Grubbs' observation of a lack of hydrogen isotope effect during polymerization in the presence of **78** would argue for the Cossee mechanism and against that of Green and Rooney.

Although the lack of a kinetic isotope effect argued against the Green-Rooney pathway, the conclusion was based upon negative evidence rather than direct support for the Cossee scheme. Later, Grubbs[92] reported an ingenious experiment that clearly supported the Cossee mechanism and excluded the other one. Schemes **10.16a** and **b** show the experiment in some detail. The starting material was alkenyl titanocene **79**, which, in the presence of EtAlCl$_2$, cyclized to give **80** by intramolecular insertion of the remote alkene group into the Ti-C bond (equation **10.57**). The mixture of **79** and EtAlCl$_2$ was an appropriate catalyst mimic of Z-N polymerization because the system also reacted with ethene to give ethene oligomers capped with a ring (equation **10.58**).

---

[91] J. Soto, M. Steigerwald, and R.H. Grubbs, *J. Am. Chem. Soc.*, **1982**, *104*, 4479.

[92] L. Clawson, J. Soto, S.L. Buchwald, M.L. Steigerwald, and R.H. Grubbs, *J. Am. Chem. Soc.*, **1985**, *107*, 3377.

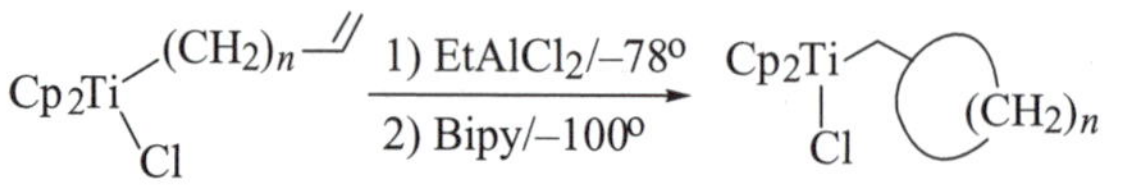

10.57

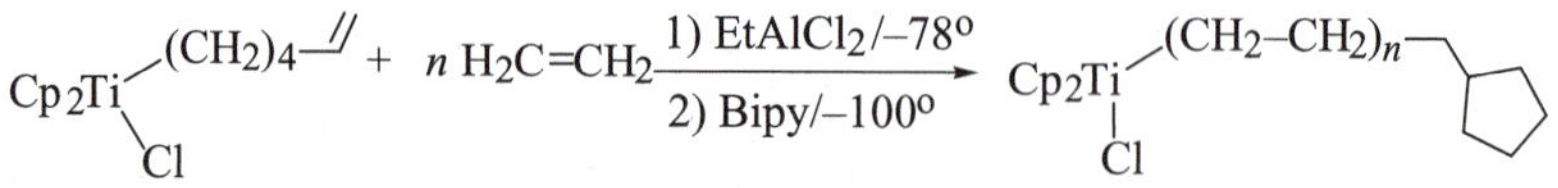

10.58

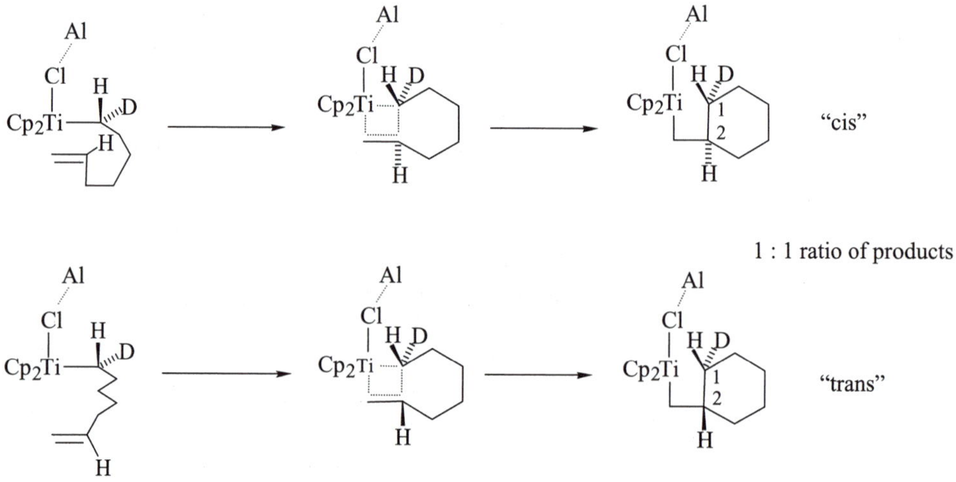

Scheme 10.16a
The Grubbs Stereochemical Isotope
Experiment: The Cossee Mechanism

Grubbs reasoned that the tethered alkene group would react at the Ti center, by either a direct insertion of the alkene into the Ti–C bond (the Cossee mechanism, Scheme **10.16a**), or by an α-hydrogen activated pathway (the Green-Rooney mechanism, Scheme **10.16b**). The experiment was designed to measure a "stereochemical" isotope effect as well as a kinetic isotope effect according to the following logic:

      If α elimination to the metal is the required first step, then it *will* make a difference whether H or D is lost at the α position. The resulting monosubstituted cyclohexanes will form according to the ease of loss of H with respect to D. In other words, the ratio of *cis* to *trans* (for example, *cis* means the D at the 1-position is *cis* with respect to H at the 2-position) will not be 1 : 1. If direct insertion occurs (the Cossee mechanism), on the other hand, the preference for H over D is irrelevant because a C–H(D) is not broken. The ratio of *cis* to *trans* should be 1 : 1. Schemes **10.16a** and **b** follow the two mechanistic pathways when $n = 5$ and for only one of the two possible α–deuterio enantiomers.

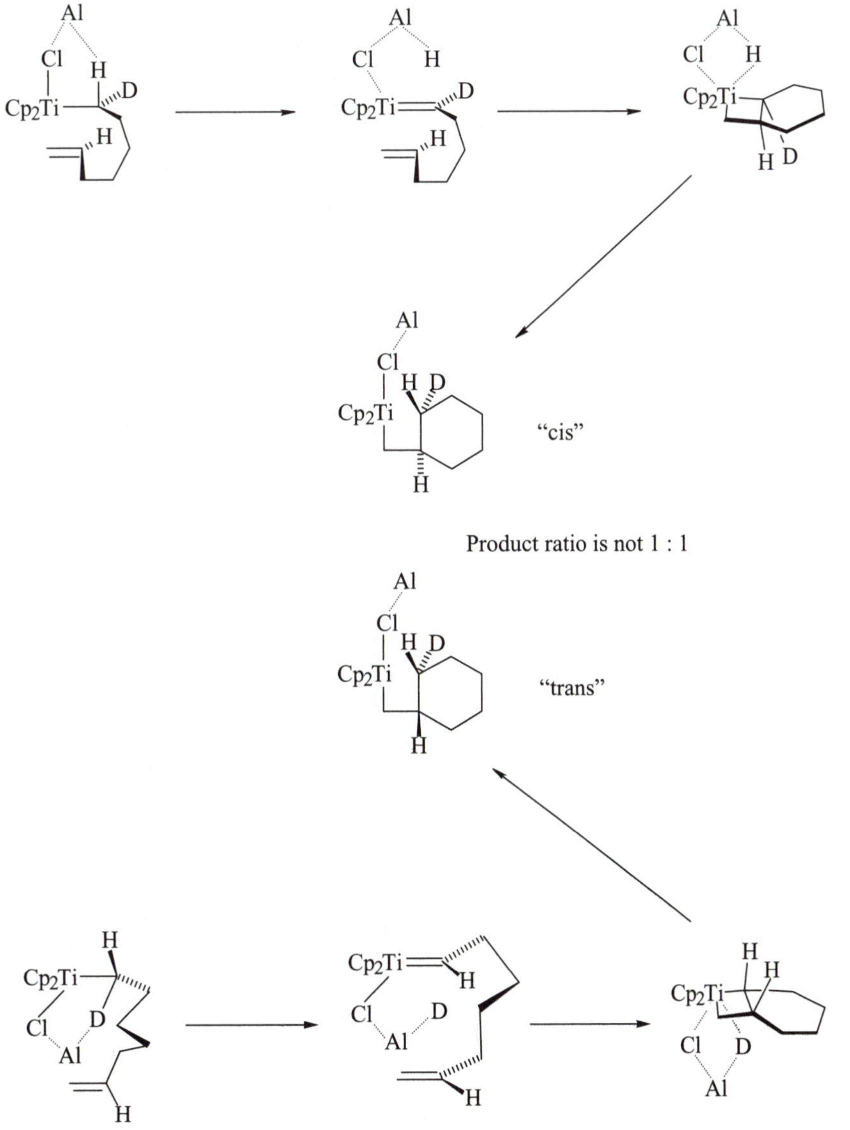

"cis"

Product ratio is not 1 : 1

"trans"

Scheme 10.16b
The Grubbs Stereochemical Isotope
Experiment: The Green-Rooney Mechanism

In all cases Grubbs found that the *cis* to *trans* ratio was always 1 : 1, thus demonstrating that α-activation does not influence the rate or stereochemistry of alkene insertion. The result of the experiment was the key piece of evidence supporting the Cossee mechanism for Z-N polymerization long sought after by chemists. The experiment allowed researchers to make a clear distinction between metathesis and Z-N polymerization, the former involving the chemistry of the M=C bond, and the latter that of the M–C bond.[93]

10-6-2 Stereochemistry of Ziegler-Natta Polymerization

Polymerization of monosubstituted alkenes introduces stereogenic centers along the carbon chain at every other position. When, for example, propene undergoes Z-N polymerization, three possible geometries are possible: *isotactic, syndiotactic,* and *atactic* (equation **10.59**).

Note that isotactic polymers have all substituent groups (called *pendant* groups) on the same side of the chain, syntdiotactic polymers have alternating stereochemistry for the pendant groups, and atactic polymers have a random stereochemical arrangement of substituents. All three types of polypropylene (PP) have been prepared, each having rather different mechanical properties. The potential existed early on to produce hundreds of polyalkenes, not only with different composition due to a change in pendant group; but also up to three different kinds of *stereoregular* polymers. In the years since Natta's first discovery of isotactic PP, progress in the design of catalysts capable of producing polymers with specific stereochemistry has been substantial.

[93]In at least one case there is an example of ethene polymerization using a Ta carbene complex in which there is strong evidence for a metathesis-based mechanism. See H.W.Turner and R.R. Schrock, *J. Am. Chem. Soc.*, **1982**, *104*, 2331, a paper that describes oligomerization of up to 35 ethene units in the presence of Ta[=CH(t-Bu)](H)(PMe$_3$)$_3$I$_2$. It seems clear that while the Cossee mechanism is operative when polymerization occurs in the presence of Z-N-type catalysts, some polymerizations may involve metathesis, especially when hydrido metal carbenes can form readily.

Even today, however, there is not complete understanding of how different stereochemistries result from different catalyst systems, especially under heterogeneous conditions. In general, Ti-Al catalysts tend to give isotactic and atactic PP, while VCl_4-$AlEt_2Cl$ gives mainly the syndiotactic polymer. Homogeneous Zr complexes, with rigid, cyclic π ligands, mixed with methylaluminoxane ($[Al(CH_3)\text{-}O]_n$–MAO) catalyze Z-N polymerization to give stereoregular polymers. Scheme **10.17** shows how formation of isotactic PP might occur using the chiral, ethano-bridged, *bis*-tetrahydroindenyl Zr complex **81**.[94] Note how propene adds to the metal in such a manner that the

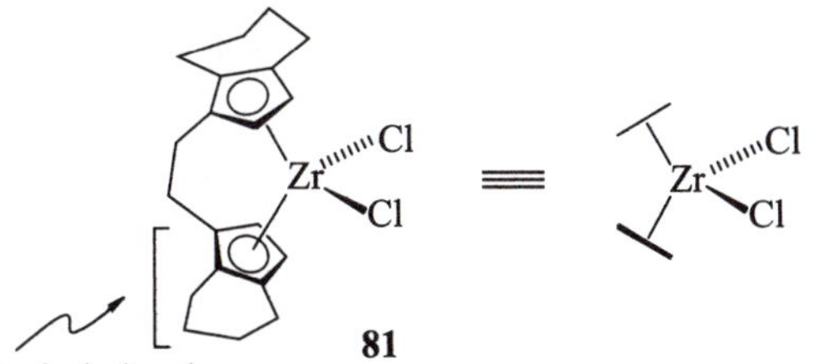

Tetrahydroindenyl group   **81**

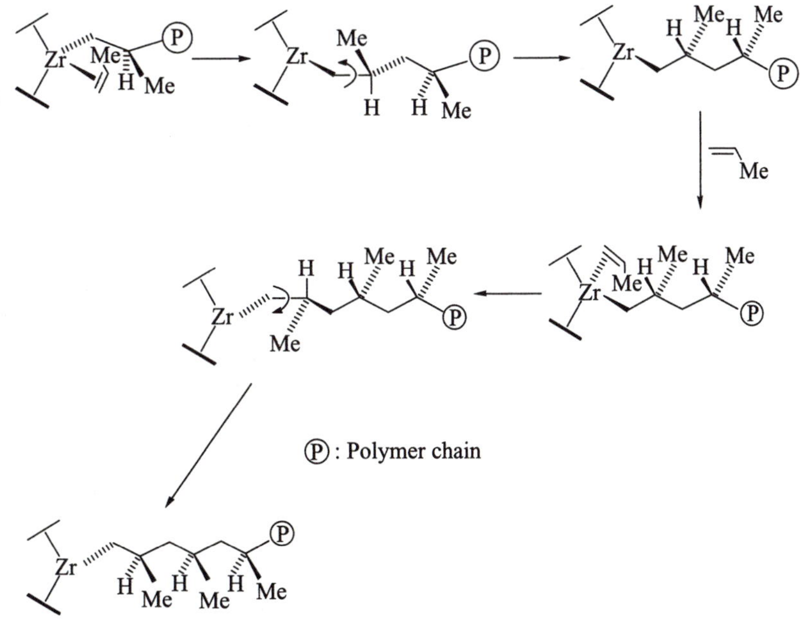

Ⓟ : Polymer chain

Scheme 10.17

A Mechanism for Isotactic Polymerization

[94]W. Kaminsky, W. K. Kulper, H.H. Brintzinger, and F.R.W.P. Wild, *Angew. Chem. Int. Ed. Eng.*, **1985**, *24*, 507.

methyl group ends up in the least sterically congested quadrant away from the cyclopentane portion of the tetrahydroindene. Upon insertion, the growing chain rotates and a second equivalent of propene complexes, again such that the methyl group is in the least sterically congested quadrant. The end result gives a chain with the pendant methyl groups on the same side.

Syndiotactic polymerization, shown in Scheme **10.18**, occurs with the Cp-fluorenyl Zr complex **82**.[95] Note that **82** has a plane of symmetry bisecting the two cyclic π ligands, while **81** has a C_2 symmetry axis,[96] but no symmetry plane. Here, the propene complexes so that the methyl group always points toward the less sterically demanding Cp ligand. Insertion, followed by rotation of the growing chain, provides a polymer with methyl groups positioned on opposite sides.[97]

Metallocene-catalyzed Z-N polymerization, similar to that described in Scheme **10.18**, is finding increased use on an industrial scale. Several different catalyst systems, consisting of combinations of metallocenes and Lewis acids, have been used to produce commercially useful polymers. Much effort at both the fundamental and applied level has occurred over the last few years to understand how these different catalysts affect the stereoregularity and molecular weight of the polymers produced.[98]

Exercise 10-15

Using the pathways described in Schemes **10.17** and **10.18** as a guide, what stereochemistry would you expect for PP when **83** (the *meso* form of **81**) is used as a co-catalyst instead of one of the enantiomers of **81**? Explain.

83

[95] J.A. Ewen, R.L. Jones, and A. Razavi, *J. Am. Chem. Soc.*, **1988**, *110*, 6255.

[96] A molecule that possesses a C_2 symmetry axis may be rotated by $360°/2$ (or $180°$) to obtain a geometry equivalent to the starting geometry. In general, a C_n axis corresponds to rotation by $360°/n$.

[97] Recently reported calculations support the Cossee mechanism as energetically feasible for Zr-based stereoregular polymerizations. See L.A. Castonguay and A.K. Rappé, *J. Am. Chem. Soc.*, **1992**, *114*, 5832.

[98] P. A. Deck and T. J. Marks, *J. Am. Chem. Soc.*, **1995**, *117*, 6128 and references therein.

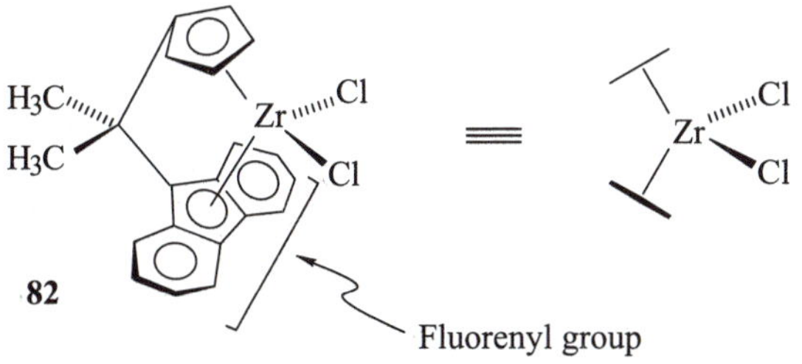
H_3C
H_3C
Zr
Cl
Cl
Fluorenyl group

82

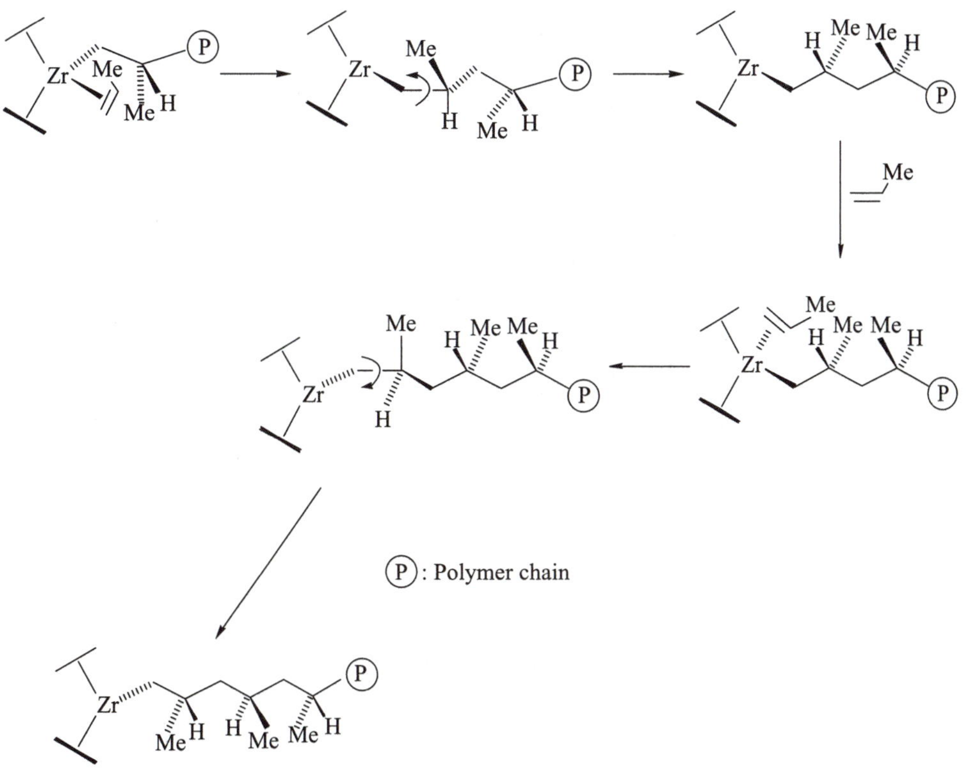
Zr
Me
H
(P): Polymer chain

Scheme 10.18
Mechanism for Syndiotactic Polymerization

10-7 Sigma (σ) Bond Metathesis

In Chapter 7 we discussed C-H and H-H activation via oxidative addition to a metal center as having great synthetic and economic importance. In recent years interest has focused on using high valent complexes of the early transition and lanthanide metals to activate σ bonds. Because these complexes are electron deficient, the OA pathway is not available. In spite of this, several groups reported that some complexes of $d^0 f^n$ early transition metals, lanthanides, and even actinides, do react with the H-H bond from dihydrogen and the C-H bonds from alkanes, alkenes, arenes, and alkynes, in a type of exchange reaction termed *sigma (σ) bond metathesis* (SBM), shown in general form in equation **10.60**.

$$L_nM\text{-}R \; + \; R'\text{-}H \; \longrightarrow \; \left[L_nM \begin{array}{c} R \\ \diagdown \\ \diagup \\ R \end{array} H \right]^{\ddagger} \; \longrightarrow \; L_nM\text{-}R' \; + \; R\text{-}H$$

10.60

So many examples of SBM have appeared in the chemical literature over the past 10 years that chemists consider it now to be another fundamental type of organometallic reaction, along with oxidative addition, reductive elimination, and others we have already discussed.

As with olefin metathesis, we may consider SBM to be formally a 2 + 2 cycloaddition, but unlike its π bond counterpart, SBM does not require the formation of a four-membered ring intermediate. Instead, a four-center, four-electron transition state occurs along the reaction coordinate, which calculations have shown to be kite-shaped. In ordinary organic reactions, 2 + 2 cycloadditions—whether they involve σ or π bonds—are disallowed[99] under thermal conditions. Organometallic metatheses, on the other hand, involve a metal that has the ability to utilize not only s and p orbitals, but also d and f (in the case of the lanthanides and actinides) orbitals, singly or in combination. Figure **10-11** shows how molecular orbitals of starting material, transition state, and products, correlate well in a pathway involving a low-energy transition state.[100] There is continuous bonding along the reaction coordinate for all the occupied orbitals, the hallmark of a concerted, thermally allowed reaction. Note how the hydrogen appears at the top of the "kite" position in the transition state. This is due to the spherical nature of the hydrogen s orbital that can best accommodate the angular overlap required at that position. Were carbon to occupy this position, the orbitals available would be shaped to point in a specific direction and thus unable to overlap smoothly with other orbitals at the kite's side positions.[101]

---

[99]The term "disallowed" means that, in order for the reaction to occur in a concerted fashion, the reaction coordinate would pass through a high-energy transition state. Lower energy paths may be possible, but these would involve a nonconcerted mechanism with radical or charged intermediates.
[100]Note that the MOs for the transition state represent a σ bond analogy to the bonding and nonbonding π MOs of the allylic system.
[101]For recent articles describing molecular orbital calculations relating to SBM, see T. Ziegler, E. Folga, and A. Berces, *J. Am. Chem. Soc.*, **1993**, *115*, 636; A.K. Rappé, *Organometallics*, **1990**, *9*, 466 and references therein.

$$Cp_2Sc-CH_3 \quad + \quad H-R \quad \longrightarrow \quad Cp_2Sc-R \quad + \quad CH_4$$

$$R = Alkyl$$

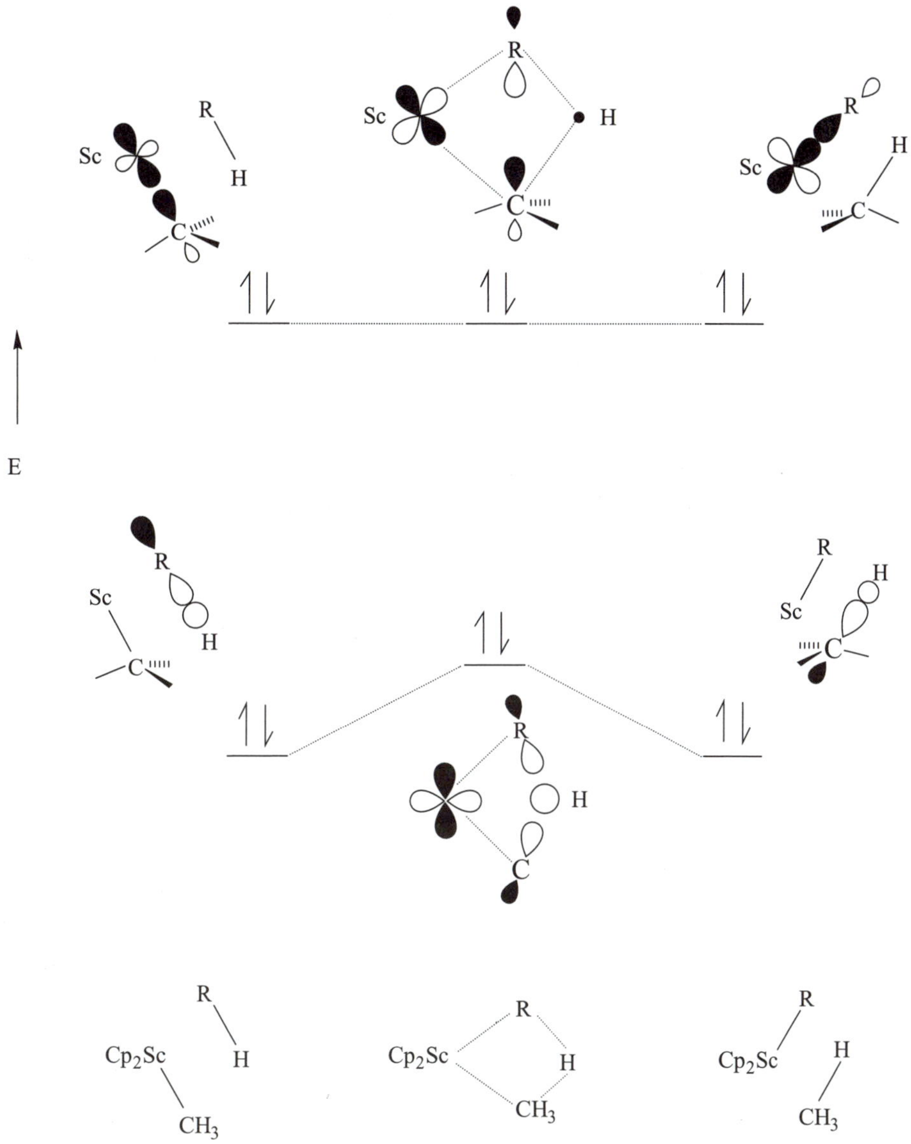

Figure 10-11
Molecular Orbital Correlation
Diagram for Sigma Bond Metathesis

Bercaw[102] first coined the term σ *bond metathesis* in 1987. For the general reaction

$$Cp^*_2Sc\text{-}R' + R\text{-}H \;\rightleftharpoons\; Cp^*_2Sc\text{-}R + R'\text{-}H$$

he found that the order of reactivity was R = R' = H >> R = H, R' = alkyl >> R-H = *sp* C-H, R' = alkyl > R-H = sp^2 C-H, R' = alkyl > R-H = sp^3 alkyl, R' = alkyl. The trend seems counter to what one might expect based on bond energies. It does, however, reflect a relationship that suggests that the greater the *s* character of a σ bond, the greater its reactivity. Equations **10.61** and **10.62** provide examples of his work with d^0 Sc-alkyl complexes.

$$Cp^*_2Sc\text{-}CH_3 \;+\; PhH \xrightarrow{\;80°\;} Cp^*_2Sc\text{-}Ph \;+\; CH_4 \qquad\qquad \textbf{10.61}$$
$$sp^2$$

$$Cp^*_2Sc\text{-}CH_3 \;+\; H\text{-}C\equiv C\text{-}CH_3 \xrightarrow{\;<0°,\ fast\;} Cp^*_2Sc\text{-}C\equiv C\text{-}CH_3 \;+\; CH_4$$
$$sp \qquad\qquad\qquad\qquad\qquad\qquad \textbf{10.62}$$

Exercise 10-16

Some alkynes undergo M–C insertion as well as metathesis. Provide a mechanism for the following transformation:
$$Cp^*_2ScC\equiv C\text{-}Me + 2H\text{-}C\equiv C\text{-}Me \rightarrow Cp^*_2ScC\equiv C\text{-}Me + H_2C=C(Me)\text{-}C\equiv C\text{-}Me$$

[Hint: The reaction involves two major steps.]

SBM may also occur intramolecularly, as equation **10.63** shows.[103]

$$\textbf{10.63}$$

[102]J.E. Bercaw, *et al.*, *J. Am. Chem. Soc.*, **1987**, *109*, 203.
[103]L.R. Chamberlain, I.P. Rothwell, and J.C. Huffman, *J. Am. Chem. Soc.*, **1986**, *108*, 1502; see also I.P. Rothwell, *Accts. Chem. Res.*, **1988**, *21*, 153 for more examples of intramolecular SBM.

The involvement of SBM in alkene polymerization is another area of current research interest. Although chain growth in Z-N polymerization occurs by 1,2-insertion into an M–C bond, chain termination may involve SBM. For example, H_2 is sometimes added to curtail the length of polymer chains. Watson[104] observed that H_2 reacts with Lu-alkyls according to equation **10.64**. The reaction is undoubtedly a SBM, and as such, provides a good model for what actually occurs when H_2 is used to limit polyalkene chain length.

$$Cp^*_2Lu-(CH_2CH)_nR' \quad \xrightarrow{\ H_2\ } \quad Cp^*_2Lu-H \ + \ H-(CH_2CH)_nR'$$
$$\underset{R}{|} \qquad\qquad\qquad\qquad\qquad\qquad\qquad \underset{R}{|}$$

10.64

Exercise 10-17

Besides β-elimination and the addition of dihydrogen, another mechanism to limit chain length during Z-N polymerization is called chain transfer (see Scheme **10.14a**). In this process the growing polymer detaches from the metal center by exchanging places with monomer or another free polymer chain. Could a mechanism involving SBM rationalize this phenomenon? Explain.

The chemistry of carbene (and the analogous carbyne) complexes is an area of great interest to organometallic chemists. Metathesis reactions that involve carbene complexes as intermediates have many useful industrial applications, especially in the area of ring opening metathesis polymerization. While we now know that the production of stereoregular polymers by Ziegler-Natta catalysis does not involve carbene complexes, the chemistry associated with this polymerization is linked to carbene complex chemistry through early mechanistic investigations. The last section of this chapter demonstrated that another type of metathesis exists in which σ M–C bonds, especially those involving the early transition metals, interact with other σ bonds. Sigma bond metathesis now ranks with oxidative elimination and 1,2-insertion as a fundamental organometallic reaction type.

[104]P.L. Watson and G.W. Parshall, *Acc. Chem. Res.*, **1985**, *18*, 51 and references therein.

Problems

10-1   Assuming that the carbene complex shown below reacts similarly to a ketone and that phosphorus ylides are good carbon nucleophiles, predict the products in the following reaction:

$$(CO)_5W=C(Ph)(Ph) + Ph_3P=CH_2 \quad \rightarrow \quad +$$

10-2   The chemistry outlined below was used to produce a cyclic arene-chromium carbene complex.[105] Note: MeOTf is a methylating agent and is equivalent to Me_3O+.

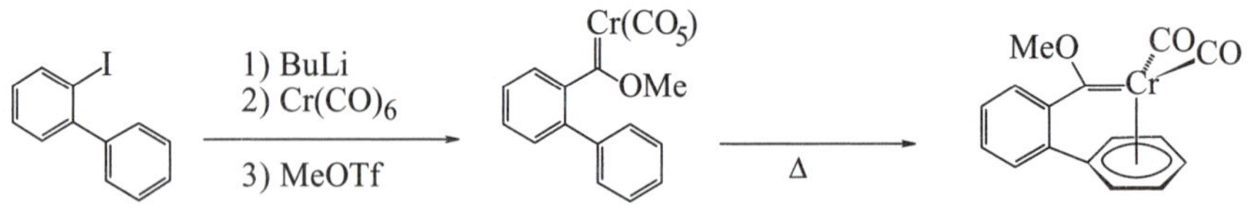

Cyclic arene-Cr carbene complex

a.   Describe in detail what happens in each reaction step.

b.   The carbene complex above was interesting because it was unreactive toward nucleophiles under conditions that cause ordinary Fischer-type carbene complexes to react readily (for example, it did not undergo the chemistry shown in equations **10.18** to **10.20**). X-ray analysis showed that the M-C$_{carbene}$ bond distance was 195.3 pm and the bond distance between the metal and carbonyl carbon (M–CO) was 184 pm. This contrasts with M-C$_{carbene}$ and M–CO bond distances of 200 to 210 pm and 180 pm respectively for Fischer-type carbene complexes such as $(CO)_5Cr=C(OCH_3)(Ph)$. Explain why the cyclic arene-Cr carbene complex is not electrophilic in its chemical behavior.

10-3   Propose a reasonable mechanism for the transformation shown below. Assume that the first step is protonation at C$_{carbyne}$.

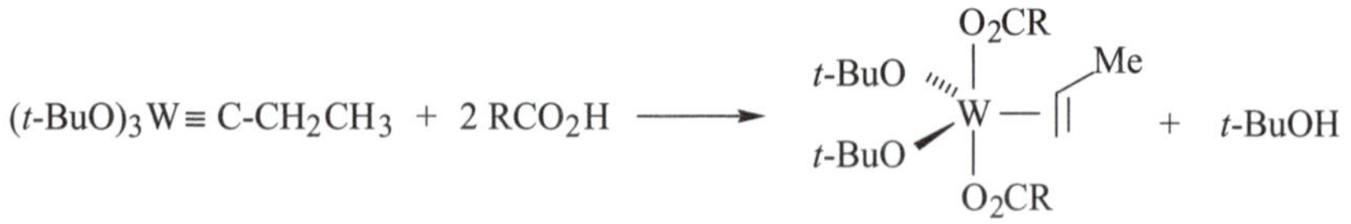

$$(t\text{-BuO})_3W \equiv C\text{-}CH_2CH_3 + 2\ RCO_2H \longrightarrow$$

10-4   Metallacyclobutadienes are proposed as intermediates in metathesis reactions of alkynes. They formally may be considered to be derived from a metal and an allyl group:

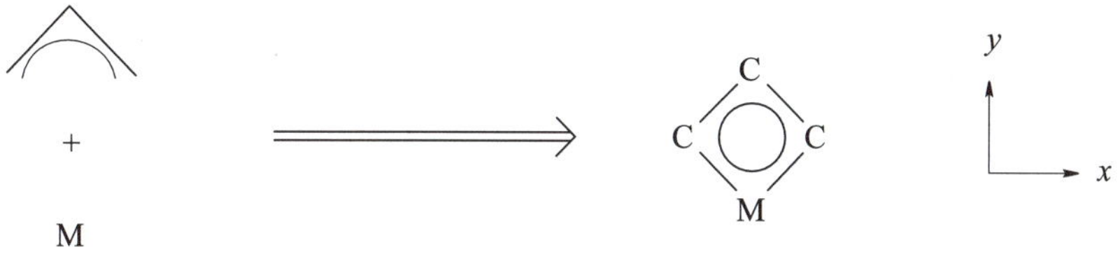

  a.   Sketch the π orbitals of the allyl group. Also indicate the relative energies of these orbitals.
  b.   For each of the π orbitals, determine the metal orbitals suitable for interaction.
  c.   Which orbital on the metal would you expect to be most strongly involved in σ bonding with the end carbons?

10-5   Propose a mechanism that would explain the following transformation. Note that no other reagents are required for the reaction.

10-6   The reaction conditions below were used to produce a polymer by ROMP. Predict what the structure of the polymer will be.[106]

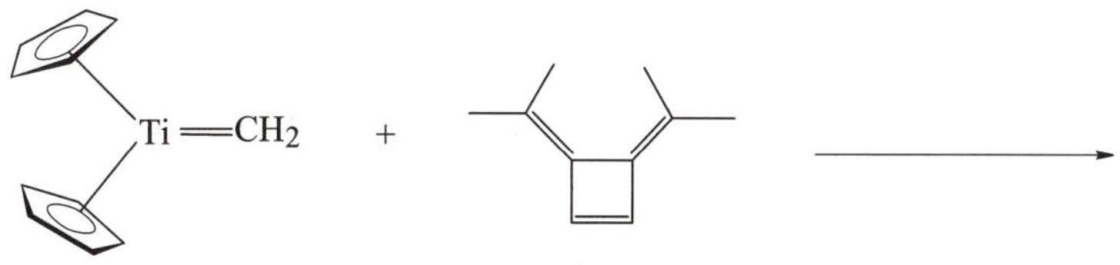

         (Generated *in situ*)

10-7   Propose a reasonable catalytic cycle for the Ni-catalyzed oligomerization of ethene that occurs during the first stage of the Shell Higher Olefin Process. Assume that the active catalyst is L_nNi-H.

[106]T.M. Swager and R.H. Grubbs, *J. Am. Chem. Soc.*, **1987**, *109*, 894.

10-8   When $Cp^*_2Sc–Me$ was allowed to react with d_6–benzene (C_6D_6) at 125°, a 1 : 1 : 1 : 1 mixture of CH_3D, CH_4, **A**, and **B** was obtained. Explain.

$$Cp^*_2\, Sc(C_6D_5)$$

A

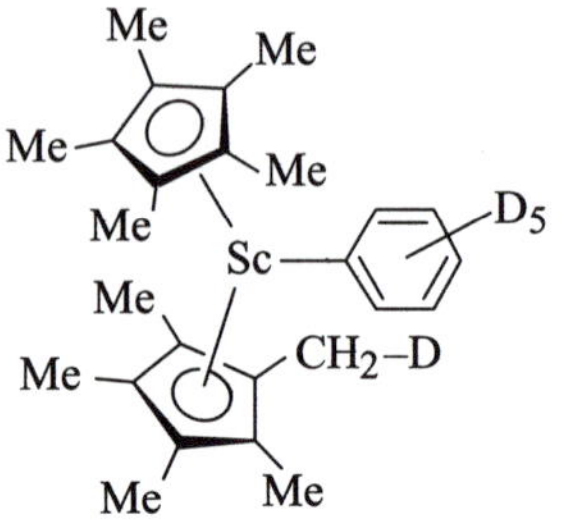

B

11

Applications of Organometallic Chemistry to Organic Synthesis

Organic synthesis is the science *and* art practiced by chemists who concern themselves with the construction of carbon-containing molecules, many of which possess biological significance. Nature, of course, constitutes the premier laboratory for the creation of organic compounds. Natural products—representing targets for laboratory synthesis—have provided chemists with tremendous challenges, which have been difficult, even impossible to meet using yesterday's technology. The last decade has seen an explosion of papers reporting new synthetic methods and descriptions of successful total syntheses of highly complex molecules. Of considerable assistance to synthesis chemists has been the recent development of new reagents and reaction conditions involving organometallic compounds—particularly complexes of the transition metals.

There are several approaches to describing the application of organotransition metal chemistry to organic synthesis. For example, one could examine the field—metal-by-metal—reporting on uses for each element. Another strategy might be to scrutinize several syntheses of interesting molecules, each of which includes the use of organometallic reagents in key steps. The approach of this chapter, however, will be different. It will build upon the coverage in previous chapters of fundamental reaction types, catalytic processes, and carbene chemistry. We will then use that knowledge to discuss a few basic synthetic transformations. These transformation types are not necessarily tied to the use of only one of the transition metals; usually several different ones could work. While coverage of the entire field of synthetic applications is impossible in one chapter,[1] it is possible for the reader to appreciate the utility of transition metals by considering the following kinds of transformations, listed in approximate order of increasing complexity and accompanied by a typical example:

Enantioselective functional group interconversions:

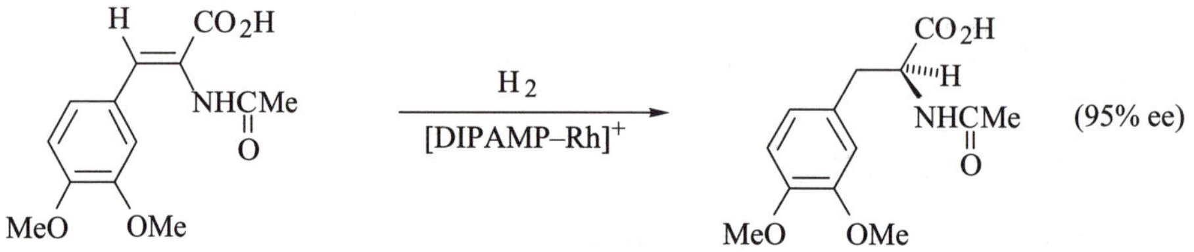

Use of organotransition metal complexes as protecting and activating groups:

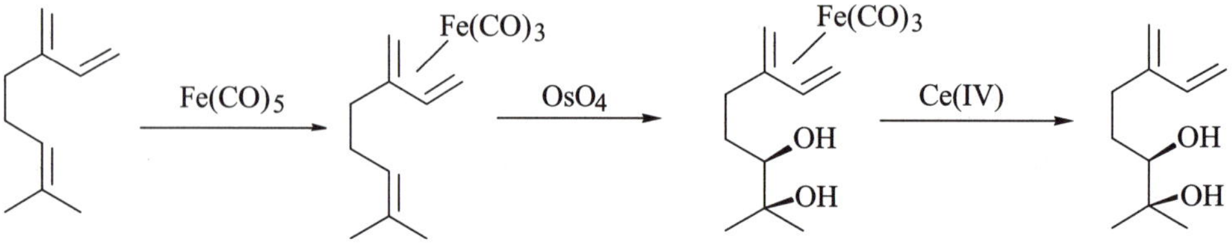

Carbon–carbon bond formation via carbonyl and alkene insertion:

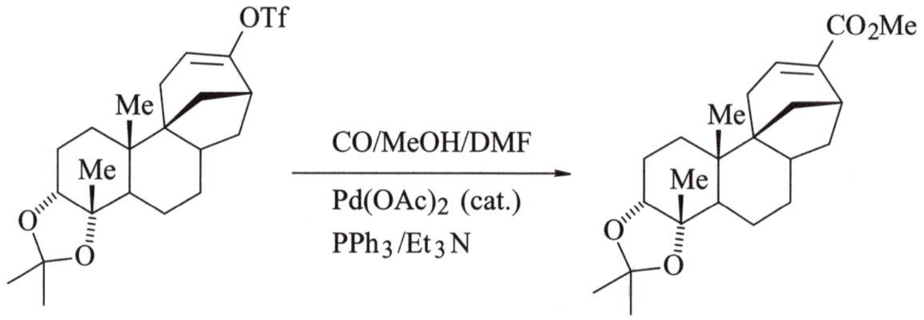

[1]Entire books or major parts of some books cover the application of organometallic chemistry to organic synthesis. For some examples, see S. G. Davies, *Organotransition Metal Chemistry: Applications to Organic Synthesis*, Pergamon Press, Oxford, 1982; J.P. Collman, *et al.*, *Principles and Applications of Organotransition Metal Chemistry*, University Science Books, Mill Valley, CA, 1987, chap. 13-20; P.J. Harrington, *Transition Metals in Total Synthesis*, Wiley, New York, 1990; F.J. McQuillan, D.G. Parker, and G.R. Stephenson, *Transition Metal Organics for Organic Synthesis*, Cambridge University Press, Cambridge, 1991; and L.S. Hegedus, *Transition Metals in the Synthesis of Complex Organic Molecules*, University Science Books, Mill Valley, CA, 1994.

Carbon-carbon bond formation via transmetallation reactions:

F—⟨benzene ring⟩—Br + ⤳—CH(OEt)—Sn(*n*-Bu)₃ ⟶ F—⟨benzene ring⟩—CH₂—CH=CH—OEt

Carbon-carbon bond formation through cyclization reactions:

Me–C≡C–H   +   (CO)₅Cr=C(OMe)(CH=CH₂) —Δ→ HO—⟨benzene ring, Me⟩—OMe

11-1   Enantioselective Functional Group Interconversions

The conversion of one functional group into another represents a fundamental process in organic synthesis. We can convert ketones to alcohols and alcohols to ketones by a reduction in the former case and an oxidation in the latter. The double bond in an alkene is rich in its chemistry, and undergoes transformation to alcohols, alkyl halides, and alkanes depending upon reaction conditions. Although C=O and C=C bonds are planar and provide an achiral reaction site, the interaction of these functional groups with specific reagents often creates one or more stereogenic centers in the reaction product, as reaction **11.1** shows.

⟨reaction scheme⟩ O=⟨ketone⟩ —NaBH₄/MeOH→ HO,H⟨alcohol⟩ + H,OH⟨alcohol⟩   **11.1**

1   :   1

In this case the achiral reagents reacting via an achiral (or racemic) intermediate in an achiral solvent should produce racemic product. Years ago this is all we could expect of such a reaction in terms of stereoselectivity. If the purpose of the transformation was to obtain one or the other enantiomer, then special and often tedious methods were required to resolve the racemate. Today the goal of synthesis chemists often is to produce molecules that are not only chiral, but also enantiomerically pure. Biologically active molecules are typically chiral and exist in only one enantiomeric form. Efficacious drugs are also usually chiral, and, in order to interact properly with a chiral active site, must themselves be administered—not as a racemic modification, but as a single enantiomer. A racemic drug is a mixture of stereoisomers, 50% of which is efficacious. The other 50% is at best worthless and at worst toxic, sometimes severely so.[2]

With the goal in mind of mimicking nature or producing materials even more efficacious than those naturally occurring, chemists have discovered several methods for obtaining a particular enantiomer of a chiral compound. Among these are the following:

1. Chemical resolution of racemic modifications
2. Chiral chromatography
3. Use of chiral natural products as starting materials
4. Stoichiometric use of chiral auxiliaries
5. Asymmetric catalysis

Method 1 represents the oldest technique for producing selectively one enantiomer, and readers should already be familiar with it.[3] A chromatography column normally is an achiral environment; elution of a racemic modification through the column should result in no separation into enantiomers. In Method 2, however, columns are modified by attaching chiral groups to the solid support. Now a chiral environment does exist. Chiral column chromatography can sometimes resolve racemic modifications, providing either enantiomer with a high degree of purity.[4]

Allowing nature to do part of the work is the central theme of Method 3. Numerous chiral molecules, isolated from natural sources and often available commercially, already contain much of the appropriate stereochemistry required in an enantioselective synthesis. These compounds are called the "chiral pool." One reported synthesis of biotin (**2**), a molecule involved in enzymatic transfer of CO_2, used the methyl ester of the amino acid cysteine (**1**) as starting material.[5] Note how **1** possesses a key stereocenter that later appears in biotin.

           1   **2**

[2]The classic example of a racemic drug, one enantiomer of which is beneficial and the other toxic, is thalidomide. The efficacious enantiomer is a tranquilizer and the other isomer causes severe birth defects when given to pregnant women. Section **9-6-2** in Chapter 9 describes the stereoselective synthesis of the analgesic drug (*S*)-naproxen; the (*R*)-isomer, also analgesic, is a liver toxin at its therapeutically effective dose (which is much higher than the dose for the (*S*)-isomer).

[3]See J. McMurry, *Organic Chemistry*, 4th ed. Brooks/Cole, Pacific Grove, CA, 1996, 313-315.

[4]For a discussion on chiral chromatography, see E. Juaristi, *Introduction to Stereochemistry and Conformational Analysis*, Wiley, New York, 1991, 132-136.

[5]E.J. Corey and M.M. Mehrotra, *Tetrahedron Lett.*, **1988**, *29*, 57.

Attaching a chiral group to a reagent and then performing a reaction that goes through two possible diastereomeric transition states is the basis for Method 4. The presence of a *chiral auxiliary* provides an environment in which two pathways—diastereomerically related—are possible between reactant and product. One pathway is usually lower in energy due to steric hindrance, and is more favorable. The result, upon removal of the chiral auxiliary, is selection for one enantiomer over the other. Absence of the chiral auxiliary during the transformation would produce a racemic modification because the transition state would be achiral. Scheme **11.1** shows an enantioselective alkylation of a species that is the equivalent of an enolate of a carboxylic acid. The acid is first converted to a derivative, known as an oxazoline (**3**), which is chiral. Treatment

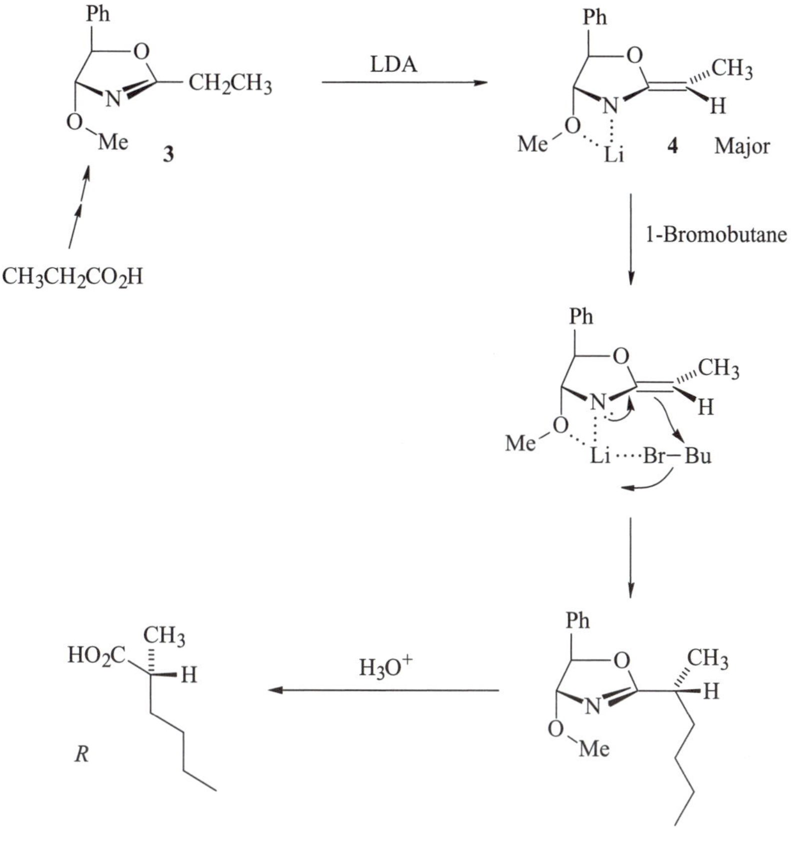

Scheme 11.1
Enantioselective Synthesis Using a Chiral Auxiliary

of **3** with lithium diisopropyl amide (LDA), a very strong base, provides the enolate (**4**) that is then allowed to react with an electrophile (1–bromobutane in this case). The lithium ion can coordinate not only to the nitrogen and to the methoxy group, but also to the halide ion of the incoming electrophile (if it attacks on the bottom face of the enolate). The bulky phenyl group, pointing upward, also tends to prevent attack on the top face of **4**. These two factors lower the energy of the transition state that involves bottom face attack. The result, upon removal of the chiral auxiliary by hydrolysis, is selection for the *R*-enantiomer in this case. There are several examples of use of Method 4 in the chemical literature,[6] but the technique does suffer from one disadvantage—one reaction is required to add the chiral auxiliary and another to remove it.

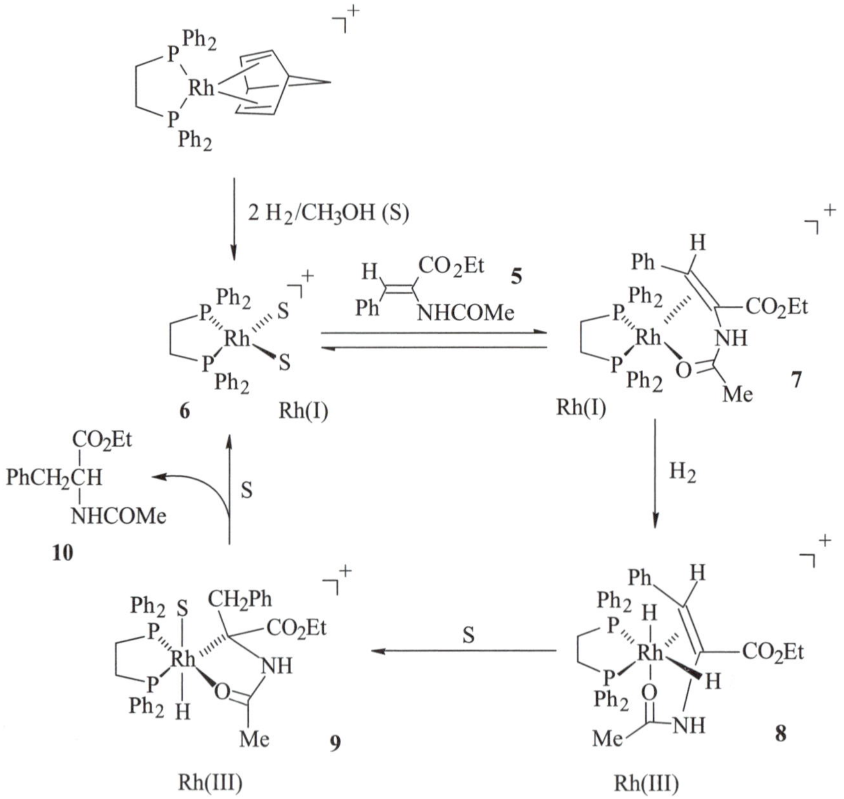

Scheme 11.2
Rh-DIPHOS-Catalyzed Hydrogenation of EAC

Method 5 is similar to the fourth technique except that asymmetric induction occurs *catalytically* without requisite separate stoichiometric steps for adding and later removing a chiral auxiliary. In nature, enzymes serve as giant chiral auxiliaries that bind substrates to active sites where chemical transformation occurs enantioselectively before release of product. Organotransition metal catalysts in some cases duplicate the high stereoselectivity of enzymic systems, and the remainder of Section **11-1** (see also some examples in Chapter 9, Sections **9-4-2** and **9-6-2**) will describe some of the most successful examples of asymmetric catalysis using organometallic complexes.

11-1-1 Asymmetric Hydrogenation Using Rhodium Complexes

Section **9-4-2** in Chapter 9 discussed an example of asymmetric hydrogenation in the synthesis of L-Dopa. We now have a good understanding of how Rh (I) complexes can catalyze the enantioselective addition of H_2 across an unsymmetrical C=C bond, thanks to the work of Halpern.[7] Scheme **11.2** shows the catalytic cycle for hydrogenation of ethyl (Z)-1-acetamidocinnamate (EAC, **5**) using a cationic Rh(I)-DIPHOS complex (**6**) as the catalyst. Since DIPHOS is an achiral ligand, the environment at the initial stages in the cycle is also achiral, and no enantioselectivity is possible. The most common ligands used in asymmetric hydrogenation are, however, chiral diphosphines that usually possess a C_2 symmetry axis; Figure **11-1** shows just a few of the dozens of diphosphine ligands that have been reported in the literature.

Halpern reasoned that, if he could work out the essential features of the catalytic cycle using readily available DIPHOS, he could then apply this knowledge to hydrogenations using chiral ligands. Employing a combination of techniques, he was able to determine the rate constant for each step in the cycle and to characterize all intermediates except dihydride **8**, formed during the rate-determining step. The formation of **8** does seem reasonable, however, based upon previous studies encountered already in Chapter 7 on the oxidative addition of H_2 to square planar Rh complexes. The other steps in the cycle should be quite familiar by now: **6** to **7** (ligand binding), **8** to **9** (1,2-insertion), and **9** to **6** (reductive elimination).

[6]For a review on the use of oxazoline chiral auxiliaries, see K.A. Lutonski and A.I. Meyers, *Asymmetric Syntheses via Chiral Oxazolines,* in *Asymmetric Syntheses,* J.D. Morrison, Ed., Academic Press, San Diego, CA, 1984, vol. 3, 213–274.

[7]J. Halpern, *Science,* **1982,** *217,* 401.

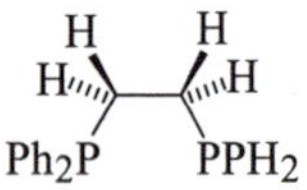

DIPHOS
(achiral)

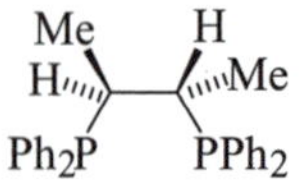

(S,S)-CHIRAPHOS

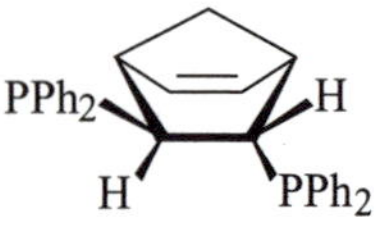

(S,S)-NORPHOS

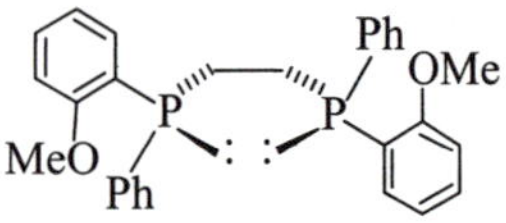

(R,R)-DIPAMP

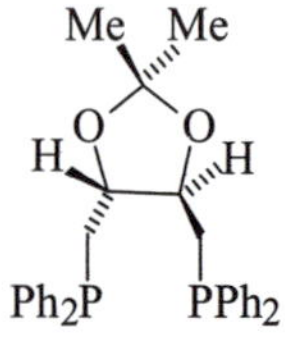

(S,S)-DIOP

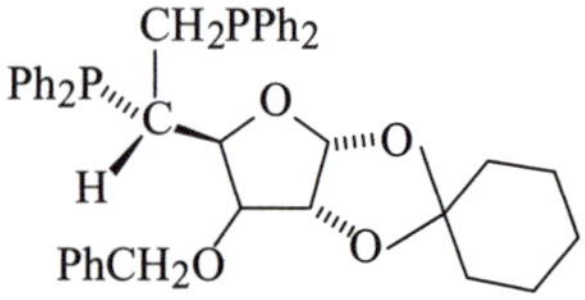

GLUCOPHOS

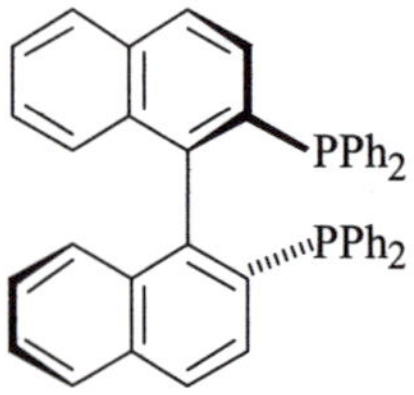

(R)-BINAP

Figure 11-1
Chiral Phosphine Ligands

The presence of a chiral ligand, such as CHIRAPHOS, complicates the cycle because now there are two parallel pathways that are diastereomerically different,[8] as shown in Scheme **11.3** (on the next page). Halpern was able to isolate **7′** and obtain its crystal structure.

If hydrogen adds to **7′** in accord with the mechanism depicted in Scheme **11.2**, then the final hydrogenation product should be N-acetyl-(S)-phenyl-alanine ethyl ester (**10′**, Scheme **11.3**). Halpern found, however, that the predominant product in the presence of CHIRAPHOS was the R-enantiomer (**10″**, Scheme **11.3**)! Based on this result and other evidence, it was possible for Halpern to say that **7′** and **7″** form as an equilibrium mixture rapidly and reversibly from reaction of **5** and **6**. Although **7′** is more stable than **7″**, the minor isomer (**7″**) reacts much faster during rate-determining oxidative addition of H₂, eventually leading to the R-amino acid.

---

[8]Since the CHIRAPHOS ligand possesses two stereocenters, the intermediates **7′** and **7″** are not mirror images and not superimposable.

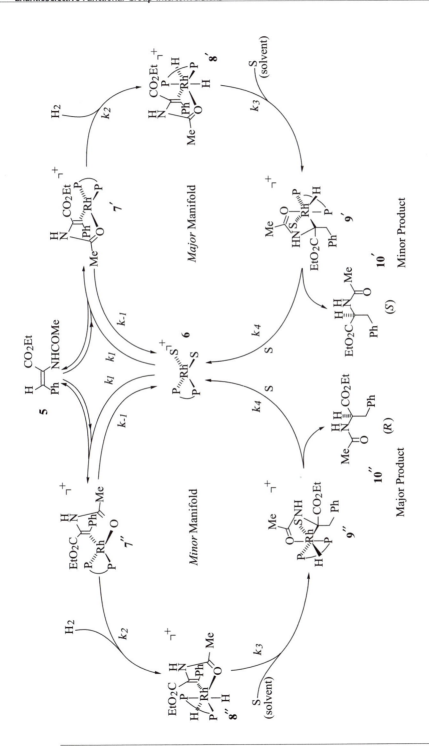

Scheme 11.3
Mechanism of Asymmetric Hydrogenation Using Rh-CHIRAPHOS

Exercise 11-1

Based on the stereochemistry associated with 1,2-insertion and reductive elimination, verify that in Scheme **11.3**, **8′** will transform ultimately into the S-amino acid and **8″** into the *R*-enantiomer.

The origin of enantioselectivity is thus the lower $\Delta G^{\ddagger}$ (free energy of activation) for the path **7″** to **8″** than for **7′** to **8′**. Figure **11-2** below shows a reaction coordinate-free energy diagram for the rate-determining step.

Calculations show that **7′** is probably more stable than **7″** because the alkene ligand fits better into the pocket formed by the phenyl groups of the CHIRAPHOS ligand in **7′**, as shown in Figure **11-3**.

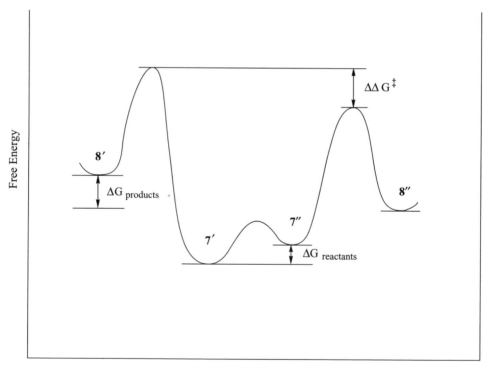

Reaction Coordinate

Figure 11-2

Reaction Coordinate-Free Energy Diagram for the
Rate-Determining Step of Asymmetric Hydrogenation

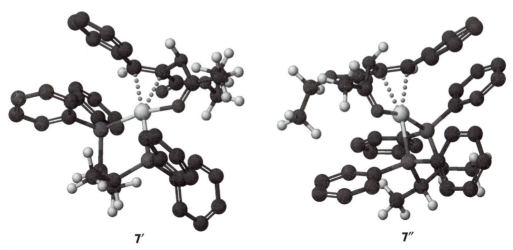

7′   **7″**

Figure 11-3
Comparison of **7′** and **7″** Showing Quality of Fit of the Alkene Ligand in the Pocket Provided by CHIRAPHOS

The difference in free energy alone between **7′** and **7″** is insufficient to explain the relative-rate difference leading to an enantiomeric excess (ee)[9] of *R* vs *S*-amino acid of 96%, that is if one assumes that ΔG is approximately equal to ΔΔG‡. In order to explain an ee of 96%, Halpern also speculated that there was a reversal in stability of the two products **8′** and **8″**, with **8″** being more stable than **8′**. More recent calculations show this to be the case.[10] As Figure **11-2** shows, the higher energy of **7″** *and* the lower energy of **8″**, when compared to their diastereomers, act synergistically to provide a much lower pathway for **7″** going to **8″** than for **7′** to **8′** (this is in agreement with the Hammond Postulate; see Chapter 7, Section **7-1-3**).

Later, Halpern conducted the same experiments using a Rh-DIPAMP complex to catalyze hydrogenation of methyl-(*Z*)-1-acetamidocinnamate (MAC), and the results were entirely analogous to the CHIRAPHOS system.[11] The important lesson learned in Chapter 9, Section **9-4-2** again applies in the mechanism for asymmetric hydrogenation—isolable intermediates such as **7′** are typically *not* the active species involved in a catalytic cycle.

Exercise 11-2

Based upon the mechanism depicted in Schemes **11.2** and **11.3**, what would the effect on enantioselectivity be if the pressure of H₂ were increased?

[9]See footnote 53, Chapter 9.
[10]Calculations performed by the authors of this textbook.
[11]C.R. Landis and J. Halpern, *J. Am. Chem. Soc.*, **1987**, *109*, 1746.

Table **11-1** lists several examples of asymmetric hydrogenations. The highest enantioselectivity seems to result when the substrate is a (Z)-α-amidocinnamic acid derivative, as shown below:

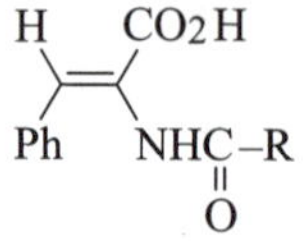

(Z)-α-Amidocinnamic acid derivative

Divergence from this "ideal" substrate tends to lower ee (entries 10 and 11, Table **11-1**). In order for catalysis to be efficient, it is necessary to have a group attached to the alkene double bond that can bind to the metal (e.g., the amide carbonyl group). This secondary binding helps to lock the C=C in a rigid conformation in the presence of the chiral diphosphine ligand thus enabling stereoselection to occur. Ordinary unsymmetrical alkenes, with only alkyl substituents attached to the double bond, undergo hydrogenation with much less enantioselectivity. Substituents attached to the carboxyl and amido groups may have some effect on the overall enantioselectivity, but their impact is relatively difficult to predict. In spite of the limited number of alkenes that undergo

Table 11-1 Asymmetric Rh-Catalyzed Hydrogenation of Alkenes

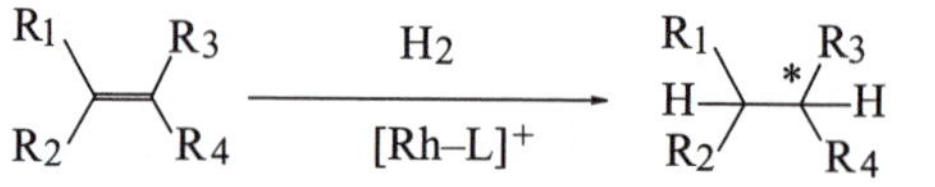

Entry No.	R_1	R_2	R_3	R_4	L	% ee	Configuration at * position
1[a]	H	Ph	CO_2H	NHCOMe	(S,S)-CHIRAPHOS	99	R
2[a]	H	Ph	CO_2H	NHCOMe	(R,R)-DIPAMP	96	S
3[a]	H	Ph	CO_2H	NHCOMe	(S,S)-NORPHOS	95	S
4[b]	H	Ph	CO_2H	NHCOPh	(R)-BINAP	100	S
5[a]	H	Ph	CO_2H	NHCOMe	(S,S)-DIOP	82	S
6[b]	Ph	H	CO_2H	NHCOPh	(S)-BINAP	87	S
7[a]	H	H	CO_2H	NHCOMe	(S,S)-CHIRAPHOS	92	R
8[a]	H	H	CO_2H	NHCOMe	(R,R)-DIOP	71-88	R
9[a]	H	Ph	CO_2Me	NHCOMe	(R,R)-DIPAMP	97	S
10[a]	Me	Me	CO_2Me	NHCOMe	(R,R)-DIPAMP	55	S
11[a]	H	Ph	CO_2H	Me	GLUCOPHOS	61	S

[a] Data taken from K.E. Koenig, *The Applicability of Asymmetric Homogeneous Catalytic Hydrogenation in Asymmetric Syntheses*, J.D. Morrison, Ed., Academic Press, San Diego, CA, 1985, vol. 5, 71-101.
[b] Data taken from S.L. Blystone, *Chem. Rev.*, **1989**, 89, 1663.

highly stereoselective hydrogenation, it is remarkable that in some cases enantioselectivity can be greater than 99%. This selectivity is possible under mild conditions (ca. 1 atm. H_2 and 25°-50°) at rates that are quite high for transition metal catalysis (100 turnovers/sec). Such results are comparable to those obtained from enzymatic systems.

Recent work on new Rh-based catalysts indicates that even better enantioselectivity is possible during hydrogenation of α-amidoacrylic acid derivatives. Workers at DuPont synthesized several chiral diphosphines (**11a-c**)—collectively called DuPHOS ligands—that promote high enantioselectivity when bound to Rh(I). When R = Et (**11b**), amino acids of either absolute configuration (obtained with an ee of at least 99%) result depending on whether the (S,S)- or the (R,R)-form of DuPHOS is used (equation **11.2**).

11

a = Me, **b**= Et, **c** = i-Pr

11.2

Rh-DuPHOS complexes also selectively catalyze hydrogenation of enol acetates, as Scheme **11.4** shows. Since it is possible to convert ketones directly to enol acetates,[12] the transformation is the equivalent of asymmetric hydrogenation of a ketone, a transformation that we will discuss in the next section.

11-1-2 Asymmetric Hydrogenation Using Ru-Based Catalysts

For many years the field of asymmetric hydrogenation was dominated by the use of Rh-based catalysts. More recent work using Ru(II) complexes has been fruitful, leading to enantioselective reduction of a number of compounds containing C=C and C=O bonds. By far the most efficient catalyst system

[12]M.J. Burk, *J. Am. Chem. Soc.*, **1991**, *113*, 8518. For a list of reaction conditions used in converting ketones to enol acetates, see R.C. Larock, *Comprehensive Organic Transformations*, VCH Publishers, New York, 1989, 743.

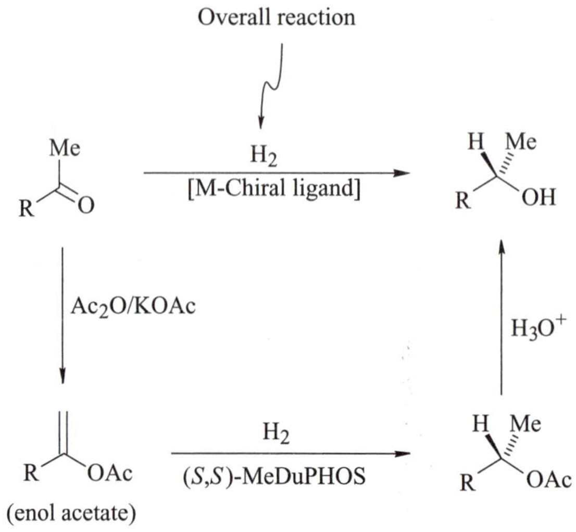

Overall reaction

Me
R—C(=O)   →[H₂][M-Chiral ligand]→   H Me / R—C—OH

Ac₂O/KOAc ↓   ↑ H₃O⁺

R—C(=CH₂)—OAc
(enol acetate)
→[H₂][(S,S)-MeDuPHOS]→
H Me / R—C—OAc

R	% ee	(Configuration)
Ph	89	(S)
1-Naphthyl	93	(S)
CO₂Et	99	(S)
CF₃	94	(S)

Scheme 11.4
Asymmetric Hydrogenation of
Enol Acetates Using Rh-DuPHOS

involves complexation of Ru(II) salts with (S)- or (R)-BINAP (see Figure 11-1). Equations **11.3–11.5** (on page 391) describe the steps involved in producing a number of Ru-BINAP catalysts.[13]

We encountered Rh-BINAP as a catalyst for double-bond isomerization in Chapter 9, Section **9-6-2**. As the following examples will show, Ru-BINAP catalysts are indeed even more versatile and selective than corresponding Rh complexes. Some[14] have even termed them "second-generation chiral catalysts." In general, like Rh(I) catalysts, Ru(II) catalysts must have a chelating het-

[13]H. Takaya, T. Ohta, and K. Mashima, "New 2,2'-Bis(diphenylphosphino)-1,1'-binaphthyl-Ru(II) Complexes for Asymmetric Catalytic Hydrogenation," in W.R. Moser and D.W. Slocum, Eds., *Homogeneous Transition Metal Catalyzed Reactions*, American Chemical Society, Washington, D.C., 123-142.
[14]I. Ojima, N. Clos, and C. Bastos, *Tetrahedron*, **1989**, *45*, 6901.

$$[RuCl_2(COD)]_n + (S)\text{-BINAP} \xrightarrow[PhCH_3]{Et_3N} Ru_2Cl_4[(S)\text{-BINAP}]_2 \bullet NEt_3 \;\; (Ru\text{-}(S)\text{-BINAP}) \quad \mathbf{11.3}$$

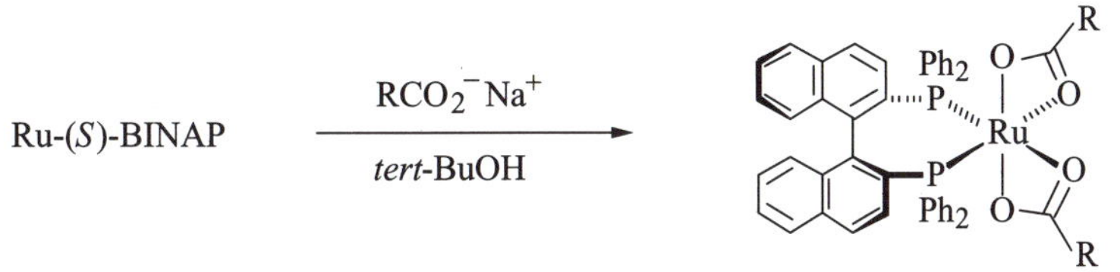
Ru-(S)-BINAP
RCO2⁻ Na⁺
tert-BuOH
Ph2
Ph2
P
P
Ru
O
R

11.4

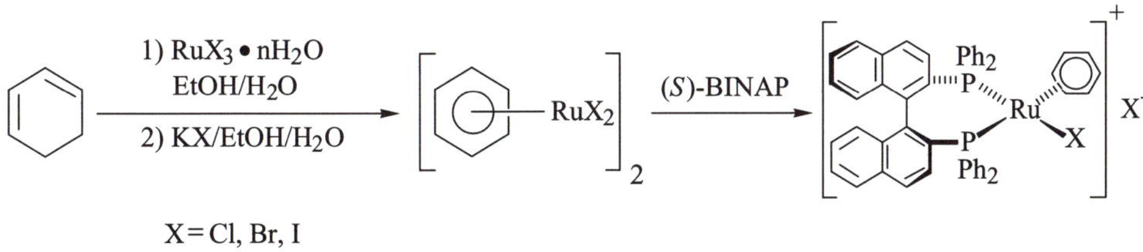
1) RuX3 • nH2O
EtOH/H2O
2) KX/EtOH/H2O
RuX2
2
(S)-BINAP
Ph2
P
Ru
X
X⁻
+
X = Cl, Br, I

11.5

eroatom (O or N) positioned close to the C=C or C=O bond undergoing saturation in order for effective stereoselection to occur. Neither the overall mechanism of hydrogenation nor the reason for enantioselection is well understood at this time. Since the stereoselectivity of the reaction tends not to decrease with increasing partial pressure of H_2, the mechanism is probably not analogous to Rh(I)-catalyzed hydrogenation. There is some evidence for the involvement of a Ru-H species before complexation of substrate. If this is the case, then the mechanism may parallel the "monohydride" pathway discussed in Chapter 9, Section **9-4-1**.

Besides α-amidoacrylic acid derivatives, Ru-BINAP complexes hydrogenate disubstituted acrylic acids, substrates not well tolerated by Rh(I) catalysts. We have already seen an example of this process in Chapter 9, Section **9-6-2** where asymmetric hydrogenation is a key step in the synthesis of (*S*)-naproxen. Equation **11.6** shows the scope of Ru-BINAP-catalyzed acrylic acid hydrogenation.[15]

[15]H. Takaya, R. Noyori, *et al.*, *J. Org. Chem.*, **1987**, *52*, 3174.

$$\underset{Y}{\overset{X}{\diagdown}}C=\underset{Z}{\overset{CO_2H}{\diagup}} \quad \xrightarrow[\text{Ru(II)-}(R)\text{-BINAP}]{H_2 \,(4\text{-}135\text{ atm})} \quad H-\underset{Y}{\overset{X}{\underset{|}{\overset{|}{C}}}}\overset{*}{\underset{Z}{\overset{CO_2H}{\underset{|}{\overset{|}{C}}}}}-H$$

(Acrylic acid: X, Y, Z = H)

Substrates:

$$\underset{Me}{\overset{H}{\diagdown}}C=\underset{Me}{\overset{CO_2H}{\diagup}} \qquad 91^{\bullet}\,(R)^{\ddagger}$$

$$\underset{Me}{\overset{Ph}{\diagdown}}C=\underset{H}{\overset{CO_2H}{\diagup}} \qquad 85\,(S)$$

$$\underset{H}{\overset{H}{\diagdown}}C=\underset{Ph}{\overset{CO_2H}{\diagup}} \qquad 92\,(R)$$

11.6

$$\underset{HO\text{–}CH_2}{\overset{Me}{\diagdown}}C=\underset{H}{\overset{CO_2H}{\diagup}} \qquad 92\,(R)$$

$$\underset{HO\text{–}(CH_2)_2}{\overset{Me}{\diagdown}}C=\underset{H}{\overset{CO_2H}{\diagup}} \qquad 93\,(S)$$

$^{\bullet}$ % ee
‡ Configuration of stereogenic center (*) in the product of hydrogenation

Allylic and homoallylic[16] alcohols undergo asymmetric hydrogenation (Scheme **11.5**).[17] The $Ru(O_2C\text{-}CF_3)_2$-BINAP catalyst used in these reactions is extremely specific in many respects. Both enantiomers of citronellol result with high ee starting with either geraniol or nerol, depending on whether (S)- or (R)-BINAP is the ligand. The reaction shows double-bond face selectivity, since the same catalyst, Ru-(S)-BINAP, transforms geraniol into (R)-citronellol and nerol into (S)-citronellol. Note that the remote double bond is unreactive to the hydrogenation conditions indicating that the allylic alcohol group provides necessary secondary chelation during the reaction. Further demonstration of this chelation is also shown in Scheme **11.5**, whereby homogeraniol reacts to give (R)-homocitronellol. Extending the carbon chain by just one more carbon, however, results in no reaction.

The remarkable selectivity of Ru–BINAP complexes is also evident when it is used to effect kinetic resolution[18] of enantiomers. Behaving much the same as enzymes, these complexes react with the two enantiomers of

---

[16]Homoallylic systems have one more carbon atom than allylic systems. For example, $CH_2=CH\text{-}CH_2$- is allylic, while $CH_2=CH\text{-}CH_2\text{-}CH_2$- is homoallylic.
[17]H. Takaya, R. Noyori, et al., J. Am. Chem. Soc., **1987**, 109, 1596.

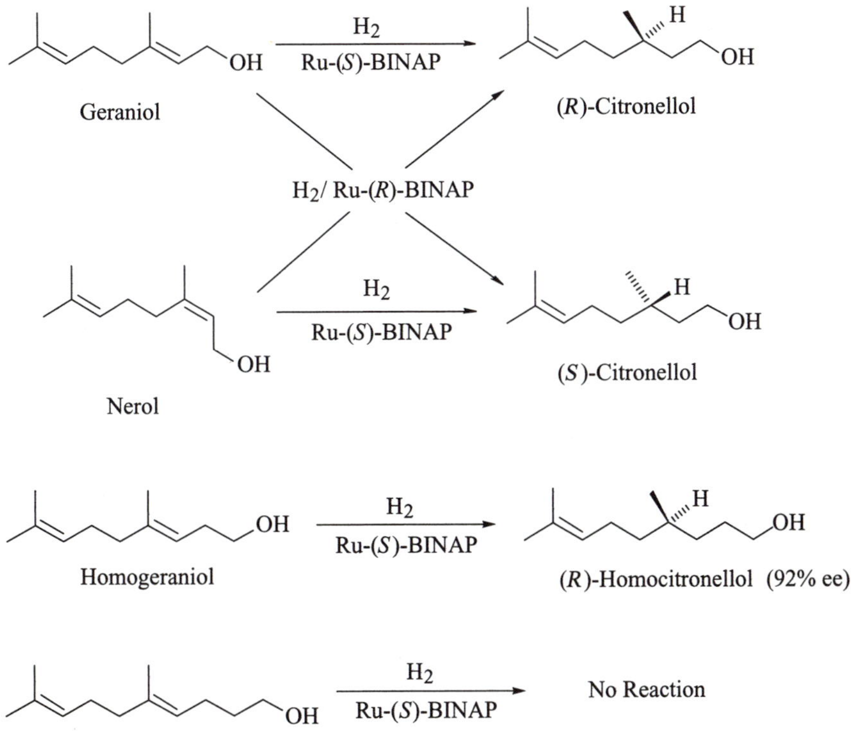

Scheme 11.5
Ru-BINAP Catalyzed Hydrogenation
of Allylic Alcohols

racemic allylic alcohols at different rates. Therefore, when racemic **12** (equation **11.7** below) reacts with H$_2$ in the presence of Ru-(S)-BINAP, the (R)-alcohol reacts preferentially, leaving the (S)-enantiomer unreacted.[19]

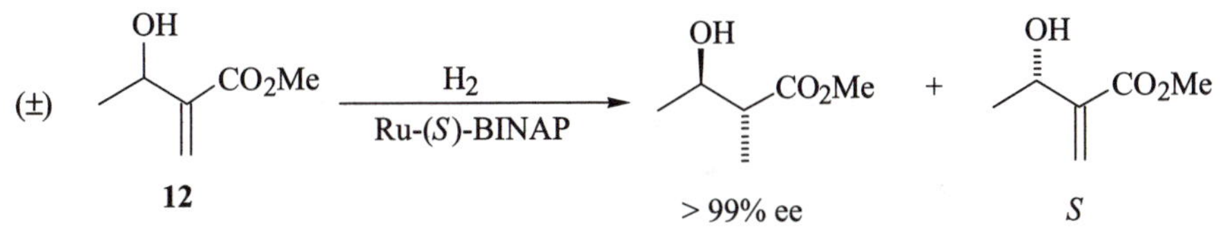

[18]Kinetic resolution represents a sixth method for separating enantiomers. A chiral reagent such as an enzyme or transition metal complex may react faster with one enantiomer than the other. If the reaction is quenched before completion, the resulting mixture consists of reaction product from one enantiomer as well as the unreacted other enantiomer, both of which could be enantiomerically pure.

[19]R. Noyori, et al., *J. Org. Chem.*, **1988**, *53*, 708.

Ketones also undergo asymmetric hydrogenation as long as a chelating heteroatom is present near the carbonyl group. Equation **11.8** shows the scope of functionality allowed in this particular facet of Ru–BINAP hydrogenation:

$$\underset{R}{\overset{O}{\|}}\underset{C_n}{\diagup}X \quad \xrightarrow[\text{Ru-BINAP}]{H_2} \quad \underset{R}{\overset{OH}{\underset{*}{\|}}}\underset{C_n}{\diagup}X$$

n = 1-3

X = N,O

C = sp^2 or sp^3 hybridized

* = New stereogenic center

11.8

Equation **11.9** below shows a particularly remarkable example of carbonyl reduction. Depending upon whether (*S*)- or (*R*)-BINAP is chosen, either enantiomer of the β-hydroxyester (with an ee greater than 99%) may be prepared.[20] These results are comparable to the enantioselectivity achieved using enzymes from baker's yeast!

$$\underset{Et}{\overset{O\quad O}{\|\quad\|}}OMe \quad \xrightarrow[R\text{- or }S\text{-BINAP}]{H_2} \quad \underset{Et}{\overset{OH\quad O}{\underset{*}{\|}\quad\|}}OMe$$

R or *S*

>99% yield (99% ee)

11.9

The chemical literature is replete with numerous examples of asymmetric functional group transformations catalyzed or promoted by transition metals. Yet, excepting a few spectacular examples, none of these transformations is as stereoselective as asymmetric hydrogenation. Later in this chapter we will encounter asymmetric transformations that involve more than simple functional group interconversion.

11-2   Organotransition Metal Complexes as Protecting and Activating Groups

The accumulation of a variety of different functional groups is an integral part of most syntheses of complex organic molecules. Quite often chemoselectivity is impossible during a step in a synthesis because two or more functional groups react with the same reagent. For example, if we wish to allow a

---

[20]R. Noyori, S. Akutagawa, *et al.*, *J. Am. Chem. Soc.*, **1987**, *109*, 5856.

Grignard reagent to react with a ketone to form a tertiary alcohol, the presence of an alcohol group at some remote site on the substrate would interfere with nucleophilic attack of the organomagnesium reagent on the carbonyl group. This is because Grignard reagents are strong Brønsted-Lowry bases that react preferentially with protic functional groups, resulting in deprotonation of the alcohol rather than attack at the carbonyl carbon. In order to ensure that attack did occur only at the C=O bond, we would first have to protect the alcohol with a group (such as a THP ether) before reaction with the Grignard reagent could occur (Scheme **11.6** below).

Inevitably the protecting group must leave; thus any good protecting group must be easy to put on, it must be chemically resilient in the face of a range of different reaction conditions, *and* it must be easy to take off when the need arises. Protection-deprotection necessarily involves two steps, so in multi-step syntheses, therefore, it is advisable to keep such operations to a minimum.[21] In spite of these necessary conditions and limitations, it is very difficult to design a synthesis of a complex molecule that does not involve protection-deprotection at some stage. In this section we will see how organotransition metal complexes serve as good protecting groups, allowing transformations to

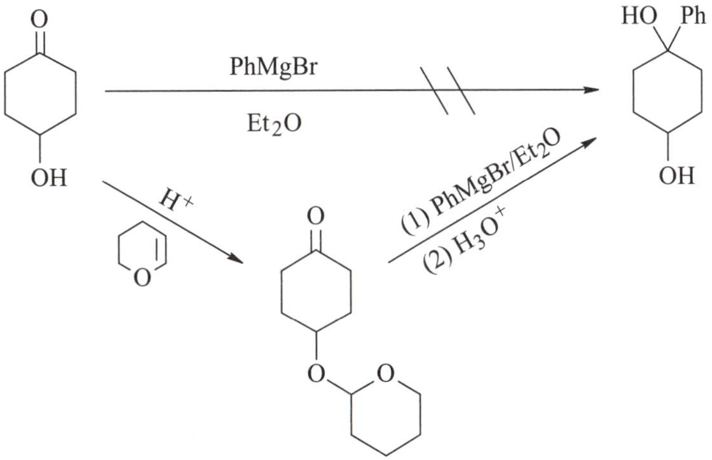

Scheme 11.6
Protection-Deprotection During a
Grignard Reaction on a Hydroxy Ketone

[21]Even if both protection and deprotection are high-yield processes, the overall yield of final product rapidly diminishes as the number of synthetic steps increases.

proceed at other sites on a molecule. We will also consider how some transition metal complexes can activate functional groups, sometimes rather subtly, to undergo transformations that would be impossible without the metal complex.

11-2-1 Use of Iron Reagents as Protecting Groups

Complexes of Fe with alkenes and polyenes tend to be quite stable, so much so that their use in catalytic reactions is limited. This inherent stability, however, is advantageous when Fe is used to complex with olefinic groups. The resulting complexes are not only readily isolable and characterizable, they are inert to electrophiles or other reagents that normally add across a π system. $[Cp(CO)_2Fe]^+$ (abbreviated as Fp^+ and pronounced "fip") coordinates to unhindered, isolated C=C bonds and serves as an excellent protecting group. Equations **11.10** to **11.12** below show how Fp^+-olefin complexes are made and some of the reactions they undergo.[22]

$$11.10$$

$$11.11$$

$$11.12$$

[Fp-isobutylene]$^+$ (equation **11.10**) usually serves as a convenient source of Fp^+ during olefin exchange to induce protection at the less hindered monosubstituted alkene (equation **11.11**). Once formed, a reaction such as hydrogenation may occur at some remote site on the molecule (equation **11.12**). Finally, nucleophilic attack of I$^-$ directly on the metal atom liberates the free alkene.

[22]K.N. Nicholas, *J. Am. Chem. Soc.*, **1975**, *97*, 3254.

Equation **11.13** below shows electrophilic aromatic substitution of bromine on an arene in which the C=C bond is first protected to prevent bromine addition at that bond.

Dienes react with $Fe(CO)_5$ or $Fe_2(CO)_9$ to give $Fe(CO)_3$-diene complexes (equation **11.14**). These are stable to a variety of reaction conditions, such as reaction with dienophiles in a Diels–Alder reaction.[23]

Scheme **11.7** shows a synthesis of a precursor to diHETE[24] that starts with Fe-diene complex **13**. Reaction of the aldehyde group of **13** with an ylide via the Wittig reaction gives **14**, the uncomplexed double bond of which then reacts with OsO_4 to give diol **15**. Hydrogenation of the alkyne group with Lindlar catalyst gives a Z-double bond in **16**. Treatment of **16** with 1,1′-carbonyldiimidazole provides carbonate **17**. The next step removes the $Fe(CO)_3$ fragment by oxidizing the metal,[25] providing a precursor to diHETE that possesses the required stereochemistry for most of the C=C bonds found in the target molecule.

Complexation of a $Fe(CO)_3$ fragment to a cyclodiene influences the stereochemistry of subsequent reactions on the ring. While the iron fragment is not acting as a protecting group, it is acting as a directing group that influences how additional functionality adds to the ring itself. Use of the $Fe(CO)_3$ frag-

[23]A.J. Pearson, *Accts. Chem. Res.*, **1980**, *13*, 463 and references therein.
[24]A. Gigou, J-P. Beaucourt, J-P. Lellouche, and R. Gree, *Tetrahedron Lett.*, **1991**, *32*, 635.
DiHETE stands for dihydroxyeicosatetraenoic acid, a biologically active metabolite derived from the metabolism of arachidonic acid. It has been the subject of great interest over the last few years.
[25]An even more specific reagent for removing $M(CO)_x$ fragments is Me_3N-O. It specifically attacks carbonyl groups, liberating the diene ligand, one equivalent of CO_2, Me_3N, and Fe salts of unknown composition. This is the preferred reagent when oxidizable groups are present on the organic substrate. See Y. Shvo and E. Hazum, *J. Chem. Soc., Chem. Commun.*, **1974**, 336.

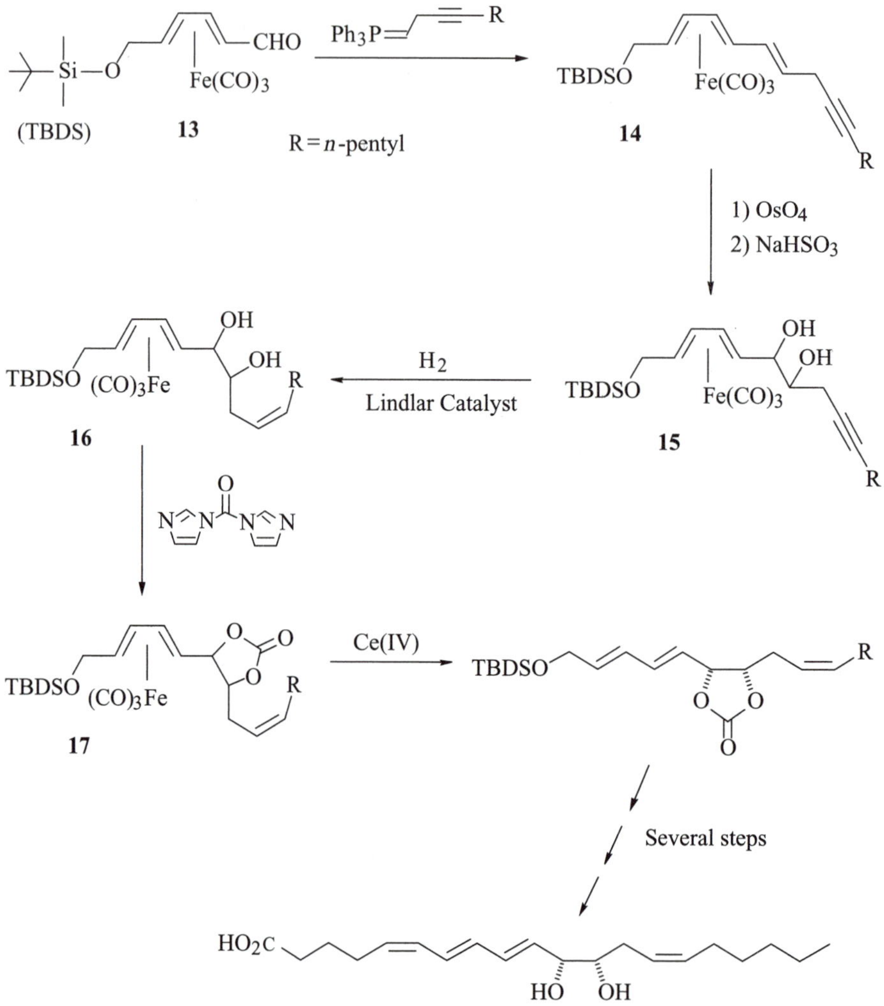

Scheme 11.7
Synthesis of a Precursor to diHETE   diHETE

ment in this way resembles a protecting group operation, because one reaction is required to add the directing group and one to remove it. Scheme **11.8** shows how cycloheptadienone complex **18** reacts with LDA to form an enolate **19**, which then undergoes alkylation by methyl iodide. This step is repeated to give the *cis*-dimethylcycloheptanone complex **20**. Note that the methyl groups were introduced *anti* to the metal because of the steric hindrance exhibited by the Fe(CO)$_3$ fragment. Reduction of **20** with NaBH$_4$ gives the corresponding alcohol **21a**. We might expect that the hydride would also be

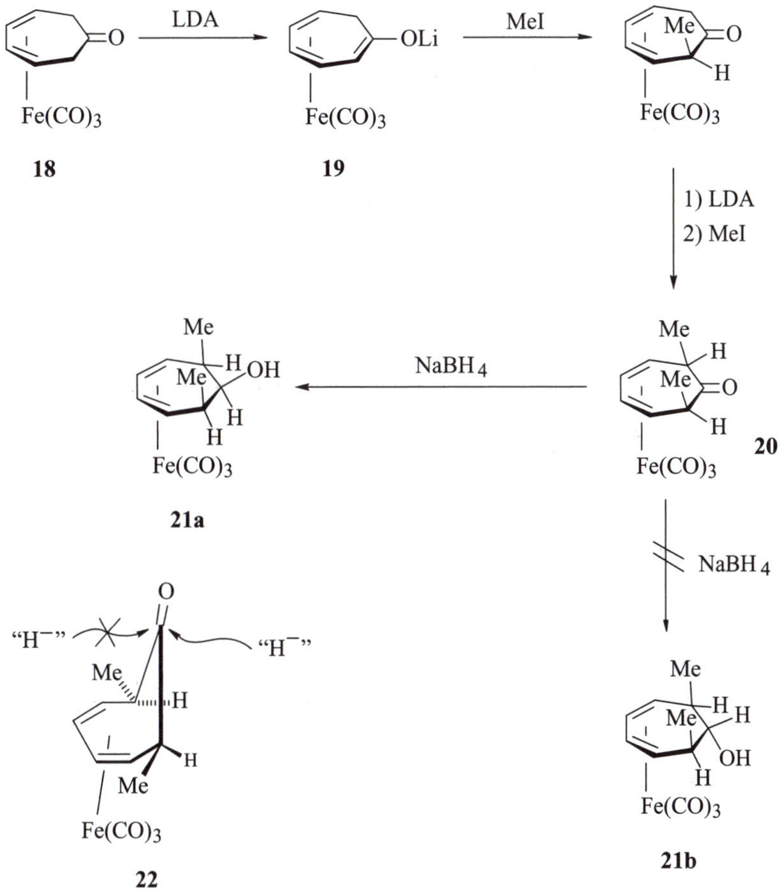

Scheme 11.8
Directing Influence of the Metal Fragment
on the Chemistry of Fe-Cyclodiene Complexes

delivered *anti* to the metal, thus giving stereoisomer **21b**. X-ray analysis of Fe-cycloheptadiene complexes indicates, however, that the ring is boat shaped (**22**). Thus the carbonyl group points upward, away from the iron fragment and almost perpendicular to a plane composed of the remaining ring atoms. The less sterically hindered approach of hydride is now from what appears—in a two-dimensional representation showing a flat seven-membered ring—to be the bottom face of the ring or *syn* to the metal.[26] Such an approach, however, avoids the steric bulk of the methyl groups.

[26]A.J. Pearson and K. Chang, *J. Chem. Soc., Chem. Commun.*, **1991**, 394; see also *J. Org. Chem.*, **1993**, *58*, 1228 for more recent examples.

Exercise 11-3

Based on the chemistry in Scheme **11.8**, predict the stereochemistry of the following reaction:

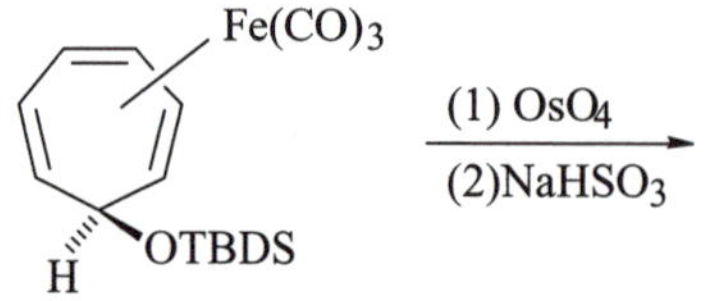

$$\xrightarrow[\text{(2)NaHSO}_3]{\text{(1) OsO}_4}$$

11-2-2 Iron Complexes Serving as Masked Functional Groups

Iron complexes with π ligands not only to protect them, but also to coerce them into a pattern of reactivity that is completely unavailable to the free ligand. A metal fragment containing iron or some other metal, when complexed to a π ligand provides a "mask" that hides the true identity of the ligand until specific reaction conditions are applied to the metal-ligand complex. For example, Fp$^+$-vinyl ether complexes[27] (**23**) are actually "masked" complexes of vinyl cations, species that are extremely reactive and appear in uncomplexed form only as transient reactive intermediates.

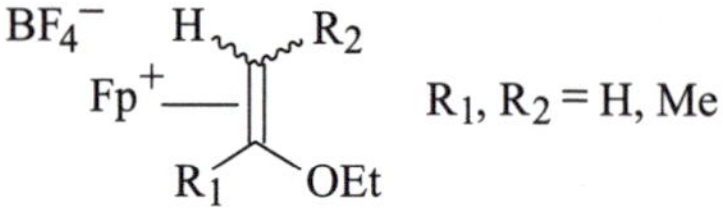

$R_1, R_2 = H, Me$

23

The organic starting material for a Fp$^+$-vinyl ether complex is an α-haloacetal or ketal. Equation **11.15** shows the scope of the preparation of these complexes. The reaction sequence in **11.15** begins with nucleophilic attack by Fp$^-$ on the acetal followed by electrophilic abstraction by H$^+$ to provide the Fp-vinyl ether complex. Once formed, the vinyl ether ligand undergoes addition by a nucleophile, typically an enolate ion (equation **11.16**), followed by another electrophilic abstraction with H$^+$; removal of the Fp$^+$ with NaI then occurs in the usual way. The overall reaction amounts to vinylation of a ketone

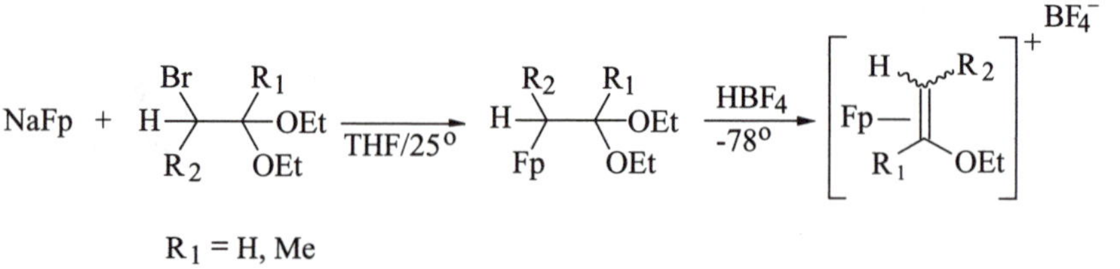

11.15

$R_1 = H, Me$

$R_2 = H, Me$

[27]T.C.T. Chang, M. Rosenblum, *et al.*, *Organometallics*, **1987**, *6*, 2394.

at the α position by a vinyl cation (equation **11.17**), a reaction impossible using conventional organic chemistry and another good example of *umpolung*:

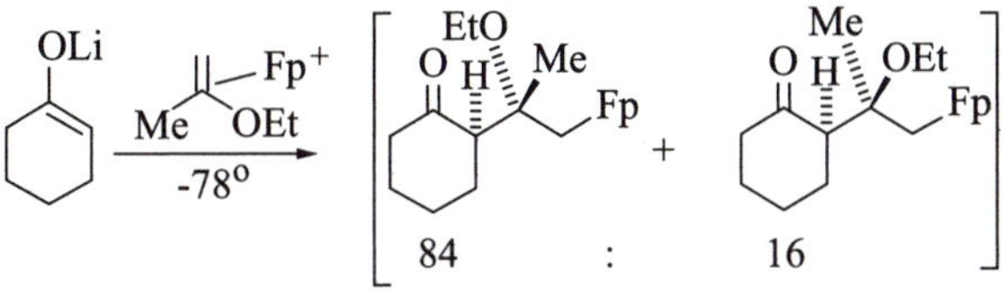

11.16

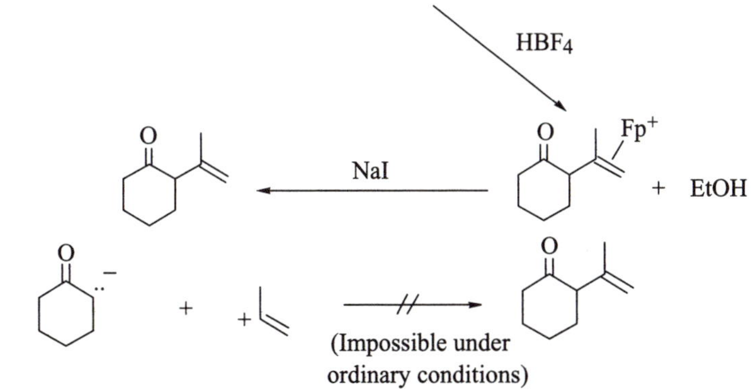

11.17

Dienylenol ethers (such as **24**) are quite reactive with protic acids and readily convert to the corresponding conjugated ketone (**25**).

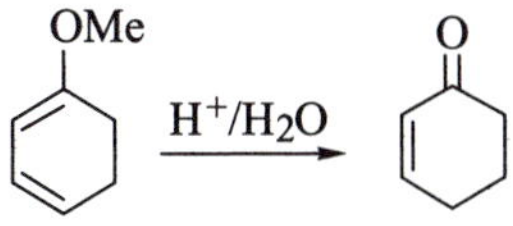

Fe(CO)$_3$ complexes of these conjugated ethers, in contrast, are quite stable to acid, but will undergo electrophilic hydride abstraction to give 18-electron η^5-dienyl cationic complexes (equation **11.18** below).

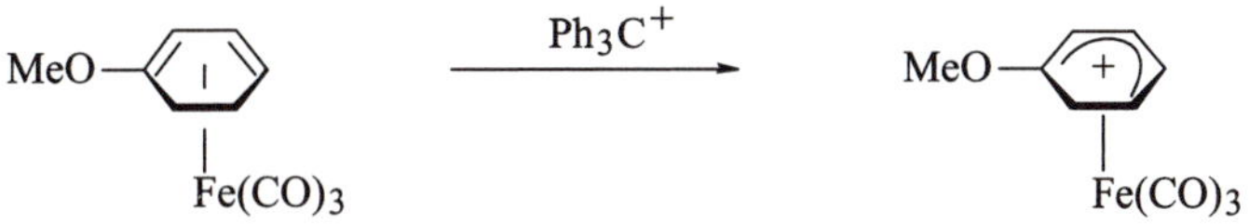

11.18

Such complexes are reactive with nucleophiles and are actually the masked equivalent of a cationic enone (**26**)—extremely unstable in the free state.

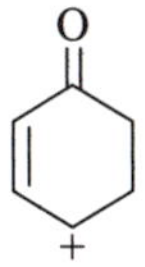

26

Scheme **11.9** shows how a strategy of masking a cationic enone led to the synthesis of an alkaloid known as O-methyljoubertamine.[28] The scheme begins with nucleophilic attack of the enolate on the masked cationic enone, which stitches together two six-membered rings. Reduction of the ketone to an alcohol with diisobutylaluminum hydride precedes acid-catalyzed dehydration of the alcohol and hydrolysis of the enol ether group of the bottom ring. These two transformations provide a conjugated ketone. The Fe(CO)$_3$ is then removed using Me$_3$NO (see footnote 25), resulting in a precursor for O-methyljoubertamine:

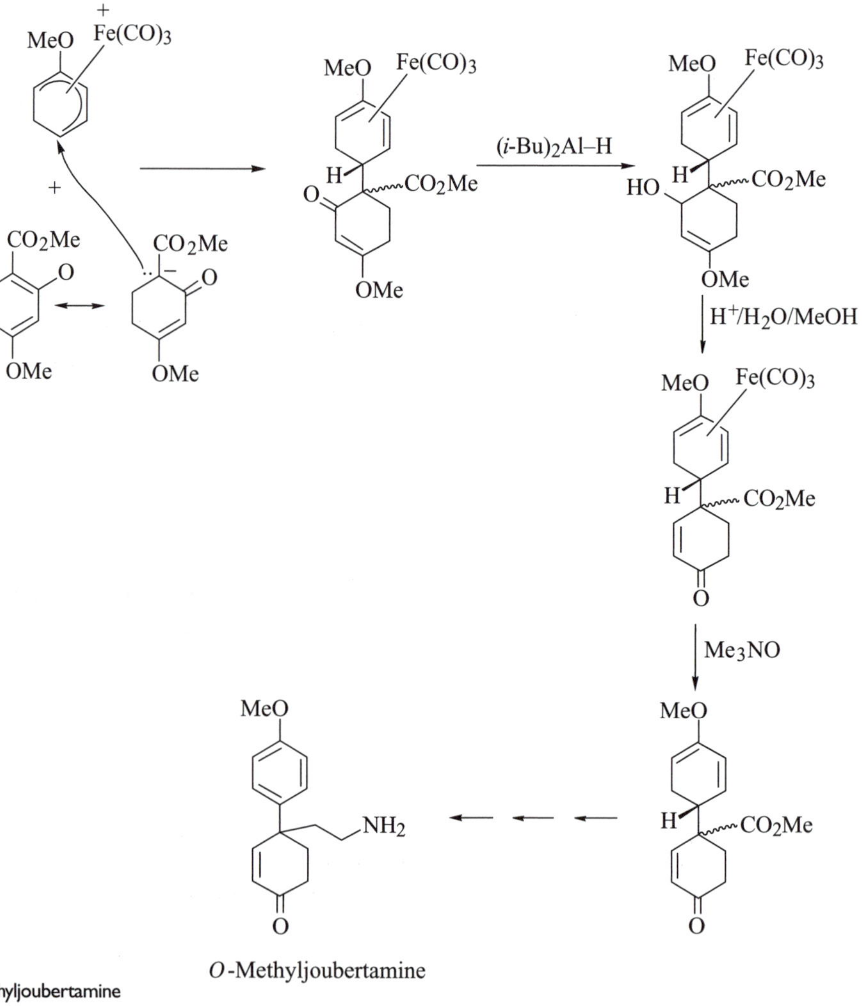

Scheme 11.9   O-Methyljoubertamine
Synthesis of O-Methyljoubertamine
Using η^5- and η^4-Fe Complexes

[28]A.J. Pearson, I.C. Richards, and D.V. Gardner, *J. Chem. Soc., Chem. Commun.*, **1982**, 807.

11-2-3 Palladium-η^3-π-Allyl Complexes: Allyl Cation Equivalents

In organic chemistry, allylic substrates are relatively reactive toward some nucleophiles, as shown in equation **11.19**. The reaction suffers from a number of disadvantages including unpredictable stereochemistry, the possibility of carbon-skeleton rearrangements, and the limited types of nucleophiles that react readily. In contrast, η^3-π-allyl metal complexes—especially cationic ones—are more reactive toward a variety of nucleophiles, usually with predictable stereochemistry. Carbon nucleophiles, moreover, react with these complexes forming the all-important C-C bond. Palladium is by far the metal of choice to bind to η^3-allyl ligands, forming stable (but not too stable for catalytic reactions) complexes that are isolable.

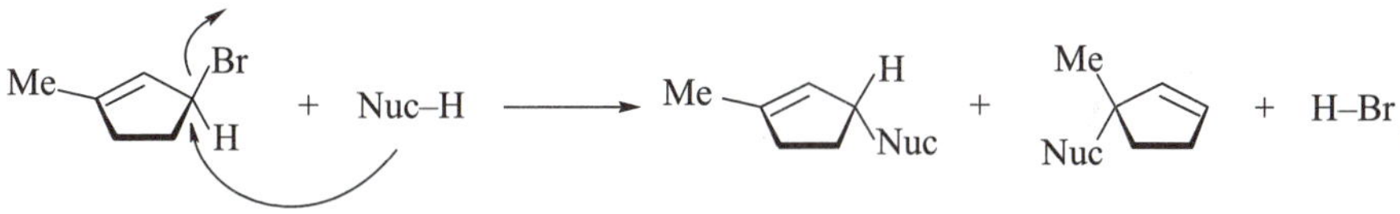

Equations **11.20** and **11.21** illustrate two general methods for preparing η^3-π-allyl metal complexes of Pd:

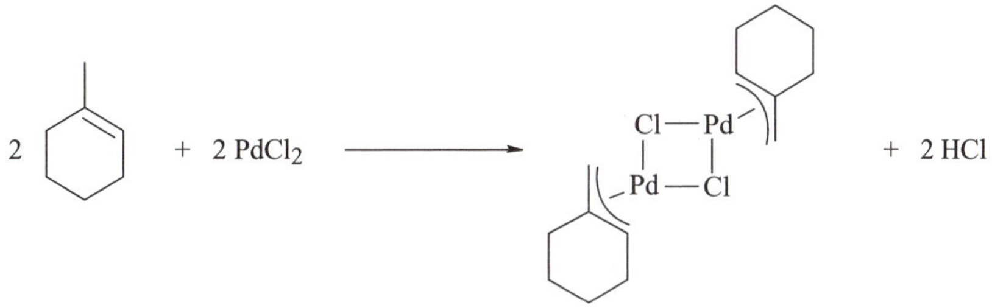

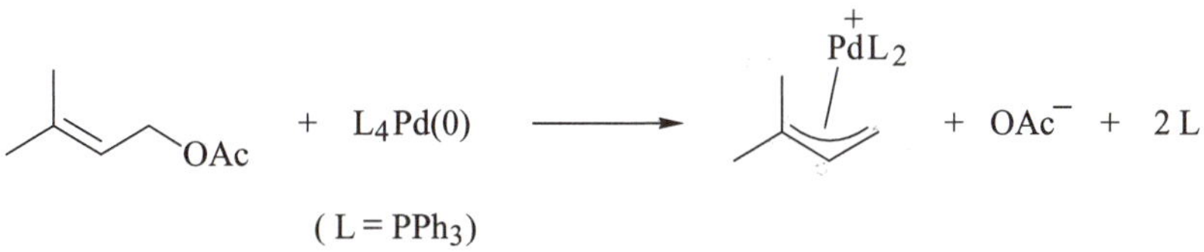

Isolated double bonds react in the first route with $PdCl_2$ to form neutral, dimeric η^3-allyl complexes. The second process involves an alkene with a leaving group (usually acetate, OAc) at the allylic position. In this case, the allylic C-OAc bond reacts with Pd(0) by oxidative addition and the remaining C=C bond becomes a π ligand to Pd, resulting in a cationic η^3-allyl complex. Both the neutral and cationic Pd complexes react with a variety of nucleophiles; of particular interest are types such as ⁻CHRR′ (where R, R′ = keto, carboxyl, or sulfonyl) and I° or II° amines.

Exercise 11-4

Propose a plausible pathway for the preparation of Pd-allyl complexes shown in equation **11.20**.

Formation of Pd-allyl complexes via the second route has some advantages over the first. Lack of regioselectivity could be a problem when isolated C=C bonds react with Pd(II) salts if there is more than one allylic position sub-

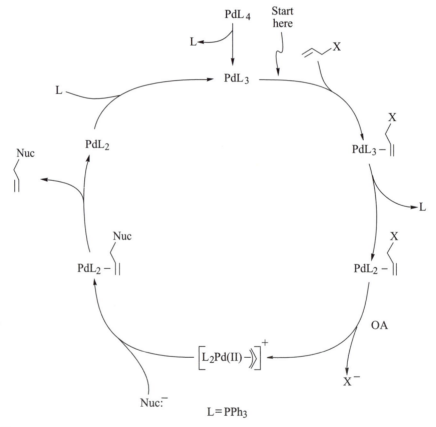

Scheme 11.10

Catalytic Formation of η³-π-Allyl Pd(II)
Complexes and Subsequent Reaction with Nucleophiles

stituted with hydrogen. The second route ensures regioselectivity because one allylic position is already activated by a leaving group; the Pd(0) complex reacts only at the allylic C–OAc bond. Since the complex resulting from the first procedure is neutral, it is not as reactive toward nucleophiles as the cationic complex formed by the second. Finally, the second procedure is *catalytic* in the use of Pd; Scheme **11.10** (on the previous page) shows a plausible catalytic cycle.

Some other examples of the use of both types of Pd complexes in synthesis follow. Equation **11.22** below shows a stoichiometric use of a Pd-allyl complex (**27**) that reacts with nucleophile **28** to give **29**, a precursor to farnesol (**30**).[29]

30(farnesol)   **11.22**

When the allyl acetate and the precursor to the nucleophile are contained in the same molecule, an intramolecular reaction to form a ring occurs, as shown in equation **11.23** below:[30]

11.23

The stereochemistry of nucleophilic attack on Pd-allyl complexes is usually predictable. Equation **11.24** shows attack of a nucleophile on complex **31**, formed by reaction of the corresponding alkene with PdCl$_2$. The nucleophile attacks at the face of the allyl ligand *anti* to the Pd.[31]

[29]B.M. Trost, *Accts. Chem. Res.*, **1980**, *13*, 385 and references therein. Farnesol (actually farnesyl pyrophosphate) is a key intermediate in the synthesis of steroids (e.g., cholesterol) and sesquiterpenes, which are secondary metabolites produced by plants.
[30]Loss of the carbethoxy group and conversion of CN to COOH in the product leads ultimately to synthesis of chrysanthemic acid, derivatives of which serve as mild, but effective insecticides known as pyrethrins. J.P. Genet and F. Piau, *J. Org. Chem.*, **1981**, *46*, 2414.
[31]D.J. Collins, W.R. Jackson, and R.N. Timms, *Tetrahedron Lett.*, **1976**, 495.

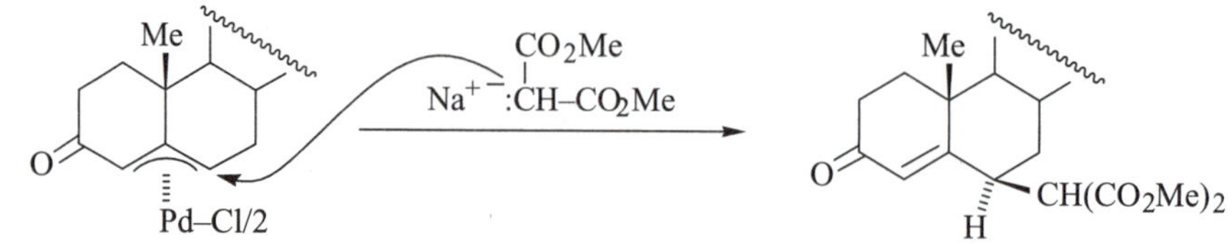

11.24

31

Allylic acetates undergo Pd-catalyzed nucleophilic substitution with net retention of configuration[32] with respect to the stereochemistry of the acetate group as shown in equations **11.25** and **11.26**. Net retention results from two inversions, the first during oxidative addition with Pd(0) (see Chapter 7, Section **7-2-2**), and the second as a result of attack of the nucleophile on the face of the allyl ligand *anti* to the metal (see Chapter 8, Section **8-2-2**), both steps shown schematically in equation **11.27** below:

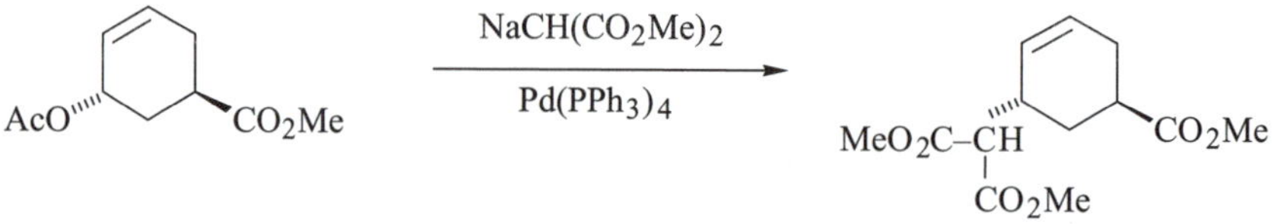

11.25

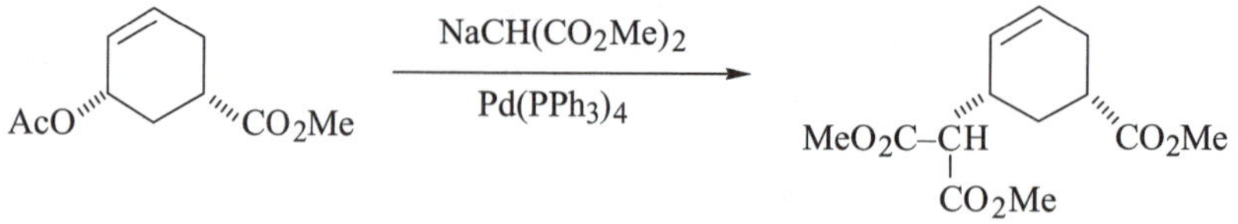

11.26

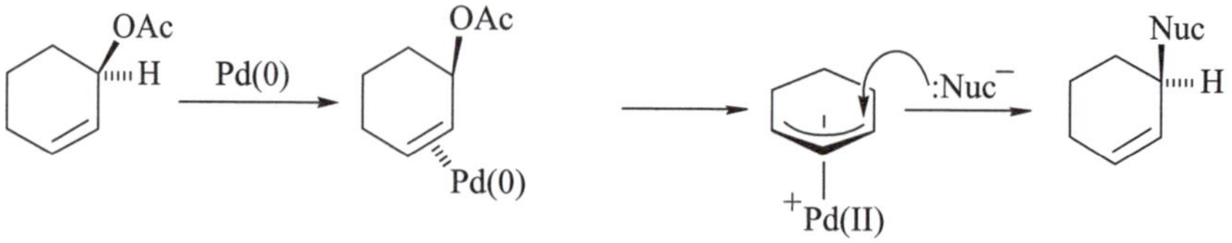

11.27

Equation **11.28** demonstrates a clever transfer of the chirality at one stereocenter to another during a Pd(0)-catalyzed nucleophilic attack on an allyl ligand.[33] The lactone (**32**) is really a cyclic acetate that undergoes cleavage of the C–O bond to give intermediate **33**, which then undergoes attack by the nucleophile to give **34**.

[32] B.M. Trost and T.R. Verhoeven, *J. Org. Chem.*, **1976**, *41*, 3215.
[33] B.M. Trost and T.P. Klun, *J. Am. Chem. Soc.*, **1979**, *101*, 6756.

$$11.28$$

Exercise 11-5

Consider equation **11.28**. Try to correlate the stereochemistry of the starting material **32** with that of the final product **34**. Is the stereochemistry of the new stereocenter in **34** the result of overall retention or inversion of configuration? Explain.

In this section we have seen a few examples of how organotransition metal complexes interact with π systems either to protect them from reaction with reagents toward which they would normally be reactive or to modify their reactivity entirely in ways that would be impossible in an uncomplexed state. The literature abounds with reports of such interactions. Section **8-2-2** in Chapter 8 contains references to other π ligand-metal interactions (especially Cr-arene) that are potentially useful in organic synthesis.

11-3   Carbon-Carbon Bond Formation via Carbonyl and Alkene Insertion

The construction of a C-C bond is probably the most important operation in organic synthesis, especially if it can be accomplished with chemo-, regio-, and stereoselectivity. The complexity of most of today's synthesis targets demands that selective, yet general, reliable, and high-yield means of introducing new C-C bonds be available. As discussed in Chapter 8, carbonyl insertion (more properly called alkyl migration to a carbonyl ligand) and 1,2-alkene insertion are general, stereospecific methods of forming such bonds in the presence of transition metals.

11-3-1 Carbonyl Insertions

Reaction of Schwartz's reagent [$(Cp)_2Zr(Cl)H$] with alkenes provides the corresponding σ-alkyl complexes by 1,2-insertion (Chapter 8, Section **8-1-2**). At room temperature under a low pressure blanket of CO, it is possible to intercept the organozirconium complex to yield acylzirconium compounds (equation **11.29**):[34]

[34]C.A. Bertelo and J. Schwartz, *J. Am. Chem. Soc.*, **1975**, *97*, 228.

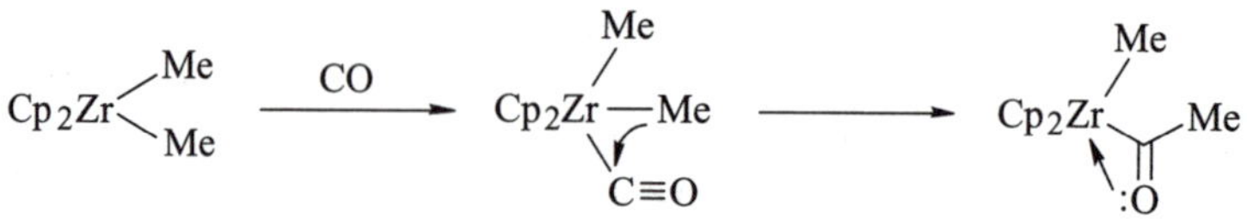

$$Cp_2Zr\overset{Me}{\underset{Me}{<}} \xrightarrow{\text{CO}} Cp_2Zr\overset{Me}{\underset{C\equiv O}{-}}Me \longrightarrow Cp_2Zr\overset{Me}{\underset{:O}{<}}Me \qquad \textbf{11.29}$$

Once formed, these complexes react with a variety of reagents under oxidizing conditions. Scheme **11.11** shows some transformations that convert alkenes ultimately to useful organic functional groups.[35] Noteworthy is the conversion of an alkene to an aldehyde possessing one more carbon atom than the starting material, a reaction equivalent to Rh-catalyzed hydroformylation (Section **9-2** in Chapter 9). Unlike the Rh system, however, the overall transformation from alkene to aldehyde is regiospecific; only the terminal aldehyde forms. The

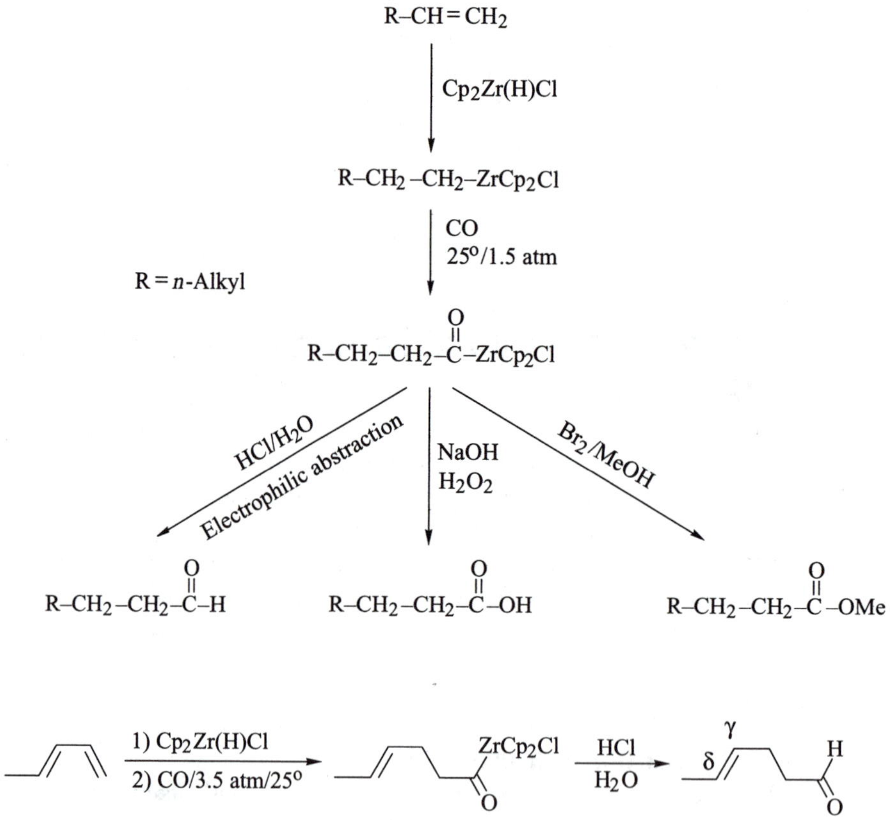

$$R-CH=CH_2$$

$$\downarrow Cp_2Zr(H)Cl$$

$$R-CH_2-CH_2-ZrCp_2Cl$$

$$\downarrow \begin{array}{c} CO \\ 25°/1.5 \text{ atm} \end{array}$$

R = n-Alkyl

$$R-CH_2-CH_2-\overset{O}{\overset{\|}{C}}-ZrCp_2Cl$$

HCl/H₂O
Electrophilic abstraction   NaOH
H₂O₂   Br₂/MeOH

$$R-CH_2-CH_2-\overset{O}{\overset{\|}{C}}-H \qquad R-CH_2-CH_2-\overset{O}{\overset{\|}{C}}-OH \qquad R-CH_2-CH_2-\overset{O}{\overset{\|}{C}}-OMe$$

1) Cp₂Zr(H)Cl
2) CO/3.5 atm/25°   ZrCp₂Cl   HCl / H₂O

Scheme 11.11
Transformations of Acylzirconium Complexes

[35]For a good review on the utility of organozirconium reagents in organic synthesis, see E-i. Negishi and T. Takahashi, *Aldrichimica Acta*, **1985**, *18*, 31.

conversion of 1,3-pentadiene ultimately into a γ, δ-unsaturated aldehyde is remarkable because only the sterically less hindered double bond reacts with Schwartz's reagent.[36]

Exercise 11-6

Show all the steps involved in the conversion of 1-hexene to heptanal. Propose a mechanism for the last step involving treatment of the acylzirconium complex with HCl.

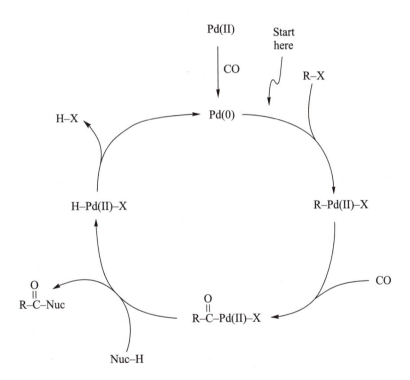

R = Alkyl, Alkenyl, Aryl

Nuc-H = R′O-H, R′R″N-H   (R′, R″= Alkyl)

X = Cl, Br, I

Scheme 11.12
Catalytic Insertion of CO into a Pd-C Bond

[36]The reaction seems to be general for a number of 1,3-dienes; C.A. Bertelo and J. Schwartz, *J. Am. Chem. Soc.*, **1976**, *98*, 262.

Sigma-alkyl (lacking β-hydrogens), -alkenyl, and -aryl complexes of Pd readily undergo carbonyl insertion. In principle, any species with a C-X (X = leaving group) bond that can undergo oxidative addition with Pd(0) is capable of conversion to a product containing a C(C=O)Y (Y = nucleophile) functionality. This extremely versatile reaction is catalytic in Pd and occurs both inter- and intramolecularly. Scheme **11.12** (on the previous page) depicts the likely catalytic cycle involving a series of steps that should be familiar.[37]

Although Pd(0) complexes are catalytically active as is, Pd(II) salts have also been used; the reducing atmosphere of CO converts Pd(II) to Pd(0) *in situ*. A "proton sponge" such as Et_3N also needs to be present to tie up the acidic by-product, HX. Equations **11.30**,[38] **11.31**,[39] and **11.32**[40] below provide a few examples of Pd-catalyzed carbonylation.

$$Pd(OAc)_2/PPh_3 \qquad CO/Et_3N/MeOH$$

11.30

93%

$$PdCl_2(PPh_3)_2 \qquad K_2CO_3/CO \text{ (4 atm)} \qquad 50°$$

11.31

78%

$$PdCl_2(PPh_3)_2 \qquad CO/R'_3N$$

11.32

83%

$$R = -(CH_2)_6-N\underset{\diagdown}{\overset{\diagup}{}}N \qquad R' = \text{Isobutyl}$$

[37]J.P. Collman, *et al., Principles and Applications of Organotransition Metal Chemistry, op. cit.,* 722.
[38]S.K. Thompson and C.H. Heathcock, *J. Org. Chem.,* **1990**, *55,* 3004.
[39]A. Cowell and J.K. Stille, *J. Am. Chem. Soc.,* **1980**, *102,* 4193.
[40]J.W. Tilley, D.L. Coffen, B.H. Schaer, and J. Lind, *J. Org. Chem.,* **1987**, *52,* 2469.

Carbonyl insertion occurs in iron complexes. A particularly versatile reagent for effecting CO insertion is $Na_2Fe(CO)_4$, known as Collman's reagent.[41] We have already seen an example of the use of this complex in converting alkyl halides to esters via acyl halides (Scheme **8.8** in Chapter 8). Scheme **11.13** gives a more complete rendition of the utility of Collman's reagent. The ferrate complex is a strong nucleophile and reacts with either alkyl or acyl halides to give **35** or **36**, respectively. Complex **35** will undergo CO insertion to give **36** in the presence of CO or a phosphine. Either complex is a potential precursor to ketones upon reductive elimination, and **36** also leads to the corresponding aldehyde. As with carbonylations we have encountered earlier in this section, a variety of carboxylic acid derivatives also result from reaction with Collman's reagent.

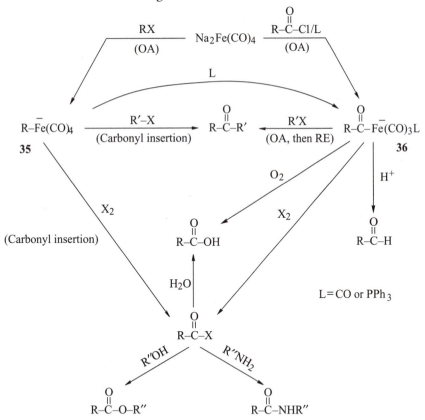

Esters, ketones, nitriles, and vinyl groups may be present in R, R′, and R″

Scheme 11.13
Carbonylation Reactions Using
$Na_2Fe(CO)_4$ (Collman's Reagent)

[41] J.P. Collman, *Acc. Chem. Res.*, **1975**, *8*, 342 and references therein.

An interesting use of Collman's reagent is the transformation of ω-substituted-1-alkenes[42] to cyclic ketones as shown in equation **11.33** below. The reaction seems to involve insertion of a C=C bond into an Fe–acyl bond. Equation **11.34** shows another illustration of this insertion in which Collman's reagent was used in a key step in the synthesis of aphidicolin, a diterpenoid[43] antiviral compound.[44]

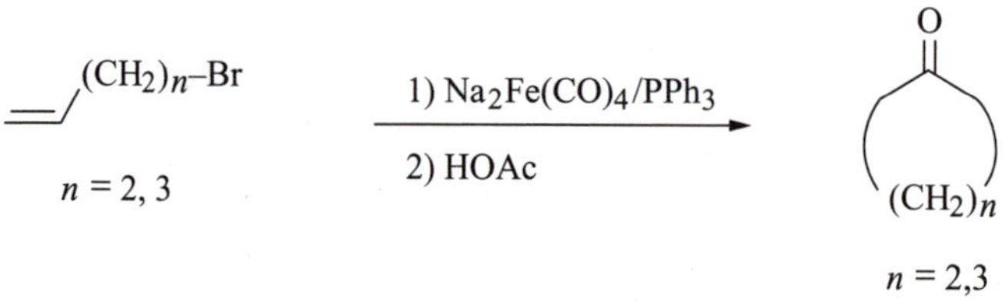

11.33

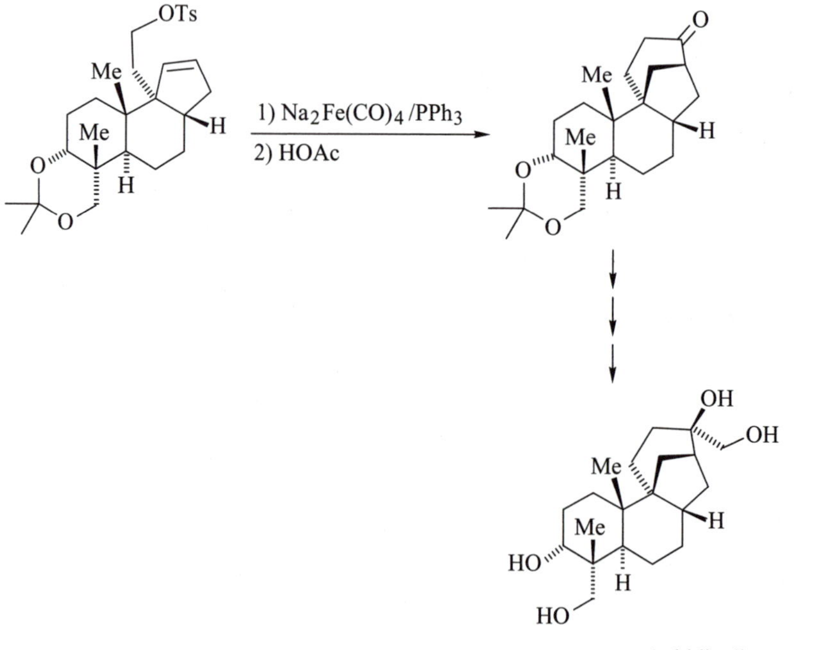

11.34

Aphidicolin

[42]The Greek letter ω is used as a generic designation for the terminal position of a carbon chain.
[43]Terpenes are natural products produced mainly by plants but also by insects and microorganisms. Terpenes result from junction of various numbers of isoprene (C_5) units. Terpenes containing two isoprene units are called monoterpenes (C_{10}), those containing three units are called sesquiterpenes (C_{15}), and those with four isoprenes are termed diterpenes (C_{20}). Aphidicolin is derived from a C_{20} precursor, hence the descriptor *diterpenoid*.
[44]J.E. McMurry, et al., *J. Am. Chem. Soc.*, **1979**, *101*, 1330.

Exercise 11-7

Propose a synthesis of **B** starting with compound **A**.

A   B

11-3-2 Carbon-Carbon Double-Bond Insertion: The Heck Reaction

Chapters 8 and 9 emphasized that there are relatively few illustrations of insertion of a C=C bond into an M-C bond, with Ziegler–Natta polymerization serving as the most significant example. Such insertion, nevertheless, is a powerful tool in the construction of rings, as equations **11.33** and **11.34** showed. Perhaps the example of C=C insertion most useful to synthesis chemists is the Heck reaction (equation **11.35** below). The transformation is similar in scope to Pd-catalyzed carbonylation, only instead of CO insertion, introduction of a C=C group into a Pd-C bond occurs.

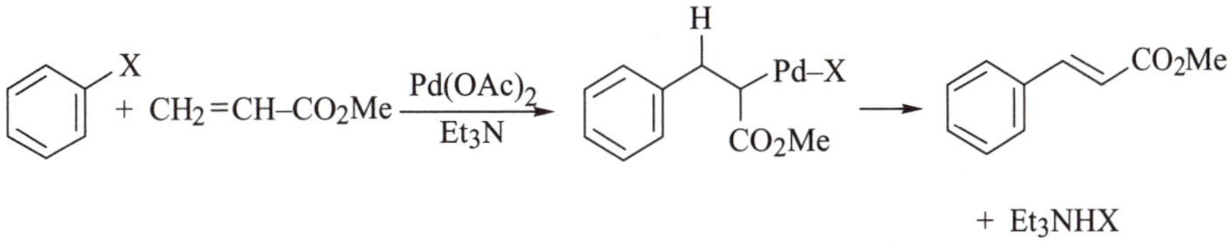

$$+ \; Et_3NHX$$

11.35

The reaction involves, typically, the oxidative addition of R-X, where R = aryl, vinyl, benzyl, or allyl (i.e., a substrate lacking β-hydrogens substituted at an sp^3 hybridized carbon), followed by alkene complexation and 1,2-insertion of an alkene. The last step is β-elimination. The Heck reaction is catalytic in Pd, and Scheme **11.14** shows what is considered to be the catalytic cycle. Complete understanding of some aspects of the mechanism awaits further experimentation.[45]

---

[45]Two reports of some of the early work on the Heck reaction that include summaries of its scope are R.F. Heck, *Acc. Chem. Res.*, **1979**, *12*, 146 and R.F. Heck, *Org. React.*, **1982**, *27*, 345.

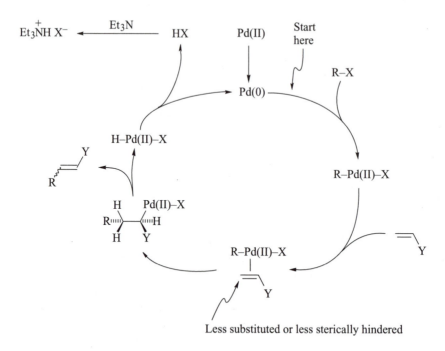

Y= H, R, Ph, CO$_2$R, CN, OMe, OAc, NHAc

Scheme 11.14
Catalytic Olefin Insertion:
The Heck Reaction

The rate of reaction and regioselectivity are sensitive to steric hindrance about the C=C bond of the vinylic partner. For simple aryl halides reacting with alkenes, the rate of reaction as a function of alkene substitution varies[46] according to the following:

$$CH_2=CH_2>CH_2=CH\text{-}OAc>CH_2=CH\text{-}Me>CH_2=CH\text{-}Ph>CH_2=C(Me)Ph$$

k_{rel}:   14,000   970   220   42   1

Figure **11-4** shows the regioselectivity of addition for a number of alkenes. The arrows with numbers attached indicate the relative tendency of the aryl group to add to the two carbons of the alkene. Although electronic effects may play a small role in directing attack by the metal in the insertion step, steric effects seem to be the dominant factor by far.

[46]S.G. Davies, *Organotransition Metal Chemistry: Applications to Organic Synthesis*, Pergamon Press, New York, 1982, 225.

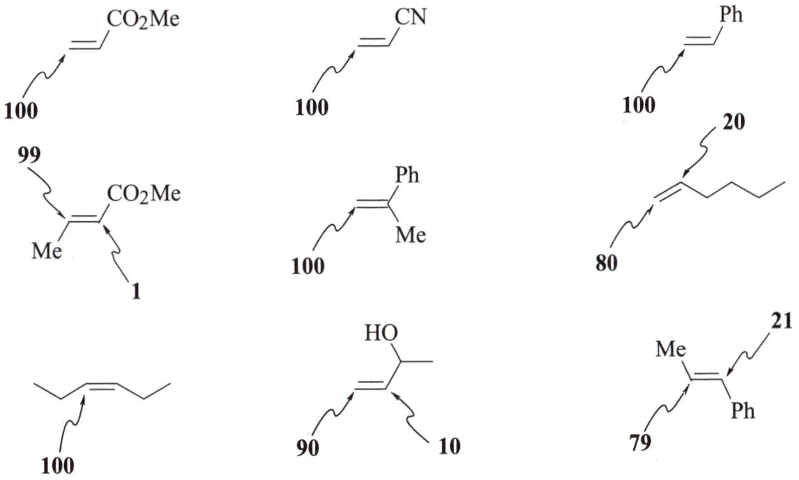

Numbers indicate the percent substitution of the aryl group at a vinylic position.

Figure 11-4
Regioselectivity of Addition
of an Aryl Halide to Various Alkenes
During Heck Olefination

The stereochemistry associated with the Heck reaction is consistent with what we now expect with a process involving 1,2-insertion and β-elimination – *syn* addition to the olefin and *syn* elimination to form a new alkene. When monosubstituted alkenes react with aryl halides, the β-hydrogen removed (Scheme **11.14**) is the one that leads preferentially to an *E*-double bond (equation **11.36**).[47]

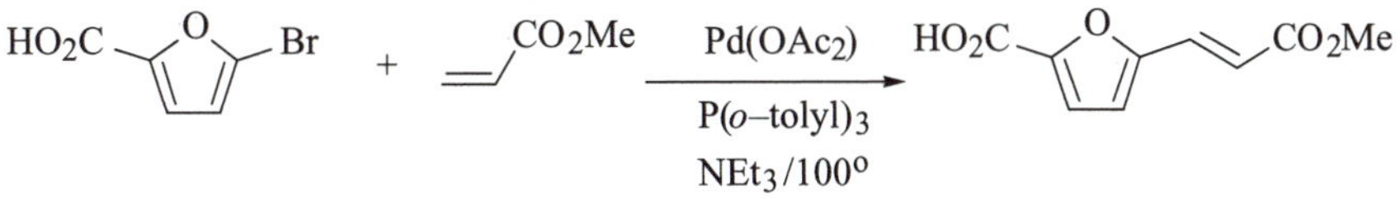

$$11.36$$

Exercise 11-8

Consider the mechanistic pathway shown in Scheme **11.14**, and show that the (*E*)-ester ought to result as product in equation **11.36**, as shown.

[47]R.P. Ghandi, *et al.*, *Synthetic Comm.*, **1989**, *19*, 1759.

1,2-Disubstituted alkenes insert to give a mixture of E- and Z-trisubstituted olefins with usually the more stable isomer predominating. If a choice of β-hydrogens is available, the most acidic is lost in the β-elimination step.

Although Pd(0) is the active oxidation state for catalysis, typical Heck olefination procedures utilize Pd(II) salts [such as PdCl$_2$, Pd(OAc)$_2$, or Na$_2$PdCl$_4$] and a reducing agent to generate Pd(0) *in situ*. Usually a stoichiometric excess of an amine such as Et$_3$N is part of the reaction mixture, serving both as a base to trap the HX given off as by-product and as a reducing agent for Pd. Scheme **11.15** below shows a possible mechanism for conversion of Pd(II) to Pd(0).[48]

Since its discovery around 20 years ago, several modifications of Heck reaction conditions have been reported. Use of DMF as a solvent is usually preferable to that originally employed, CH$_3$CN.[49] The presence of Bu$_4$NCl, especially accompanied by KOAc, increases reaction rate and allows the reaction to proceed at room temperature instead of 80 to 130°, as normally required.[50] Originally thought to serve as a phase transfer catalyst,[51] the quaternary ammonium salt probably serves also as a source of Cl$^-$, which tends to bind to Pd and stabilize intermediate metal complexes involved in the catalytic cycle.

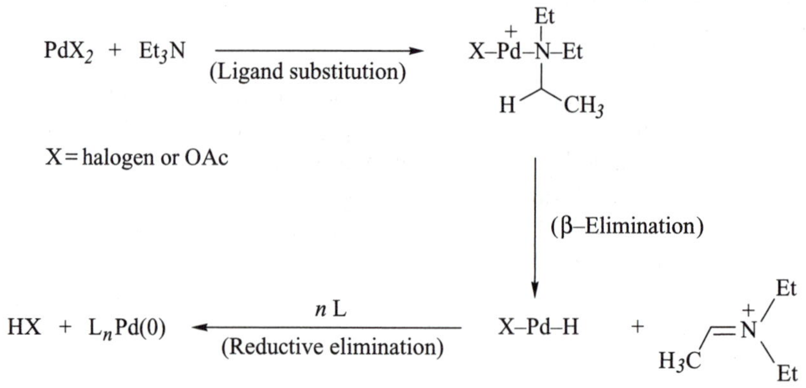

Scheme 11.15
Reduction of Pd(II) to Pd(0) with Et$_3$N

[48]J.P Collman, *et al.*, *op. cit.*, 725.
[49]A. Spencer, *J. Organomet. Chem.*, **1984**, *270*, 115, which includes references to his earlier work.
[50]T. Jeffery, *Tetrahedron Lett.*, **1985**, *26*, 2667; *J. Chem. Soc., Chem. Commun.*, **1984**, 1287; and R. Larock and B.E. Baker, *Tetrahedron Lett.*, **1988**, *29*, 905.
[51]A phase transfer catalyst is typically a quaternary ammonium salt with medium-to-long hydrocarbon chains attached to the nitrogen. The hydrocarbon region is hydrophobic, and thus compatible with non-polar organic solvents. The charged nitrogen end of the molecule is hydrophilic or water-compatible. Because of these properties, phase transfer catalysts function as anion carriers capable of transporting nucleophiles or bases from an aqueous phase to an organic liquid phase. Reactions that require the presence of the salt of a strong nucleophile, base, or oxidizing agent—normally not soluble in a non-polar solvent—proceed considerably faster in the presence of such catalysts. See J. March, *Advanced Organic Chemistry*, 4th ed. Wiley, New York, 1992, 362–365.

Equations **11.37**[52] and **11.38**[53] show two recent examples of intramolecular Heck olefination. In the first case, the Pd catalyst also acts to isomerize a C=C bond to give a final product that has the C=N bond in conjugation with the aromatic ring. The second example demonstrates how asymmetric induction is possible in the presence of BINAP.

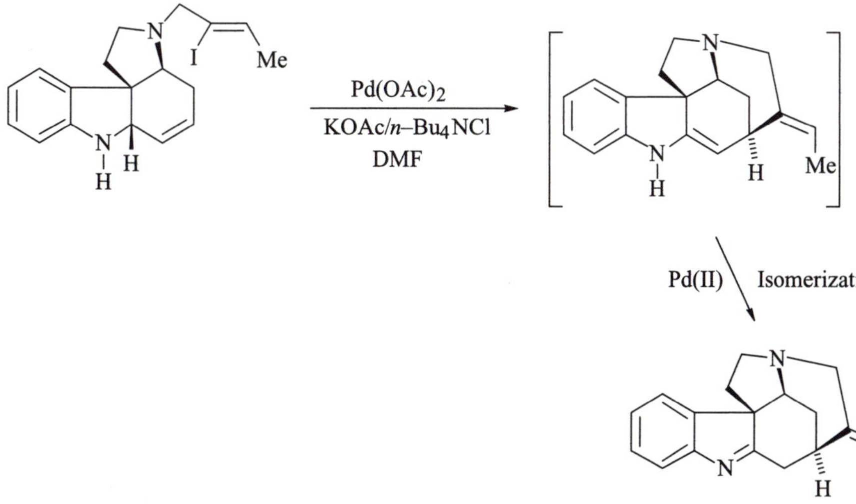

11.37

Pd(II)   Isomerization

11.38

Pd$_2$(dba)$_3$

(R)–BINAP/80°

71% ee

dba:

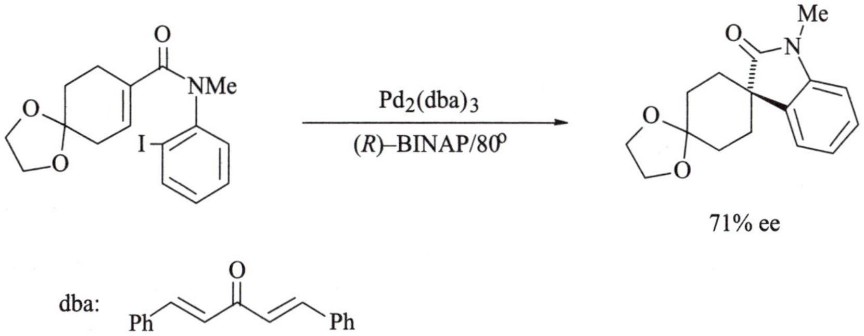

[52]V.H. Rawal and C. Michoud, *J. Org. Chem.*, **1993**, *58*, 5583.
[53]A. Ashimori and L.E. Overman, *J. Org. Chem.*, **1992**, *57*, 4571.

11-4 Carbon-Carbon Bond Formation via Transmetallation Reactions

Transmetallation according to general equation **11.39** is an excellent way of introducing σ-bonded hydrocarbon ligands into the coordination sphere of transition metals. The equilibrium is thermodynamically favorable from left to right if the electronegativity of M (usually the transition metal) is greater than M′ (often a main group or early transition metal), and kinetically favorable if an empty orbital is available on both metals.

$$M'\text{-}R + M\text{-}X \rightleftharpoons M\text{-}R + M'\text{-}X \qquad\qquad \textbf{11.39}$$

$$via \qquad \begin{matrix} M' \bullet\bullet\bullet\bullet R \\ \bullet \qquad\qquad \bullet \\ \bullet \qquad\qquad \bullet \\ X \bullet\bullet\bullet\bullet\bullet M \end{matrix}^{\ddagger}$$

37

The reaction appears to be concerted, involving a four-center transition state (**37**) that leads to transfer of the organic group to M with retention of configuration. Even if transfer is not favorable thermodynamically, the reaction may still be useful if it is possible to trap the equilibrium concentration of M-R and then utilize it in a subsequent irreversible step.[54] Table **11-2** shows a portion of the Periodic Table with electronegativities listed below each element.

Table 11-2 Electronegativities[a] of Selected Elements Useful in Synthesis

I[b]	II	III	IV	II	II	II	II	II	II	II	II	III	IV
Li												B	
0.98												2.04	
Na	Mg											Al	Si
0.93	1.31											1.61	1.90
K	Ca	Sc	Ti	V	Cr	Mn	Fe	Co	Ni	Cu	Zn		
0.82	1.00	1.36	1.54	1.63	1.66	1.55	1.83	1.88	1.91	1.90	1.65		
			Zr		Mo		Ru	Rh	Pd	Ag			Sn
			1.33		2.16		2.2[c]	2.38	2.20	1.93			1.96
					W			Ir	Pt	Au	Hg	Tl	Pb
					2.36			2.20	2.28	2.54	2.00	2.04	2.33

[a] Pauling electronegativities taken from A.L Allred, *J. Inorg. Nucl. Chem.*, **1961**, *17*, 215.
[b] The Roman numerals at the top of the table indicate the oxidation state of the metal. Electronegativity generally increases for a metal as its oxidation state increases.
[c] Estimated

[54]E-i. Negishi, *Organometallics in Organic Synthesis*, Wiley, New York, 1980, 54-56.

Organolithium and magnesium reagents are the most reactive transmetallation reagents (containing M′), but their use is disadvantageous because they often react with transition metal ligands. Less-reactive organozirconium, zinc, tin, and aluminum compounds often possess just the right amount of reactivity to be useful in transferring σ-bonded hydrocarbon ligands to transition metals without affecting the ligands already present.

11-4-1 Transmetallation Involving Zirconium

We saw earlier (Chapter 8, Section **8-1-2**) that hydrozirconation of alkenes and alkynes proceeds readily with predictable regio- and stereoselectivity. Although the resulting d^0 organozirconium complexes undergo electrophilic cleavage to give corresponding halides or undergo carbonylation followed by nucleophilic attack to give carbonyl derivatives, they are too stable to undergo other useful chemistry. Scheme **11.16** shows the use of a zirconium reagent to transfer an alkene ligand to Al, which then reacts with acyl halides.[55] This is an excellent route to unsymmetrical ketones.

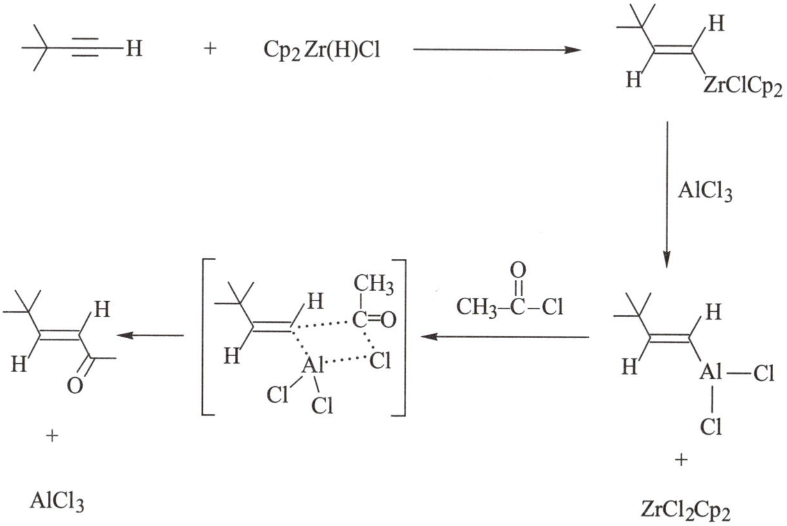

Scheme 11.16
Zr-Al Exchange

[55]D.B. Carr and J. Schwartz, *J. Am. Chem. Soc.*, **1977**, *99*, 638; **1979**, *101*, 3521.

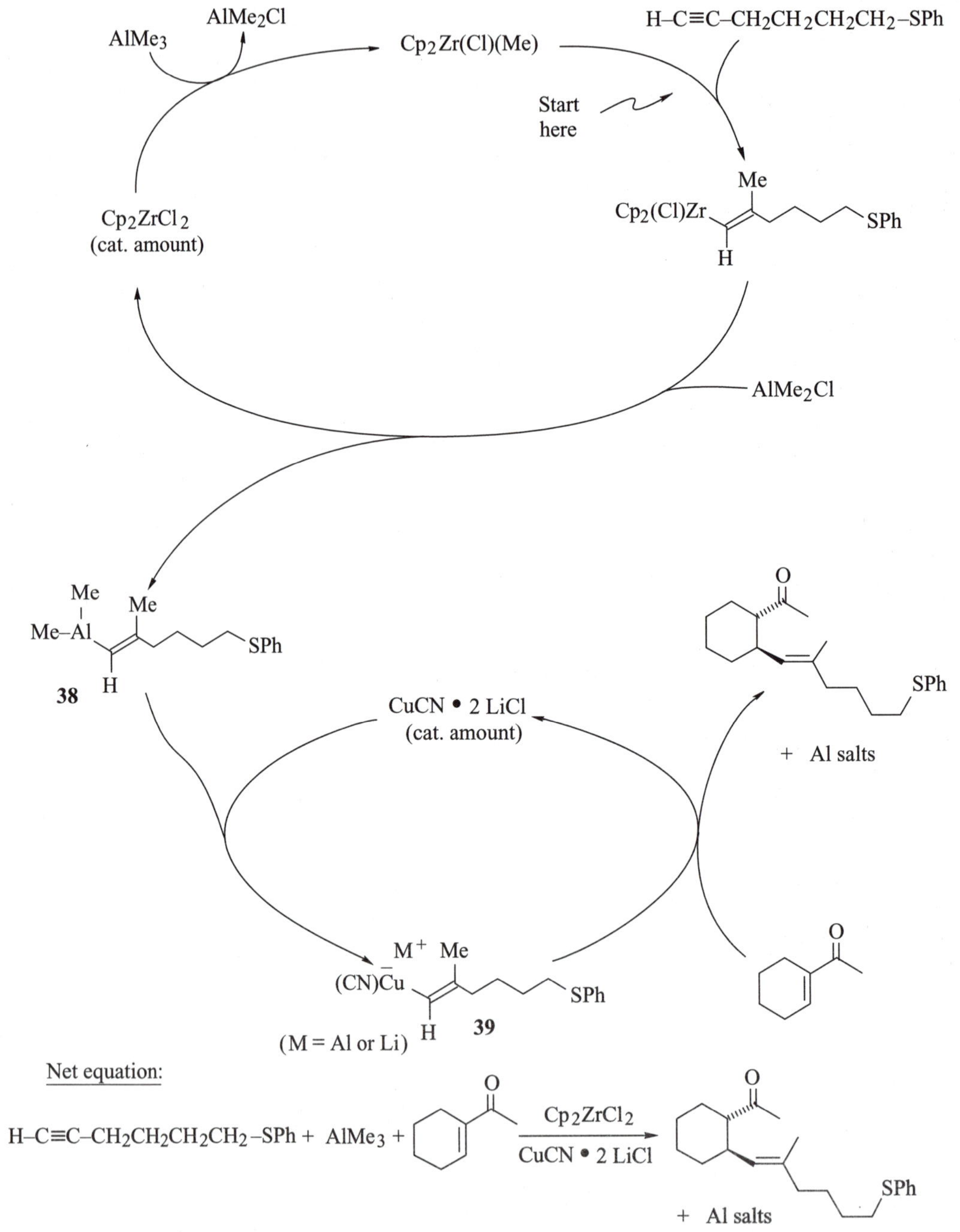

Scheme 11.17
Multiple Transmetallations

Scheme **11.17** shows how two transmetallations were used to connect a monosubstituted alkyne to an enone. The process begins with methylation of a catalytic amount of Cp_2ZrCl_2 using Me_3Al. The resulting $Cp_2Zr(Cl)(Me)$ then reacts with a terminal alkyne to give a *syn* 1,2-insertion of the triple bond into a Zr-Me bond. Metal-metal exchange with Al yields the alane **38**, which then undergoes another metal-metal exchange with a catalytic amount of Cu to give mixed cuprate **39**. Cuprate **39** finally transfers stereoselectively its vinyl group via 1,4-addition to the β–position of the enone, creating a new C-C bond and giving the ultimate product. The mechanism of this last step is not well understood and may involve radical intermediates. This β-alkenylation is general for a variety of alkynes and enones.[56]

Sigma-Hydrocarbyl[57] zirconium complexes undergo transmetallation with other metals such as Pd and Ni. Because Zr is relatively electropositive compared to most of the transition metals, and therefore will react with most transition metals, a great deal of useful synthetic chemistry involving its transmetallation could develop with further research.

Exercise 11-9

Consider the transformation below that is a *stoichiometric* variation of Scheme **11.17**. Provide the detailed steps for each major transformation.

11-4-2 Transmetallation Involving Palladium

Palladium is one of the most versatile metals used by synthesis chemists because it promotes a myriad of transformations. In this section we introduce yet another example of such versatility in which a variety of different partners combine to form a new C-C bond via transmetallation to Pd. Metals such as Si, Zn, Zr, B, and Al transmetallate readily with Pd, but organotin compounds (called stannanes) seem to be the most useful for several reasons. First, these compounds are relatively easy to synthesize by a number of different routes; some are even available commercially. Second, stannanes, once formed, are stable because the Sn-C bond is relatively robust, containing ca. 50 kcal/mole

---

[56]B.H. Lipschutz and S.H. Dimock, *J. Org. Chem.*, **1991**, *56*, 5761.
[57]Hydrocarbyl is a general term for any hydrocarbon ligand that attaches to the metal with one σ bond (e.g., alkyl, alkenyl, alkynyl, or aryl).

of bond energy. Such bond strength makes the Sn–C bond resistant to rupture in the presence of both oxidizing and reducing reagents; several functional groups, moreover, are compatible with Sn. Finally, organotin compounds are not particularly air or moisture sensitive, which reduces considerably the need for use of sophisticated laboratory techniques.

Equations **11.40** to **11.42** (below) outline the general types of coupling transformations that are possible using organotin reagents in the presence of Pd. Stille and his co-workers performed most of the original work in developing the scope and utility of this coupling reaction, so the reaction is often called Stille coupling.[58]

$$R\text{-}X + R''\text{-}SnR'_3 \xrightarrow[\text{PdL}_2]{} R\text{-}R'' + X\text{-}SnR'_3 \qquad\qquad \textbf{11.40}$$

$$R(CO)X + R''\text{-}SnR'_3 \xrightarrow[\text{PdL}_2]{} R(CO)R'' + X\text{-}SnR'_3 \qquad\qquad \textbf{11.41}$$

$$R\text{-}X + R''\text{-}SnR'_3 \xrightarrow[\text{PdL}_2 \,/\, CO]{} R(CO)R'' + X\text{-}SnR'_3 \qquad\qquad \textbf{11.42}$$

As with the Heck olefination or Pd-catalyzed CO insertion, most any organic bromide or iodide (without hydrogens attached to an sp^3-hybridized β carbon, of course) will serve as an electrophile and undergo coupling with the organotin reagent (equation **11.40**). Usually the stannane is designed so that one group will transfer preferentially over the rest. The order of transfer is as follows:

$$RC{\equiv}C\text{-} > RCH{=}CH\text{-} > Aryl > allyl \sim benzyl > CH_3(CO)CH_2\text{-} > alkyl$$

Methyl or *n*-butyl groups typically comprise the remaining, nontransferable groups in the organotin reagent (the R' groups in equations **11.40** to **11.42**).[59] Acyl chlorides react to give ketones (equation **11.41**) , or, alternatively, R–X reacts in the presence of CO with the same result (**11.42**). Use of the alternate path to ketones works when the acid chloride is not readily available or if the presence of an acyl chloride group is incompatible with co-attachment of remote protic functional groups such as OH or NH_2. Table **11-3** summarizes the broad scope of Stille coupling. In general, any electrophile listed in the first column will couple with any stannane from the second column.

[58]For an excellent summary of C–C bond formation using organotin reagents in the presence of Pd, see J.K. Stille, *Angew. Chem. Int. Ed. Eng.*, **1986**, *25*, 508.

[59]Tetraalkyltin reagents will react—albeit more slowly than unsymmetrical stannanes—if the transfer of a simple alkyl group is desired.

Table 11-3 Electrophiles and Organotin Reagents Suitable for Coupling Reactions[a]

Electrophile	Organotin Reagent
R(CO)Cl	H-SnR$_3$
R'R''C=CR'''-CH2-X (allyl)	R'''C=C-SnR$_3$
Ar-CH$_2$-X (benzyl)	R'R''C=CR'''-SnR$_3$
R'R''C=CR'''-X	Ar-SnR$_3$
Ar-X	R'R''C=CR'''-CH$_2$-SnR$_3$
R'-C(H)(X)-CO$_2$R''	Ar-CH2-SnR$_3$
	R'-SnR$_3$ (R,R' = alkyl)

[a]X = Br, I; R=Me, n-Butyl.

The reaction is catalytic in Pd, the active form of which is Pd(0). Pd(PR$_3$)$_4$ or PdCl$_2$(MeCN)$_2$ are useful catalysts for Stille coupling [the latter Pd(II) complex is likely reduced *in situ*]. Scheme **11.18** shows the catalytic cycles for direct coupling and coupling with carbonyl insertion. The mechanism of catalysis is not known completely, but the steps in Scheme **11.18** seem reasonable,

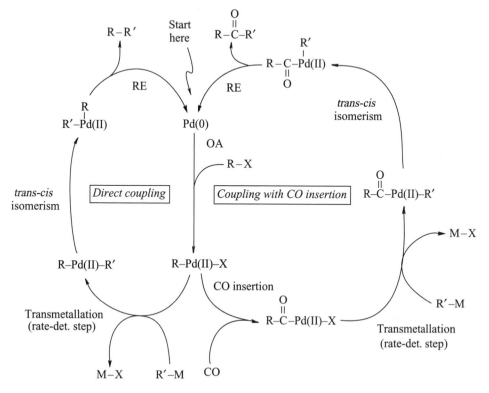

Scheme 11.18
Catalytic Cycles for Stille Coupling

based upon previous work and the study of the stoichiometric version of Stille coupling. The rate-determining step seems to be transmetallation, so transferable groups attached to Sn may have β–hydrogens attached to sp^3 carbons, since steps subsequent to transmetallation are rapid.

Stille coupling occurs not only with chemo- and regioselectivity, but also with stereoselectivity. Equation **11.43** illustrates the occurrence of retention of configuration at the substituted position of both the vinyl halide and the vinyl tin reagent. The exceptions are allyl or benzyl halides, which undergo inversion of configuration at the halogen-substituted position, suggesting an S_N2-type mechanism (Chapter 7, Section **7-2-2**) for oxidative addition of RX.

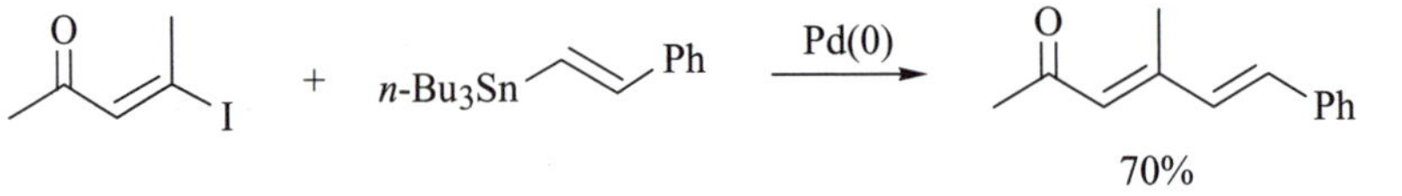

11.43

70%

Exercise 11-10

Complete the following reaction and specify the stereochemistry of the coupled product:

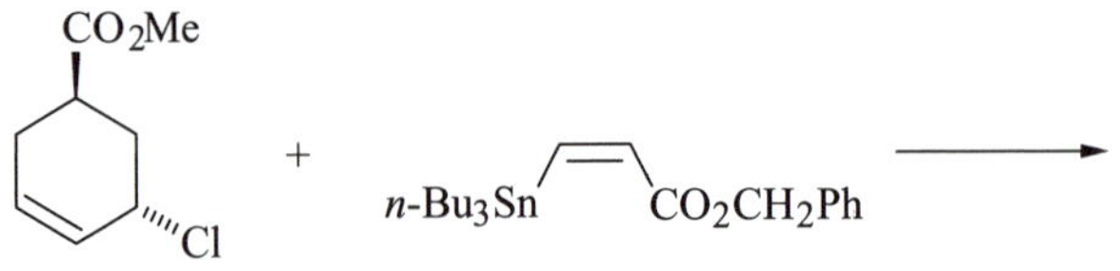

After the pioneering work of Stille, the discovery that enol triflates[60] (**40**, equation **11.44**) undergo Pd-catalyzed coupling with organotin complexes provided an added dimension to the scope of the reaction[61]. Equation **11.44** shows one method for the formation of enol triflates from a ketone; such derivatives constitute, therefore, a stable form of an enol. Though the triflate group stabilizes enols, it is an excellent leaving group for oxidative addition to Pd. Equation **11.45** hints at the versatility of coupling enol triflates with organotin reagents, the net result of which is the conversion of a ketone into a substituted alkene.[62] Usually LiCl is added to the reaction mixture to complex

[60]The triflate group, further abbreviated as OTf, is actually trifluoromethanesulfonate (CF_3SO_3). Organic chemists consider triflate to be an excellent leaving group—even better than tosylate, OTs.

[61]A summary of Pd-catalyzed coupling using triflates appeared recently: W.J. Scott and J.E. McMurry, *Acc. Chem. Res.*, **1988**, *21*, 47.

[62]W.J. Scott and J.K. Stille, *J. Am. Chem. Soc.*, **1986**, *108*, 3033.

with Pd ($^-$OTf is a very poor ligand) after oxidative addition. The chloride ion then attaches to the Sn during transmetallation (Scheme **11.18**). Analogous to Stille coupling in the presence of CO (shown in equation **11.42**), vinyl triflates also couple with stannanes in the presence of CO to give conjugated ketones and aldehydes.

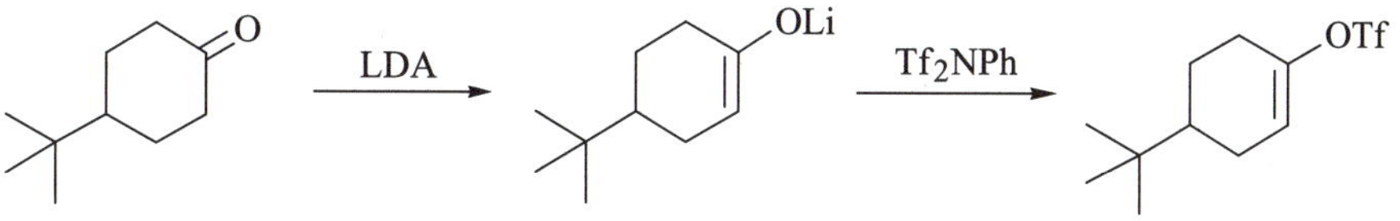

11.44

40

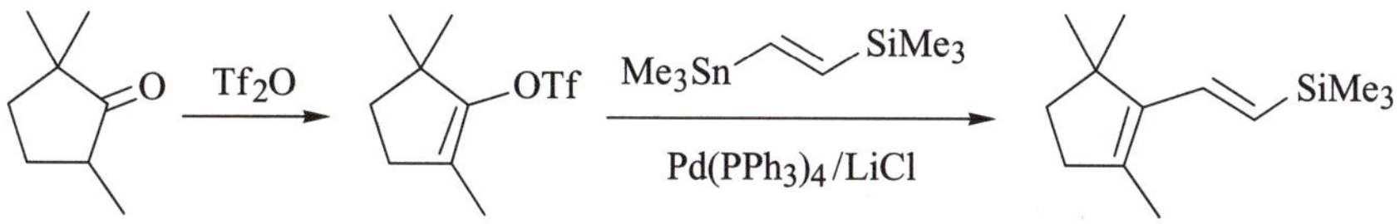

11.45

Exercise 11-11

What are the starting materials required to synthesize compound **41**?

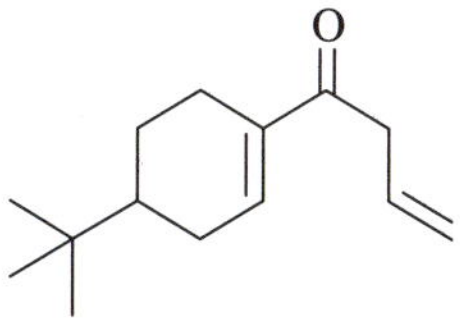

41

Countless reports of how Stille coupling played a key role in the synthesis of complex molecules have appeared recently, attesting to the great utility of the reaction. Equation **11.46** shows intramolecular coupling as the ring-closing step in the synthesis of zearelenone (**42**), a fungal metabolite possessing antibiotic activity.[63] Intermolecular coupling of **43** and **44** (equation **11.47**) provides the side chain on **45**, which contains the carbon backbone essential for construction of three rings in Lepicidin A (**46**), a structurally complex microbial product with insecticidal activity.[64]

[63]A. Kalivretenos, J.K. Stille, and L.S. Hegedus, *J. Org. Chem.*, **1991**, *56*, 2883.
[64]D.A. Evans and W.C. Black, *J. Am. Chem. Soc.*, **1993**, *115*, 4497.

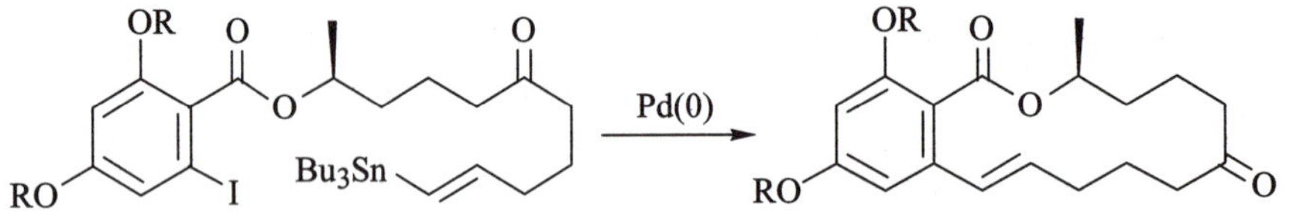

R = Protecting group

Pd(0)

R = H   **42**

11.46

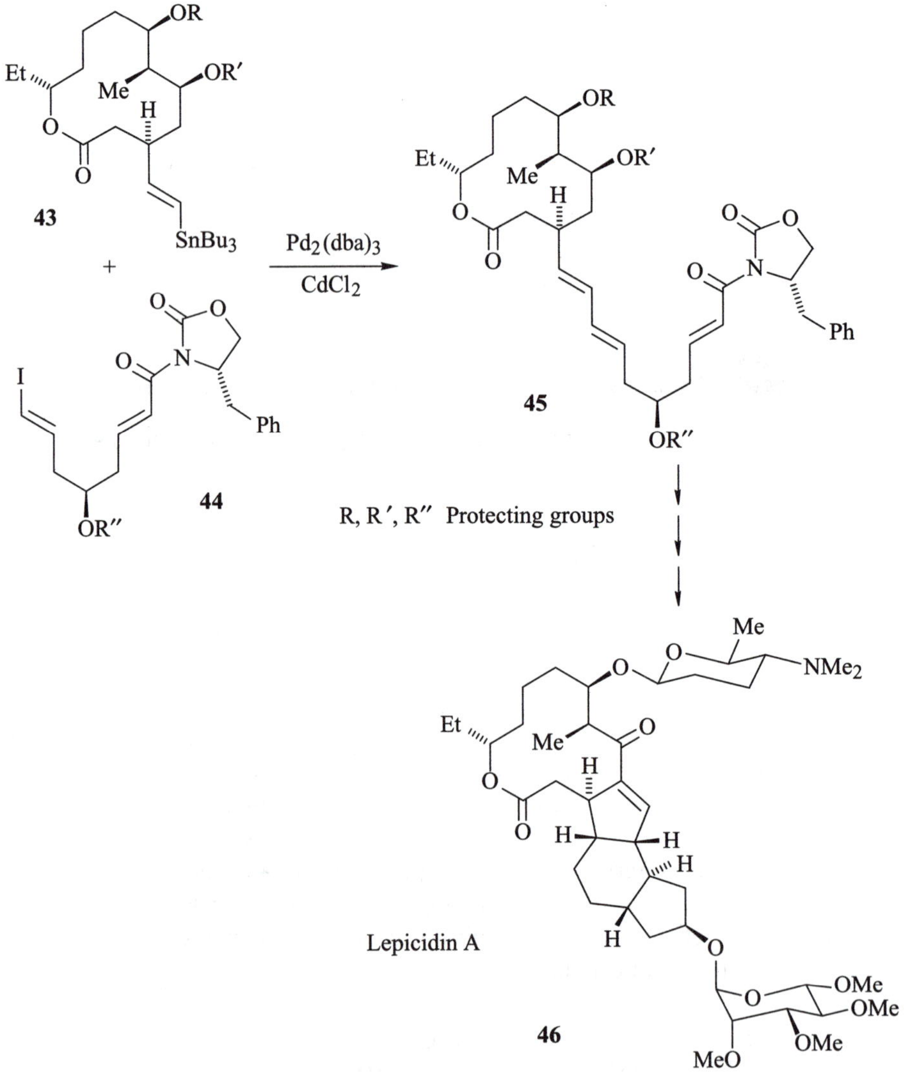

43

+

44

$Pd_2(dba)_3$

$CdCl_2$

45

R, R′, R″ Protecting groups

Lepicidin A

46

11.47

11-5   Carbon-Carbon Bond Formation through Cyclization Reactions

Although several examples of cyclization reactions appeared in earlier sections of this chapter—none has involved ring formation accompanied by formation of more than one C-C bond. In Section **11-5** we will show a limited number of cases where organotransition metal complexes effectively promote the construction of rings where at least two bond connections occur during the same transformation. Some examples will be extensions of reactions already covered, while others will entail "new" chemistry.

11-5-1 Cyclizations Involving Fischer-Type Carbene Complexes

Fischer-type chromium carbene complexes have numerous applications to organic synthesis. Dötz,[65] Wulff,[66] and several other groups developed a procedure for synthesizing substituted phenols, naphthols, and higher polycyclic homologs that amounts to a $[3 + 2 + 1]$ cyclization involving three partners: a Cr carbene complex (contributing three ring atoms), an alkyne (contributing two ring atoms), and CO (contributing the sixth ring atom). Equation **11.48** shows a schematic representation of the cyclization, indicating how the reactant fragments fit together. Note that the methoxy group attached to $C_{carbene}$ ends up *para* to the phenolic OH group.

11.48

Scheme **11.19** shows a plausible mechanism for the cyclization. Section **10-3-1** in Chapter 10 pointed out that Fischer-type carbene complexes readily lose a CO ligand, which in this case provides an open coordination site for the incoming alkyne and formation of **47**. Rearrangement of **47** gives the chromacyclobutene (**48**), a reaction that should be familiar from Chapter 10. Next carbonyl insertion (or migration of an alkenyl carbon atom in the ring to CO) occurs to relieve ring strain and to provide intermediate **49**. Rearrangement of

---

[65]For a summary of work on the use of Cr carbenes to synthesize arenes, see K.H. Dötz, *Angew. Chem. Int. Ed. Eng.*, **1984**, *23*, 587.
[66]W.D. Wulff, *et al.*, *Tetrahedron*, **1985**, *41*, 5813.

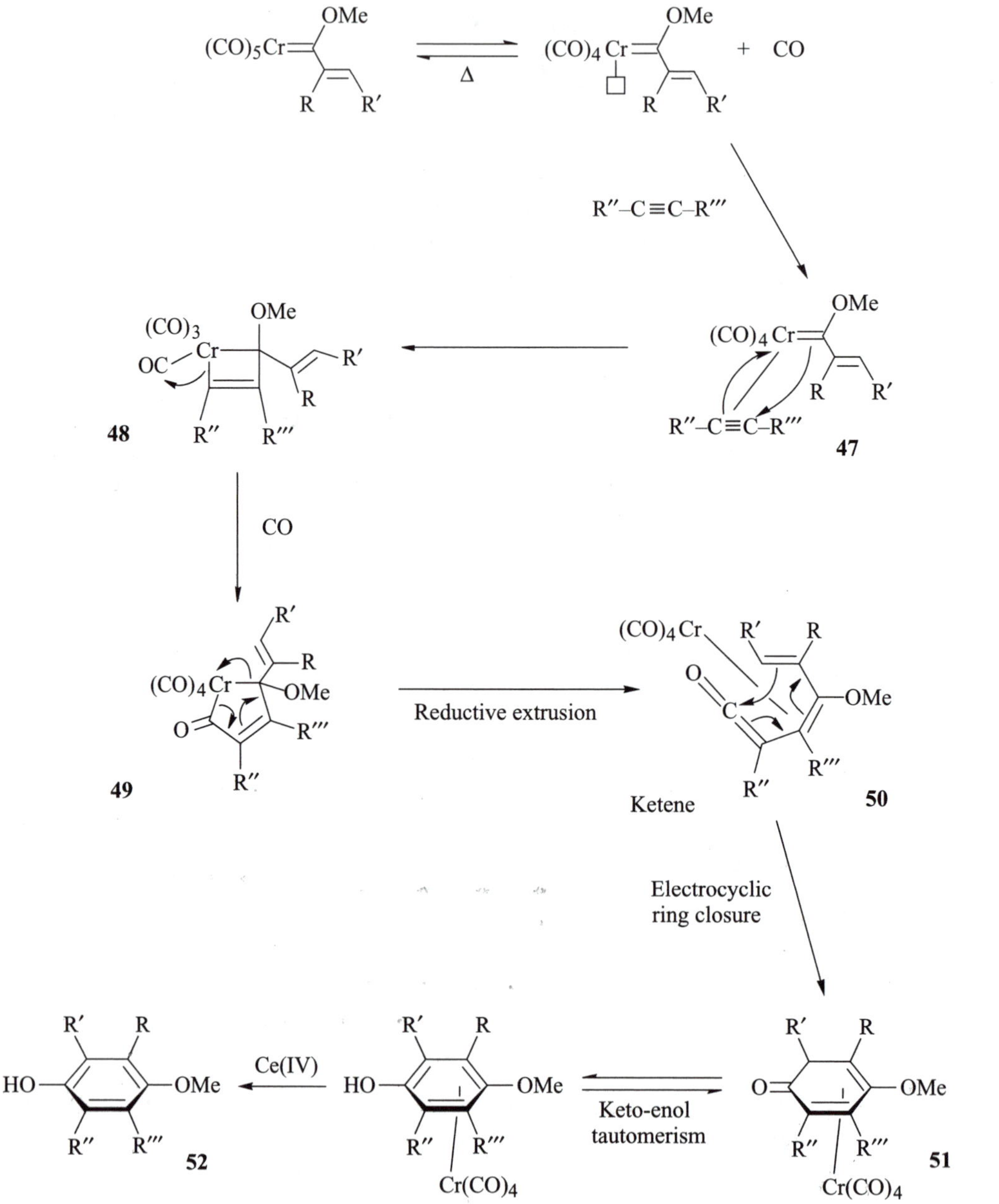

Scheme 11.19
Proposed Mechanism of the Dötz-Wulff Cyclization

49 by reductive extrusion[67] of the metal fragment proceeds to yield ketene complex **50**. The transformation of **50** into **51** is an electrocyclic ring closure, examples of which are numerous in the literature of organic chemistry.[68] Intermediate **51** then undergoes keto-enol tautomerism to give the final product **52** upon removal of the metal fragment. Several methods exist that will remove the metal from the π ligand; these include oxidative treatment with air or Ce(IV) and carbonylation of Cr in an atmosphere of CO.

Exercise 11-12

For unsymmetrical alkynes, there are two orientations possible for the alkyne substituents in the final six-membered ring product. Usually the predominant regioisomer is that in which the bulkier of the two substituents is *ortho* to the phenol group.[69] Assume that 1-hexyne reacts with $(CO)_5Cr=C(OMe)(Ph)$ to give a naphthol derivative. Follow the reaction through according to Scheme **11.19**, and determine the predominant regioisomer based upon the difference in steric hindrance of the substituents attached to the triple bond of the alkyne.

[67]Reductive extrusion and its reverse reaction, oxidative coupling (shown in the equation below), may occur when β elimination in the metallacycle is not feasible geometrically. We may think of the reaction as going through a bis-metallacyclopropane intermediate. Thus, in the forward direction, the oxidation state of the metal decreases by two units. The reverse reaction is formally an oxidation.

bis-metallacyclopropane

The dotted line signifies an extra bond between the two carbon atoms if required.

For a discussion of reductive extrusion - oxidative coupling, see R.H. Crabtree, *The Organometallic Chemistry of the Transition Metals*, 2nd ed. Wiley, New York, 1994, 155-158.
[68]For additional information on electrocyclic ring openings and closures, see J.E. McMurry, *Organic Chemistry*, 4th ed. Brooks/Cole, Pacific Grove, CA, 1996, chap. 31, especially 1215–1221 and F.A. Carey and R.J. Sundberg, *Advanced Organic Chemistry, Part A*, 3rd ed. Plenum, New York, 1990, chap. 11, especially 596-609.
[69]Electronic effects also play a role in directing regiochemistry. Electron withdrawing substituents end up *ortho* to the phenol group whereas electron donating groups (such as ethers) prefer the *meta* position. See F.J. McQuillan, D.J. Parker, and G.R. Stephenson, *op. cit.*, 419-420.

The Dötz–Wulff arene synthesis is an excellent technique for construction of complex arenes. Equations **11.49**[70] and **11.50**[71] show two examples of the cyclization procedure, the former illustrating formation of a naphthol and the latter leading to an *N*-methylindole.

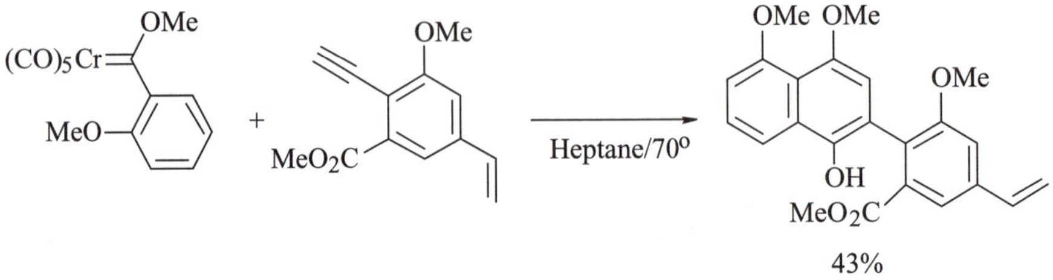

11.49

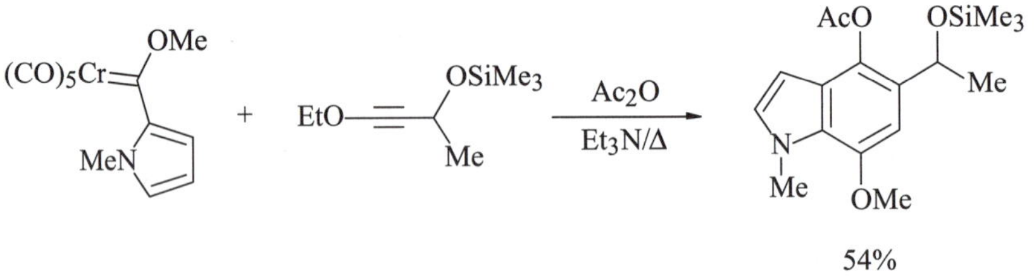

43%

11.50

54%

Cr carbene complexes are also useful in the synthesis of β-lactams (**53**), cyclic amides that appear in the structure of the penicillin (**54**) and cephalosporin (**55**) antibiotics.

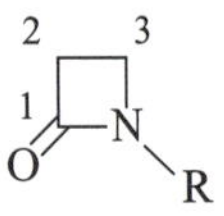

53 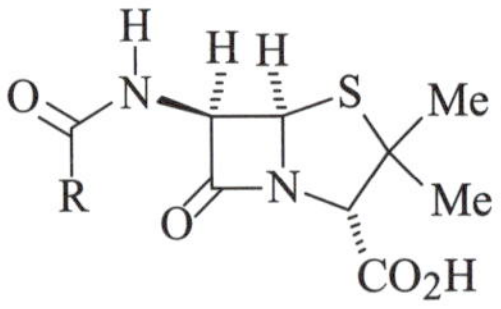   54 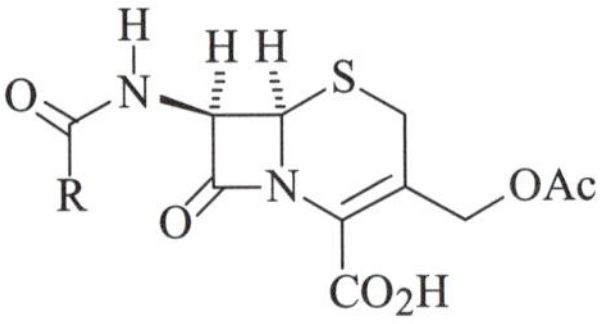   55

When carbene complexes such as $(CO)_5Cr=CRR'$ (R = H or Me; R' = OR or NR) undergo irradiation with light (λ = 340 to 450 nm), they absorb in such a manner that electron density transfers from the metal to $C_{carbene}$. Scheme **11.20** shows the result of this absorption. The first step is a reversible CO insertion into the M=C bond to give intermediate **56**, one resonance form of which is represented as a complexed ketene (**56b**). In the Dötz–Wulff cyclization, the ketene intermediate underwent intramolecular closure to form a six-membered ring. In photochemical ketene generation,

[70]K.A. Parker and C.A. Coburn, *J. Org. Chem.*, **1991**, *56*, 1666.
[71]A. Yamashita, A. Toy, and T.A. Scahill, *J. Org. Chem.*, **1989**, *54*, 3625.

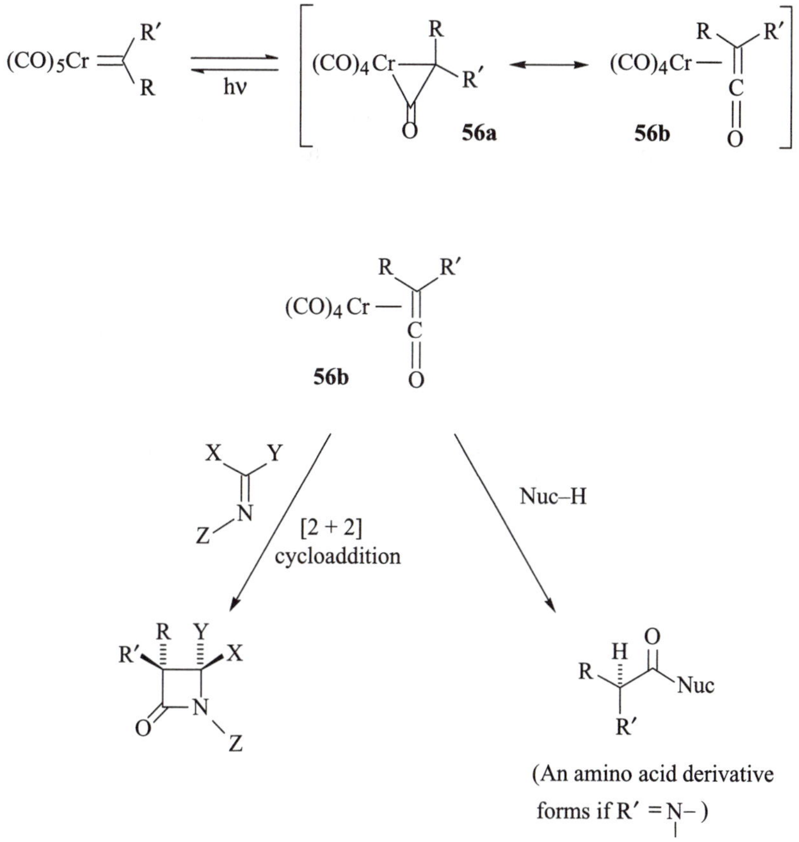

Scheme 11.20
Photochemical-Induced
Reactions of Cr Carbenes

reaction with an external reactant occurs instead because of the absence of a conjugated π system capable of undergoing such ring closure. Ketenes are extremely reactive and will suffer attack by nucleophiles at the carbonyl carbon or undergo [2 + 2] cycloaddition with a π system such as a C=C bond, or in the case of interest here, an imine.

The functionality of the lactam portion of penicillins and cephalosporins includes a substituted acylamine group and a hydrogen attached to C-2 (structure **53**). Hegedus[72] and co-workers discovered a general route to

---

[72]For a summary of the chemistry of ketenes that are generated photochemically from Cr carbenes, see L.S. Hegedus, *Pure Appl. Chem.*, **1990**, *62*, 691 and references therein.

aminocarbene complexes (**59**) shown in Scheme **11.21** below. The reaction involves using potassium-graphite to reduce $Cr(CO)_6$ to $Cr(CO)_5^{2-}$, which then reacts with formamide derivatives (**57**). Trimethylsilyl chloride reacts with intermediate **58** to form the corresponding silyl ether, and loss of that group finally yields **59**. If the amide used contains a chiral oxazoline moiety (Scheme **11.1**) that can be removed later, chiral carbenes (**60**) (and ultimately ketenes upon photolysis) result.

Equation **11.51** shows the general procedure Hegedus used to produce several β lactams. When imine substituents R_2 and R_3 were different (i.e., one was alkyl and the other H), high stereoselectivity resulted at the C-2 and C-3 positions. For example, if R_2 = Me and R_3 = H, the composition of the final product mixture was > 97% *trans* and < 3% *cis*. Use of the (*R*)-chiral auxiliary gave *R* stereochemistry at C-2, and the (*S*)-auxiliary gave *S* stereochemistry (same as natural penicillins and cephalosporins) at that position.

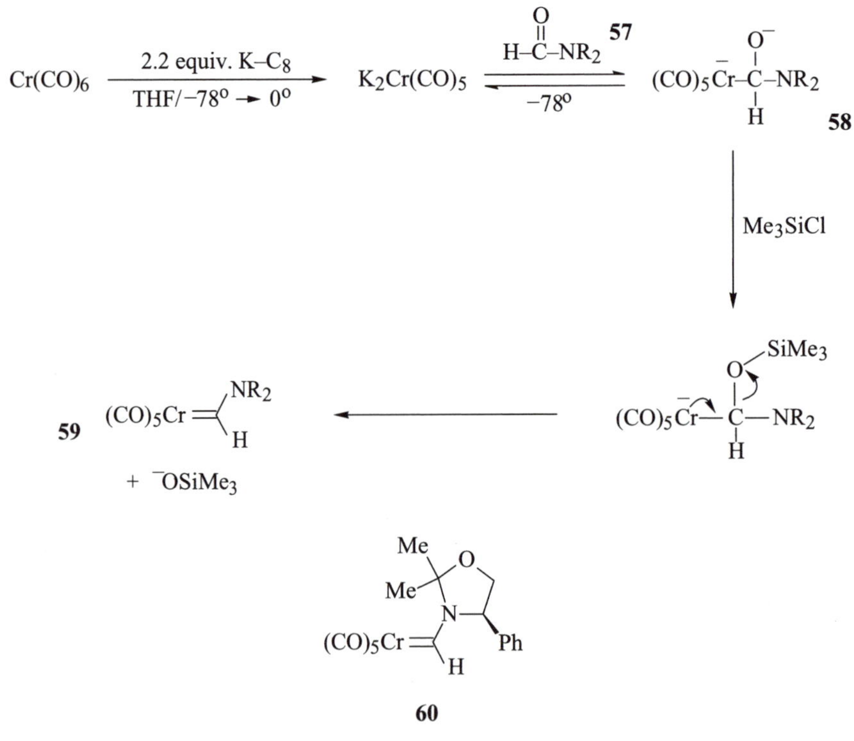

Scheme 11.21
Synthesis of Amino Carbene Complexes

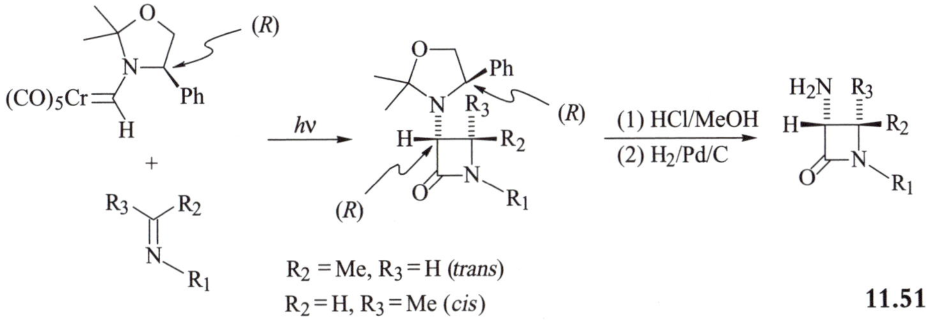

$$R_2 = Me, R_3 = H \ (trans)$$
$$R_2 = H, R_3 = Me \ (cis)$$

11.51

In equation **11.52** we see an extension of the photocycloaddition into the synthesis of cephalosporins. Reaction of racemic imine **62** with **61** gave **63** as a 1:1 mixture of diastereomers.

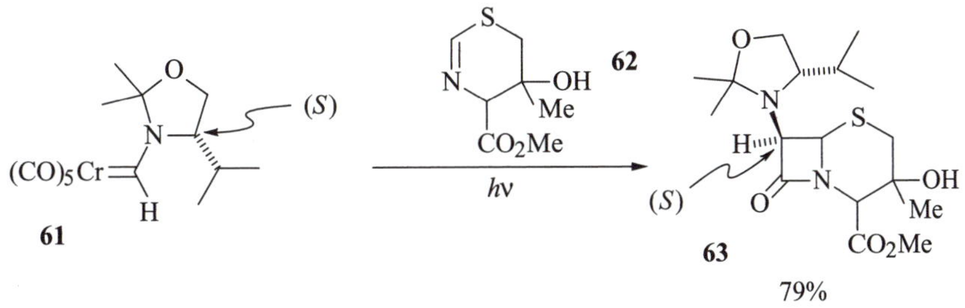

79%

11.52

Photochemical generation of ketenes from Cr carbene complexes is an excellent means of producing four-membered rings. The application of this chemistry does not stop, however, with the synthesis of these ring systems. The ability of ketenes to react readily with a variety of nucleophiles complements the capability of ketenes to undergo cycloaddition with π systems. Since photolysis of Cr carbene complexes is a convenient method for generating ketenes, many more applications of this chemistry will undoubtedly appear in the literature.

11-5-2 Cyclizations Involving Palladium

Discovered over 50 years ago[73] and analogous to the renowned Diels-Alder reaction, the *ene reaction* (equation **11.53**) is a concerted transformation that brings together two partners—one of which is a simple π system, called the *ene*, (typically an isolated C=C or C=O bond with at least one hydrogen at the

[73]K. Alder, F. Pascher, and A. Schmitz, *Chem. Ber.*, **1943**, 76, 27.

allylic position) and the other an isolated π bond, termed the *enophile* (enophiles with electron withdrawing groups attached to the π system work best). The reaction resembles a cycloaddition, proceeding through a six-electron, six-membered-ring transition state (**64**), and both intra- and inter-molecular versions are known. Because the reaction results in the formation of a new C-C bond with the stereoselectivity attendant concerted processes, the ene reaction has great potential as a useful synthetic procedure. Intramolecular versions, moreover, produce rings. The utility of the transformation was limited for many years, however, because of the high temperatures required compared to those used for its more famous analog, the Diels–Alder reaction.

$$11.53$$

The discovery that Lewis acids catalyze ene reactions greatly increased the utility of the transformation.[74] Equations **11.54**[75] and **11.55**[76] below demonstrate the effect of a Lewis acid, ZnBr$_2$, in not only decreasing the severity of reaction conditions required, but also in increasing the stereoselectivity observed.

$$11.54$$

$$11.55$$

[74]For a review of acid-catalyzed ene reactions, see B.B. Snider, *Acc. Chem. Res.*, **1980**, *13*, 426.
[75]S.K. Ghosh and T.K. Sarkar, *Tetrahedron Lett.*, **1986**, *27*, 525.
[76]L.F. Tietze, V. Beifuss, and M. Ruther, *J. Org. Chem.*, **1989**, *54*, 3120.

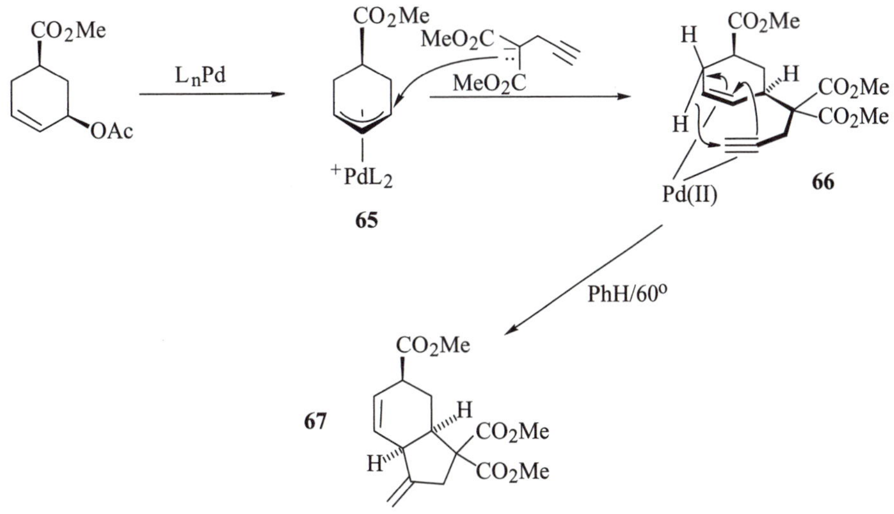

Scheme 11.22
Trost's Pd-Catalyzed Ene Reaction

85% yield

Palladium is effective as a catalyst for the ene reaction. Two research groups, Trost's[77] in the United States and Oppolzer's[78] in Switzerland, have used slightly different approaches to synthesize five-membered rings via a Pd-catalyzed ene reaction. Scheme **11.22** shows an example of Trost's procedure. The starting material for the ene reaction is a 1,6-enyne (**66**), which results from a nice application of chemistry we have already seen—attack of a stabilized carbanion on a Pd-allyl complex (**65**) (Section **11-2-3**). Once formed, **66** undergoes ene reaction in the presence of Pd(II) to give **67**. The arrows shown in the scheme indicate the bond breaking and making that occurs in an ene reaction.

Research on the mechanism of Pd(II) catalysis points to a cycle shown in Scheme **11.23** that involves hydropalladation of the alkyne group. The cycle begins with complexation of Pd(II) to both π systems to give **68**. 1,2-Insertion of the alkyne provides **69**, and then insertion of the η^2-coordinated double bond occurs to yield **70**. The last step is β-elimination providing two possible regioisomers (**71** and **72**).

Several applications of the Trost ene reaction have appeared in the literature. A spectacular extension of this reaction involves first using the ene reaction to construct a five-membered ring and then utilizing the remaining functionality to initiate a subsequent ene reaction followed by an electrocyclic ring closure, a reaction sequence called a cascade.[79] Scheme **11.24** below describes this transformation in which an acyclic molecule is transformed in one operation into a tricyclic species.

[77]B.M. Trost, *Acc. Chem. Res.*, **1990**, *23*, 34 and references therein.
[78]W. Oppolzer, *Angew. Chem. Int. Ed. Engl.*, **1989**, *28*, 38.
[79]B.M. Trost and Y. Shi, *J. Am. Chem. Soc.*, **1992**, *114*, 791.

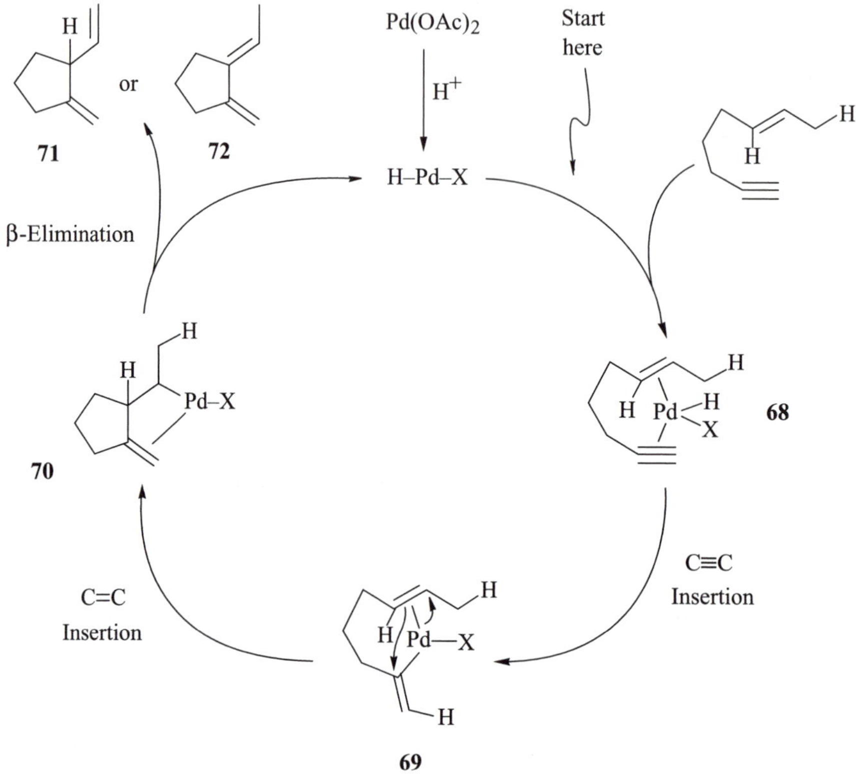

Scheme 11.23
Hydropalladation Mechanism for Ene Reaction

Exercise 11-13

The reaction below is another impressive example of cascading intramolecular ene reactions. Show how the transformation occurs by writing down all the intermediates formed after each ene reaction.

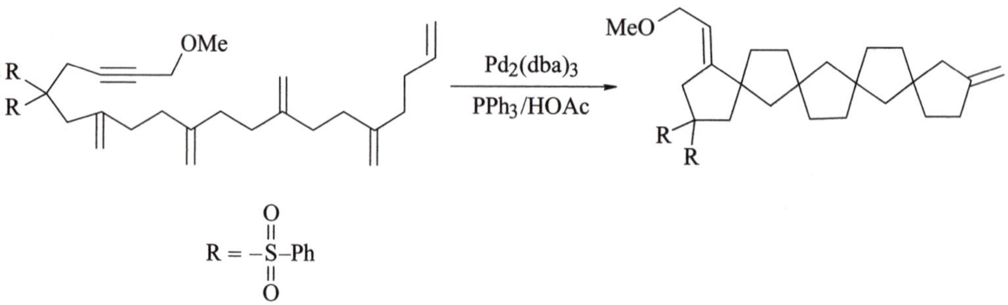

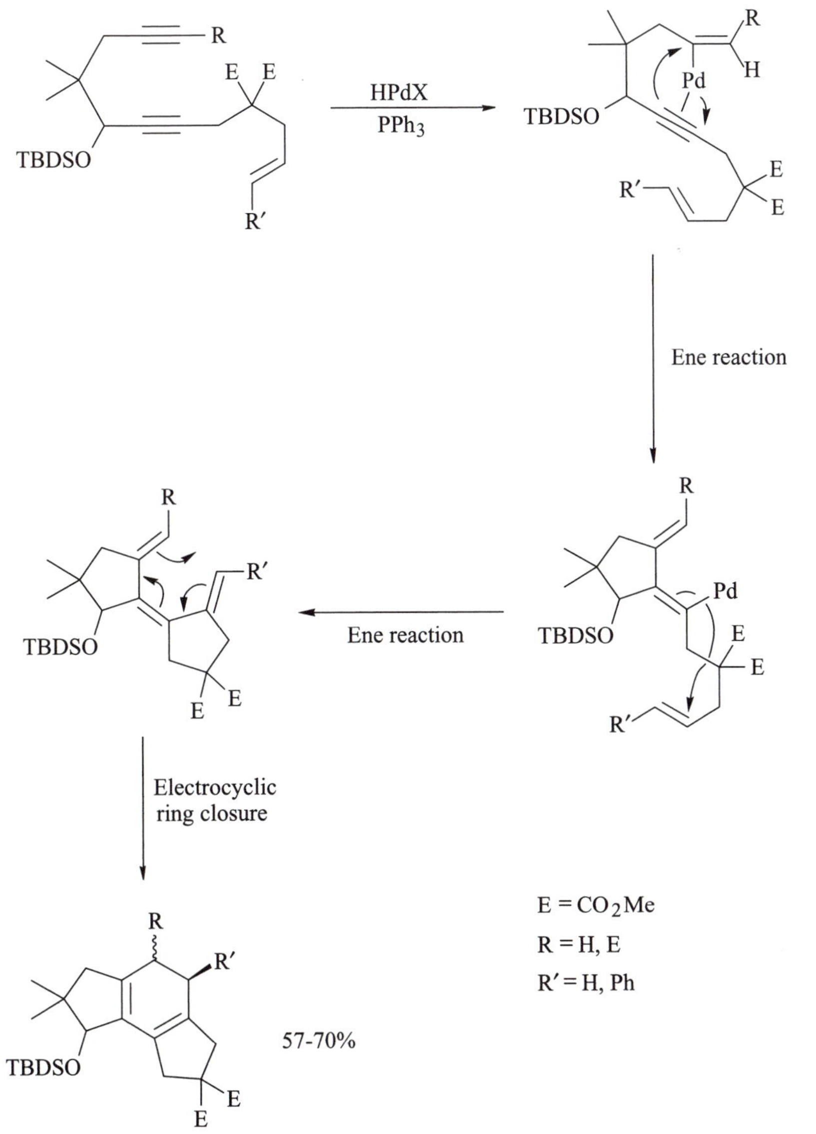
R
E E
TBDSO
R′
HPdX
PPh3
R
Pd
H
TBDSO
E
R′
E
Ene reaction
R
Pd
TBDSO
E
E
R′
Ene reaction
R
R′
TBDSO
E
E
Electrocyclic
ring closure
R
R′
TBDSO
E
E
57-70%
E = CO2Me
R = H, E
R′ = H, Ph

Scheme 11.24
Pd-Catalyzed Ene Cascade

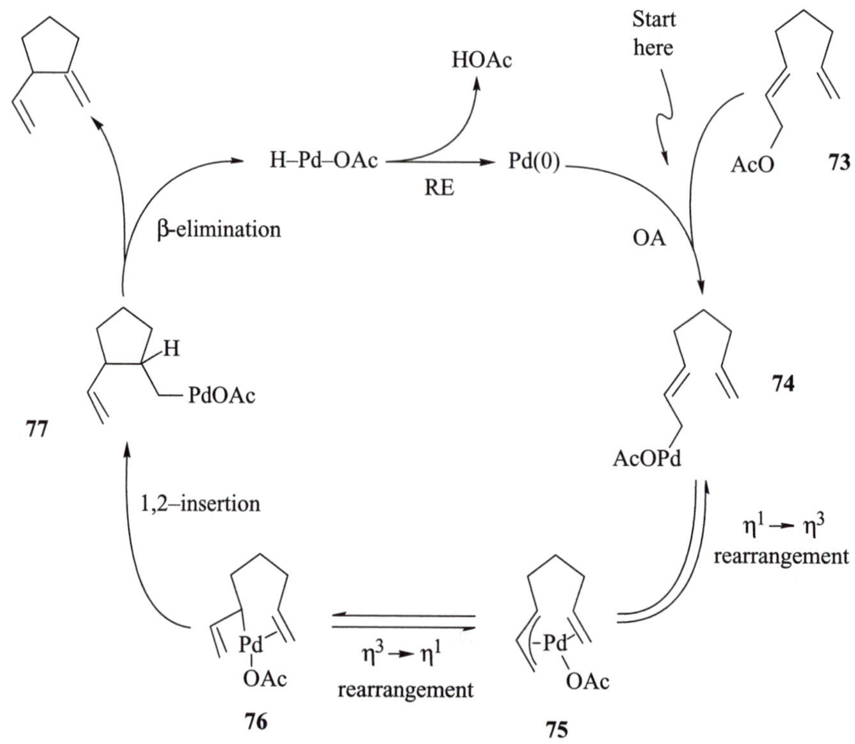

Scheme 11.25
Oppolzer Pd-Catalyzed Ene Reaction

Oppolzer's approach to Pd-catalyzed ene reactions starts with an allyl acetate that is part of a 1,6-diene system. Scheme **11.25** details a plausible mechanism that involves a Pd(0) species as the active catalyst. The cycle begins with oxidative addition of the allyl acetate portion of **73** to give **74**. Next, η^1 to η^3-allylic rearrangement leads to **75**. Allylic rearrangement again gives **76**. 1,2-C=C bond insertion follows, yielding **77**. The last step, as in the Trost procedure, is β-elimination.

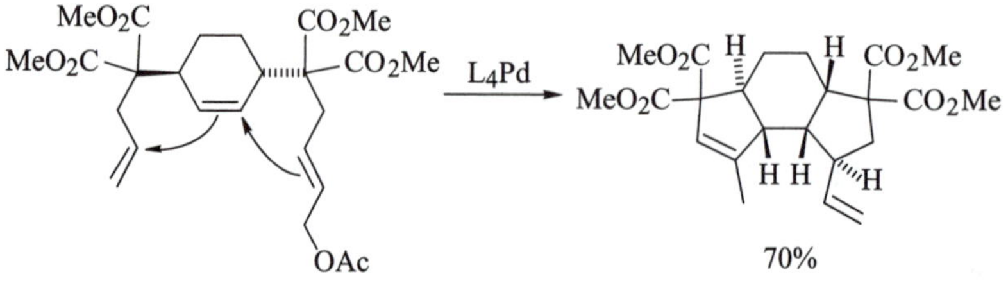

11.56

Equation **11.56** shows a recent application of Oppolzer's approach in which two five-membered rings result from consecutive ene reactions.[80]

Whether one uses the Trost or Oppolzer approach, Pd-catalyzed ene reactions stereoselectively provide good yields of compounds containing five-membered rings under moderate reaction conditions. This is yet another example of the superb ability of Pd to catalyze a large number of synthetically useful transformations.

11-5-3 Cobalt-Promoted Formation of Five-Membered Rings: The Pauson-Khand Reaction

Alkynes react with $Co_2(CO)_8$ to form stable complexes according to equation **11.57**. When these dicobalt complexes react at elevated temperature with alkenes, a cyclopentenone forms as shown in equation **11.58**. The transformation, formally a [2 + 2 + 1] cycloaddition (the connecting fragments are shown in **11.58**), is known as the Pauson-Khand (P-K) reaction.[81]

Unlike most of the reactions involving organotransition metals that we have already mentioned, this transformation is *stoichiometric* in Co, not catalytic. The mechanism is unknown at this time; Scheme **11.26** shows a possible pathway for the intermolecular version of the P-K reaction that begins with complexation of $Co_2(CO)_8$ with the alkyne followed by complexation of the alkene fragment.[82] 1,2-Insertion of the alkene may precede carbonyl insertion.

[80]W. Oppolzer and R.J. DeVita, *J. Org. Chem.*, **1991**, *56*, 6256. Oppolzer has also used Rh(I) complexes to catalyze the ene reaction. Yields are similar but the stereochemical results are opposite to those obtained with Pd catalysts. This difference may be a function of the octahedral geometry of Rh(III) intermediates, presumably generated during the catalytic cycle. See W. Oppolzer and A. Fürstner, *Helv. Chim. Acta.*, **1993**, *76*, 2339.

[81]I.U. Khand, P.L. Pauson, *et al.*, *J. Chem. Soc., Perkin Trans. I*, **1973**, 977.

[82]N.E. Schore, *Chem. Rev.*, **1988**, *88*, 1081.

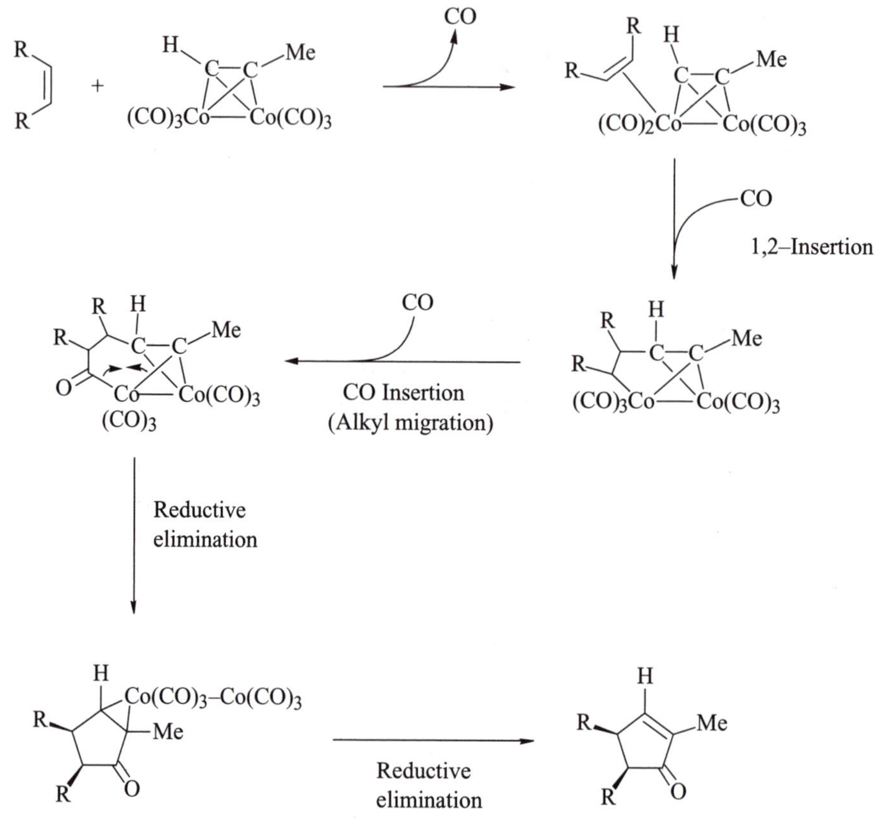

Scheme 11.26
Possible Mechanism for the Pauson-Khand Reaction

Reductive elimination forms a cobaltocyclopropane (**78**), which then undergoes a second reductive elimination to form cyclopentenone **79**. The reaction is regioselective for cycloaddition of monosubstituted alkynes, yielding a cyclopentenone in which the alkyne substituent is adjacent to the keto group. Stereochemistry about 1,2-disubstituted alkenes is usually preserved upon cycloaddition, although disubstituted olefins with bulky substituents tend to be unreactive.

The intramolecular version of the P-K reaction is probably more useful as a means of ring synthesis than the intermolecular procedure, especially in producing polycyclic systems possessing five-membered rings. Several improvements in the intramolecular transformation have resulted from continued research efforts over the last 20 years. These include a substantial increase in rate with the addition of silica gel[83] to the reaction mixture (adsorption of the alkyne-Co complex onto silica may restrict molecular motion, allowing the ene-yne system to interact more readily); the use of tertiary amine N-oxides;[84,85] and the use of acetonitrile as a solvent (rather than those of the non-polar and moderately-polar, aprotic variety). The reaction conditions tolerate the presence of a number of different organic functional groups; substrates containing alkenes or alkynes with electron withdrawing groups directly attached, however, usually do not undergo P-K reaction efficiently.[86]

Equation **11.59** shows how the P-K reaction produces a [3.3.0] bicyclic system **80** with high stereoselectivity.[87] Formation of a complex, tricyclic molecule with three 5-membered rings fused together is straightforward using the P-K reactions according to equation **11.60**.

$$\text{11.59}$$

74%

$$\text{11.60}$$

45%   6%

Numerous other examples of this useful transformation have been reported.[88]

[83]W.A. Smit, R. Caple, *et al., Tetrahedron Lett.*, **1986**, *27*, 1245.

[84]S. Shambayani, W.E. Crowe, and S.L. Schreiber, *Tetrahedron Lett.*, **1990**, *31*, 5289.

[85]Remember that N-oxides remove CO ligands by formation of CO_2 (footnote 25, this chapter); this probably opens a site on Co for complexation of the alkene fragment.

[86]A recent report indicates that compounds such as $R\text{-}C\equiv C(CO)\text{-}C(Me)_2\text{-}CH_2\text{-}CH=CH_2$ will readily undergo P-K ring closure in acetonitrile. The *gem* dimethyl group seems to restrict molecular motion, coercing the two π systems to coordinate with Co. See T.R. Hoye and J.A., Suriano, *J. Org. Chem.*, **1993**, *58*, 1659.

[87]W.A. Smit, *et al., Izv. Akad. Nauk. SSSR, Ser. Khim.*, **1985**, 2650.

[88]For a comprehensive review of the P-K reaction, see N.E. Schore, *Org. React.*, **1991**, *40*, 1.

Research on the application of organotransition metal chemistry to organic synthesis remains an active and fruitful endeavor. At present, no examples of complex molecule total synthesis exist in which every step along the path to the target involves use of a transition metal. Hundreds of reports, on the other hand, have appeared in the literature where the use of organotransition metal complexes at key points in the synthetic strategy has meant the difference between success and failure.

Problems

11-1   Calculate the difference in free energies of activation, $\Delta\Delta G^{\ddagger}$, for conversion of **7′** to **8′** and **7″** to **8″** that would be required to lead to an enantiomeric excess (ee) of 96% for the *R*-amino acid ester, **10″** (Scheme **11.3**).

11-2   Using (*E*,*E*)-3,5-heptadienal as a starting material and making use of your knowledge of Fe–diene chemistry, the Wittig reaction, and formation of cyclopropanes with Fischer-type carbene complexes, design a synthesis of following compound:

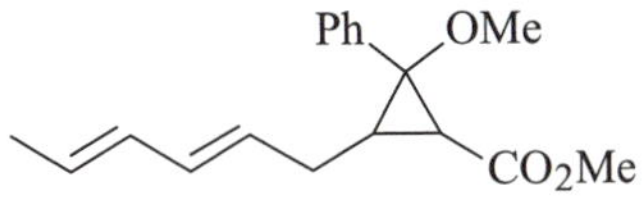

11-3   Propose a mechanism for the transformation shown.

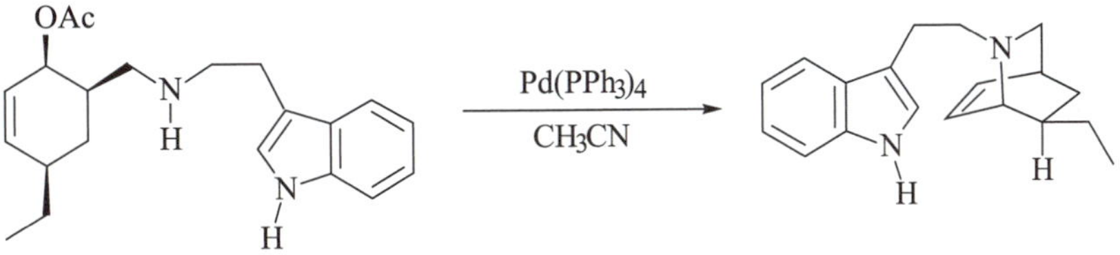

11-4   What are the individual steps that are likely to occur in the chemistry described by equation **11.33**? Why is PPh$_3$ present in the first major step and HOAc in the second?

11-5   When the alkene reactant is an allyl alcohol, a Heck reaction often yields an aldehyde or ketone. For the intramolecular olefination shown below, provide a pathway that accounts for the formation of the ketone.

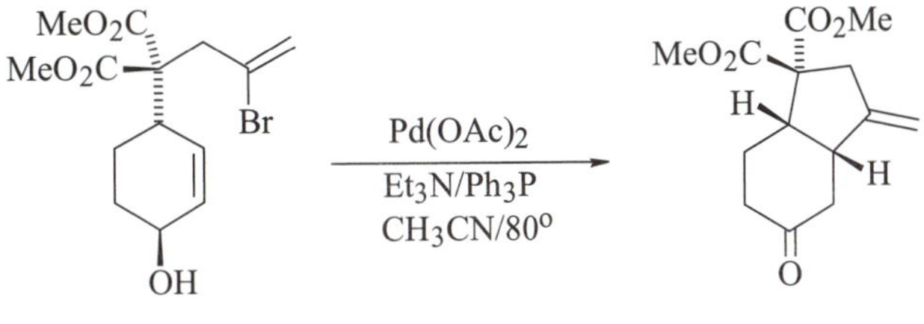

11-6   Transmetallation and CO insertion reactions play key roles in the following Pd-catalyzed transformation, shown below in general form.[89]

    a.   Propose a mechanism for this transformation.
    b.   Starting with **A** and **B**, propose a synthesis of compound **C**.

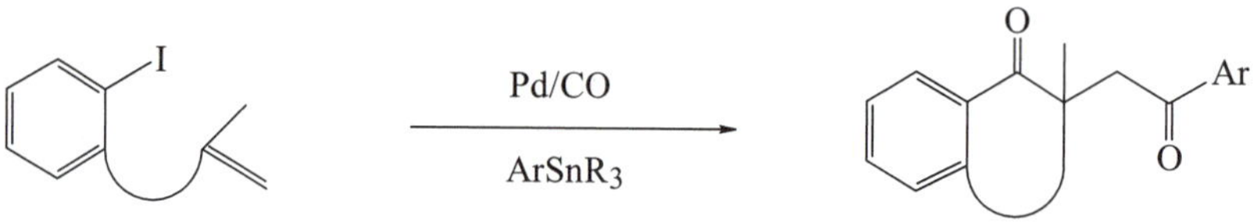

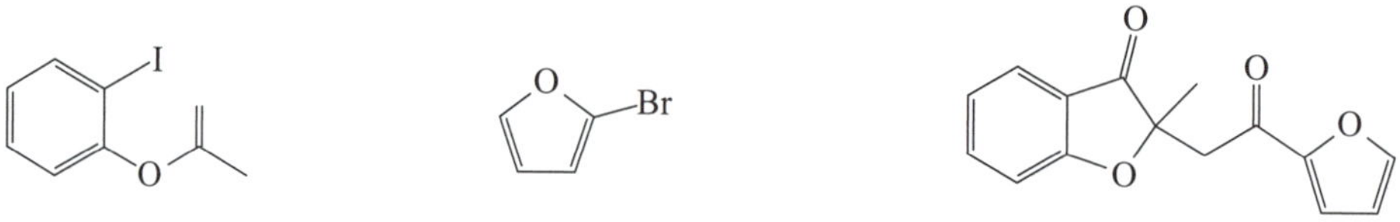

          A   **B**   **C**

11-7   Scheme **11.20** alludes to the application of Hegedus aminocarbene photochemistry to the synthesis of α-amino acids. Assume that you could make **D**. Show the steps you would use to convert **D** into alanine.

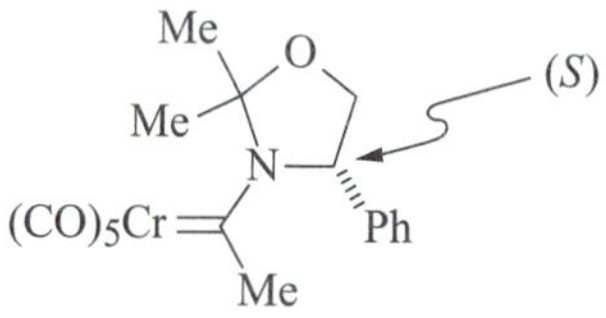

$H_2N-CH(CH_3)CO_2H$

alanine

          D

11-8   The transformation below demonstrates an interesting approach to the synthesis of 1,2–disubstituted aromatic rings. It combines aspects of Cr carbene complex chemistry outlined in Schemes **11.19** and **11.20**.[90]

---

[89]R. Grigg, J. Redpath, V. Sridharan, and D. Wilson, *Tetrahedron Lett.*, **1994**, *35*, 7661.
[90]C.A. Merlic, D. Xu, and B.G. Gladstone, *J. Org. Chem.*, **1993**, *58*, 538.

a. Propose a mechanism for this transformation.
b. Starting with **E**, propose a synthetic pathway to **F**.

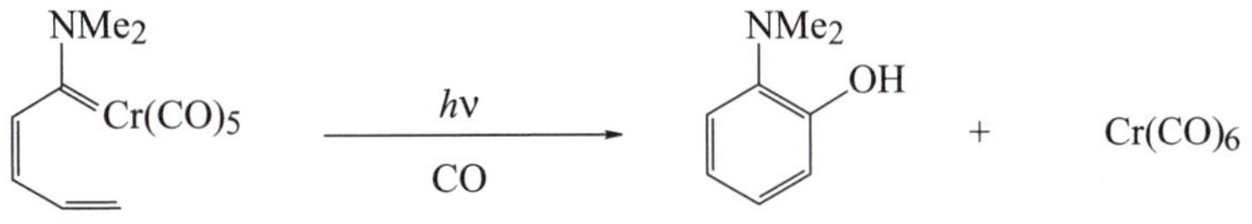

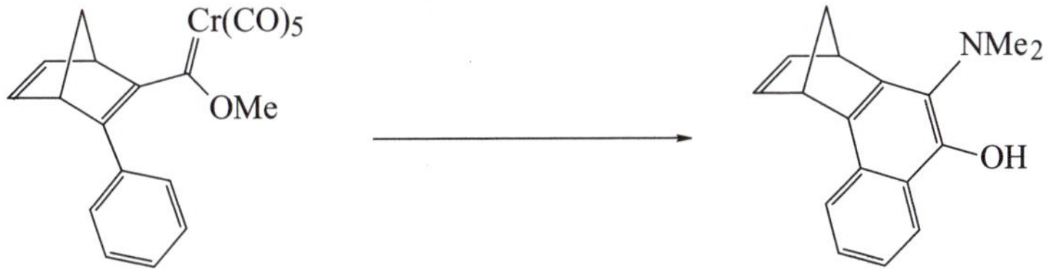

E		**F**

11-9 The ene reaction was used to convert acyclic compound **G** to bicyclic ketone **I**, as shown below. The reaction is catalytic in Pd(0). Propose a catalytic cycle that accounts for the formation of two fused five-membered rings. Assume that one of the intermediates in the cycle is **H**.[91]

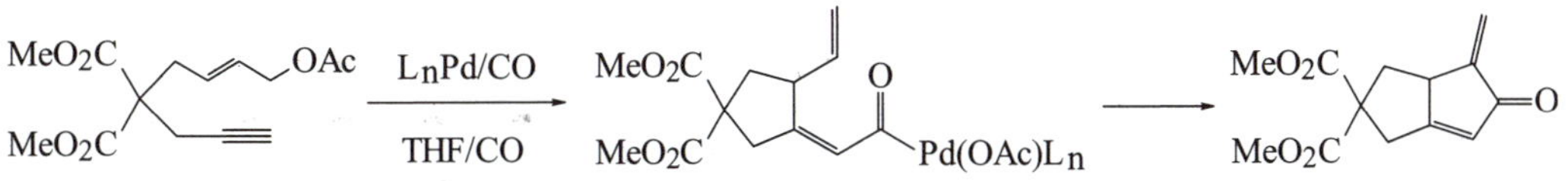

G	**H**	**I**

11-10 For the reaction below, predict the most likely product and indicate its stereochemistry:

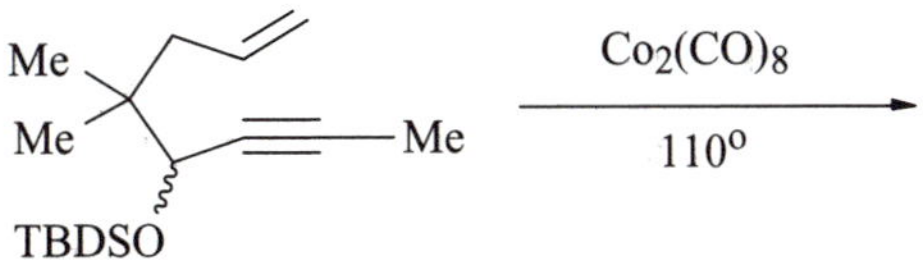

[91]N.H. Ihle and C.H. Heathcock, *J. Org. Chem.*, **1993**, *58*, 560.

12

Isolobal Groups and Cluster Compounds

12-1   Isolobal Analogy

An important contribution to the understanding of parallels among organic, inorganic, and organometallic chemistry has been the concept of isolobal molecular fragments. In his 1982 Nobel lecture, Hoffmann described molecular fragments as isolobal,

> if the number, symmetry properties, approximate energy and shape of the frontier orbitals and the number of electrons in them are similar—not identical, but similar.[1]

To illustrate this definition, it is useful to compare fragments of methane with fragments of an octahedrally coordinated transition metal complex, ML_6. For simplicity, we will consider only σ interactions between the metal and the ligands in this complex.[2] The fragments to be discussed are shown in Figure **12-1**.

The parent compounds have filled valence-shell electron configurations,

---

[1] R. Hoffmann, *Angew. Chem. Int. Ed. Eng.*, **1982**, *21*, 711.

[2] Pi interactions can also be considered; see reference 1 and H-J. Krause, *Z. Chem.*, **1988**, *28*, 129.

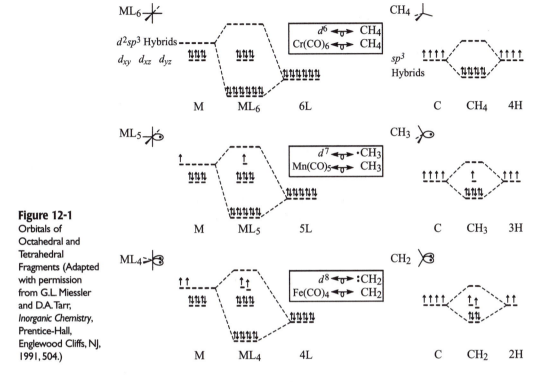

Figure 12-1
Orbitals of
Octahedral and
Tetrahedral
Fragments (Adapted
with permission
from G.L. Miessler
and D.A. Tarr,
Inorganic Chemistry,
Prentice-Hall,
Englewood Cliffs, NJ,
1991, 504.)

an octet for CH_4, and 18 electrons for ML_6. Methane may be considered to use sp^3 hybrid orbitals in bonding, with eight electrons occupying bonding pairs formed from interactions between the hybrids and $1s$ orbitals on hydrogen. The metal in ML_6, by similar reasoning, uses d^2sp^3 hybrids to bond to the ligands, with 12 electrons occupying bonding orbitals and six nonbonding electrons occupying d_{xy}, d_{xz}, and d_{yz} orbitals.

Molecular fragments containing fewer ligands than the parent polyhedra can now be described; for the purpose of the analogy, these fragments will be assumed to preserve the original geometry of the remaining ligands.

For example, in the 7-electron fragment CH_3, three of the sp^3 orbitals of carbon are involved in σ bonding with the hydrogens. The fourth hybrid is singly occupied and at higher energy than the σ bonding pairs of CH_3, as shown in Figure **12-1**. The frontier orbitals of the 17-electron fragment $Mn(CO)_5$ are similar to those of CH_3. The σ interactions between the ligands and Mn in this fragment may be considered to involve five of the metal's d^2sp^3 hybrid orbitals. The sixth hybrid is singly occupied and at higher energy than the five σ bonding orbitals.

As shown in Figure **12-1,** each of these fragments has a single electron in a hybrid orbital at the vacant site of the parent polyhedron. These orbitals are sufficiently similar to meet Hoffmann's isolobal definition. Using Hoffmann's symbol ⟵σ⟶ to designate groups as isolobal, we may therefore write the following:

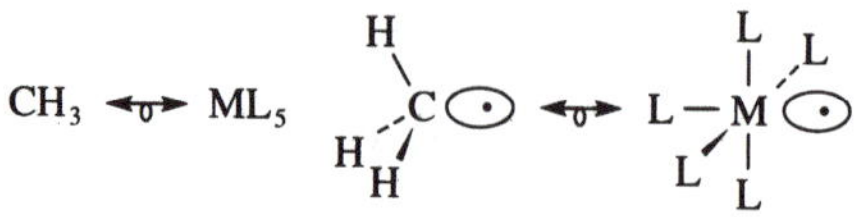

$$CH_3 \;\longleftrightarrow\; ML_5$$

Similarly, 6-electron CH_2 and 16-electron ML_4 are isolobal. Each of these fragments represents the parent polyhedron with single electrons occupying hybrid orbitals at otherwise vacant sites; each fragment also has two electrons less than the filled shell octet or 18-electron configurations. Absence of a third ligand from the parent polyhedra also gives a pair of isolobal fragments, CH and ML_3 as shown below:

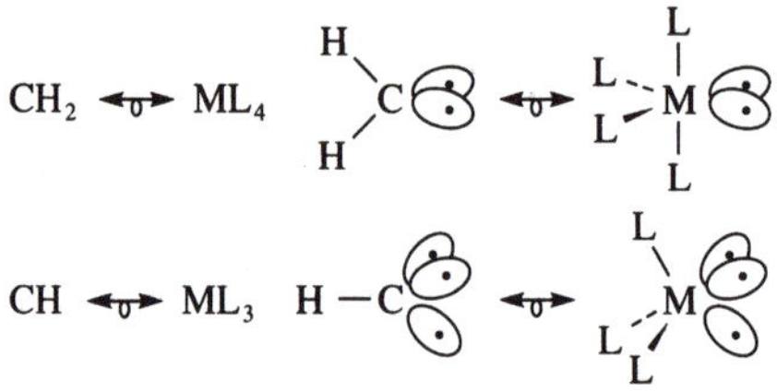

$$CH_2 \;\longleftrightarrow\; ML_4$$

$$CH \;\longleftrightarrow\; ML_3$$

These relationships are summarized in Table **12-1.**

Isolobal fragments can be formally combined into molecules, as shown in Figure **12-2.** For example, two CH_3 fragments, when linked, form ethane,

Table 12-1 Isolobal Fragments

	Organic	Inorganic	Organo-metallic Example[a]	Vertices Missing from Parent Polyhedron	Electrons Short of Filled Shell
Parent:	**CH₄**	**ML₆**	**Cr(CO)₆**	0	0
Fragments:	CH₃	ML₅	Mn(CO)₅	1	1
	CH₂	ML₄	Fe(CO)₄	2	2
	CH	ML₃	Co(CO)₃	3	3

[a] These examples involve π interactions between the metal and the ligands (affecting the energies of the d_{xy}, d_{xz}, and d_{yz} orbitals); however, the overall electronic structures of the fragments shown are still consistent with their being considered isolobal.

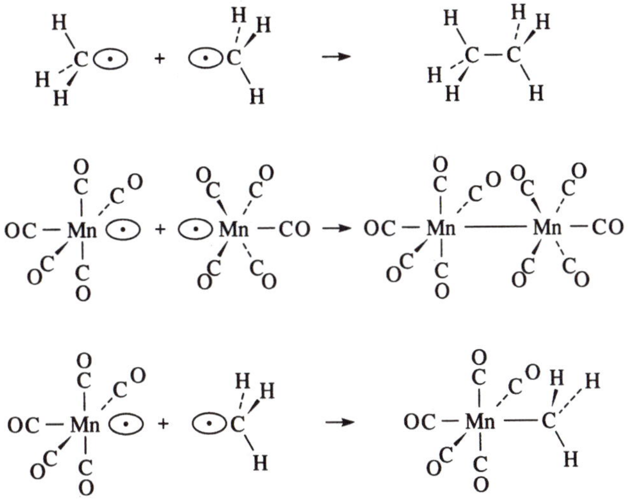

Figure 12-2
Molecules Resulting from the Combination of Isolobal Fragments (Adapted with permission from G.L. Miessler and D.A. Tarr, *Inorganic Chemistry*, Prentice-Hall, Englewood Cliffs, NJ, 1991, 506.)

and two $Mn(CO)_5$ fragments form the dimeric $(OC)_5Mn–Mn(CO)_5$. Furthermore, organic and organometallic fragments can be intermixed; an example is $H_3C–Mn(CO)_5$, also a known compound.

Organic and organometallic parallels are not always this complete. For example, while two six-electron CH_2 fragments form ethylene, $H_2C=CH_2$, the dimer of the isolobal $Fe(CO)_4$ is not nearly as stable; it is known as a transient species obtained thermally or photochemically from $Fe_2(CO)_9$.[3] Both CH_2 and $Fe(CO)_4$, however, form three-membered rings: cyclopropane, and $Fe_3(CO)_{12}$. Although cyclopropane is a trimer of CH_2 fragments, $Fe_3(CO)_{12}$ has two bridging carbonyls and is therefore not a perfect trimer of $Fe(CO)_4$. The isoelectronic $Os_3(CO)_{12}$, on the other hand, is a trimeric combination of three $Os(CO)_4$ fragments (which are isolobal with both $Fe(CO)_4$ and CH_2), and can correctly be described as $[Os(CO)_4]_3$. These structures are shown in Figure **12-3**.

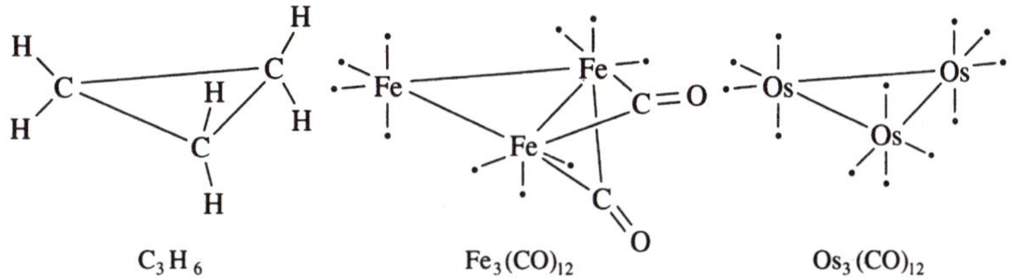

Figure 12-3
Isolobally Related Three-Membered Rings (Adapted with permission from G.L. Miessler and D.A. Tarr, *Inorganic Chemistry*, Prentice-Hall, Englewood Cliffs, NJ, 1991, 506.)

C_3H_6 $Fe_3(CO)_{12}$ $Os_3(CO)_{12}$
                         ($\cdot$ = terminal carbonyl)

[3]M. Poliakoff and J.J. Turner, *J. Chem. Soc. (A)*, **1971**, 2403.

The 15-electron fragment $Ir(CO)_3$ forms $[Ir(CO)_3]_4$, which has an Ir_4 core in the shape of a regular tetrahedron. In this complex all the carbonyl groups are terminal. The isoelectronic complexes $Co_4(CO)_{12}$ and $Rh_4(CO)_{12}$ have nearly tetrahedral arrays of metal atoms, but three carbonyls bridge one of the triangular faces in each of these clusters. Interestingly, these compounds are also known having a central tetrahedral core, with one or more $Co(CO)_3$ fragments [isolobal and isoelectronic with $Ir(CO)_3$] replaced by the isolobal CR fragment (shown in Figure **12-4**).

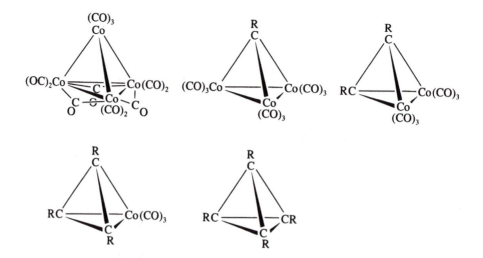

Figure 12-4
Isolobally Related Tetrahedral Molecules (Adapted with permission from G.L. Miessler and D.A. Tarr, *Inorganic Chemistry*, Prentice-Hall, Englewood Cliffs, NJ, 1991, 507.)

12-1-1 Extensions of the Analogy

The concept of isolobal fragments can be extended beyond the examples given so far, to include charged species, a variety of ligands other than CO, and organometallic fragments based on structures other than octahedral. Some of the ways of extending the isolobal parallels can be summarized as follows:

1. The isolobal definition may be extended to isoelectronic fragments having the same coordination number. For example,

$$\text{Since } Mn(CO)_5 \;\overset{\leftrightarrow}{}\; CH_3, \quad \begin{matrix} Re(CO)_5 \\ [Fe(CO)_5]^+ \\ [Cr(CO)_5]^- \end{matrix} \;\overset{\leftrightarrow}{}\; CH_3$$

2. Gain or loss of electrons from two isolobal fragments yields isolobal fragments. For example,

Since $Mn(CO)_5$ ⟷ CH_3 , $[Mn(CO)_5]^+$ ⟷ CH_3^+
[17-electron   [7-electron   [16-electron   [6-electron
fragment]   fragment   fragment]   fragment]

                                        $[Mn(CO)_5]^-$ ⟷ CH_3^-
                                        [18-electron   [8-electron
                                        fragment]   fragment]

3. Other two-electron donors are treated similarly to CO:

$Mn(CO)_5$ ⟷ $Mn(PR_3)_5$ ⟷ $Mn(NCR)_5$ ⟷ $[MnCl_5]^{5-}$ ⟷ CH_3

4. η^5-Cp is considered to occupy three coordination sites and to be a six-electron donor (as $C_5H_5^-$):

$(\eta^5$-Cp$)Fe(CO)_2$ ⟷ $[Fe(CO)_5]^+$ ⟷ $Mn(CO)_5$ (17-electron fragments) ⟷ CH_3

$(\eta^5$-Cp$)Mn(CO)_2$ ⟷ $[Mn(CO)_5]^+$ ⟷ $Cr(CO)_5$ (16-electron fragments) ⟷ CH_2

Examples of isolobal fragments containing CO and η^5-Cp ligands are given in Table **12-2**.

Table 12-2 Examples of Isolobal Fragments

	Number of Electrons Short of Parent Configuration (8 or 18)			
	0	**1**	**2**	**3**
Neutral hydrocarbon Isolobal fragments	CH_4 $Cr(CO)_6$ $[Mn(CO)_6]^+$ $CpMn(CO)_3$	CH_3 $Mn(CO)_5$ $[Fe(CO)_5]^+$ $CpFe(CO)_2$	CH_2 $Fe(CO)_4$ $[Co(CO)_4]^+$ $CpCo(CO)$	CH $Co(CO)_3$ $[Ni(CO)_3]^+$ $CpNi$
Anionic hydrocarbon fragments[a] Isolobal fragments	CH_3^- $Fe(CO)_5$	CH_2^- $Co(CO)_4$	CH^- $Ni(CO)_3$	
Cationic hydrocarbon fragments[b] Isolobal fragments		CH_4^+ $V(CO)_6$	CH_3^+ $Cr(CO)_5$	CH_2^+ $Mn(CO)_4$

[a] Anionic hydrocarbon fragments were obtained by removing H^+ from the neutral hydrocarbons at the top of the columns.
[b] Cationic hydrocarbon fragments were obtained by adding H^+ to the neutral hydrocarbons at the top of the columns.

Exercise 12-1

For the following, propose examples of isolobal organometallic fragments other than those given so far in this chapter:

     a.   A fragment isolobal with CH_2^+
     b.   A fragment isolobal with CH^-
     c.   Three fragments isolobal with CH_3

Exercise 12-2

Give organic fragments isolobal with each of the following:

     a. $(\eta^5\text{-Cp})Ni$
     b. $(\eta^6\text{-C}_6\text{H}_6)Cr(CO)_2$
     c. $[Fe(CO)_2(PPh_3)]^-$

Analogies are by no means limited to organometallic fragments of octahedra; similar arguments can be used to derive fragments of different polyhedra. For example, $Co(CO)_4$, a 17-electron fragment of a trigonal bipyramid, is isolobal with $Mn(CO)_5$, a 17-electron fragment of an octahedron:

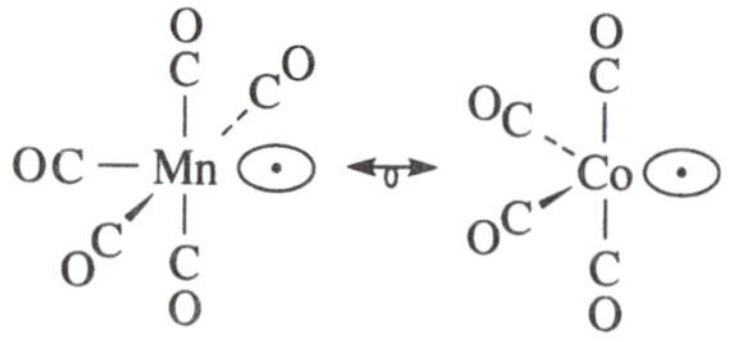

Examples of electron configurations of isolobal fragments of polyhedra having five through eight vertices are given in Table **12-3**.[4]

---

[4]For an analysis of the energies and symmetries of fragments of a variety of polyhedra, see R. Hoffmann, *op. cit.*; M. Elian and R. Hoffmann, *Inorg. Chem.*, **1975**, *14*, 1058; and T. A. Albright, R. Hoffmann, J. C. Thibeault, and D. L. Thorn, *J. Am. Chem. Soc.*, **1979**, *101*, 3801.

Table 12-3 Isolobal Relationships for Fragments of Polyhedra

Organic Fragment	Coordination Number of Transition Metal for Parent Polyhedron				Valence Electrons of Fragment
	5	6	7	8	
CH_3	d^9-ML_4	d^7-ML_5	d^5-ML_6	d^3-ML_7	17
CH_2	d^{10}-ML_3	d^8-ML_4	d^6-ML_5	d^4-ML_6	16
CH		d^9-ML_3	d^7-ML_4	d^5-ML_5	15

12-1-2 Examples of Applications of the Analogy

The isolobal analogy can, in principle, be extended to *any* molecular fragment having frontier orbitals of suitable size, shape, symmetry, and energy. In many cases the characteristics of frontier orbitals are not as easy to predict as in the examples cited above, and calculations are necessary to determine the symmetry and energy of molecular fragments. For example, $Au(PPh_3)$, a 13–electron fragment, has a single electron in a hybrid orbital pointing away from the phosphine.[5] This electron is in an orbital of similar symmetry but of somewhat higher energy than the singly occupied hybrid in the $Mn(CO)_5$ fragment.

The $Au(PPh_3)$ fragment can be combined with the isolobal $Mn(CO)_5$ and CH_3 fragments to form $(OC)_5Mn-Au(PPh_3)$ and $H_3C-Au(PPh_3)$.

Even a hydrogen atom, with a single electron in its $1s$ orbital, can in some cases be viewed as a fragment isolobal with such species as CH_3, $Mn(CO)_5$, and $Au(PPh_3)$. Hydrides of the first two (i.e., CH_4 and $HMn(CO)_5$) are very well known. In addition, in some cases $Au(PPh_3)$ and H show surprisingly similar behavior, such as in their ability to bridge the triosmium clusters shown on p. 454.[6,7]

[5]D.G. Evans and D.M.P. Mingos, *J. Organomet. Chem.*, **1982**, *232*, 171.

[6]A.G. Orpen, A.V. Rivera, E.G. Bryan, D. Pippard, G. Sheldrick, and K.D. Rouse, *J. Chem. Soc., Chem. Commun.*, **1978**, 723.

[7]B.F.G. Johnson, D.A. Kaner, J. Lewis, and P.R. Raithby, *J. Organomet. Chem.*, **1981**, *215*, C33.

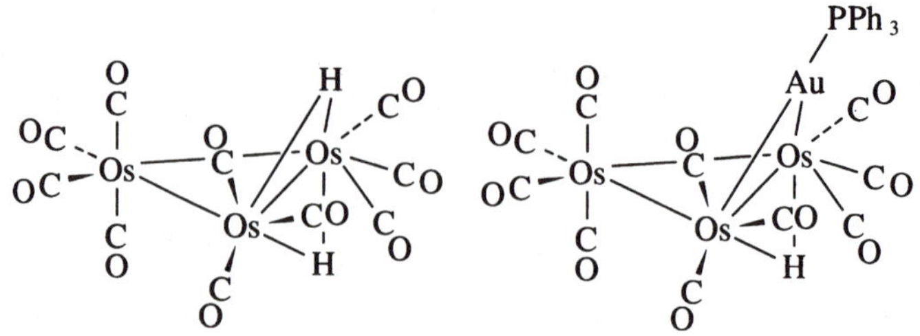

Potentially the greatest practical use of isolobal analogies is in suggesting syntheses of new compounds. For example, CH_2 is isolobal with 16-electron $Cu(\eta^5\text{-}Cp^*)$ (Extension 4 of the analogy, as described previously), and with 14-electron PtL_2 (L = PR_3, CO);[8] all three of these fragments are two ligands and two electrons short of their parent polyhedra. Recognition that these fragments are isolobal has been exploited in preparing new organometallic compounds composed of fragments isolobal with fragments of known compounds.[9] Two of the compounds obtained in these studies are shown below.

Previously Known Compounds	**New Compounds Composed of Isolobal Fragments**

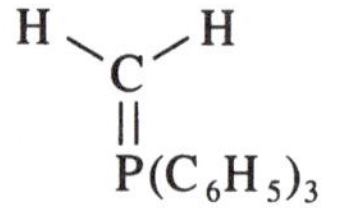

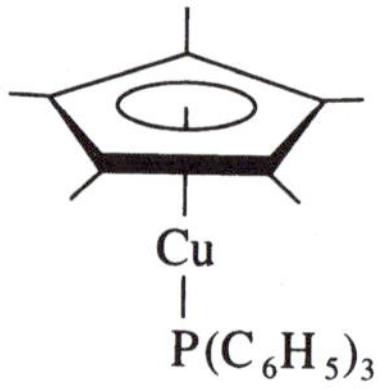

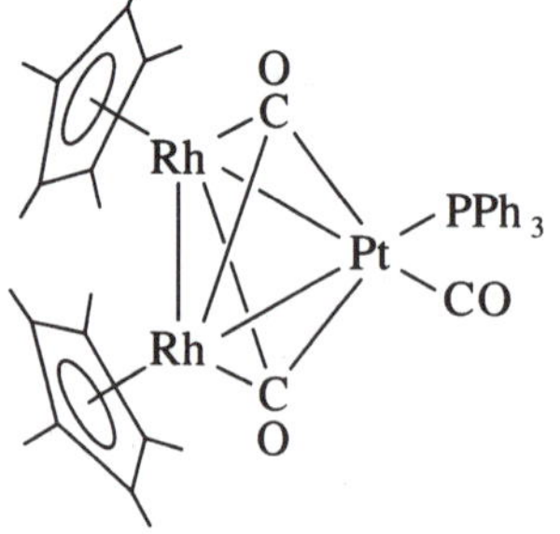

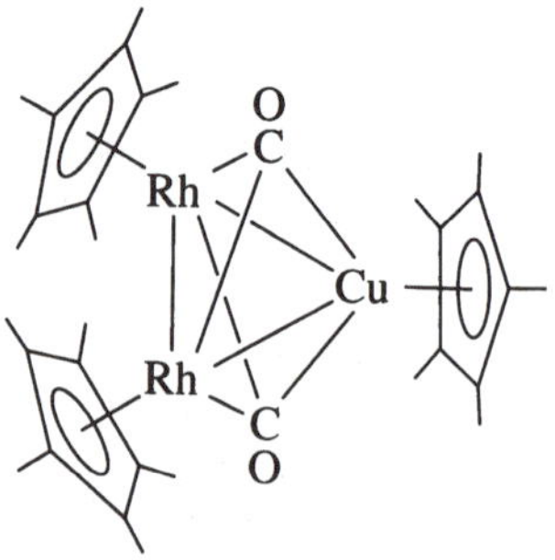

[8]In a manner similar to the relationship between the "18-electron rule" for octahedra and other structures, and the "16-electron rule" for square planar complexes, isolobal relationships between 16-electron fragments of octahedra and 14-electron fragments of square planar structures can also be demonstrated; see footnote 4 for references.

[9]G.A. Carriedo, J.A.K. Howard, and F.G.A. Stone, *J. Organomet. Chem.*, **1983**, *250*, C28.

12-2   Cluster Compounds

Examples of cluster compounds have been cited previously in this text. Transition metal cluster chemistry has developed rapidly in recent years. Beginning with simple dimeric molecules such as $Co_2(CO)_8$ and $Fe_2(CO)_9$,[10] chemists have developed syntheses of far more complex clusters, some having interesting and unusual structures and chemical properties. Large clusters have been studied with the objective of developing catalysts that may duplicate or improve upon the properties of heterogeneous catalysts; the surface of a large cluster may in these cases mimic in some degree the behavior of the surface of a solid catalyst.

Before considering transition metal clusters in more detail, we will consider compounds of boron, which has an extremely detailed cluster chemistry. Some of these compounds exhibit similarities in their bonding and structures to transition metal clusters, especially clusters containing carbonyls.

12-2-1 Boranes

There are a great many neutral and ionic species composed of boron and hydrogen, far too numerous to mention in this text.[11] For the purposes of illustrating parallels between these species and transition metal clusters, we will first consider one category of boranes, *closo* ("cage-like") boranes which have the formula $B_nH_n^{2-}$. These boranes are closed polyhedra having n corners and all triangular faces (triangulated polyhedra). Each corner is occupied by a BH group; an example, $B_6H_6^{2-}$, is shown below.

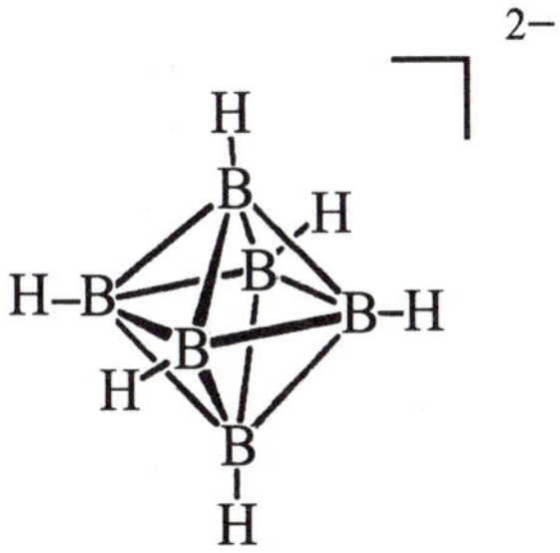

$$B_6H_6^{2-}$$

Molecular orbital calculations have shown that *closo* boranes have $2n + 1$ bonding molecular orbitals. These orbitals may be classified into two types: n B–H σ bonding orbitals, and $n + 1$ bonding orbitals in the central core

---

[10]Some chemists define clusters as having at least three metal atoms.

[11]For a more detailed introduction to borane chemistry, see N. N. Greenwood and A. Earnshaw, *Chemistry of the Elements*, Pergamon Press, Elmsford, NY, **1984**, 171-202.

(described as "framework," or "skeletal," bonding orbitals).[12] Electrons in these central core orbitals are primarily responsible for holding the core of the cluster together.

A useful example is $B_6H_6^{2-}$, which has octahedral geometry. In this ion each boron has four valence orbitals that can participate in bonding (s, p_x, p_y, and p_z), giving a total of 24 boron valence orbitals for the cluster. Of these, 13 orbitals ($2n + 1$) are bonding and can be classified into the two sets (B–H σ bonding and framework bonding), described above. A useful reference coordinate system for $B_6H_6^{2-}$ is shown in Figure **12-5**, with the z axis of each boron chosen to point toward the center of the octahedron.

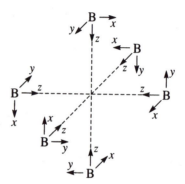

Figure 12-5
Coordinate System for Bonding in $B_6H_6^{2-}$ (Adapted with permission from G.L. Miessler and D.A. Tarr, *Inorganic Chemistry*, Prentice-Hall, Englewood Cliffs, NJ, 1991, 512.)

The p_z and s orbitals of the borons collectively have the same symmetry and may therefore be considered to combine to form sp hybrid orbitals. These hybrid orbitals on each boron point out toward the hydrogen atoms and in toward the center of the cluster. Six ($=n$) of the hybrids form σ bonds with the $1s$ orbitals of the hydrogens. The remaining hybrids and the unhybridized $2p$ orbitals (p_x and p_y) of the borons remain to participate in interactions within the B_6 core.

Seven orbital combinations ($=n + 1$) lead to bonding interactions within the B_6 core; these are shown in Figure **12-6**. Constructive overlap of all six hybrid orbitals at the center of the octahedron yields one framework bonding orbital.[13] Additional bonding interactions are of two types: overlap of two sp hybrid orbitals with parallel p orbitals on the remaining four boron atoms (three such interactions), and overlap of p orbitals on four boron atoms within the same plane (three interactions). The remaining orbitals form antibonding molecular orbitals, or are nonbonding.

---

[12]K. Wade, *Electron Deficient Compounds*, Thomas Nelson & Sons, London, 1971.
[13]A_{1g} designates the highly symmetric interaction that occurs when all sp hybrids point toward the center of the octahedron.

To summarize:

From the 24 valence atomic orbitals of the boron atoms the following are formed:

13 bonding orbitals (=2n+1), consisting of:
   6 boron-hydrogen bonding orbitals (=n)
   7 framework molecular orbitals (=n+1):
      1 bonding orbital from overlap of sp hybrid orbitals
      6 bonding orbitals from overlap of p orbitals of boron with sp
        hybrid orbitals or with other boron p orbitals
11 antibonding or nonbonding orbitals

Figure 12-6
Bonding in $B_6H_6^{2-}$ (Adapted with permission from G.L. Miessler and D.A. Tarr, *Inorganic Chemistry*, Prentice-Hall, Englewood Cliffs, NJ, 1991, 513.)

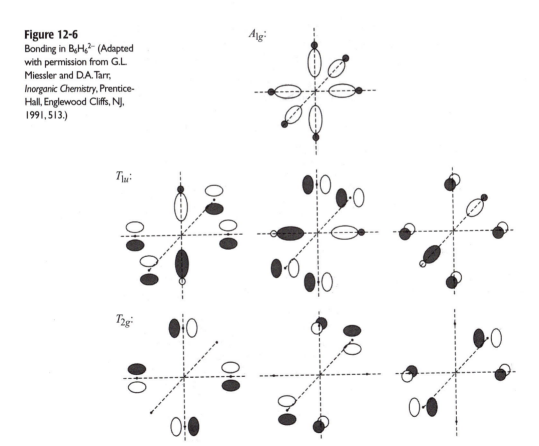

Table 12-4 Bonding Pairs for *Closo* Boranes

Formula	Total Valence Electron Pairs	Framework Bonding Pairs			B–H Bonding Pairs
		A Symmetry[a]	Other Symmetry		
$B_6H_6^{2-}$	13	1	6		6
$B_7H_7^{2-}$	15	1	7		7
$B_8H_8^{2-}$	17	1	8		8
$B_nH_n^{2-}$	$2n+1$	1	n		n

[a]These are bonding pairs occupying the highly symmetric orbital that results from the combination of all *sp* hybrid orbitals pointing directly toward the center of the polyhedron. The actual designation (such as A_{1g}) depends on the overall symmetry of the cluster.

Similar descriptions of bonding can be derived for other *closo* boranes. In each case one particularly useful similarity can be found: there is one more framework bonding pair than the number of vertices in the polyhedron. The extra framework bonding pair is in a highly symmetric orbital resulting from overlap of atomic (or hybrid) orbitals at the center of the polyhedron. In addition, there is a significant gap in energy between the highest bonding orbital (HOMO) and the lowest nonbonding orbital (LUMO).[14] The numbers of bonding pairs for some common geometries are shown in Table **12–4**.

Together, the *closo* structures make up only a very small fraction of all known borane species. Additional structural types can be obtained by removing one or more corners from the *closo* framework. Removal of one corner yields a *nido* ("nest-like") structure, removal of two corners an *arachno* ("spiderweb-like") structure, and removal of three corners a *hypho* ("net-like") structure. Examples of three related *closo*, *nido*, and *arachno* borane structures are shown in

Figure 12-7
Closo, Nido, and *Arachno* Borane Structures (Adapted with permission from G.L. Miessler and D.A. Tarr, *Inorganic Chemistry,* Prentice-Hall, Englewood Cliffs, NJ, 1991, 514.)

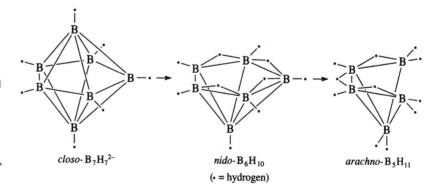

closo-$B_7H_7^{2-}$ *nido*-B_6H_{10} *arachno*-B_5H_{11}

(• = hydrogen)

[14]K. Wade, "Some Bonding Considerations" in B.F.G. Johnson, Ed., *Transition Metal Clusters*, John Wiley & Sons, New York, **1980**, 217.

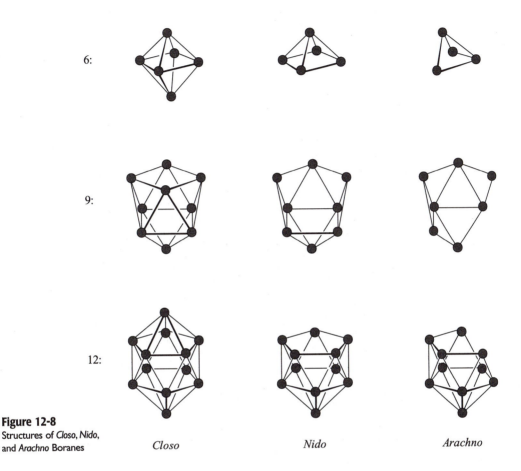

Figure 12-8
Structures of *Closo*, *Nido*, and *Arachno* Boranes

Closo   *Nido*   *Arachno*

6:

9:

12:

Figure **12–7**, and the structures of the boron core of selected additional boranes are shown in Figure **12–8**.

The classification of structural types can often be done more conveniently on the basis of valence electron counts. Various schemes for relating electron counts to structures have been proposed, with most proposals based on the set of rules formulated by Wade in 1971.[15] A classification scheme based on these rules is summarized in Table **12–5**.

[15]K. Wade, *Adv. Inorg. Chem. Radiochem.*, **1976**, *18*, 1.

Table 12-5 Classification of Cluster Structures

Structure Type	Corners Occupied		Pairs of Framework Bonding Electrons	Empty Corners
closo	n	Corners of n-cornered polyhedron	$n+1$	0
nido	$n-1$	Corners of n-cornered polyhedron	$n+1$	1
arachno	$n-2$	Corners of n-cornered polyhedron	$n+1$	2
hypho	$n-3$	Corners of n-cornered polyhedron	$n+1$	3

In addition, it is sometimes useful to relate the total valence electron count in boranes to the structural type. In *closo* boranes the total number of valence electron pairs is equal to the sum of the number of vertices in the polyhedron (at each boron, one electron pair is involved in boron-hydrogen bonding) and the number of framework bond pairs. For example, in $B_6H_6^{2-}$ there are 26 valence electrons, or 13 pairs ($=2n + 1$, as mentioned previously). The number of vertices in the parent polyhedron (an octahedron) is six, and the number of framework bond pairs is seven. The total of 13 pairs corresponds to the six pairs involved in bonding to the hydrogens (one per boron), and the seven pairs involved in framework bonding. The *closo* structure is the parent polyhedron of the other structural types. These electron counts are summarized for several examples of boranes in Table **12–6**.

Table 12-6 Examples of Electron Counting in Boranes

Vertices in Parent Polyhedron	Classification	Boron Atoms in Cluster	Valence Electrons	Framework Electron Pairs	Examples	Formally Derived from
6	closo	6	26	7	$B_6H_6^{2-}$	$B_6H_6^{2-}$
6	nido	5	24	7	B_5H_9	$B_5H_5^{4-}$
6	arachno	4	22	7	B_4H_{10}	$B_4H_4^{6-}$
12	closo	12	50	13	$B_{12}H_{12}^{2-}$	$B_{12}H_{12}^{2-}$
12	nido	11	48	13	$B_{11}H_{13}^{2-}$	$B_{11}H_{11}^{4-}$
12	arachno	10	46	13	$B_{10}H_{15}^{-}$	$B_{10}H_{10}^{6-}$

12-2-2 Method for Classifying Structures

From a practical standpoint, it is useful to have a classification scheme based on molecular formulas rather than parent polyhedra (which may not immediately be obvious). In addition, such a classification scheme should ideally be adaptable to other clusters, whether or not they involve the element boron. One such scheme can be stated as follows:

Classification	Formally derived from[16]
closo	$B_nH_n^{2-}$
nido	$B_nH_n^{4-}$
arachno	$B_nH_n^{6-}$
hypho	$B_nH_n^{8-}$

For boranes, the matching formulas above can be obtained by formally subtracting a sufficient number of H^+ ions (such that the number of boron atoms becomes equal to the number of hydrogen atoms) from the actual formulas (except for the *closo* clusters, the number of hydrogen atoms in general is greater than the number of boron atoms). For example, to classify B_5H_{11}, subtract six H^+ ions from the formula:

$$B_5H_{11} - 6\ H^+ \rightarrow B_5H_5^{6-}$$

The resulting formula matches the *arachno* classification.

Example 12-1

The following examples illustrate this method for classifying boranes according to structural type:

a.	$B_{10}H_{14}$	$B_{10}H_{14} - 4\ H^+ = B_{10}H_{10}^{4-}$	Classification: *nido*
b	$B_2H_7^-$	$B_2H_7^- - 5\ H^+ = B_2H_2^{6-}$	Classification: *arachno*
c.	B_8H_{16}	$B_8H_{16} - 8\ H^+ = B_8H_8^{8-}$	Classification: *hypho*

Exercise 12-3

Classify the following boranes by structural type:
a. $B_5H_8^-$ b. $B_{11}H_{11}^{2-}$ c. $B_{10}H_{18}$

12-2-3 Heteroboranes

The electron counting schemes can be extended to isoelectronic species such as the carboranes (also known as carbaboranes), clusters containing both carbon

---

[16]This is a formalism only and does not, except for the *closo* classification, imply that ions of the given formulas (such as $B_nH_n^{8-}$) actually exist.

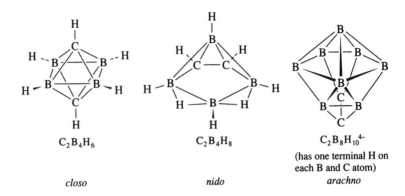

Figure 12-9

Examples of Carboranes (Adapted with permission from G.L. Miessler and D.A. Tarr, *Inorganic Chemistry*, Prentice-Hall, Englewood Cliffs, NJ, 1991, 517.)

$C_2B_4H_6$

closo

$C_2B_4H_8$

nido

$C_2B_8H_{10}^{4-}$

(has one terminal H on each B and C atom)

arachno

and boron as framework atoms. The CH^+ unit is isoelectronic with BH; many compounds are known in which one or more BH groups have been replaced by CH^+ (or by C, which has the same number of electrons as BH). For example, replacement of two BH groups in *closo*-$B_6H_6^{2-}$ with CH^+ yields *closo*-$C_2B_4H_6$, a neutral compound. *Closo, nido,* and *arachno* carboranes are all known, most commonly containing two carbon atoms; examples are shown in Figure **12-9**. Examples of chemical formulas corresponding to these designations are:

Type	Borane	Example	Carborane	Example
closo	$B_nH_n^{2-}$	$B_{12}H_{12}^{2-}$	$C_2B_{n-2}H_n$	$C_2B_{10}H_{12}$
nido	B_nH_{n+4} [17]	$B_{10}H_{14}$	$C_2B_{n-2}H_{n+2}$	$C_2B_8H_{12}$
arachno	B_nH_{n+6} [18]	B_9H_{15}	$C_2B_{n-2}H_{n+4}$	$C_2B_7H_{13}$

Carboranes may be classified structurally using the same method as for boranes. Since a carbon atom has the same number of valence electrons as a boron atom plus a hydrogen atom, each C can be formally converted to BH in the classification scheme. For example, for a carborane having the formula $C_2B_8H_{10}$:

$$C_2B_8H_{10} \rightarrow B_{10}H_{12}$$
$$B_{10}H_{12} - 2\ H+ \rightarrow B_{10}H_{10}^{2-}$$

The classification of $C_2B_8H_{10}$ is therefore *closo*.

[17]*Nido* boranes may also have the formulas $B_nH_{n+3}^-$ and $B_nH_{n+2}^{2-}$.

[18]*Arachno* boranes may also have the formulas $B_nH_{n+5}^-$ and $B_nH_{n+4}^{2-}$.

Many derivatives of boranes containing other main group atoms (hetero-atoms) are also known. These "heteroboranes" may be classified by formally converting the heteroatom to a BH_x group having the same number of valence electrons, then proceeding as in previous examples. For some of the most common heteroatoms, the following substitutions can be used:

Heteroatom	Replace with
C, Si, Ge, Sn	BH
N, P, As	BH_2
S, Se	BH_3

Exercise 12-4

Determine formulas of boranes isoelectronic with the following:

a. $closo$-$C_2B_3H_5$

b. $nido$-CB_5H_9

c. SB_9H_9 (classify as $closo$, $nido$, or $arachno$)

d. $CPB_{10}H_{11}$ (classify as $closo$, $nido$, or $arachno$)

While it may not be surprising that the same set of electron counting rules can be used to satisfactorily describe such similar compounds as boranes and carboranes, it is of interest to examine how far the comparison can be extended. Can Wade's rules, for example, be used effectively on compounds containing organometallic fragments bonded to boranes or carboranes? Can the rules be extended even further, to describe the bonding in polyhedral metal clusters?

12-2-4 Metallaboranes and Metallacarboranes

The CH group of a carborane is isolobal with 15-electron fragments of an octahedron such as $Co(CO)_3$. Similarly, BH, which has four valence electrons, is isolobal with 14-electron fragments such as $Fe(CO)_3$ and $Co(\eta^5\text{-Cp})$. These organometallic fragments have been found in substituted boranes and carboranes in which the organometallic fragments substitute for the isolobal CH and BH groups. For example, the organometallic derivatives of B_5H_9 shown in Figure **12–10** have been synthesized.

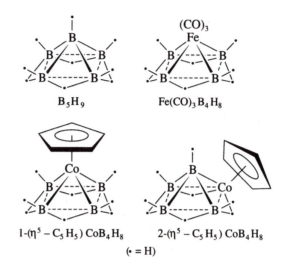

Figure 12-10

Organometallic Derivatives of B_5H_9 (Adapted with permission from G.L. Miessler and D.A. Tarr, *Inorganic Chemistry*, Prentice-Hall, Englewood Cliffs, NJ, 1991, 519.)

Theoretical calculations on the iron derivatives have supported the view that $Fe(CO)_3$ bonds in a manner isolobal with BH.[19] In both fragments the orbitals involved in framework bonding within the cluster are similar (Figure **12-11**). In BH the orbitals participating in framework bonding are an sp_z hybrid pointing toward the center of the polyhedron, and p_x and p_y orbitals tangential to the surface of the cluster. In $Fe(CO)_3$ an $sp_z d_{z^2}$ hybrid points toward the center, and pd hybrid orbitals are oriented tangentially to the cluster surface.

Examples of metallaboranes and metallacarboranes are numerous. Selected examples with *closo* structures are given in Table **12-7**.

Figure 12-11

Orbitals of Isolobal Fragments BH and $Fe(CO)_3$ (Adapted with permission from G.L. Miessler and D.A. Tarr, *Inorganic Chemistry*, Prentice-Hall, Englewood Cliffs, NJ, 1991, 520.)

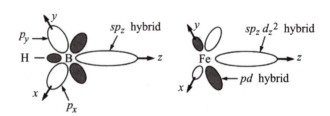

[19]R.L. DeKock and T.P Fehlner, *Polyhedron*, **1982**, *1*, 521.

Table 12-7 Metallaboranes and Metallacarboranes with *Closo* Structures

Number of Framework Atoms	Shape		Examples
6	Octahedron		$B_4H_6(CoCp)_2$ $C_2B_3H_5Fe(CO)_3$
7	Pentagonal bipyramid		$C_2B_4H_6Ni(PPh_3)_2$ $C_2B_3H_5(CoCp)_2$
8	Dodecahedron		$C_2B_4H_4[(CH_3)_2Sn]CoCp$
10	Bicapped square antiprism		$[B_9H_9NiCp]^-$ $CB_7H_8(CoCp)(NiCp)$
12	Icosahedron		$C_2B_7H_9(CoCp)_3$ $C_2B_9H_{11}Ru(CO)_3$

Anionic boranes and carboranes can also act as ligands toward metals in a manner resembling that of cyclic organic ligands. For example, *nido* carboranes of formula $C_2B_9H_{11}^{2-}$ have *p* orbital lobes pointing toward the "missing" site of the icosahedron (remember that the *nido* structure corresponds to a *closo* structure [which in this case is the 12-vertex icosahedron] with one vertex missing). This arrangement of *p* orbitals can be compared with the *p* orbitals of the cyclopentadienyl ring, as shown in Figure **12-12**.

The similarity between these ligands is sufficient that $C_2B_9H_{11}^{2-}$ can bond to iron to form a carborane analog of ferrocene, $[Fe(\eta^5\text{-}C_2B_9H_{11}^{2-})_2]^{2-}$. A mixed ligand sandwich compound containing one carborane and one cyclopentadienyl ligand, $[Fe(\eta^5\text{-}C_2B_9H_{11})(Cp)]$, has also been made (Figure **12-13**). Numerous other examples of boranes and carboranes serving as ligands to transition metals are also known.[20]

Figure 12-12
Comparison of $C_2B_9H_{11}^{2-}$ with $C_5H_5^-$ (Adapted with permission from G.L. Miessler and D.A. Tarr, *Inorganic Chemistry*, Prentice-Hall, Englewood Cliffs, NJ, 1991, 521.)

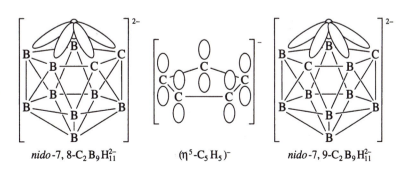

nido-7, 8-$C_2B_9H_{11}^{2-}$ $(\eta^5\text{-}C_5H_5)^-$ *nido*-7, 9-$C_2B_9H_{11}^{2-}$

[20]K.P. Callahan and M.F. Hawthorne, *Adv. Organomet. Chem.*, **1976**, *14*, 145.

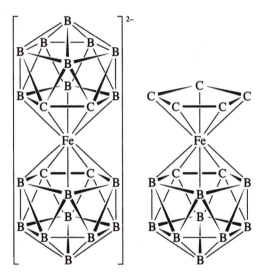

Figure 12-13
Carborane Analogs of
Ferrocene (Adapted with
permission from G.L.
Miessler and D.A. Tarr,
Inorganic Chemistry,
Prentice-Hall, Englewood
Cliffs, NJ, 1991, 521.)

Metallaboranes and metallacarboranes can be classified structurally by a procedure similar to that for boranes and their main group derivatives. In this scheme, the valence electron count of the metal-containing fragment is first determined, and then compared with the requirement of the 18-electron rule. This fragment can then be considered equivalent to a BH_x fragment needing the same number of electrons to satisfy the octet rule. For example, a 14-electron fragment such as $Co(\eta^5\text{-}Cp)$ is four electrons short of 18; this fragment may be considered the equivalent of the four-electron fragment BH, which is four electrons short of an octet. Examples of organometallic fragments and their corresponding BH_x fragments are given in Table **12-8**.

Table 12-8 Organometallic and Borane Fragments

Valence Electrons in Organometallic Fragment	Example	Valence Electrons in Borane Fragment	Borane Fragment
13	$Mn(CO)_3$	3	B
14	$Co(\eta^5\text{-}Cp)$	4	BH
15	$Co(CO)_3$	5	BH_2
16	$Fe(CO)_4$	6	BH_3

Example 12-2

The following examples illustrate that the method described previously for classifying boranes can also be used for metallaboranes:

$B_4H_6(CoCp)_2$

$$B_4H_6(CoCp)_2 \rightarrow B_4H_6(BH)_2 = B_6H_8$$
$$B_6H_8 - 2\,H^+ = B_6H_6{}^{2-} \qquad \text{Classification: } \textit{closo}$$

$B_3H_8Re(CO)_3$

$$B_3H_8Re(CO)_3 \rightarrow B_3H_8(B) = B_4H_8$$
$$B_4H_8 - 4\,H^+ = B_4H_4{}^{4-} \qquad \text{Classification: } \textit{nido}$$

$B_3H_7[Fe(CO)_3]_2$

$$B_3H_7[Fe(CO)_3]_2 \rightarrow B_3H_7[BH]_2 = B_5H_9$$
$$B_5H_9 - 4\,H^+ = B_5H_5{}^{4-} \qquad \text{Classification: } \textit{nido}$$

Exercise 12-5

Classify the following metallacarboranes by structural type:
a. $C_2B_7H_9(CoCp)_3$ b. $C_2B_4H_6Ni(PPh_3)_2$

12-2-5 Carbonyl Clusters

Many carbonyl clusters have structures similar to boranes; it is therefore of interest to determine to what extent the approach used to describe bonding in boranes may also be applicable to bonding in carbonyl clusters.

According to Wade, in addition to the obvious relation shown in equation **12.1** below, the valence electrons in a cluster can be assigned to framework and metal-ligand bonding, shown in equation **12.2**:[21]

Total number of valence electrons in cluster	=	Number of valence electrons contributed by metal atoms	+	Number of valence electrons contributed by ligands	**12.1**

[21]K. Wade, *Adv. Inorg. Chem. Radiochem.*, **1976**, *18*, 1.

$$\begin{matrix}
\text{Total number of} \\
\text{valence electrons} \\
\text{in cluster}
\end{matrix} = \begin{matrix}
\text{Number of electrons} \\
\text{involved in framework} + \\
\text{interactions}
\end{matrix} \begin{matrix}
\text{Number of electrons} \\
\text{involved in metal-} \\
\text{ligand interactions}
\end{matrix} \qquad \textbf{12.2}$$

As we have seen previously, the number of electrons involved in framework interactions in boranes is related to the classification of the structure as *closo*, *nido*, etc. Rearranging equation **12.2** gives the following:

$$\begin{matrix}
\text{Number of electrons} \\
\text{involved in framework} \\
\text{interactions}
\end{matrix} = \begin{matrix}
\text{Total number of} \\
\text{valence electrons} \\
\text{in cluster}
\end{matrix} - \begin{matrix}
\text{Number of electrons} \\
\text{involved in metal-} \\
\text{ligand interactions} \qquad \textbf{12.3}
\end{matrix}$$

For a borane, two electrons (one pair) are assigned to each boron-hydrogen bond (including each three-center, two-electron bond for bridging hydrogens). For a transition metal carbonyl complex, Wade suggests that 12 electrons per metal are either involved in metal-carbonyl bonding (to all carbonyls on a metal), or are nonbonding and therefore unavailable for participation in framework bonding. The result is that there is a net difference of 10 electrons per framework atom in comparing boranes with transition metal carbonyl clusters. A metal carbonyl analogue of *closo*-$B_6H_6^{2-}$, which has 26 valence electrons, would therefore need a total of 86 valence electrons to adopt a *closo* structure. An 86-electron cluster that satisfies this requirement is $Co_6(CO)_{16}$. Like $B_6H_6^{2-}$, $Co_6(CO)_{16}$ has an octahedral framework. As in the case of boranes, *nido* structures correspond to *closo* geometries from which one vertex is empty; *arachno* structures lack two vertices, and so on. The valence electron counts corresponding to the various structural classifications for main group and transition metal clusters are summarized in Table **12-9**.

Examples of *closo*, *nido*, and *arachno* borane and transition metal clusters are given in Table **12-10**. Transition metal clusters formally containing seven metal-metal framework bonding pairs are among the most common; examples illustrating the structural diversity of these clusters are given in Table **12-11**[22] and Figure **12-14**.

Table 12-9 Electron Counting in Main Group and Transition Metal Clusters

Structure Type	Main Group Cluster	Transition Metal Cluster
closo	$4n + 2$	$14n + 2$
nido	$4n + 4$	$14n + 4$
arachno	$4n + 6$	$14n + 6$
hypho	$4n + 8$	$14n + 8$

[22]K. Wade, "Some Bonding Considerations," in B.F.G. Johnson, Ed., *Transition Metal Clusters*, Wiley, **1980**, 232.

Table 12-10 *Closo*, *Nido*, and *Arachno* Borane and Transition Metal Clusters

Atoms in Cluster	Vertices in Parent Polyhedron	Framework Electron Pairs	Valence Electrons (Boranes)				Valence Electrons (Transition Metal Clusters)			
			closo	*nido*	*arachno*	Example	*closo*	*nido*	*arachno*	Example
4	4	5	18				58			
	5	6		20		$B_4H_7^-$		60		$Co_4(CO)_{12}$
	6	7			22	B_4H_{10}			62	$[Fe_4C(CO)_{12}]^{2-}$
5	5	6	22			$C_2B_3H_5$	72			$Os_5(CO)_{16}$
	6	7		24		B_5H_9		74		$Os_5C(CO)_{15}$
	7	8			26	B_5H_{11}			76	$[Ni_5(CO)_{12}]^{2-}$
6	6	7	26			$B_6H_6^{2-}$	86			$Co_6(CO)_{16}$
	7	8		28		B_6H_{10}		88		$Os_6(CO)_{17}[P(OMe)_3]_3$
	8	9			30	B_6H_{12}			90	

Table 12-11 Clusters That Formally Contain Seven Metal-Metal Framework Bond Pairs

Number of Framework Atoms	Cluster Type	Shape	Examples
7	Capped *closo*[a]	Capped octahedron	$[Rh_7(CO)_{16}]^{3-}$
6	*closo*	Octahedron	$Rh_6(CO)_{16}$
6	Capped *nido*[b]	Capped square pyramid	$H_2Os_6(CO)_{18}$
5	*nido*	Square pyramid	$Ru_5C(CO)_{15}$
4	*arachno*	Butterfly	$[HFe_4(CO)_{13}]^{-c}$

[a]A capped *closo* cluster has a valence electron count equivalent to neutral B_nH_n.
[b]A capped *nido* cluster has the same electron count as a *closo* cluster, equivalent to $B_nH_n^{2-}$.
[c]This complex has an electron count matching a *nido* cluster, but it adopts the butterfly structure expected for *arachno*. This is one of many examples in which structures of metal complexes are not predicted accurately by Wade's rules.

Predicted structures of transition metal carbonyl complexes using Wade's rules are often, but not always, accurate.[23] For example, the clusters $M_4(CO)_{12}$ (M = Co, Rh, Ir) have 60 valence electrons and are predicted to be *nido* complexes ($14n+4$ valence electrons). A *nido* structure would be a trigonal bipyramid (the parent structure) with one position vacant. X-ray crystallographic studies, however, have shown these complexes to have tetrahedral metal cores.[24]

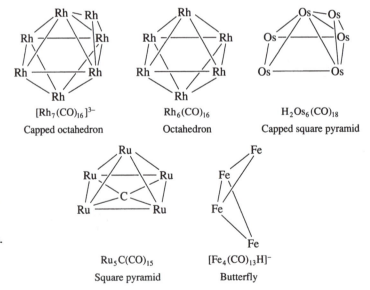

Figure 12-14
Metal Cores for Clusters Containing Seven Framework Bond Pairs (Adapted with permission from G.L. Miessler and D.A. Tarr, *Inorganic Chemistry*, Prentice-Hall, Englewood Cliffs, NJ, 1991, 525.)

$[Rh_7(CO)_{16}]^{3-}$
Capped octahedron

$Rh_6(CO)_{16}$
Octahedron

$H_2Os_6(CO)_{18}$
Capped square pyramid

$Ru_5C(CO)_{15}$
Square pyramid

$[Fe_4(CO)_{13}H]^-$
Butterfly

[23]Limitations of Wade's rules are discussed in R. N. Grimes, "Metallacarboranes and Metallaboranes," in G. Wilkinson, F.G.A. Stone, and W. Abel, Eds., *Comprehensive Organometallic Chemistry*, vol. 1, Pergamon Press, Elmsford, NY, 1982, 473.
[24]The metal cores of $Co_4(CO)_{12}$ and $Rh_4(CO)_{12}$ are slightly distorted; as mentioned earlier in this chapter (see Figure 12-4), these complexes have three bridging carbonyls on one triangular face.

12-2-6 Carbide Clusters

In recent years, many compounds have been synthesized, often fortuitously, in which one or more atoms have been partially or completely encapsulated within metal clusters. The most common of these cases have been the carbon-centered clusters, with carbon exhibiting coordination numbers and geometries not found in classical organic structures. Other nonmetals such as nitrogen have also been found encapsulated in clusters. Examples of these unusual coordination geometries are shown in Figure **12-15**.

Figure 12-15
Carbide Clusters and a Nitride Cluster (Adapted with permission from G.L. Miessler and D.A. Tarr, *Inorganic Chemistry*, Prentice-Hall, Englewood Cliffs, NJ, 1991, 526.)

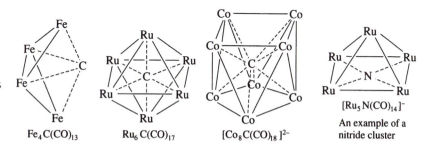

$Fe_4C(CO)_{13}$ $Ru_6C(CO)_{17}$ $[Co_8C(CO)_{18}]^{2-}$ $[Ru_5N(CO)_{14}]^-$
An example of a nitride cluster

Encapsulated atoms contribute their valence electrons to the total electron count. For example, carbon contributes its four valence electrons in $Ru_6C(CO)_{17}$ to give a total of 86 electrons, corresponding to a *closo* electron count.

Exercise 12-6

Classify the clusters shown in Figure **12-15** (other than $Ru_6C(CO)_{17}$) as *closo*, *nido*, or *arachno*.

How is it possible for carbon (or nitrogen), with only four valence orbitals, to form bonds to more than four surrounding transition metal atoms? $Ru_6C(CO)_{17}$ is again a useful example. The octahedral Ru_6 core has framework molecular orbitals (Figure **12-16**) similar to those of $B_6H_6^{2-}$ (Figure **12-6**). The key point is that carbon is not restricted to forming bonds with individual atoms (as is commonly described in organic compounds of carbon), but can participate in the formation of molecular orbitals extending from the carbon to the surrounding metals. Figure **12-6** shows the orbitals from which four of the most important bonding orbitals in the central Ru_6C core are derived. The 2*s*

Figure 12-16

Bonding Interactions Between
Central Carbon and
Octahedral Ru_6 (Adapted
with permission from G.L.
Miessler and D.A. Tarr,
Inorganic Chemistry, Prentice-
Hall, Englewood Cliffs, NJ,
1991, 526.)

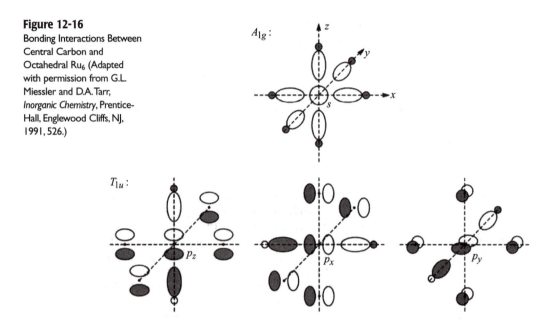

orbital of carbon can interact with six metal orbitals (hybrid orbitals, involving
significant *d* character) in a σ fashion, as shown in the top diagram in the figure. In addition, each of the carbon's 2*p* orbitals can participate in more complex interactions, as shown in the three lower diagrams in the figure. The net
result of these interactions is the formation of four C–Ru bonding orbitals,
occupied by electron pairs to hold the carbon in the center of the "cage."

Problems

12-1 Propose organic fragments isolobal with the following:
- **a.** $[Re(CO)_4]$
- **b.** $Tc(CO)_4(PPh_3)$
- **c.** $(\eta^5\text{-}Cp)Ir(CO)$
- **d.** $(\eta^4\text{-}C_4H_4)Co(PMe_3)_2$
- **e.** The ligands in the carbon-centered cluster $[C(SiMe_3)_2(AuPPh_3)_3]^+$, an example of pentacoordinate carbon.

12-2 Propose an organometallic fragment, not mentioned in this chapter, isolobal with:
- **a.** CH_3^+
- **b.** CH_2
- **c.** CH_2^-
- **d.** CH_3^-

12-3 On the basis of the isolobal analogy, propose a synthesis for the following compound: $(\eta^5\text{-}Cp)(CO)_2Fe{-}Mn(CO)_5$

12-4 Organoimido ligands (NR^-) may be viewed as isolobal with the cyclopentadienyl ligand C_5H_5 or other neutral organometallic fragments (alternatively, NR^{2-} may be considered isolobal with $C_5H_5^-$). Thus, a transition metal complex containing C_5H_5 may have chemical parallels with an organoimido complex of the same charge involving a metal from the following group in the Periodic Table. (For example, $[WCp]^{5+}$ would be analogous with $[Re(NR)]^{5+}$.) On this basis, predict formulas of organoimido complexes isolobal with Cp_2Zr and with $Cp_2Nb(NR)$.[25]

12-5 The 15-electron $Co(CO)_3$ fragment has a number of chemical parallels with the phosphorus atom.
- **a.** Propose structures for the compounds formed by replacing one or more phosphorus atoms of the tetrahedral P_4 cluster with $Co(CO)_3$ fragments.
- **b.** Like $Co(CO)_3$, the $(\eta^5\text{-}Cp)Mo(CO)_2$ fragment may substitute for phosphorus atoms in P_4. Propose formulas for the clusters that would result from such substitution.[26]

[25] J. Sundermeyer and D. Runge, *Angew. Chem. Int. Ed. Engl.*, **1994**, *33*, 1255.
[26] T. Kilthau, B. Nuber, and M.L. Ziegler, *Chem. Ber.*, **1995**, *128*, 197.

12-6   Classify the following as *closo*, *nido*, or *arachno*:
   a. $B_{10}H_{14}^{2-}$
   b. $C_3B_5H_7$
   c. $PCB_{10}H_{11}$
   d. $AsCB_9H_{11}^-$
   e. $B_4H_6(RhCp)_2$
   f. $C_2B_9H_{11}Os(CO)_3$

12-7   The following azaboranes have recently been reported.[27] Classify them as *closo*, *nido*, or *arachno*.
   a. $B_3H_2(CMe_3)_3(NCMe_3)$
   b. $B_3H_2(CMe_3)_2(NEt_2)(NCMe_3)$

12-8   The complex $Cp_2Zr(CH_3)_2$ reacts with the highly electrophilic borane $HB(C_6F_5)_2$ to form a product having stoichiometry $(CH_2)[HB(C_6F_5)_2]_2(ZrCp_2)$; the product is an example of pentacoordinate carbon.[28]
   a.   Propose a structure for this product. [Useful information: The ^1H NMR spectrum has singlets at $\delta = 5.23$ (relative area = 5) and 2.29 ppm (rel. area = 1) and a broad signal at $\delta = -2.05$ ppm (rel. area = 1); the ^{19}F NMR spectrum has three resonances; and the ^{11}B NMR has a single peak.]
   b.   An isomer of this product, $[Cp_2ZrH]^+[CH_2\{B(C_6F_5)_2\}_2(\mu\text{-}H)]^-$, has been proposed as a potential Ziegler–Natta catalyst. Suggest a mechanistic pathway by which this isomer might serve as a catalyst for the polymerization of ethylene.[29]

[27]M. Müller, T. Wagner, U. Englert, and P. Paetzold, *Chem. Ber.*, **1995**, *128*, 1.
[28]R.E. von H. Spence, D.J. Parks, W.E. Piers, M-A. MacDonald, M.J. Zaworotko, and S.J. Rettig, *Angew. Chem. Int. Ed. Eng.*, **1995**, *34*, 1230.
[29]See also L. Jia, X. Yang, C. Stern, and T.J. Marks, *Organometallics*, **1994**, *13*, 3755.

13

Other Applications of Organometallic Chemistry

Bioorganometallic Chemistry, Organolanthanide Chemistry, Solid-State and Surface Chemistry, and Fullerene-Metal Complexes

In this chapter we will look briefly at several other areas of chemical science where the carbon–transition metal bond plays an important role. These areas are often interdisciplinary and exist at the frontier of our knowledge of chemistry.

The number of biologically active organotransition metal complexes has, until recently, been quite limited. Indeed, the only well-characterized, naturally-occurring example is Coenzyme B_{12} and its homologs, the *cobalamins* (Figure **13-1**).[1] Several reports in recent years, however, have appeared on the use of transition metal complexes in biological systems; in addition to discussing the chemistry of cobalamins, Section **13-1** will explore some of these applications.

The chemistry of organolanthanide complexes has some resemblance to, but also some striking differences with, that of organotransition metal compounds. Section **13-2** will introduce the reader to some of the rich chemistry involving the lanthanide–carbon bond.

[1] A recent report provides the first direct evidence for a molybdenum–carbon bond found in transient species resulting from interaction of formaldehyde with xanthine oxidase enzyme; see B. D. Howes, *et al.*, *J. Am. Chem. Soc.*, **1994**, *116*, 11624.

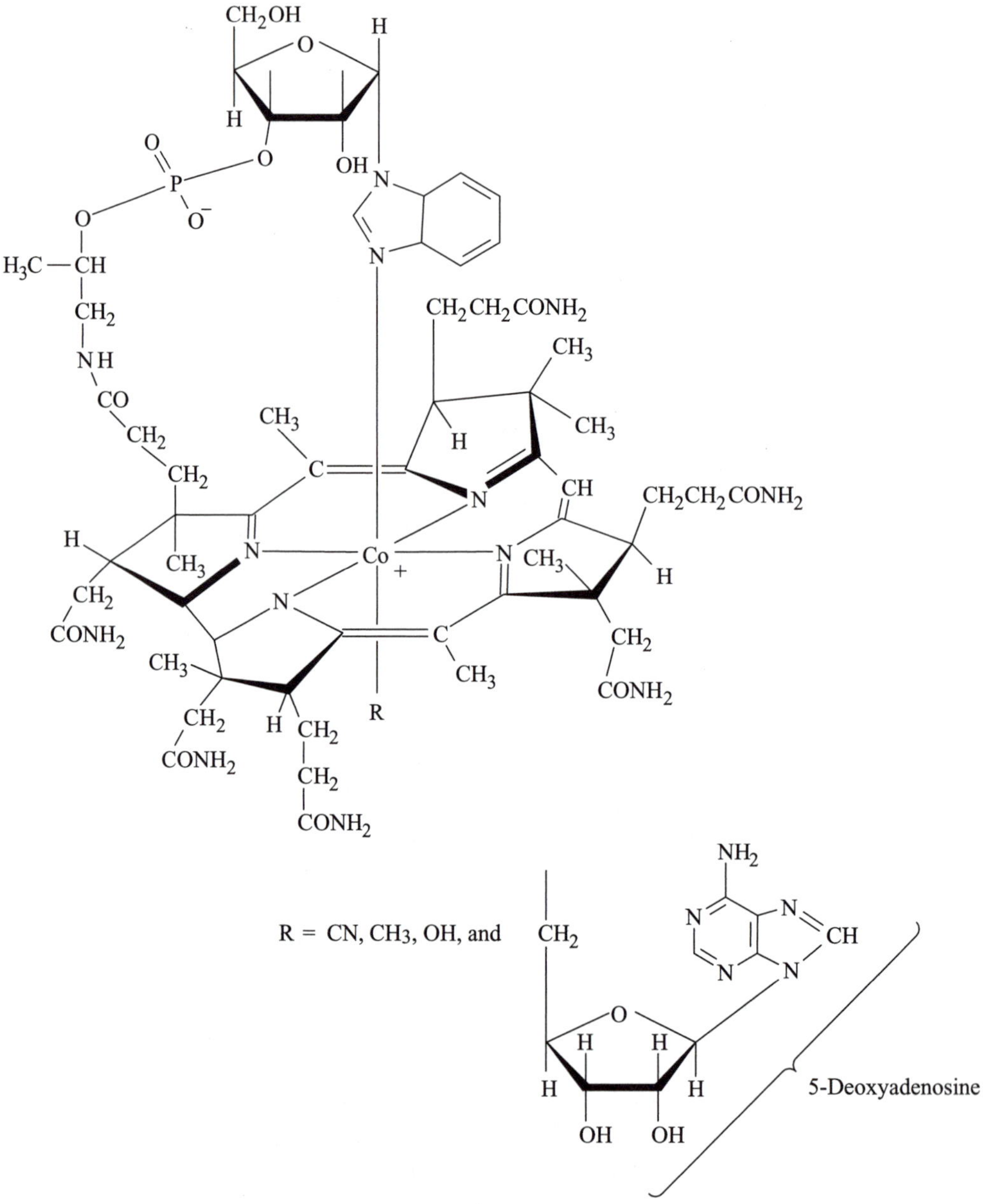

Figure 13-1
Vitamin B$_{12}$ Coenzyme and Homologs

R = CN, CH3, OH, and

5-Deoxyadenosine

Solid-state and surface chemistry are broad areas encompassing the disciplines of chemistry and physics. Transition metals are often present in materials that exhibit either electrical superconductivity or thermal stability at extremely high temperatures. Proper understanding of the mechanisms of heterogeneous catalysis requires an understanding of surface science. Section **13-3** will briefly explore a few aspects of these fields where transition metal cluster compounds serve as models for chemical reactions that occur on a solid support.

A new form of carbon was discovered just a few years ago in which 60 carbon atoms join to form a ball-like structure (Figures **13-2** and **13-10**).

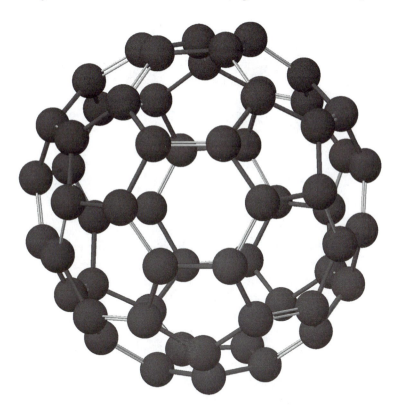

Figure 13-2
Buckminsterfullerene
(C_{60} Fullerene)

The carbon orb, resembling two geodesic domes fused together, was named *buckminsterfullerene* (or "buckyball," in honor of the late Buckminster Fuller). Since the discovery of buckyball, several other fullerenes containing different numbers of carbon atoms have been reported. The interaction of transition metals with fullerenes is an interesting facet to an exciting new development in chemistry; Section **13-4** will consider aspects of the preparation and reactions of some metal-fullerene complexes.

13-1   Bioorganometallic Chemistry

13-1-1 Organocobalt Complexes; Vitamin B$_{12}$ and Homologs

Complexes containing transition metals such as Fe, Mo, Cu, and Zn play an important role as co-enzymes in many biological systems. These complexes are, however, not truly organometallic because no metal–carbon bond is present. Of all the transition metals only cobalt bonds directly to carbon to form the only known, stable organotransition compounds existing in nature. Figure **13-1** shows a closely related family of biologically active compounds called *cobalamins*. A tetradentate ligand known as a *corrin*[2] takes up the equatorial positions of the complex, and at the top position a relatively unusual ribonucleotide resides that contains 1,2-dimethylbenzimidazole as the heterocyclic base fragment. The position at the bottom of the complex is variable, as shown in the following list (also see Figure **13-1**):

Bottom ligand (R)	Compound name
5'-deoxyadenosine	Coenzyme B$_{12}$
CN	Vitamin B$_{12}$ (cyanocobalamin)
CH$_3$	methylcobalamin
OH	hydroxycobalamin

The lability of the Co–C bond in coenzyme B$_{12}$, with a measured bond strength of only 26 kcal/mole,[3] thwarted early attempts to isolate the molecule. This bond is easily broken in the presence of visible light, leading to subsequent molecular decomposition during isolation steps. Proper care to exclude light and the addition of CN$^-$ during the extraction process leads to the deep red cyanocobalamin,[4] commonly called vitamin B$_{12}$. Hodgkin reported the crystal structure of structurally interesting vitamin B$_{12}$ in 1955 and was subsequently awarded a Nobel Prize for her efforts.[5] In 1972, a team of approximately 100 chemists, led by Woodward at Harvard University and Eschenmoser in Switzerland, described their successful total synthesis of cyanocobalamin.[6] The effort required **11 years**, truly a crowning achievement in organic chemistry at that time.

---

[2]The corrin ring system is identical to the porphyrins except that porphyrins have four =CR-bridges (R = H or alkyl) between the five-membered rings, whereas corrins have only three. Porphyrins are found in hemoglobin, chlorophyll, and the cytochromes.
[3]J. Halpern, S-H. Kim, and T.W. Leung, *J. Am. Chem. Soc.*, **1984**, *106*, 8317.
[4]Cyanocobalamin is not present in biological systems but is an artifact of the isolation process.
[5]D. C. Hodgkin, *Science*, **1965**, *150*, 979.
[6]A news article in *Science*, **1973**, *179*, 266 summarizes the original disclosure of the synthetic effort presented at a symposium in the fall of 1972.

Vitamin B_{12} and Pernicious Anemia

The inability in humans to absorb cobalamin results in a condition called *pernicious anemia*. If left untreated, the disease results in the potentially fatal formation of abnormal blood cells. Animals and plants lack the ability to synthesize cobalamins; only a few species of bacteria (primarily anaerobic) possess this capability. Since these bacteria are present in animals consumed as food by humans, sufficient cobalamin is usually available from a normal diet. Even if this is the case, however, a special protein—called an *intrinsic factor*—needs to be present in the lower intestine for proper uptake of the vitamin. The treatment for humans who lack the intrinsic factor is usually direct injection of large amounts of cyanocobalamin.

Coenzyme B_{12} (in conjunction with particular enzymes) is involved in two important functions in biological systems: permutation of two adjacent substituents (outlined in equation **13.1**), and methylation (equation **13.2**).

$$
\begin{array}{c}
| \ \ | \\
-C-C- \\
| \ \ | \\
H \ \ X
\end{array}
\xrightarrow{\text{Coenzyme } B_{12}}
\begin{array}{c}
| \ \ | \\
-C-C- \\
| \ \ | \\
X \ \ H
\end{array}
\qquad \textbf{13.1}
$$

$$
L_3X_2Co(III)\text{-}CH_3 + Acceptor \rightarrow L_nX_mCo(I, II, \text{ or } III) + CH_3\text{-}Acceptor
$$

$$\textbf{13.2}$$

Equation **13.3** shows an example of the first function, in which methylmalonyl-CoA[7] isomerizes to succinyl-CoA, a compound that undergoes further degradation via the citric acid cycle. Methylmalonyl-CoA arises from metabolism of the amino acid leucine and also from metabolic breakdown of propanoic acid.

Scheme **13.1** shows a catalytic cycle (after isomerization of methylmalonyl-CoA to succinyl-CoA) that involves coenzyme B_{12}.[8] Note that the

Methylmalonyl-CoA   Succinyl-CoA   **13.3**

---

[7]Note that the Co in CoA does *not* stand for cobalt. CoA designates the *coenzyme A* group, which for our purposes is an ethyl group in which one of the hydrogen atoms is replaced with a highly complex substituent. Thioesters with CoA as the alkyl group are reactive in biological systems, releasing a high amount of free energy upon cleavage of the acyl-S bond.

[8]A. L. Lehninger, D. L. Nelson, and M.M. Cox, *Principles of Biochemistry*, 2nd ed. Worth Publishers, New York, 1993, *495*; see also S. Wollowitz and J. Halpern, *J. Am. Chem. Soc.*, **1984**, *106*, 8319.

postulated mechanism shows the cobalamin undergoing homolytic Co-C bond cleavage (step **a**). In keeping with the observation that no incorporation of hydrogen atom occurs by donation from the solvent water, the adenosyl ligand abstracts a hydrogen atom from the substrate in step **b** and then returns it to substrate in step **d**.[9]

In order to understand the second function of coenzyme B_{12}, we need to quickly review the redox chemistry of cobalamins in biological systems. The Co in coenzyme B_{12} (Figure **13-1**) possesses 18 electrons and is in the +3 oxidation state. It is possible to reduce cobalamins (and coenzeme B_{12} in particular) stepwise from +3 to +1 as outlined in equation **13.4**.

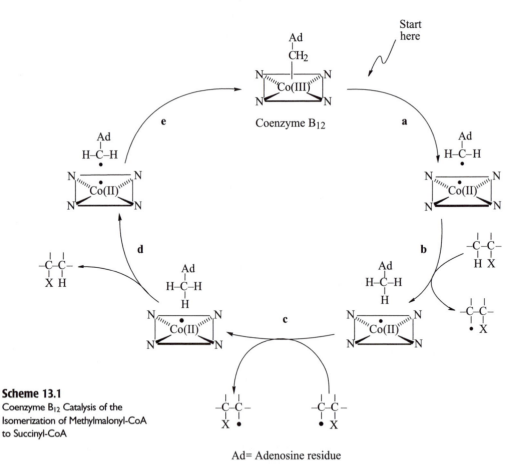

Scheme 13.1
Coenzyme B_{12} Catalysis of the
Isomerization of Methylmalonyl-CoA
to Succinyl-CoA

Ad= Adenosine residue

[9]See C. Walsh, *Enzymatic Reaction Mechanisms*, W. H. Freeman, San Francisco, 1979, 640–665 and H. Dugas and C . Penney, *Bioorganic Chemistry*, Springer-Verlag, New York, 1981, 369–386 for additional examples of 1,2-rearrangements that occur in biological systems.

$$[\,B_{12}\,Co(III)\,]^{+} \xrightarrow{\;1\,e^{-}\;} [\,B_{12}\,Co(II)\,]^{\bullet} \xrightarrow{\;1\,e^{-}\;} [\,B_{12}\,Co(I)\,]^{\bullet\bullet\,-}$$

$$B_{12r} \qquad\qquad B_{12s}$$

13.4

The reduced form, B_{12r}, is a four- or five-coordinate, neutral radical, postulated as an intermediate in Scheme **13.1**. Further reduction gives the superreduced form, B_{12s}, which is a carbanion and a powerful nucleophile. It is probably the B_{12s} form that is involved in the biosynthesis of coenzyme B_{12}, according to equation **13.5**. Here adenosine triphosphate (ATP) undergoes S_N2 attack by B_{12s} at the CH_2 group with concurrent displacement of triphosphate.

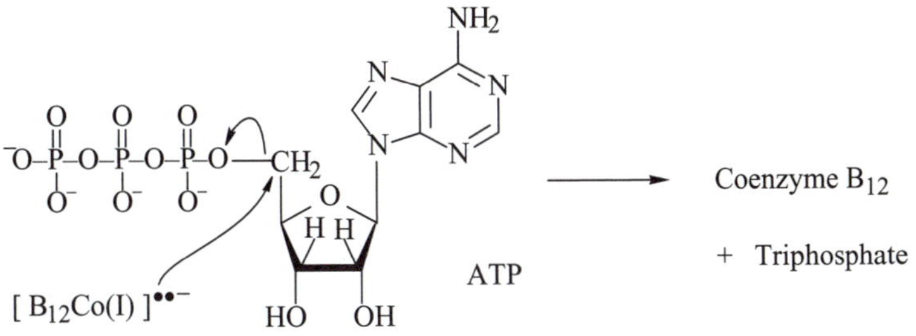

ATP → Coenzyme B_{12} + Triphosphate

13.5

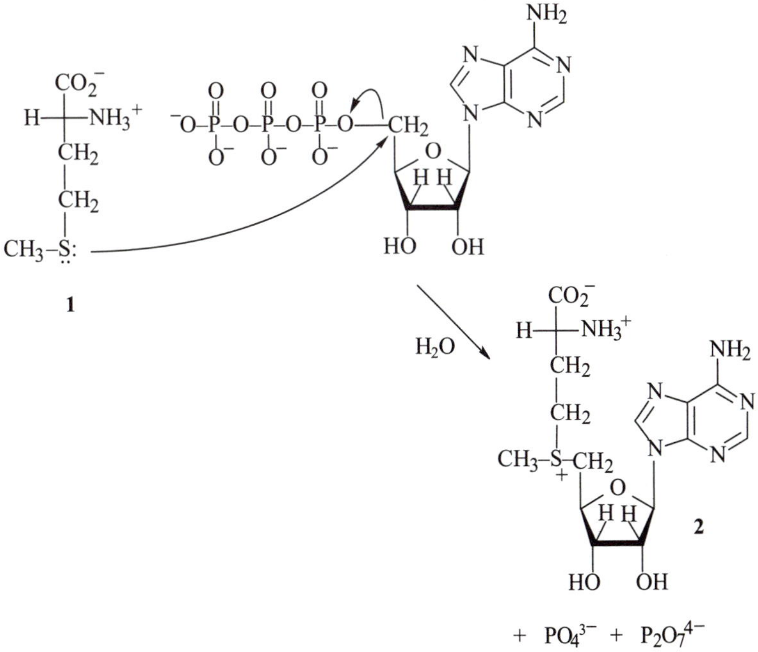

13.6

$+ \; PO_4{}^{3-} \; + \; P_2O_7{}^{4-}$

The most common methyl transfer agent in biochemical systems is S-adenosylmethionine, species **2** in equation **13.6**, formed from reaction of the amino acid methionine (**1**) with ATP. Once **2** transfers a methyl group to an acceptor (equation **13.7**), the products are the methylated acceptor, and S-adenosylhomocysteine, which subsequently breaks down to adenosine (**3**), and homocysteine (**4**). Methionine (and later **2**) can be regenerated by transfer of a methyl group from methylcobalamin (equation **13.8**). This last reaction, and isomerization of methylmalonyl-CoA to succinyl-CoA (equation **13.3**), are the only reactions dependent on some form of coenzyme B_{12} that are known to occur in mammals.

Acceptor:⟶ [structure of species **2**] ⟶ Acceptor–CH_3 + [structure of S-Adenosylhomocysteine]   **13.7**

S-Adenosylhomocysteine

[structure of adenosine **3**] + [structure of homocysteine **4**]

$$[B_{12}Co(III)\text{–}CH_3]^+ \;+\; [\text{structure } \mathbf{4}] \;\longrightarrow\; [B_{12}Co(I)]^{\bullet\bullet-} \;+\; [\text{structure } \mathbf{1}] \qquad \mathbf{13.8}$$

4   **1**

Exercise 13-1

Equation **13.8** provides a good example of what fundamental reaction type discussed in Chapter 8?

For many years, most scientists believed that mercury salts were chemically inert in the environment. It is now clear that some bacteria possess the capability of "mobilizing" inorganic mercury by converting salts such as $HgCl_2$ to CH_3Hg^+. Methyl mercury compounds are readily bioaccumulated in fat tissues of animals, leading to mercury intoxication. The classic example of mercury poisoning in humans occurred in Minamata Bay, Japan, where bacterial action on industrial discharges of mercury salts converted these compounds to methyl mercury. In this form, the metal was eventually incorporated into the edible flesh of shellfish. The local population consumed the seafood and later suffered greatly from the effects of mercury intoxication, a condition that some now call Minamata disease. Methylcobalamin seems to be the methyl transfer agent involved; equations **13.9** and **13.10** show a likely sequence of steps for the transferral process. Unlike the previous example of methyl transfer in which the equivalent of a "CH_3^+" is involved, the species transferred to Hg from methylcobalamin is equivalent to a methyl carbanion.

$$[\,B_{12}Co(III)\!-\!CH_3\,]^+ \;+\; H_2O \;\longrightarrow\; [\,B_{12}\underset{\underset{\displaystyle OH_2}{|}}{Co(III)}\!-\!CH_3\,]^+ \;\equiv\; [\,B_{12}\underset{\underset{\displaystyle OH_2}{|}}{Co(III)^{\delta+}}\,{}^{\delta-}CH_3\,]^+ \qquad \textbf{13.9}$$

$$[\,B_{12}\underset{\underset{\displaystyle OH_2}{|}}{Co(III)^{\delta+}}\,{}^{\delta-}CH_3\,]^+ \;+\; HgCl_2 \;\longrightarrow\; [\,B_{12}Co(III)(H_2O)\,]^+\,Cl^- \;+\; CH_3HgCl \qquad \textbf{13.10}$$

Exercise 13-2

The overall process shown in equations **13.9** and **13.10** is an example of what fundamental reaction type discussed in Chapter 8?

While their presence in biological systems is limited, the chemically interesting cobalamins serve crucial roles. The capability of Co to take on three different oxidation states, and the ability of cobalamins to transfer alkyl groups as cations, radicals, and carbanions, has stimulated vigorous research efforts worldwide.

13-1-2 Organometallic Compounds with Anti-Tumor Activity

Researchers have either isolated from natural sources, or synthesized, literally hundreds of compounds that possess anti-tumor activity. Many of these are used to treat cancer in humans; one of the most widely prescribed of these is cisplatin (**5**). Although extremely effective, severe side effects often accompany the administration of **5** during a chemotherapy regimen. Since cisplatin was introduced about 20 years ago, chemists have prepared several analogs of **5** that have been more palatable, yet still efficacious. Carboplatin (**6**) is prominent among these analogs. The *cis* relationship between the Cl ligands (or X-type ligands in general) is essential to the anti-tumor activity of all the Pt-based chemotherapy drugs.[10]

Cisplatin and its relatives inhibit tumor cell growth by covalently interacting with one of the two strands of DNA in the nucleus of a cancerous cell. This event occurs as the result of two ligand substitution reactions in which chloride ligands (or X-type ligands occupying analogous *cis*-positions) are displaced by neighboring guanine bases, both in one strand (equation **13.11**). The resulting covalent cross-link bends the double-stranded DNA helix. This leads to reactions that enhance the toxicity of cisplatin to tumor cells.[11] As with the use of any anti-tumor drug, the hope is that the agent will attack tumor cells preferentially to normal tissue. Toxic side effects in varying degrees accompany all chemotherapy regimens.

Complexes of the early and middle transition metals have been at the center of the search for non-platinum anti-tumor agents. While cisplatin and its homologs are not truly organometallic complexes because they contain no Pt-C bonds, several more recently discovered compounds with anti-tumor activity are indeed organotransition metal complexes. These include metallocenes involving Ti (**7**) and Fe (**8**).[12]

---

[10]For a brief summary of recent research on platinum-based anti-tumor agents, see M. J. Abrams and B. A. Murrer, *Science*, **1993**, *261*, 725.

[11]P.M. Takahara, A.K. Rosenzweig, C.A. Frederick, and S. J. Lippard, *Nature*, **1995**, *377*, 649.

[12]For a summary of work on metallocenes as anti-tumor agents, see P. Köpf-Maier and H. Köpf, *Chem. Rev.*, **1987**, *87*, 1137 and P. Köpf-Maier, *Eur. J. Clin. Pharmacol.*, **1994**, *47*, 1.

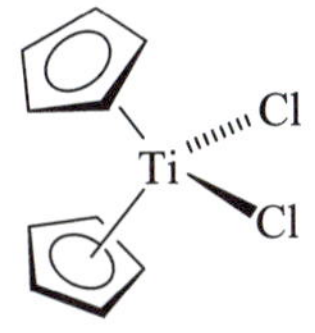

7

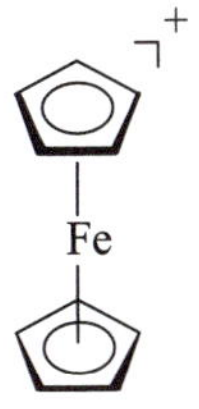

8

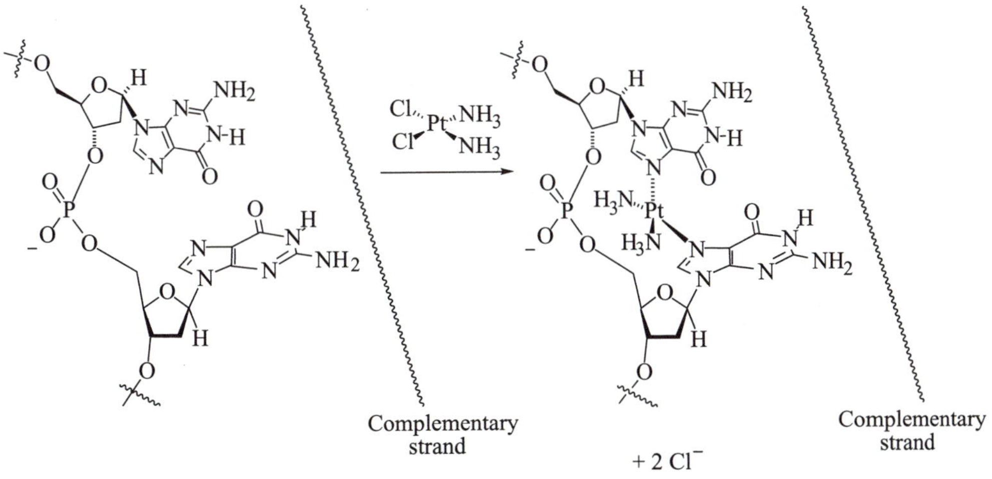

Complementary strand

Complementary strand

+ 2 Cl⁻

13.11

The titanocene derivatives again possess the *cis* relationship between the X-type ligands analogous to cisplatin. Whether that ligand arrangement is crucial to anti-tumor activity is unclear, because the ferrocenium ion complexes possess no X-type ligands in the coordination sphere. Yet both families of compounds exhibit a similar spectrum of anti-tumor activity comparable to cisplatin and other structurally unrelated chemotherapeutic agents. The metallocenes exhibit different toxic side effects than platinum-based drugs, but at levels that could still make them clinically useful. The modes of action of both **7** and **8** seem to be similar, yet are not completely understood. They do appear to inhibit DNA replication, but the manner in which they bind to the nucleic acid may be different than that of cisplatin. Recent studies indicate that binding of the metal to DNA occurs on the outer surface of the double helix.[13] Work on developing clinically useful metallocene anti-tumor agents continues to be active in the hope that some of these compounds will soon be effective in treating human cancers.

13-1-3 Organotransition Metal Complexes as Biochemical Tracers

Radioisotopes have traditionally served as labels used to follow the fates of molecules in metabolic pathways. Radioisotopes provide the chemist with at least two distinct advantages when they are used to probe biochemical reaction mechanisms. First, they are detectable in extremely small amounts (nmol and even pmol), and second, their presence does not significantly affect the chemistry of a molecule. In spite of the real advantages of radioisotopes and their widespread use as reaction probes, problems are associated with their use. These include: exposure to hazardous radiation, the need for elaborate laboratory facilities, and the difficulty of disposing of radioactive waste material.

In order to overcome these difficulties, there has recently been much effort put in to designing molecular tags that are just as sensitive to detection as radioisotopes yet do not have the problems associated with their use. Two types of biochemical interactions have been analyzed traditionally using radiolabels. These are antigen–antibody and hormone–receptor associations, both of which are high-affinity, specific interactions involving relatively small molecules binding to a much larger proteic entity.

Antigens **9**, **10**, and **11** are metal carbonyl derivatives of three important antiepileptic drugs—phenobarbital, carbamazepine, and phenytoin, respectively. Measurements using Fourier-transform infrared spectroscopy (FT-IR)[14] showed that the labeled drugs were able to bind specifically to their corresponding antibodies with an affinity comparable to that obtained with radiolabeled antiepileptics. The metal carbonyl fragment turns out to be an excellent molecular tag because the IR spectrum of proteins is usually transparent in the region of 1850-2200 cm^{-1}, whereas metal carbonyls absorb quite strongly in this region. By repeated scanning in the metal carbonyl region of the IR spectrum,

---

[13]T. J. Marks, *et al., J. Am. Chem. Soc.,* **1991**, *113*, 9027.

[14]Fourier transform IR spectroscopy has several advantages over conventional, continuous wave IR methods. The ability to scan the entire spectrum in less than one second, and then to accumulate the information provided by each scan, means that samples present in amounts undetectable by conventional IR methods are amenable to analysis by FT-IR. See R. M. Silverstein, C. C. Bassler, and T. C. Morrill, *Spectrometric Identification of Organic Compounds*, 5th ed. Wiley, New York, 1991, 98-99 for a discussion of the FT-IR method.

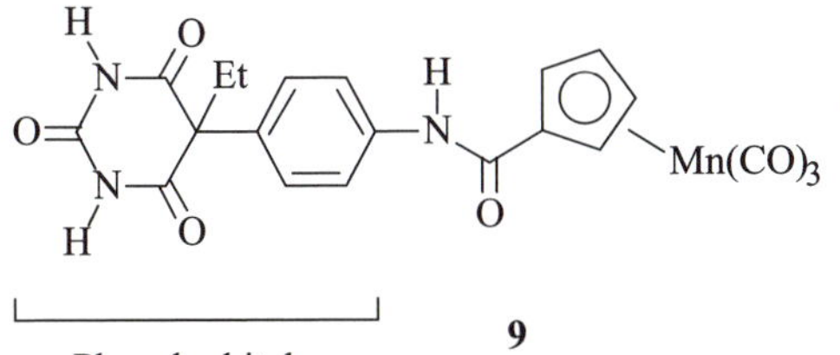

Phenobarbital

9

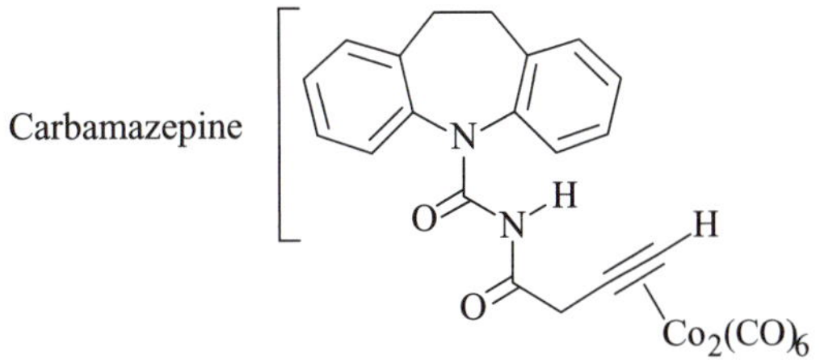

Carbamazepine

10

Phenytoin

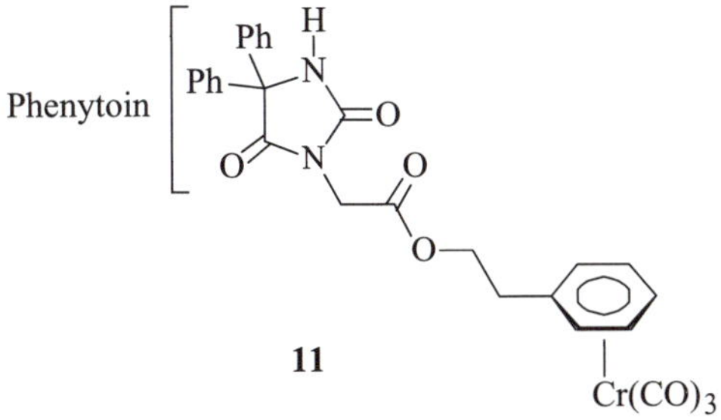

11

quantities of antigen–antibody complex as low as a fraction of a picomole are detectable. This is easily comparable in sensitivity to radioisotopes. Depending on the metal used, each different carbonyl fragment displays its own characteristic IR spectrum. By labeling different antigens with different metal carbonyl fragments, it is possible that a mixture of antigens could be analyzed simultaneously and quantitatively. Early efforts in this endeavor appear promising.

Hormone–receptor binding involves the association of a relatively small molecule (the hormone) with a sterically compatible, complementary site (the receptor), which is typically a protein attached to a membrane system. Binding is the first step in a hormonal response, and is followed by a cascade of events that lead to some regulatory action in the organism. For example, epinephrine binding triggers the release of energy for muscle contraction. FT-IR measurements of binding efficiencies of several metal carbonyl-estradiol derivatives

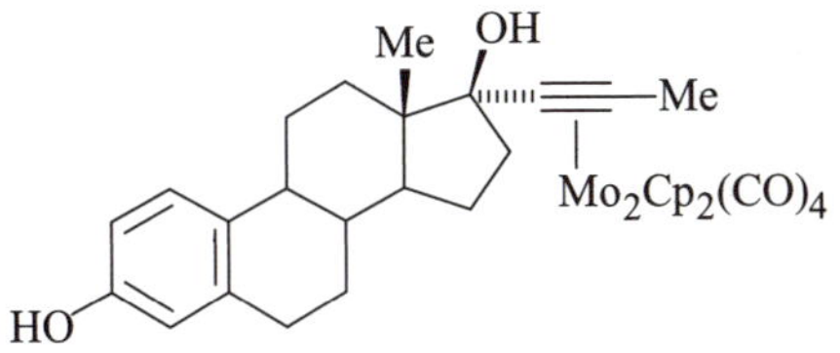

12

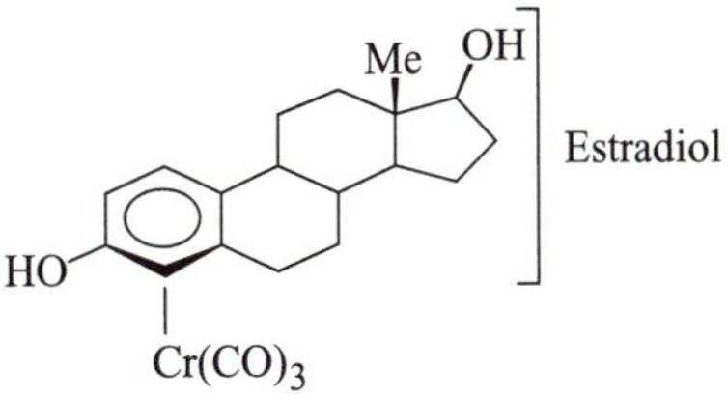

Estradiol

13

Table 13-1 Comparison of Some Properties of Lanthanides and Transition Metal Ions

Property	Transition metal (d-block)	Lanthanide
Valence e⁻s	[noble gas] md^n m = 3,4,5	$[Xe]\ 4f^n$
Oxidation state	-2 to +8	+3 (most common)
Coordination number	≤ 6 (most common)	8-12
Electronegativity	low to moderate	usually very low
Ligands available for bonding to metal	Virtually any Lewis base	Cp^-, Cp^{*-}, $C_8H_8^{2-}$
Nature of ligand bonding	Often highly covalent (back bonding)	Usually highly ionic (no back bonding)
Oxophilicity[a]	High with early transition metals; low with late metals	Very high

[a]Oxophilicity is a measure of a metal's affinity for bonding to oxygen.

(e.g., structures **12** and **13**) to estrogen receptors were of crucial assistance to researchers in producing the first picture of the binding site.[15]

13-2 Organolanthanide Chemistry

Long neglected by chemists as an active field of research, organolanthanide chemistry has become an attractive area for investigation in recent years. The results of these studies show that the so-called *rare earth*[16] metals possess properties unlike any other elements, and that they undergo rich chemistry in their interactions with organic ligands. It is only natural that we should compare and contrast the properties of the lanthanides with those of other transition metals;[17] Table **13–1** summarizes some of these differences.[18]

[15]A summary on the use of FT-IR to analyze antigen–antibody and hormone–receptor interactions appears in G. Jaouen, A. Vessieres, and I. S. Butler, *Acc. Chem. Res.*, **1993**, *26*, 361.

[16]The term "rare earth" is actually a misnomer. The relative abundance of individual lanthanide metals exceeds that of such "common" elements as gold or mercury, and, in fact, the abundance of many of these elements is comparable to that of tin. See N. N. Greenwood and A. Earnshaw, *Chemistry of the Elements,* Pergamon Press, Oxford, 1984, 1496 for more information on relative elemental abundance on earth.

[17]Lanthanides and actinides are sometimes called the "*f*-block" transition metals, since most elements in these rows of the Periodic Table have a partially filled valence *f* subshell. Other elements that are normally considered transition metals are termed "*d*-block" transition metals.

[18]Several reviews have appeared describing the chemistry of organolanthanide complexes that also contrast their chemistry to that of the *d*-block transition metals. For example, see T. J. Marks and R. J. Ernst, in G. Wilkinson, F. G. A. Stone, and E. W. Abel, Eds., *Comprehensive Organometallic Chemistry*, Pergamon Press, Oxford, 1982, chap. 21; W. J. Evans, *Adv. Organomet. Chem.*, **1985**, *24*, 131; *Polyhedron*, **1987**, *6*, 803; and most recently, C. J. Schaverien, *Adv. Organomet. Chem.*, **1994**, *36*, 283.

Several aspects of Table **13-1** require further comment. The valence electrons of lanthanide ions reside in $4f$ orbitals, which have rather limited extension away from the nucleus compared to d orbitals in the transition metals. Indications are that the $4f$ orbitals do not extend much beyond the $5s^2 5p^6$ orbitals of the inert Xe core. This means that *all* lanthanides in the +3 oxidation state [symbolized as Ln(III), with Ln serving as the general symbol for all the lanthanides] possess an electron distribution that strongly resembles the closed-shell, inert gas Xe with a +3 charge attached.[19] *Such an electron configuration suggests that each lanthanide reacts in a similar manner.* This is often the case, in contrast to the transition metals, whose chemistry varies markedly going from left to right across the Periodic Table.

All lanthanides share a common oxidation state of +3. Only four of the lanthanides normally take on different additional oxidation states: Ce(IV) [$4f^0$], Sm(II) [$4f^6$], Eu(II) [$4f^7$], and Yb(II) [$4f^{14}$]. This means that two of the most common fundamental reaction types in transition metal chemistry—oxidative addition and reductive elimination—are unlikely with the lanthanides.

Organolanthanide complexes typically exhibit much larger coordination numbers than organotransition metal compounds. This is due to the higher effective ionic radius of the lanthanides—ranging from 102 pm for Ce (III) to 86.1 pm for Lu(III)—compared with 55 to 86 pm for transition metal ions.[20]

Lanthanide ions are strongly electropositive. Ligands attached to lanthanides, therefore, tend to bind ionically rather than covalently, as is the case with transition metal complexes. This characteristic limits the types of ligands available for interaction with lanthanide atoms. Carbonyl and the phosphines, the "bread and butter" ligands we have come to know so well in organotransition metal chemistry, do not stabilize lanthanide complexes and are typically not seen as ligands. Instead, ligands such as Cp, Cp*, and cyclooctatetraenyl dianion (COT^{2-}) are the most common groups found in organolanthanide complexes. Mu-Bridging ligands, such as H in structure **15** and O in **16**, are also common. Structures **14**,[21] **15**,[22] and **16**[23] serve as examples. Because ligands such as Cp are primarily ionic and electron-rich, they satisfy—through ionic bonding—the electron-poor nature of Ln(III) without requiring the back bonding stabilization that the Ln(III) complexes are incapable of providing. They also serve another purpose because of their steric bulk: a lanthanide metal, surrounded by bulky ligands and coordinatively saturated, is less susceptible to attack by other nucleophiles and is thus more stable.

---

[19]Readers should not confuse the symbol Ln with L_n, the latter meaning n L-type ligands.

[20]J. A. Dean, *Lange's Handbook of Chemistry*, 14th ed. McGraw-Hill, New York, 1992, 4.13 to 4.17.

[21]J. H. Burns and W. H. Baldwin, *J. Organomet. Chem.*, **1976**, *120*, 361 (the nitrile is an L-type ligand).

[22]W. J. Evans, *et al.*, *J. Am. Chem. Soc.*, **1982**, *104*, 2008.

[23]H. Schumann, *et al.*, *Chem. Ber.*, **1993**, *126*, 907.

14   **15**   **16**

In some respects, the organolanthanides resemble high-valent early transition metal complexes, with their high oxophilicity. Lanthanide complexes typically exhibit a high affinity for oxygen, both in elemental and combined form, requiring that they be handled in an atmosphere that excludes both oxygen and moisture.

We will divide our brief discussion of the synthesis and reactivity of organolanthanide chemistry into three parts based upon the three oxidation states of the metal that occur when bound to organic ligands: Ln(III), Ln(II), and Ln(0).

13-2-1 Ln(III) Chemistry

One of the most thoroughly studied trivalent organolanthanide complexes is Cp_3Ln (see structure **14** above), prepared according to equation **13.12** below.[24] These complexes generally possess the same chemical properties regardless of the metal. If two equivalents of Cp^- react with $LnCl_3$, Cp_2LnCl results, which can then react with alkyllithium reagents to give $Cp_2Ln\text{-}R$ (R = CH_3 or *tert*-butyl) according to equation **13.13** below.

$$LnCl_3 + 3\ NaCp \xrightarrow[THF]{} Cp_3Ln\cdot THF + 3NaCl \qquad \textbf{13.12}$$

$$LnCl_3 + 2\ NaCp \longrightarrow Cp_2Ln\text{-}Cl \xrightarrow{R\text{-}Li} Cp_2Ln\text{-}R + LiCl \qquad \textbf{13.13}$$

Steric effects are extremely important in influencing the chemistry of organolanthanide compounds. Equations **13.14** and **13.15** demonstrate that terminal groups are much more susceptible to hydrogenolysis (and attack by

[24]T. J. Marks and R. J. Ernst, *op. cit.*

other reagents) than bridged ligands.[25] The reason for this difference is presumably the exposed nature of terminal ligands in contrast to bridged ligands that are surrounded by the protective influence of the bulky Cp groups.

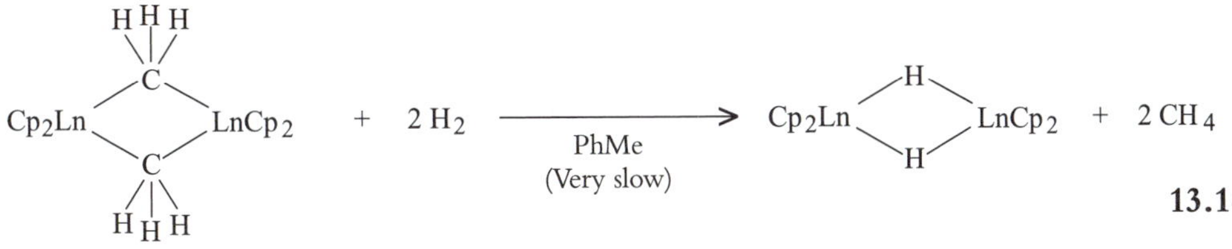

$$Cp_2Ln \underset{\underset{\overset{|}{\underset{H}{\overset{|}{C}}}{}\big/\!\!\big\backslash_{H}^{H}}{\overset{\overset{H}{\underset{\big\backslash}{\overset{H}{|}}\!\big/\,{}^{H}}{C}}}{\big\backslash}/\!\!\big/ LnCp_2 \; + \; 2\,H_2 \quad \xrightarrow[\substack{\text{PhMe} \\ \text{(Very slow)}}]{} \quad Cp_2Ln \underset{\overset{}{\underset{H}{\diagdown}}}{\overset{H\diagdown}{\diagup}} LnCp_2 \; + \; 2\,CH_4$$

$$\textbf{13.14}$$

Ln = Yb, Er, Lu

$$2\,Cp_2Ln(CMe_3)(THF) + 2\,H_2 \quad \xrightarrow[\substack{\text{PhMe} \\ \text{(rapid)}}]{} \quad [Cp_2Ln(\mu\text{-}H)(THF)]_2 + 2\,HCMe_3$$
$$Ln = Er,\,Lu$$

$$\textbf{13.15}$$

Equation **13.16** provides additional evidence that steric effects are important to the reactivity of Ln(III) complexes. Hydrogenolysis is facile for complexes in which the ionic radius of the Ln(III) is large. For Lu, the smallest of the lanthanides, the reaction is slow because the complex is now "sterically saturated" and the alkyl group is not accessible to H_2.[26]

$$2\,Cp_2Lu\text{-}Me(THF) + 2\,H_2 \quad \xrightarrow[\substack{\text{THF} \\ \text{(very slow)}}]{} \quad [Cp_2Lu(\mu\text{-}H)(thf)]_2 + 2\,CH_4$$

$$\textbf{13.16}$$

The examples shown above demonstrate a salient characteristic of organolanthanide chemistry: in spite of its inherent sameness, regardless of metal, *a subtle adjustment in the steric bulk of ligands surrounding the metal can make a significant difference in the reactivity of a particular type of complex.* This property is unique to the lanthanides. Thus chemists, through a good understanding of the relationship between degree of steric saturation and chemical reactivity, have been able in many cases to design molecules with predictable reactivity.

Nevertheless, this understanding is still not complete. For example, the first Cp^*_3Ln complex was reported[27] recently, where Ln = Sm (equation 13.17). Previous thinking proposed that the Cp^* ligand would be too bulky for three such groups to surround a metal. The existence of the Sm complex

[25]W. J. Evans, *et al.*, *J. Am. Chem. Soc.*, **1982**, *104*, 2008 and **1984**, *106*, 1291.

[26]W. J. Evans, R. Dominquez, and T. P. Hanusa, *Organometallics*, **1986**, *5*, 263.

[27]W. J. Evans, S. L. Gonzales, and J. W. Ziller, *J. Am. Chem. Soc.*, **1991**, *113*, 7423.

suggests that larger lanthanides such as Ce and Pr could also coordinate to three Cp* ligands and provide stable Cp*$_3$Ln complexes. Synthesis of these awaits further research.

$$2\,(Cp^*)_2Sm \quad + \quad \text{[cyclooctatetraene]} \quad \longrightarrow \quad (Cp^*)_3Sm \quad + \quad (Cp^*)Sm(C_8H_8)$$

13.17

Although organolanthanide complexes do not undergo oxidative addition and reductive elimination, they readily promote or catalyze other fundamental reactions such as 1,2-olefin insertion (section **10-6**), σ bond metathesis (section 10-7), and protonolysis of hydrocarbyl groups (section **8-4-1**). Equation **13.18** shows a hydroboration reaction that several metals will catalyze.

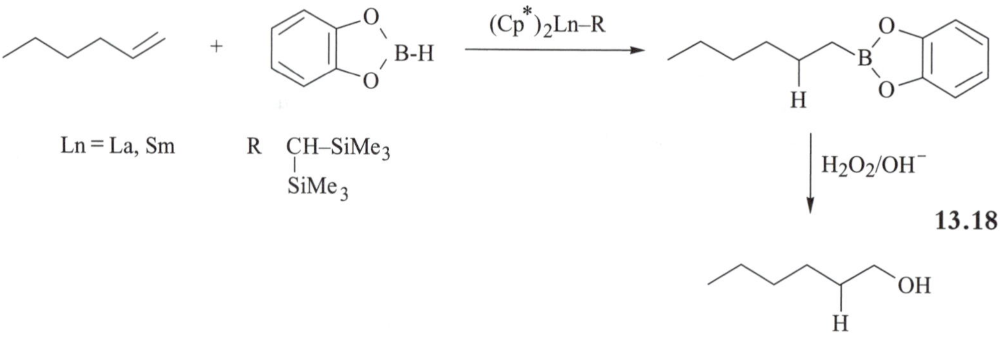

Ln = La, Sm   R   CH–SiMe$_3$
                               |
                            SiMe$_3$

13.18

Scheme **13.2** depicts the postulated catalytic cycle, one that involves an Ln(III) species at each step.[28]

Exercise 13-3

For each step in Scheme **13.2**, write a detailed mechanism. What fundamental organometallic reaction type is involved in each step?

[28]K. N. Harrison and T. J. Marks, *J. Am. Chem. Soc.*, **1992**, *114*, 9220.

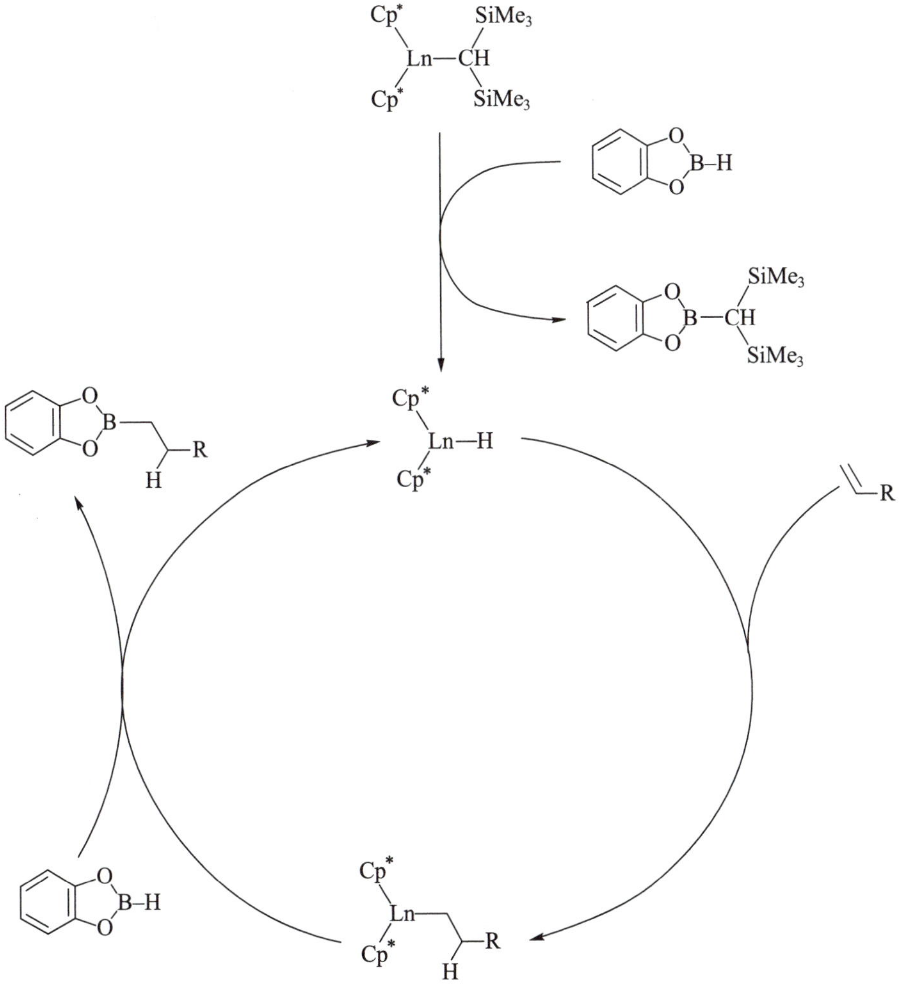

Scheme 13.2
Lanthanide-Catalyzed Hydroboration

13-2-2 Ln(II) Chemistry

Systematic study of organolanthanide complexes where the metal exists as Ln(II) began only relatively recently, and is still an active research area. Since the +2 oxidation state is readily available only for Sm, Eu, and Yb, the number of divalent organolanthanide complexes is small compared to those for Ln(III). Equations **13.19** and **13.20** show two different ways of preparing Ln(II) complexes. The first procedure shows a metal-metal exchange between Yb and Hg.[29] The second transformation generates a complex with Cp^* ligands. The extra methyl groups of Cp^* create a molecule that is more soluble than the comparable complex substituted with Cp ligands.[30]

$$Cp_2Hg + Yb \rightarrow Hg + Cp_2Yb \qquad\qquad\qquad\qquad \textbf{13.19}$$

$$SmI_2 + 2\,KCp^* \xrightarrow{\text{THF}} (Cp^*)_2Sm(THF)_2 + 2\,KI$$

$$\xrightarrow[\text{75\%/vacuum}]{} (Cp^*)_2Sm + 2\,THF \qquad \textbf{13.20}$$

Divalent organolanthanide complexes possess some interesting reactivity. Equation **13.21** shows reaction of a divalent Sm complex with diphenylacetylene. The reaction probably goes through a vinyl radical intermediate that reacts with a second equivalent of Sm complex to give **17**.[31]

$$(Cp^*)_2Sm(THF)_2 + Ph–C\equiv C–Ph \longrightarrow \underset{Ph}{\overset{(Cp^*)_2Sm}{\diagdown}}C=C\underset{\bullet}{\overset{Ph}{\diagup}} + 2\,THF \qquad \textbf{13.21}$$

$$(Cp^*)_2Sm(THF)_2 \downarrow$$

$$\underset{Ph}{\overset{(Cp^*)_2Sm}{\diagdown}}C=C\underset{Sm(Cp^*)_2}{\overset{Ph}{\diagup}}$$

17

Once formed, **17** reacts with CO to give ultimately a tetracyclic product **18**, according to Scheme **13.3**.[32]

Samarium (II) compounds show great promise as useful reagents in organic synthesis.[33] In the example below (equation **13.22**), Cp_2Sm reacts with different allyl chlorides to give an η^3-allyl samarium (III) complex (**19**), which is unstable[34] above -20° C. After it forms, **19** behaves as a Grignard reagent, and reacts with aldehydes and ketones to yield alcohols (equation **13.23**).[35]

[29]G. Z. Suleimanov, *et al.*, *J. Organomet. Chem.*, **1982**, *235*, C19.

[30]W. J. Evans, L. A. Hughes, and T. P. Hanusa, *J. Am. Chem. Soc.*, **1984**, *106*, 4270.

[31]W. J. Evans, *et al.*, *J. Am. Chem. Soc.*, **1983**, *105*, 1401.

[32]W. J. Evans, *et al.*, *J. Am. Chem. Soc.*, **1986**, *108*, 1722.

[33]For a recent review on the use of lanthanide reagents in organic synthesis, see G. A. Molander, *Chem. Rev.*, **1992**, *92*, 29.

[34]$Cp^*_2Sm(\eta^3$-allyl) complexes have been made and characterized by X-ray spectroscopy. Note here that the Cp^* analog is more stable than the corresponding Cp derivative, probably because it is closer to steric saturation than the complex with Cp ligands. W. J. Evans, T. A. Ulibarri, and J. W. Ziller, *J. Am. Chem. Soc.*, **1990**, *112*, 2314.

[35]C. Bied, J. Collin, and H. B. Kagan, *Tetrahedron*, **1992**, *48*, 3877.

Scheme 13.3
CO Insertion Reaction
Involving an Sm(II)
Complex

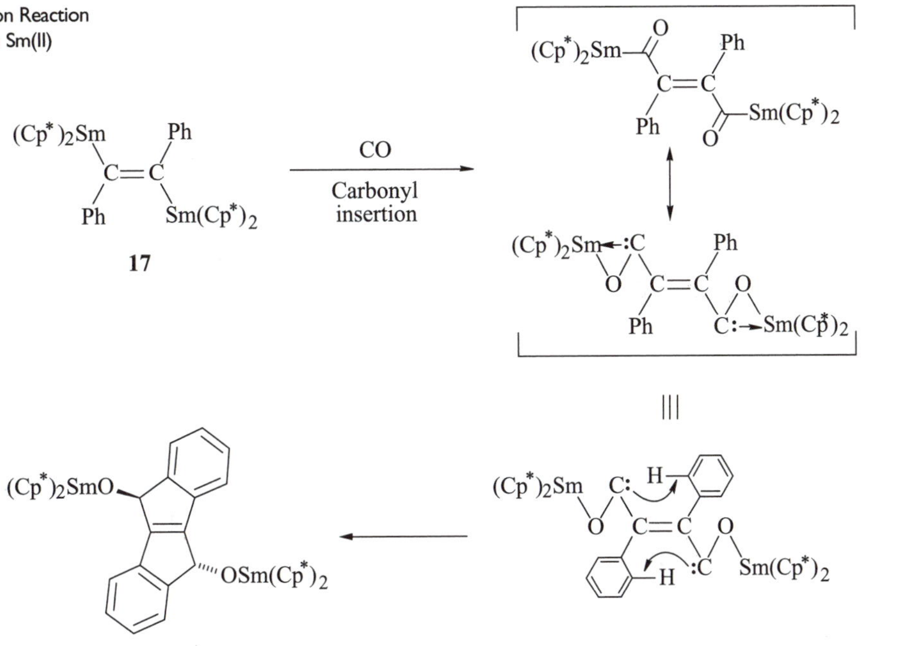

Double carbene insertion into a C–H bond

$$R = H, \quad R' = H \qquad\qquad R, R' = H$$
$$R = H, \quad R' = Ph \qquad\qquad R = H, R' = Ph \qquad\qquad \textbf{13.22}$$
$$R = H, \quad R' = Me \qquad\qquad R = H, R' = Me$$
$$R = Me, \quad R' = H$$

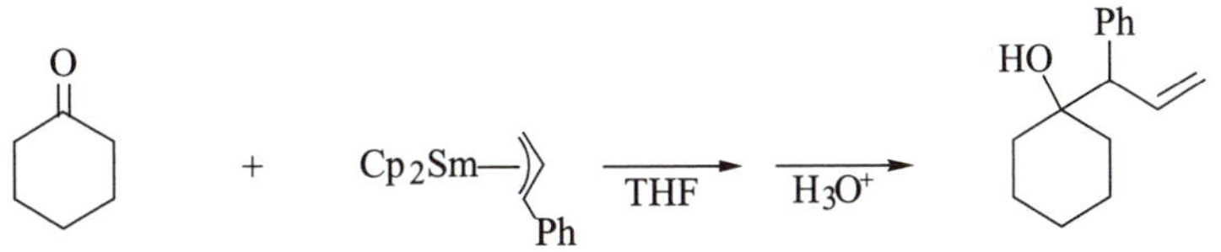

13.23

13-2-3 Ln(0) Chemistry

Transition metals form a variety of zero-valent complexes with neutral π systems (Chapter 5). The stability of these complexes depends substantially on the ability of the relatively electron-rich metal to back donate electron density to the ligand. If f orbitals, with their short extension away from the nucleus, are the outer orbitals of the lanthanides, such back donation should be difficult. Studies show that elemental lanthanides do react with neutral π systems, but most of the resulting products have not been well characterized. The reason that lanthanides react at all is probably due to the presence of empty $5d$ orbitals that are relatively close in energy to the $4f$ orbitals containing the valence electrons. Spectroscopic studies show that stable electron configurations exist in which $5d$ orbitals are populated.

One good example of a stable Ln(0) π complex is diarene **20**. Depending upon the metal, some of these sandwich compounds exhibit impressive thermal stability (they sublime at 100°).[36] The bulkiness of the trisubstituted benzene rings probably accounts for some of the chemical stability, making the metal sterically unaccessible. The thermal stabilities also correlate well[37] with the ease of promotion of an electron from a configuration $4f^n 6s^2$ to $4f^{n-1} 5d^1 6s^2$.

20

Ln= Nd, Tb, Dy, Ho, Er, Lu, Gd

There appears to be much remaining opportunity for fruitful exploration of organolanthanide chemistry. The lanthanides, with their distinctive chemical properties, offer a special brand of organometallic chemistry to the researcher that both complements and extends traditional organotransition metal chemistry.

---

[36] F. G. N. Cloke, *et al.*, *J. Chem. Soc., Chem. Commun.*, **1987**, 1668.

[37] W. A. King and T. J. Marks, *J. Am. Chem. Soc.*, **1992**, *114*, 9221. See also F. G. N. Cloke, *et al.*, *J. Chem. Soc., Chem. Commun.*, **1989**, 53.

13-3   Surface Organometallic Chemistry

Many catalytic processes involving organometallic complexes are known to occur at metal surfaces.[38] While some of these processes have been known for many years, and some are of immense industrial importance, the details of the reactions occurring at the surfaces have proven to be difficult to understand. Despite the experimental difficulties of analyzing molecules (or molecular fragments) attached to surfaces, an understanding of some of these catalytic processes is emerging. A useful approach to viewing reactions at surfaces uses the concept of frontier orbitals.

Metal surfaces can be viewed as arrays of atoms, with the potential for molecules to become attached to positions on these arrays. Such attachment can occur if the orbitals of the metal(s), and of the approaching molecules, can interact in such a way to stabilize electrons; this is similar to how we can view the formation of transition metal complexes from interactions between metals and ligands. In principle, a variety of modes of attachment is possible: attachment of a molecule to a single metal atom on the surface, to a site bridging two or three adjacent metal atoms, or to a site bridging metals that are not directly bonded in the solid; these are shown schematically in Figure **13-3**.

Each of these bonding modes has parallels in the known range of organometallic complexes. Examples of these parallels and a brief discussion of their implications for organometallic reactions at surfaces follow.

13-3-1 Single Metal (Terminal) Sites

Strictly speaking, there are considerable differences between a metal atom surrounded by ligands and a metal atom on a surface. In the former case, metal-ligand orbital interactions can be described as in earlier chapters of this text, with the result of such interactions being the formation of molecular orbitals having specific energies and shapes. In the case of metal atoms on surfaces, the metal-metal interactions do not result in discrete orbitals but rather in energy bands (each having a continuum of energy levels), with valence metal electrons able to travel throughout the metal rather than being restricted to a much more localized region (orbital).[39] Considering ways in which incoming molecules (adsorbates) can interact with such a surface might, at first view, seem hopelessly complex.

One approach to seeking parallels between transition metal complexes and metal surfaces is to divide a surface into its components (fragments consisting of metal atoms or small groups of atoms) and to study the orbitals of

[38]M. R. Albert and J. T. Yates, Jr., *The Surface Scientist's Guide to Organometallic Chemistry*, American Chemical Society, Washington, D.C., 1987; R. Hoffmann, *Solids and Surfaces*, VCH Publishers, New York, 1988.

[39]See, for example, C. Kittel, *Introduction to Solid State Physics*, Wiley, New York, 1971.

Single metal (terminal) site

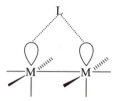

or

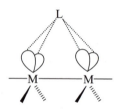

Adjacent doubly bridging site

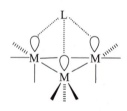

Adjacent triply bridging site

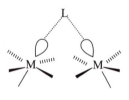

Non-adjacent doubly bridging site

designates an acceptor metal orbital lobe

Figure 13-3
Bonding Sites on Metal Surfaces

each component. This approach makes it possible to consider possible interactions between adsorbate molecules and the surface components. The result of such an analysis is to model a metal surface by considering how a fragment of this surface may be able to participate in bonding interactions with incoming molecules; if this type of interaction also has a significant effect on the bonding within the attached molecule (such as weakening a bond), the result may be to facilitate a reaction.

For example, consider how such an analysis can be applied to the interaction of H_2 with a metal surface. Such interactions are very important to understand because H_2 scission is a crucial step in a variety of transition metal-catalyzed reactions (Section **9-4-2**). Since many examples are known in which the H-H bond is broken at a single metal site in homogeneous reactions, it is reasonable to consider how a similar reaction might occur at a single metal site on a surface.

If one ligand is removed from an octahedral transition metal complex, a fragment results having a low energy (empty) acceptor orbital, as shown in Figure **13-4**. This fragment has the same geometry as a metal atom on some surfaces.[40]

Figure 13-4
Formation of a Five-Coordinate Fragment

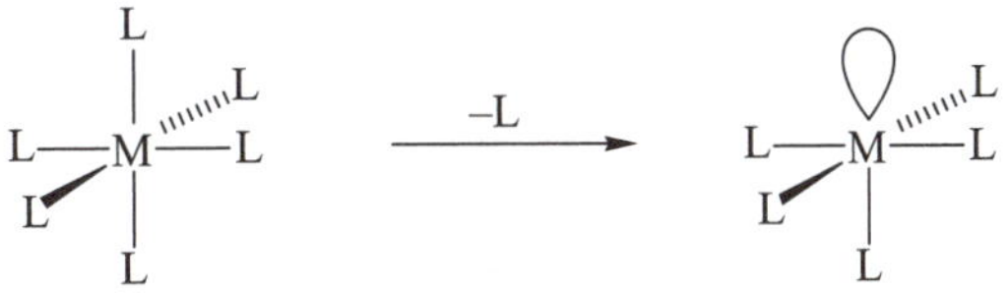

$$L-M\begin{matrix}L\\|\\\end{matrix}L \quad \xrightarrow{-L} \quad L-M\begin{matrix}L\\|\\\end{matrix}L$$

If an H_2 molecule approaches such a metal site in an orientation parallel to the surface, the σ orbital of H_2 can interact with the empty orbital of the metal fragment to stabilize the H-H bonding pair. At the same time, a filled d orbital of the metal can interact with the empty π* orbital of H_2, as shown in Figure **13-5**.

Figure 13-5
Interaction of H_2 with
Metal Atom on Surface

[40]See J. W. Lauher, *J. Am. Chem. Soc.*, **1979**, *101*, 2604. Metals in other environments on surfaces also have low energy, empty orbitals available for interaction with donor orbitals of adsorbed molecules.

The result of these two interactions is not only to bind the hydrogen atoms to the metal, but also to weaken the H-H bond. The bond weakening is a consequence of both interactions: loss of bonding electrons from the σ (bonding) orbital, and the transfer of electrons from the metal to the $\sigma*$ (antibonding) orbital of H_2. This is analogous to the weakening of the H-H bond when H_2 bonds to transition metals (Chapter 6, Section **6-2**; see Figure **6-6**), and when H_2 interacts with frontier metal orbitals in oxidative addition reactions (Chapter 7, Section **7-2-1**; see Figure **7-8**). Weakening of the H-H bond is believed to be an important step in a variety of surface-catalyzed processes. For example, if another molecule such as an alkene is also attached to a neighboring site, one of the hydrogens of the original H_2 may be able to escape from the H_2 and form a new bond to this molecule.

Exercise 13-4

Show how orbital interactions might, in principle, make it possible for ethylene to become attached to a single metal on a surface. Would you expect such a process to result in strengthening or weakening of the carbon-carbon bond?

13-3-2 Bridging Sites

As in the case of transition metal complexes, bridging ligands (molecules or molecular fragments) are believed important in a variety of processes occurring at surfaces. With two or more surface atoms involved, more types of orbital lobes are available than in the case of individual atoms, and more complex orbital interactions may therefore become possible.

Doubly Bridging Sites

One way to model a two-atom site is by using the molecular fragment $M_2(CO)_6$. Such a fragment could be derived by removing three ligands from octahedral $M(CO)_6$, then joining two $M(CO)_3$ fragments with a metal-metal bond. Each metal in $M_2(CO)_6$ would have two empty lobes, four lobes for the total fragment shown below:

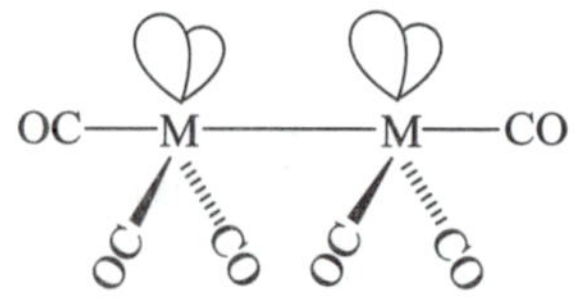

In addition, two metal-metal σ orbitals, one bonding and one antibonding, are formed. The M-M bonding orbital is similar in energy to the lobes shown above, and is capable of interactions with substrate molecules in some cases. In all, five orbitals, shown in Figure **13-6**, are potentially available for these interactions.[41]

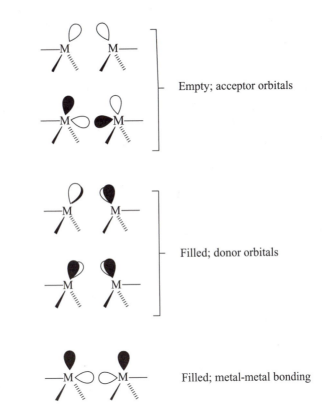

Figure 13-6
Orbitals of $M_2(CO)_6$
Available for Donor-
Acceptor Interactions

Empty; acceptor orbitals

Filled; donor orbitals

Filled; metal-metal bonding

One molecule that can become attached to $M_2(CO)_6$ (M = Co) is acetylene. Two types of interaction involving the π orbitals of acetylene are possible, one in which (filled) π orbitals act as donors, and one in which (empty) π^* orbitals act as acceptors; a pair of such interactions is shown in Figure **13-7**.

[41]The shapes of these orbitals have been determined from calculations based on linear combinations of the atomic orbitals involved. See D. L. Thorn and R. Hoffmann, *Inorg. Chem.*, **1978**, *17*, 126. This reference also discusses the relative energies of the orbitals and shows their shapes in more detail than provided here.

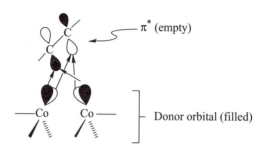

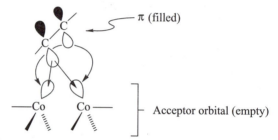

Figure 13-7
Interaction of Co$_2$(CO)$_6$
with Acetylene

Exercise 13-5

Show how another pair of donor–acceptor interactions, perpendicular to those shown in Figure **13-7**, is possible between acetylene and Co$_2$(CO)$_6$.

The complex resulting from these interactions, Co$_2$(CO)$_6$(acetylene), is stable and has been well characterized. Similar metal–acetylene interactions have been proposed for reactions on metal surfaces. One such reaction is the oligomerization of acetylene [n C$_2$H$_2$ → (C$_2$H$_2$)$_n$]. If one acetylene is attached to a metal surface (Fe in this case), an additional acetylene molecule may be able to participate in an orbital interaction of the type shown in Figure **13-8**.

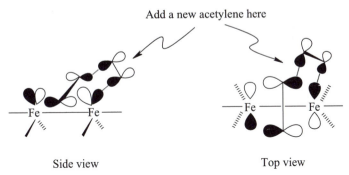

Add a new acetylene here

Side view   Top view

Figure 13-8
Possible Interaction
Between Acetylene
Orbitals on Metal Surface

Such an interaction could, in principle, lead to carbon-carbon bond formation between two acetylenes at the surface and, subsequently, to the generation of larger polyacetylene molecules.[42]

Other Bridging Sites

A third possible type of molecule-surface interaction would involve a site bridging a triangle of three metal atoms. Triply bridging ligands have been well characterized in cluster chemistry. A bonding model here can be based on an $M_3(CO)_9$ fragment, formed from three $M(CO)_3$ fragments; a known molecular fragment having this formula is $Co_3(CO)_9$, which forms complexes with triply bridging ligands such as S^{2-} and alkylidynes.

In the $M_3(CO)_9$ fragment there are three vacant acceptor lobes pointing up, as shown in Figure **13-9**. These can interact with molecules, or molecular fragments, having suitable orbitals. An example of a fragment that can interact with these lobes is the alkylidyne CR; the interactions of CR with $M_3(CO)_9$ are also shown in Figure **13-9**.

Interactions between incoming molecules and metal sites not directly linked by metal-metal bonds are also possible. These yield "A-frame" arrangements, examples of which are well known in coordination chemistry. Examination of this type of interaction is left as an exercise at the end of this chapter.

---

[42]D. L. Thorn and R. Hoffmann, *op. cit.*

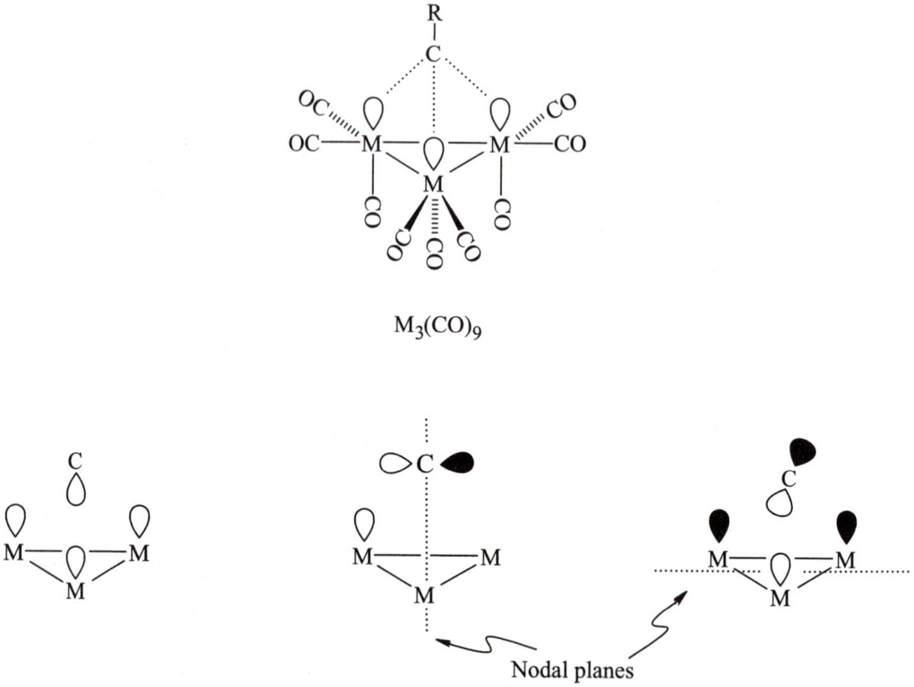

$$M_3(CO)_9$$

Nodal planes

Figure 13-9
Interaction of $M_3(CO)_9$ with Alkylidyne

13-4   Fullerene-Metal Complexes

One of the most fascinating developments in modern chemistry has been the synthesis of buckminsterfullerene,[43] C_{60}, and the related "fullerenes," molecules having near-spherical shapes resembling geodesic domes. First reported as synthesized in the gas phase by Kroto, Smalley, and co-workers in 1985,[44] C_{60}, C_{70}, and a variety of related species were soon synthesized; examples of their structures are shown in Figure **13-2** and **13-10**. Subsequent work has been extensive, and many attempts have been made to coordinate fullerenes to metals. Remarkably, roughly nine years after the first synthesis of fullerenes, natural deposits of these molecules have been found at the impact sites of ancient meteorites.[45]

[43]Often referred to as "buckyball."
[44]H. W. Kroto, J. R. Heath, S. C. O'Brien, R. F. Curl, and R. E. Smalley, *Nature*, **1985**, *318*, 162.
[45]L. Becker, J. L. Bada, R. E. Winans, J. E. Hunt, T. E. Bunch, and B. M. French, *Science*, **1994**, *265*, 642; and D. Heymann, L. P. F. Chibante, R. R. Brooks, W. S. Wolbach, and R. E. Smalley, *Science*, **1994**, *265*, 645.

Figure 13-10
Examples of Fullerene Structures
(Structures reproduced with permission
from C.J. Brabec, A. Maiti, and J. Bernholc,
Chem. Phys. Letters, **1994,** *219,* 473.)

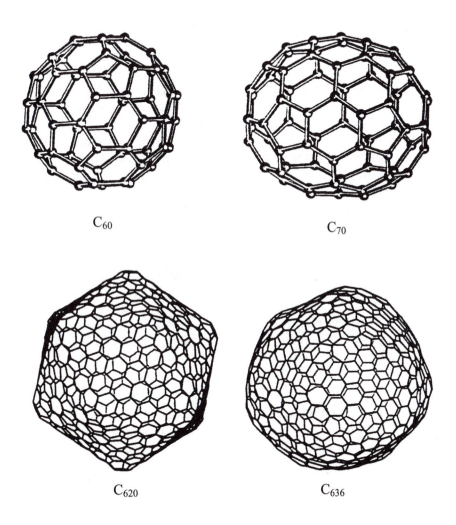

C_{60}

C_{70}

C_{620}

C_{636}

13-4-1 Structures of Fullerenes

The prototype fullerene, C_{60}, consists of fused five- and six-membered carbon rings. Each six-membered ring is surrounded, alternately, by hexagons and pentagons of carbons; each pentagon is fused to five hexagons. The consequence of this structural motif is that each hexagon is like the base of a bowl;

the three pentagons fused to this ring, linked by hexagons, force the structure to curve (in contrast to graphite, in which each hexagon is fused to six surrounding hexagons in the same plane). This phenomenon, best seen by assembling a model of C_{60}, results in a dome-like structure that eventually curves around on itself to give a structure resembling a sphere.[46] The shape resembles a soccer ball (which has an identical arrangement of pentagons and hexagons on its surface); all 60 atoms are equivalent and give rise to a single ^{13}C NMR resonance.

Although all atoms in C_{60} are equivalent, the bonds are not. Two types of bonds occur (best viewed using a model): at the fusion of two six-membered rings and at the fusion of five- and six-membered rings. X-ray crystallographic studies on C_{60} complexes have shown that the C–C bond length at the fusion of two six-membered rings in these complexes is shorter, 135.5 pm, in comparison with the comparable distance at the fusion of five- and six-membered rings, 146.7 pm.[47] This is indicative of a greater degree of π bonding at the fusion of the six-membered rings.

Surrounding each six-membered ring with two pentagons (on opposite sides) and four hexagons (with each pentagon, as in C_{60}, fused to five hexagons) gives a slightly larger, somewhat prolate structure with 70 carbon atoms (Figure 13-10). C_{70} is often obtained as a by-product of the synthesis of C_{60} and is among the most stable of the fullerenes. Unlike C_{60}, five different types of carbon are present in C_{70}, giving rise to five ^{13}C NMR resonances.[48]

This is not all! Recently synthesized extensions of the fullerene family include carbon nanotubes, long tubes of carbon capped by a half-fullerene structure;[49] carbon "onions," apparently containing concentric spheres of carbons;[50] and acetylene-bridged difullerenes.[51]

13-4-2 Fullerene-Metal Complexes

Fullerene-metal complexes[52] have now been prepared for a variety of metals. These complexes fall into several structural types:

[46]C_{60} has the same symmetry characteristics as an icosahedron.

[47]These distances were obtained for a twinned crystal of C_{60} at 110 K; S. Liu, Y. Lu, M. M. Kappes, and J. A. Ibers, *Science*, **1991**, *254*, 408.

[48]R. Taylor, J. P. Hare, A. K. Abdul-Sada, and H. W. Kroto, *J. Chem. Soc., Chem. Commun.*, **1990**, 1423.

[49]S. Iijima, *Nature*, **1991**, *354*, 56.

[50]D. Ugarte, *Nature*, **1992**, *359*, 707.

[51]H. L. Anderson, R. Faust, Y. Rubin, and F. Diederich, *Angew. Chem. Int. Ed. Engl.*, **1994**, *33*, 1366.

[52]For a review of metal complexes of C_{60} through 1991, see P. J. Fagan, J. C. Calabrese, and B. Malone, *Acc. Chem. Res.*, **1992**, *25*, 134.

- Complexes containing encapsulated metals. These may contain one, two, or three metals inside the fullerene.[53,54]

  Examples: FeC_{60},[55] UC_{60}, LaC_{82}, Sc_2C_{74}, Sc_3C_{82}

- Complexes in which the fullerene itself behaves as a ligand.[56]

  Examples: $Fe(CO)_4(\eta^2\text{-}C_{60})$, $Mo(\eta^5\text{-}C_5H_5)_2(\eta^2\text{-}C_{60})$,[57] $(C_6H_5)_3P]_2Pt(\eta^2\text{-}C_{60})$,[58]

- Adducts to the oxygens of osmium tetroxide.

  Example: $C_{60}(OsO_4)(4\text{-}t\text{-butylpyridine})_2$[59]

- Intercalation compounds of alkali metals.[60,61] These contain alkali metal ions occupying sites between fullerene clusters (interstitial sites).

  Examples: NaC_{60}, RbC_{60}, KC_{70}

These are conductive, or in some cases superconductive materials and are of great interest in the field of materials science. Since these are principally ionic, rather than covalent in nature, they are not within the scope of this text; the interested reader is encouraged to consult references 60 and 61 for additional information about these compounds.

---

[53]J. R. Heath, S. C. O'Brien, Q. Zhang, Y. Liu, R. F. Curl, H. W. Kroto, F. K. Tittel, and R. E. Smalley, *op. cit.*

[54]H. Shinohara, *et al.*, *J. Phys. Chem.*, **1993**, *97*, 4259.

[55]A compound of this stoichiometry is also known with Fe outside the cage. See T. Pradeep, G. U. Kulkarni, K. R. Kannan, T. N. Guru Row, and C. N. R. Rao, *J. Am. Chem. Soc.*, **1992**, *114*, 2272 and references therein.

[56]P. J. Fagan, J. C. Calabrese, and B. Malone, "The Chemical Nature of C_{60} as Revealed by the Synthesis of Metal Complexes," in G. S. Hammond and V. J. Kuck, Eds., *Fullerenes*, ACS Symposium Series 481, 1992, 177-186.

[57]R. E. Douthwaite, M. L. H. Green, A. H. H. Stephens, and J. F. C. Turner, *J. Chem. Soc., Chem. Commun.*, **1993**, 1522.

[58]P. J. Fagan, J. C. Calabrese, and B. Malone, *Science*, **1991**, *252*, 1160.

[59]J. M. Hawkins, A. Meyer, T. A. Lewis, S. D. Loren, and F. J. Hollander, *Science*, **1991**, *252*, 312.

[60]R. C. Haddon, A. F. Hebard, M. J. Rosseinsky, D. W. Murphy, S. H. Glarum, T. T. M. Palstra, A. P. Ramirez, S. J. Duclos, R. M. Fleming, T. Siegrist, and R. Tycko, "Conductivity and Superconductivity in Alkali Metal Doped C_{60}", in G. S. Hammond and V. J. Kuck, Eds., *Fullerenes*, American Chemical Society, Washington, DC, 1992, 71-89.

[61]R. C. Haddon, *Acc. Chem. Res.*, **1992**, *25*, 127.

Complexes with Encapsulated Metals

These complexes are interesting structural examples of "cage" organometallic complexes, in which the metal is completely surrounded by the fullerene. The first of these to be synthesized was FeC_{60}, reported by Smalley and coworkers shortly after the original synthesis of buckminsterfullerene.[62] Typically, complexes containing encapsulated metals are prepared by laser-induced vapor phase reactions between carbon and the metals; often mixtures of products (in some cases containing different numbers of encapsulated metals, as well as a variety of fullerene cages) are obtained, and separation of these into individual components may prove challenging. Although structural characterization has proven very difficult for most compounds in this category, spectroscopic measurements have supported structures containing central metal cations surrounded by a fulleride, a fullerene that has been reduced. These compounds may be quite stable.

Chemical formulas of fullerene compounds containing encapsulated metals are written using the @ symbol to designate encapsulation. Some examples include:

$Fe@C_{60}$	contains Fe surrounded by C_{60}
$K@C_{59}B$	contains K surrounded by $C_{59}B$[63]
$Sc_3@C_{82}$	contains Sc_3 (possibly in a triangular arrangement) surrounded by C_{82}[64]

This designation indicates structure only and does not include charges on ions that may occur. For example, $La@C_{82}$ is believed to contain La^{3+} surrounded by the C_{82}^{3-} ion.[65]

Fullerenes as Ligands[66]

Viewing the structure of a fullerene such as C_{60}, one is tempted to propose that it might be involved in sandwich compounds, bonding to metals in a penta-hapto fashion, through a five-membered ring (such as in ferrocene) or in a hexahapto fashion, through a six-membered ring (as in dibenzenechromium). However, experiments have yielded no evidence for these types of bonding;

[62]J. R. Heath, S. C. O'Brien, Q. Zhang, Y. Liu, R. F. Curl, H. W. Kroto, F. K. Tittel, and R. E. Smalley, *J. Am. Chem. Soc.*, **1985**, *107*, 7779.

[63]Y. Chai, T. Guo, C. Jin, R. E. Haufler, L. P. F. Chibante, J. Fure, L. Wang, J. M. Alford, and R. E. Smalley, *J. Phys. Chem.*, **1991**, *95*, 7564.

[64]H. Shinohara, *et al.*, *J. Phys. Chem.*, **1993**, *97*, 4259.

[65]K. Kikuchi, S. Suzuki, Y. Nakao, N. Nakahara, T. Wakabayashi, H. Shiromaru, K. Saito, I. Ikemoto, and Y. Achiba, *Chem. Phys. Lett.*, **1993**, *216*, 67.

[66]P. J. Fagan, J. C. Calabrese, and B. Malone, *Acc. Chem. Res.*, **1992**, *25*, 134.

as a ligand, C_{60} appears to behave as an electron–deficient alkene (or arene) and to bond in a dihapto fashion. This type of bonding was observed in the first complex to be synthesized in which C_{60} acts as a ligand toward a metal, $[(C_6H_5)_3P]_2Pt(\eta^2\text{-}C_{60})$,[67] shown in Figure **13-11**.

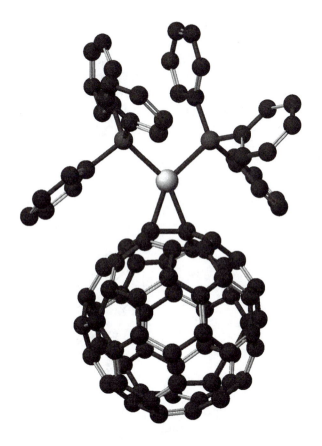

Figure 13-11
Fullerene Complex of Platinum

A common route to the synthesis of complexes involving fullerenes as ligands is displacement of other ligands, typically those weakly coordinated to metals. Examples of such ligands include ethylene (equation **13.24**) and CH_3CN (equation **13.25**):

[67]P. J. Fagan, J. C. Calabrese, and B. Malone, *op. cit.*

$$[(C_6H_5)_3P]_2Pt(\eta^2\text{-}C_2H_4) + C_{60} \rightarrow [(C_6H_5)_3P]_2Pt(\eta^2\text{-}C_{60}) \qquad \textbf{13.24}$$
$$\text{(shown in Figure 13-11)}$$

$$3\ Cp^*Ru(CH_3CN)_3{}^+X^- + C_{60} \rightarrow \{[Cp*Ru(CH_3CN)_2]_3C_{60}\}^{3+}\ (X^-)_3$$
$$\textbf{13.25}$$
$$(X^- = O_3SCF_3{}^-)$$

The C_{60} ligand tends to bond to metals in a dihapto fashion, involving the carbon-carbon bond at the fusion of two six-membered rings, as shown in Figure **13-12**. This is not surprising, since a carbon-carbon bond at the fusion of two six-membered rings is shorter, and thus has more π bonding character, than a bond at the fusion of a five- and six-membered ring.

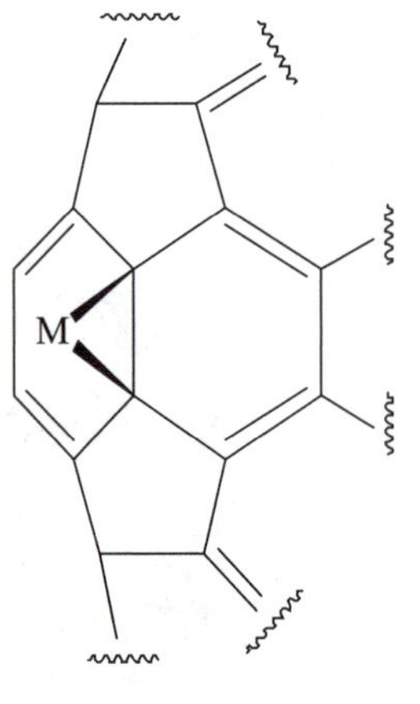

Figure 13-12
Bonding of C_{60} to Metal

Bonding of a fullerene to a metal leads to some distortion at the site of attachment. This is similar to what happens when alkenes bond to metals: the four groups attached to the C=C bond of the alkene bend back away from the metal as a consequence of the donation of d electron density into the $\pi*$ orbital of the alkene (see Chapter 5, Section **5-1-1**). In a similar fashion, the d electron density of the metal can donate to an empty antibonding orbital of a fullerene and cause distortion. In effect, this pulls the two carbons involved slightly away from the C_{60} surface. Since a C=C antibonding orbital is populated as a consequence of this interaction, the distance between these carbons is slightly elongated. This increase in C–C bond distance is analogous to the elongation that occurs when ethylene and other alkenes bond to metals, as discussed in Section **5-1**.

In some cases more than one metal can become attached to a fullerene surface. A spectacular example is $[(Et_3P)_2Pt]_6C_{60}$, shown in Figure **13-13**.[68] In this structure the six $(Et_3P)_2Pt$ units are arranged in an octahedral fashion around the C_{60}. As in the case of other complexes of fullerenes, each platinum is bonded to two carbons at the juncture of two six-membered rings; each of these pairs of carbons is pulled out slightly from the C_{60} surface.

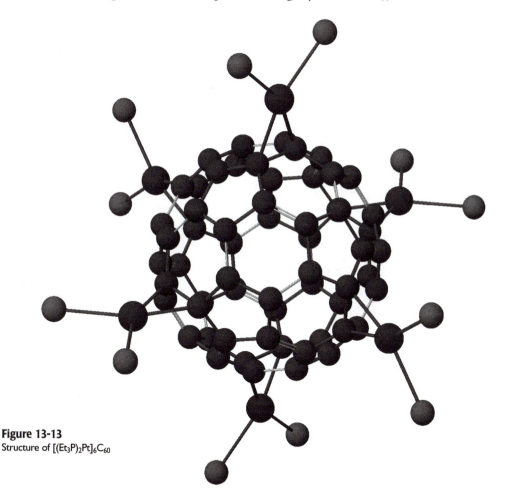

Figure 13-13
Structure of $[(Et_3P)_2Pt]_6C_{60}$

Ethyl groups are not shown in order to enhance picture clarity.

[68]P. J. Fagan, J. C. Calabrese, and B. Malone, *J. Am. Chem. Soc.*, **1991**, *113*, 9408.

More recently, a variety of additional compounds, in which C_{60} appears to function as a classical dihapto organic ligand, have been synthesized.[69] These contain transition metals from a variety of groups; examples of proposed structures are given in Figure **13-14**.

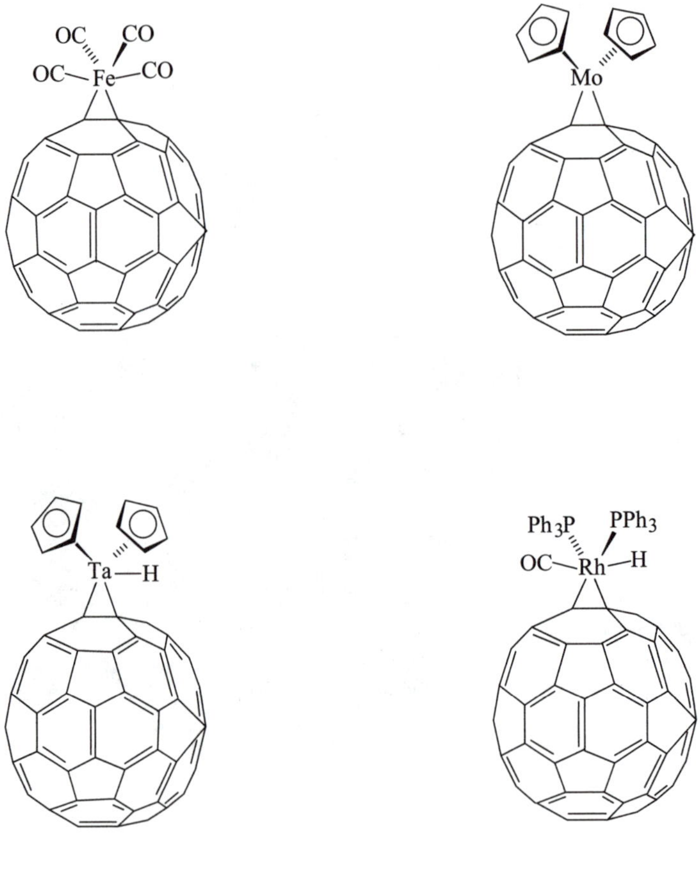

Figure 13-14
Transition Metal Complexes Containing C_{60}

These complexes are formed under very mild conditions (room temperature, atmospheric pressure) from replacement of other ligands such as CO and hydrides. Equations **13.26** and **13.27** provide examples of these reactions:

[69] R. E. Douthwaite, M. L. H. Green, A. H. H. Stephens, and J. F. C. Turner, *J. Chem. Soc., Chem. Commun.*, **1993**, 1522.

$$Fe_2(CO)_9 + C_{60} \rightarrow Fe(CO)_4(\eta^2\text{-}C_{60}) \qquad\qquad \textbf{13.26}$$

$$Ta(\eta^5\text{-}C_5H_5)_2H_3 + C_{60} \rightarrow Ta(\eta^5\text{-}C_5H_5)_2(\eta^2\text{-}C_{60})H \qquad\qquad \textbf{13.27}$$

Although complexes of C_{60} have been studied most extensively, some complexes of other fullerenes have also been prepared. An example is $(\eta^2\text{-}C_{70})Ir(CO)Cl(PPh_3)_2$, shown in Figure **13-15**.[70] As in the case of the known C_{60} complexes, bonding to the metal occurs at the fusion of two six-membered rings.

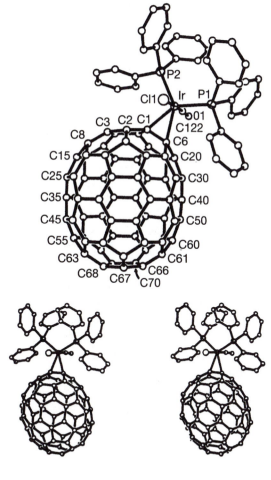

Figure 13-15
Structure of $(\eta^2\text{-}C_{70})Ir(CO)Cl(PPh_3)_2$
(Structures reproduced with permission
from A. L. Balch, V. J. Catalano, J. W. Lee,
M. M. Olmstead, and S. R. Parkin,
J. Am. Chem. Soc., **1991**, *113*, 8953,
Copyright 1991 American Chemical Society.)

[70] A. L. Balch, V. J. Catalano, J. W. Lee, M. M. Olmstead, and S. R. Parkin, *J. Am. Chem. Soc.*, **1991**, *113*, 8953.

Adducts to Oxygens of Osmium Tetroxide[71]

The first pure fullerene derivative to be prepared was $C_{60}(OsO_4)(4-t-butyl-pyridine)_2$.[72] The X-ray crystal structure of this compound provided the first direct evidence that the proposed structure for C_{60} was, in fact, correct. Osmium tetroxide, a powerful oxidizing agent, can add across the double bonds of many compounds, including polycyclic aromatic hydrocarbons. When OsO_4 was allowed to react with C_{60}, polycyclic molecule par excellence, and 4-*tert*-butylpyridine, 1:1 and 2:1 adducts[73] were formed, parallel products to those anticipated in classical organic chemistry. The 1:1 adduct was characterized by X-ray crystallography; its structure is shown in Figure **13–16**.

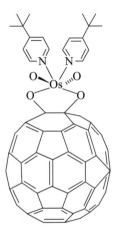

Figure 13-16
Structure of $C_{60}(OsO_4)$
(4-*tert*-butylpyridine)$_2$

As in the case of the fullerene-metal complexes described previously, attachment is at the juncture between two six-membered rings, and the two carbons at the position of attachment protrude from the surface of the C_{60}. The distance from the center of the "sphere" to these carbons is approximately 380 pm in comparison with an average of 351 pm from the center to the other carbons.

---

[71]J. M. Hawkins, *Acc. Chem. Res.*, **1992**, *25*, 150 and references therein.
[72]J. M. Hawkins, A. Meyer, T. A. Lewis, S. Loren, and F. J. Hollander, *op. cit.*
[73]The 2:1 refers to two fullerenes bound to one Os center.

Problems

13-1   Propose a structure for the product (**A**) of the following reaction:

$$Sm + C_8H_8 \ (COT) + PhS–SPh \xrightarrow[\text{THF}]{\text{I}_2 \ (\text{cat.})} A \ (\text{purple prisms})$$

(The I_2 is used to activate the Sm metal surface.)

Elemental analysis of **A** gave an empirical formula of $C_{22}H_{29}O_2SSm$. Partial ^1H-NMR data include the following peaks: δ 11.24 (singlet, 16H), 10.41 (singlet, 4H), 7.99 (singlet, 4H), and 7.83 ppm (singlet, 2H).[74]

13-2   Hydrosilation is a useful reaction that converts an alkene into a silylalkane, as shown below:

$$R–CH=CH_2 + H–SiR'_3 \longrightarrow R–CH_2–CH_2–SiR'_3$$
$$R' = H, \text{alkyl, aryl}$$

Complexes involving several different transition metals catalyze hydrosilation. A recent example (below) shows that lanthanide complexes can also catalyze this transformation.[75]

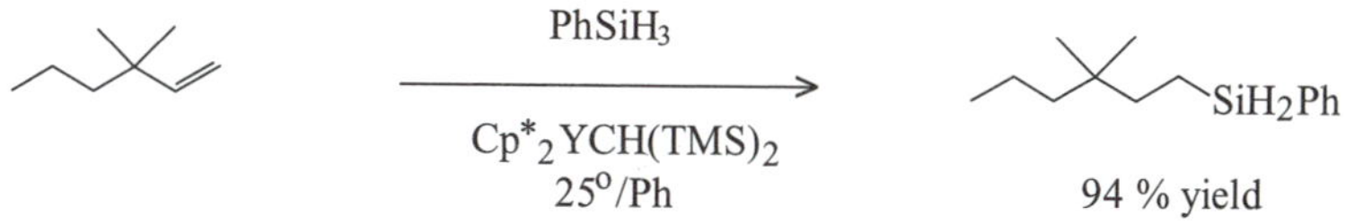

$$TMS = Si(CH_3)_3$$

The authors proposed that the active catalytic species was $Cp^*_2 YH$, formed according to:

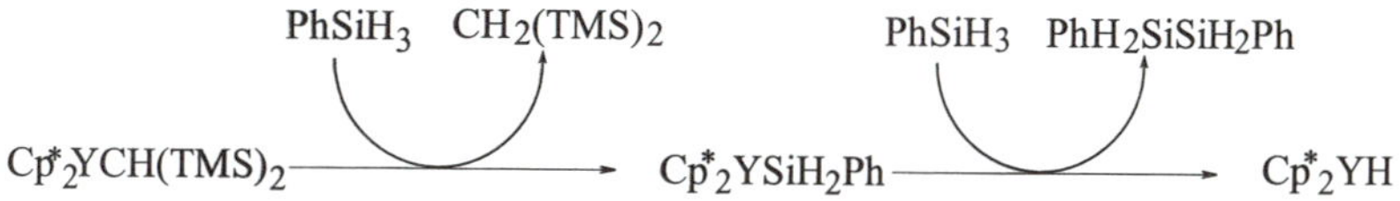

[74]K. Mashima, *et al.*, *J. Chem. Soc. Chem. Commun.*, **1993**, 1847.
[75]G. A. Molander and M. Julius, *J. Org. Chem.*, **1992**, *57*, 6347.

a.   Describe in detail how the steps for the formation of $Cp^*_2 YH$ occur. [Hint: see Chapter 10, Section **10–7**]

b.   Assuming that $Cp^*_2 YH$ is the true catalytic species, propose a catalytic cycle for hydrosilation. Use 1-octene and $PhSiH_3$ as starting materials.

c.   Based upon your answers to parts a and b, predict the product of the following reaction:

$$ + \ PhSiD_3 \quad \xrightarrow[\text{PhH/60}^\circ]{Cp^*_2\,YCH(TMS)_2} $$

13-3   In this chapter the interaction between H_2 parallel to a metal surface (exemplified by an ML_5 fragment) is described. Suppose such an interaction occurred with the H_2 perpendicular to the surface (the H–H bond pointed directly toward a single metal). Describe the types of orbital interactions that would occur. Would you expect the overall interaction to be stronger or weaker than for the case of parallel bonding?[76]

13-4   It has been proposed that a possible step in the mechanism in the preparation of surface ethylidyne ($\equiv$C–CH_3) from acetylene is the formation of surface vinylidene (=C=CH_2). This process could involve a triply bridging site on a metal surface. Show how this might be possible.[77]

13-5   Show how CO and acetylene may bridge metal positions *not* connected by metal-metal bonds to form A-frame structures.[78]

13-6   Reaction of Ir complex **B** with C_{60} gave a black solid residue **C** with the following spectral characteristics: mass spectrum: M^{+} (m/z = 1056); ^{1}H-NMR: δ 7.65 (multiplet, 2H), 7.48 (multiplet, 2H), 6.89 (triplet, 1H, J = 2.7 Hz), and 5.97 ppm (doublet, 2H, J = 2.7 Hz); IR: ν_{CO} = 1998 cm^{-1}.[79]

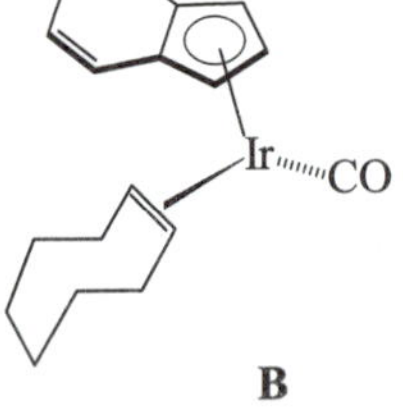

B

[76]J. Y. Saillard and R. Hoffmann, *J. Am. Chem. Soc.*, **1984**, *106*, 2006.

[77]L. L. Kesmodel, L. H. Dubois, and G. A. Somorjai, *J. Chem. Phys.*, **1979**, *70*, 2180.

[78]See D. M. Hoffman and R. Hoffmann, *Inorg. Chem.*, **1981**, *20*, 3543.

[79]R. S. Koefod, M. F. Hudgens, and J. R. Shapley, *J. Am. Chem. Soc.*, **1991**, *113*, 8957.

a.   Propose a structure for **C**.

b.   The carbonyl stretching frequency of **B** was reported as 1954 cm^{-1}. How does the electron density at Ir change in going from **B** to **C**? Explain.

c.   When **C** was treated with PPh$_3$, a new complex **D** rapidly formed along with some C$_{60}$. What is the structure of **D**?

d.   When **C** was treated with H$_2$C=CH$_2$, a new complex **E** formed along with C$_{60}$. This time, however, the rate of formation of **E** was much slower than that for **D**. What is the structure of **E**? Why does it form much more slowly than **D**?

Appendix A

List of Abbreviations

—□	Empty coordination site
↔	Isolobal
I°, II°, III°	Primary, secondary, tertiary
α	Position on carbon chain attached to metal; α denotes carbon directly attached to metal
β	Position on carbon chain attached to metal; β denotes one carbon beyond α position
η^n	Descriptor of hapticity (the number of binding sites on a ligand); the superscript designates the number of binding sites
μ	Descriptor for bridging ligand(s)
ν	Frequency
A	Associative substitution
Ar	Aryl
Bu	Butyl (previously known as n-butyl)

t-Bu	*tert*-Butyl or *tert*-Bu
CO	Carbonyl ligand
COD (or cod)	1,5-Cyclooctadiene
COT (or cot)	1,3,5,7-Cyclooctatetraene
Cp	Cyclopentadienylide ion, $C_5H_5^-$
Cp*	Pentamethylcyclopentadienylide ion, $C_5(CH_3)_5^-$
Cy	Cyclohexyl
D	Dissociative substitution
DMF	*N,N*-Dimethylformamide
DMSO	Dimethylsulfoxide
d^n	Formal *d*-electron configuration
E or E^+	Electrophile
ee	Percent enantiomeric excess
Et	Ethyl
Et_2O	Ethyl ether
EtOH	Ethanol
eu	Entropy units (cal/mole K)
en	Ethylenediamine
fac	Facial stereochemistry (three ligands occupy the same face of an octahedral complex)
Fp	$Cp(CO)_2$ Fe fragment (pronounced "fip")

HMPA	Hexamethylphosphoric triamide or commonly hexa ethylphosphoramide, $(Me_2N)_3P\text{-}O$
HOMO	Highest occupied molecular orbital
I	Interchange
IR	Infrared
L	A generalized ligand, most often a 2-e^- neutral ligand
LDA	Lithium diisopropylamide
L_nM	A generalized metal complex fragment with n L-type ligands attached
Ln	Lanthanide
LUMO	Lowest unoccupied molecular orbital
M	Central metal in a complex
Me	Methyl
MeOH	Methanol
mer	Meridional stereochemistry (three ligands arranged to define a plane through an octahedral complex)
MO	molecular orbital
NMR	Nuclear magnetic resonance
Nuc or Nuc:$^-$	Nucleophile
OA	Oxidative addition
OAc^-	Acetate ion
O.S.	Oxidation state
OMs	Mesylate, $CH_3SO_3^-$

OTf	Triflate, $CF_3SO_3^-$
OTs	Tosylate, p-CH_3–C_6H_4–SO_3^-
PE	Polyethylene
Ph	Phenyl
pm	Picometer
PP	Polypropylene
Pr	Propyl
PR_3	Phosphine
Py	Pyridine
R	Usually a generalized alkyl group
RE	Reductive elimination
ROMP	Ring opening metathesis polymerization
S	Solvent
SBM	Sigma bond metathesis
SHOP	Shell higher olefin process
THF or thf	Tetrahydrofuran
TMS	Trimethylsilyl, $(CH_3)_3Si$
X	A generalized X-type ligand (1 e^- in the neutral ligand counting method)
Z-N	Ziegler-Natta

Appendix B

Answers to Exercises

Chapter 2

2-1

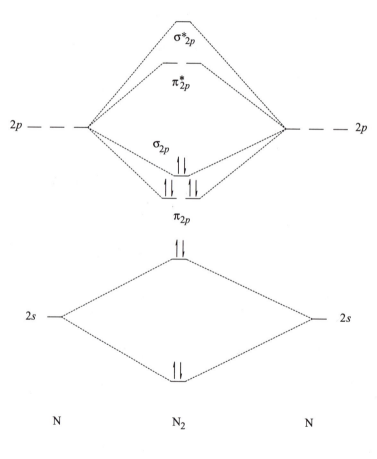

Bond order = $\frac{1}{2}(8-2)$ = 3 (a triple bond)

2-2   The MO diagram is similar to that of N_2. Isoelectronic with N_2, C_2^{2-} also has bond order = 3.

2-3 The group orbitals are the same as in CO_2 (see Figure **2-7**).

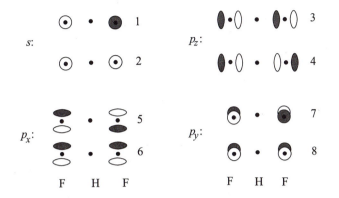

2-4 Again, the group orbitals are the same as in CO_2. Also, the central atom/group orbital interactions are the same as shown for CO_2. In the case of N_3^-, the central atom and group orbitals have the same energy, of course. For example, the $2s$ interactions (group orbital 2) are:

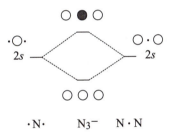

2-5

Relative Energy

Chapter 3

3-1 **a.** $[Fe(CO)_4]^{2-}$

Fe	$8e^-$
4 (CO)	$8e^-$
2^- charge	$\underline{2e^-}$
	$18e^-$

b. $[(\eta^5\text{-}C_5H_5)_2Co]^+$

Co	$9e^-$
2 $(\eta^5\text{-}C_5H_5)$	$10e^-$
+ charge	$\underline{-1e^-}$
	$18e^-$

c. $(\eta^3\text{-}C_5H_5)(\eta^5\text{-}C_5H_5)Fe(CO)$

Fe	$8e^-$
$\eta^3\text{-}C_5H_5$	$3e^-$
$\eta^5\text{-}C_5H_5$	$5e^-$
CO	$\underline{2e^-}$
	$18e^-$

3-2 **a.** $[M(CO)_3(PPh_3)]^-$

M	?
3 (CO)	$6e^-$
PPh$_3$	$2e^-$
	$\underline{1e^-}$
	$18e^-$

To give a total of 18, M must have 9 electrons outside its noble gas core: **Co.**

b. $HM(CO)_5$

M	?
H	$1e^-$
5 (CO)	$\underline{10e^-}$
	$18e^-$

M must have 7 electrons outside its noble gas core: **Mn.**

c. $(\eta^4\text{-}C_8H_8)M(CO)_3$

M	?
$\eta^4\text{-}C_8H_8$	$4e^-$
3 (CO)	$\underline{6e^-}$
	$18e^-$

M needs 8 electrons outside its noble gas core: **Fe**.

d. $[(\eta^5\text{-}C_5H_5)M(CO)_3]_2$

M	?
η^5-C_5H_5	$5e^-$
3 (CO)	$6e^-$
M–M bond	$1e^-$
	$18e^-$

M must have 6 electrons outside its noble gas core: **Cr**.

3-3 $[Ir(CO)(PPh_3)_2(Cl)(NO)]^+$ corresponds to $\mathbf{ML_4X_2^+}$ if NO is linear and to $\mathbf{ML_3X_2^+}$ if NO is bent.

EAN = 9 (Ir) + 8 (4 L-type ligands) + 2 (2 X-type ligands) − 1 (charge) = **18** (NO is linear)

or

EAN = 9 + 6 (3 L-type ligands) + 2 − 1 = **16** (NO is bent)

CN = 3 + 2 = **5** (NO always contributes 1 to CN regardless of whether it is linear or bent.)

OS = 2 + 1 = **3** regardless of whether NO is linear or bent according to the formula. It should be pointed out, however, that conversion of a linear NO complex to a bent one is considered to be a formal increase in oxidation state of +2. Therefore, the bent complex should have an oxidation state of +2 higher than the linear one.

d^n = 9 (Ir is in Group 9) - 3 (OS) = **6**

3-4

Ni	$10e^-$	Pt	$10e^-$
4 (CN)	$4e^-$	2 Cl	$2e^-$
2^-	$2e^-$	en	$4e^-$ (ethylenediamine)
	$16e^-$		$16e^-$

Rh	$9e^-$	Ir	$9e^-$
Cl	$1e^-$	CO	$2e^-$
3 (PPh₃)	$6e^-$	Cl	$1e^-$
	$16e^-$	2 (PPh₃)	$4e^-$
			$16e^-$

Chapter 4

4-1 N_2 would be expected to be a *weaker* π-acceptor than CO. In N_2, the π^* orbital is distributed evenly over both atoms; in CO, the π^* is concentrated on the C, making it more available to π-accept.

4-2 Choose whichever you like; they should all check out as 18-electron complexes.

Note: Be sure that bridging CO donates one electron to each metal bridged. For example, for each Co in the solid form of $Co_2(CO)_8$:

Co	$9e^-$
Co-Co bond	$1e^-$
3 (CO) (terminal)	$6e^-$
2 (CO) (bridging)	$\underline{2e^-}$
	$18e^-$

Chapter 5

5-1 **2-Node group orbitals**

d_{xy} none $d_{x^2-y^2}$ none

1-Node group orbitals

p_y d_{yz} p_x d_{xz}

0-Node group orbitals

s, d_{z^2} p_z

5-2

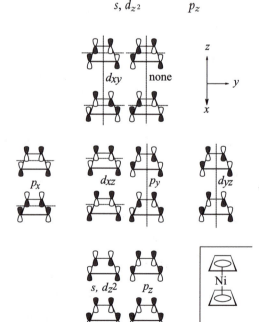

5-3

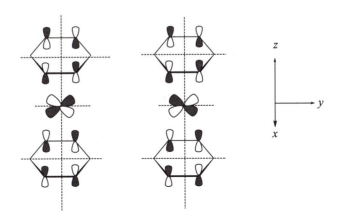

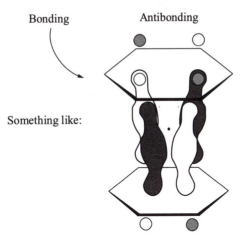

Bonding   Antibonding

Something like:

5-4 a. 2^+
         b. 1^+
         c. 2^+

5-5 a. Cr
         b. Mn

5-6 The relative energy levels of $C_8H_8^{2-}$ may be obtained using the Frost circle trick, as shown below. The MOs are obtained by adding and subtracting butadiene MOs.

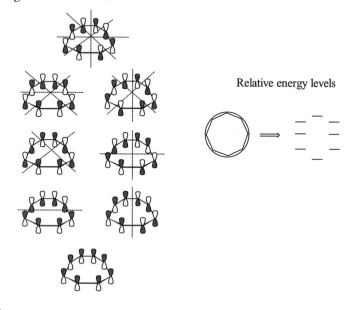

Relative energy levels

Chapter 6

6-1 **a.** Mn
  b. Fe
  c. Os

6-2 **a.** Ta
  b. Os

6-3 $Mo(PMe_3)_5H_2$ has two H ligands.
  $Mo(CO)_3(PR_3)_2H_2$ has one H_2 ligand (R = isopropyl).
  PMe_3 is a stronger donor (and weaker acceptor) than CO. The consequence is that Mo in $Mo(PMe_3)_5H_2$ has a greater concentration of electrons and a greater tendency to back bond to the hydrogens. This tendency is sufficiently strong to rupture any H–H bonds (donation to σ^* of H_2, as shown in Figure **6-6**), converting H_2 into separate H ligands.

6-4 As the compound is oxidized, the Co becomes less able to back donate to the acceptor MO of the phosphine. This means less occupation of the P–C antibonding orbital, thus a shorter P–C distance. Since there is less Co–P back bonding (a bonding interaction between Co and P), the Co–P distance increases.

Chapter 7

7-1

a. $[PtCl_3(NH_3)]^-$ $\xrightarrow{\ Br^-\ }$

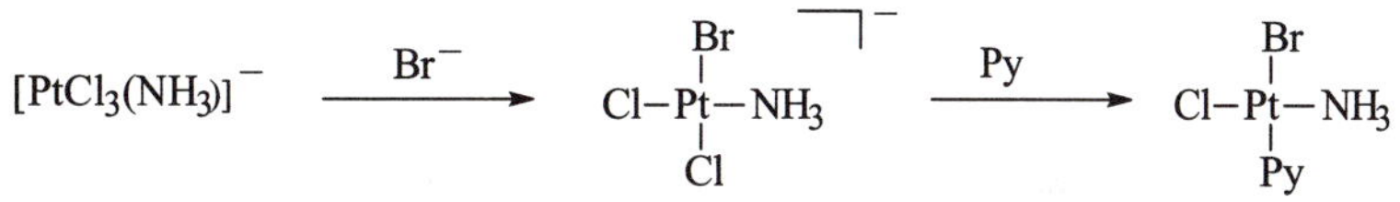

b.

$PtCl_4{}^{2-}$ $\xrightarrow{\ NO_2{}^-\ }$

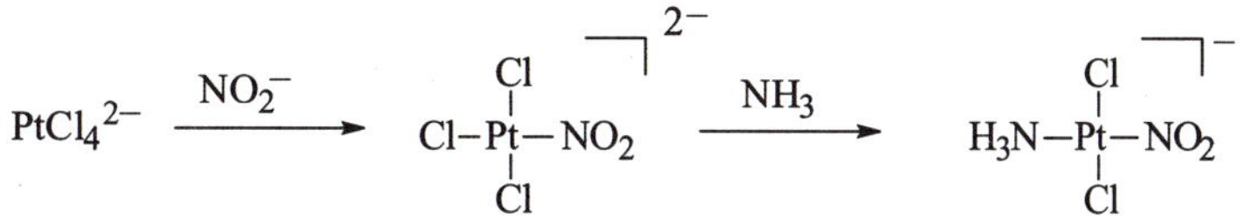

7-2   **a.**

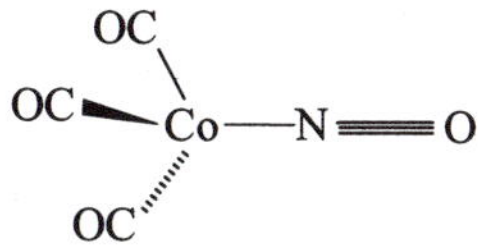

9 (Co) + 3 (linear NO) + 6 (3 CO) = 18 electrons

b. Starting material		**Transition state**	
Fe	$8e^-$	Fe	$8e^-$
3 CO	$6e^-$	3 CO	$6e^-$
		PMe$_3$	$2e^-$
Me$_2$N$_4$	$4e^-$ (an L$_2$ ligand)	Me$_2$N$_4$	$2e^-$ (an X$_2$ ligand)
	$\overline{18e^-}$		$\overline{18e^-}$

Product

Fe	$8e^-$
2 CO	$4e^-$
PMe$_3$	$2e^-$
Me$_2$N$_4$	$4e^-$ (an L$_2$ ligand)
	$\overline{18e^-}$

7-3 Equation **7.24**

Reactant: $[Ir(CO)_3(PMePh_2)_2]^+ \equiv [ML_5]^+$; OS = 0 + 1 = 1
Product: $[Ir(CO)_2(H)_2(PMePh_2)_2]^+ \equiv [MX_2L_4]^+$; OS = 2 + 1 = 3

Equation **7.26**

Reactant: $Ir(PPh_3)_3(Cl) \equiv [MXL_3]^0$; OS = 1 + 0 = 1

Product: 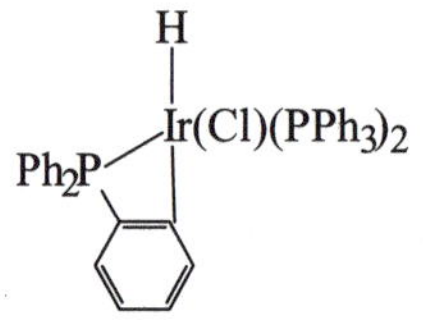
H
Ir(Cl)(PPh3)2
Ph2P

$\equiv [MX_3L_3]^0$; OS = 3 + 0 = 3

In both equations, the oxidation state has increased by two units. The coordination number has also increased in both cases. These two phenomena are characteristic of oxidative additions.

7-4 The answer lies with the *trans* effect. Once the intermediate forms, the next ligand exchange occurs *trans* to the ligand with the strongest *trans* effect; this ligand is CO. The two CO ligands that are *trans* to each other in the intermediate also labilize each other. Replacement of either with X gives the resulting *cis*-product.

7-5 The mechanism can be outlined as follows:

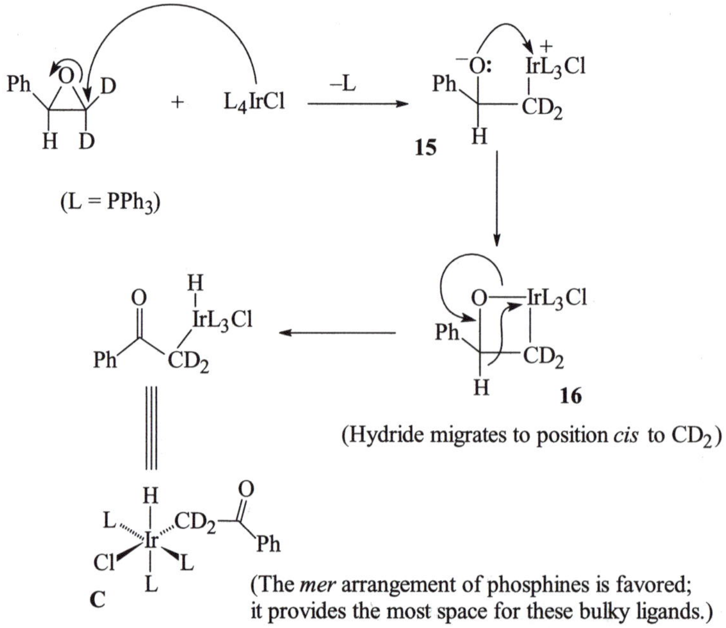
Ph
O
D
H D
+
L4IrCl
–L
(L = PPh3)
Ph
O:
+
IrL3Cl
CD2
H
15
O
H
IrL3Cl
Ph
CD2
O
IrL3Cl
Ph
CD2
H
16
(Hydride migrates to position cis to CD2)
H
L
CD2
O
Ir
Ph
Cl
L
L
C
(The mer arrangement of phosphines is favored;
it provides the most space for these bulky ligands.)

7-6

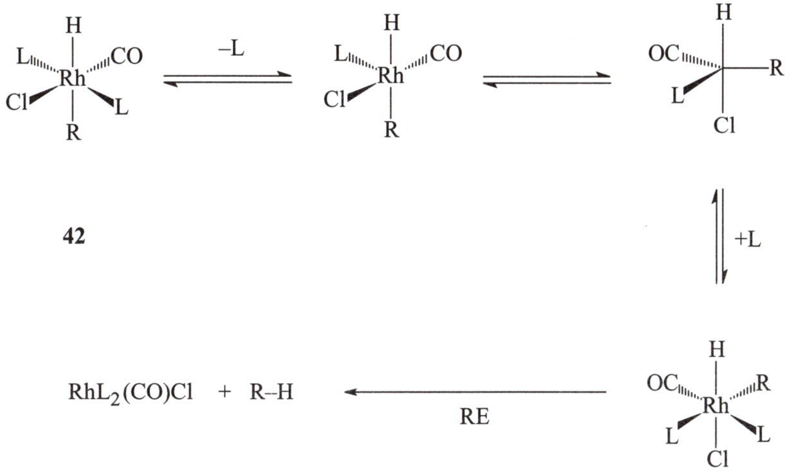

42   +L

$$RhL_2(CO)Cl \quad + \quad R\text{–}H \quad \longleftarrow \quad \underset{RE}{}$$

Chapter 8

8-1   Mechanism **1** (reverse reaction):

(* = ^{13}C)

Since the forward reaction involves direct CO insertion (from the outside) to form the acyl ligand, the reverse reaction involves loss of the acyl CO (in this case containing the ^{13}C).

8-2   *cis*-CH$_3$Mn(CO)$_4$(^{13}CO) + PR$_3$ (R = C$_2$H$_5$)

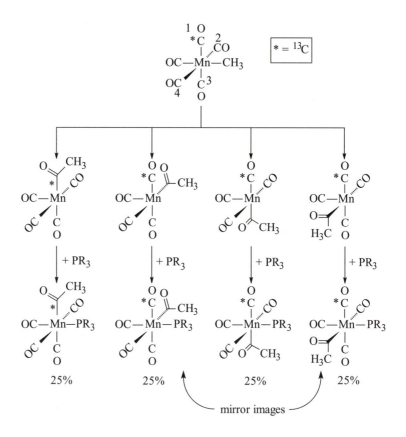

8-3

B

Note: Other isomers are possible that maintain a *cis* relationship between CO and the benzyl group.

8-4   C-F bonds in the α position make the M-C bond strong because the C-F σ* orbital is low enough in energy to accept electrons from the metal via back donation. A strong C-M σ bond is less likely to participate in β-elimination.

M–F bonds in the late transition metals are relatively weak. Even if β–elimination occurred as shown in the following, there would not be a good thermodynamic driving force toward the right-hand side of the equation.

8-5

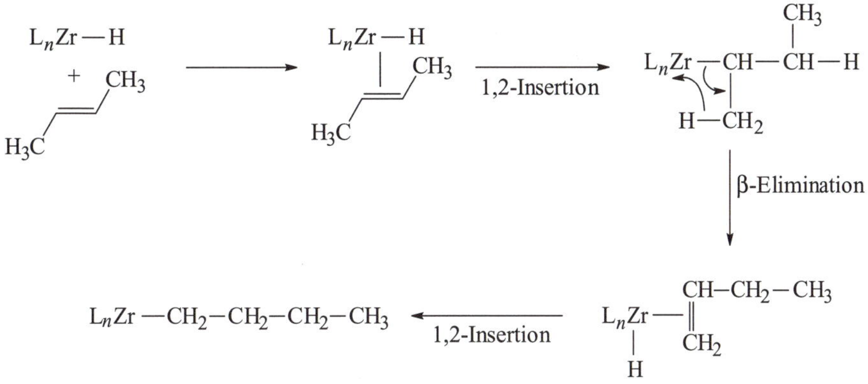

8-6

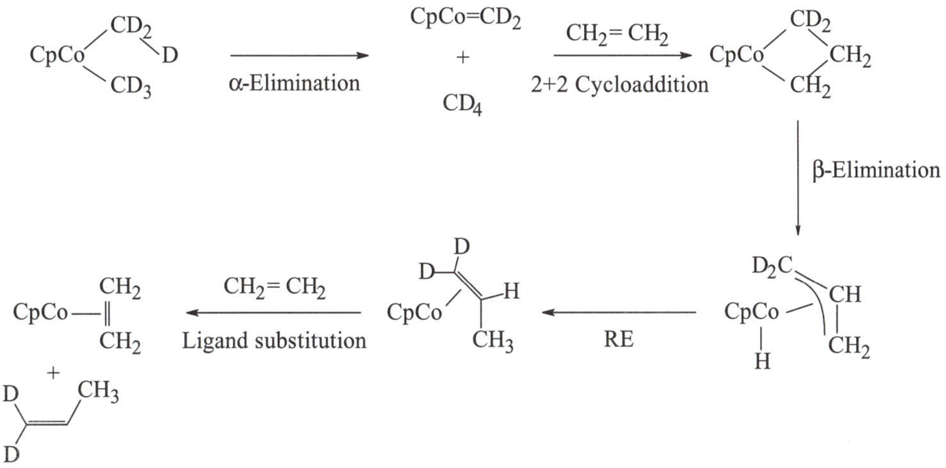

8-7

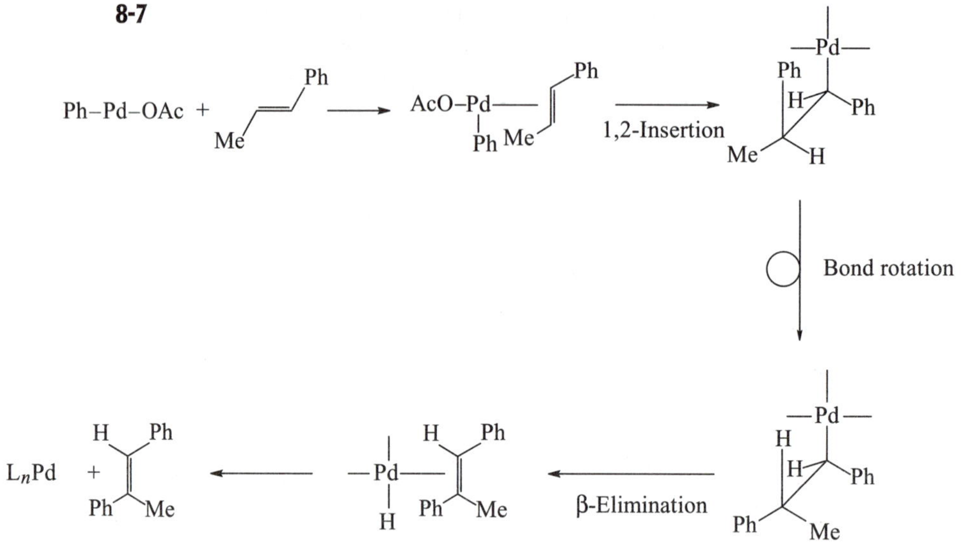

8-8

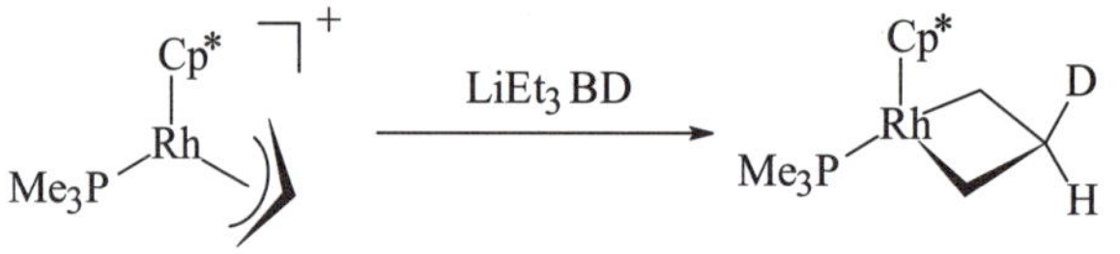

Because the complex is positively charged, it is possible that the hydride adds to the metal first. Once attached to the Rh, it could then transfer to the η^3 allyl on the same side of the allyl ligand as the metal. In order to do this, the η^5-Cp* ligand would have to slip to η^3 before final bond reorganization gives the expected product. Otherwise, attack of H$^-$ (or D$^-$) at the metal would give a 20-electron complex.

Another possibility would be for PMe$_3$ to dissociate before H$^-$ (D$^-$) attack, as shown in the following:

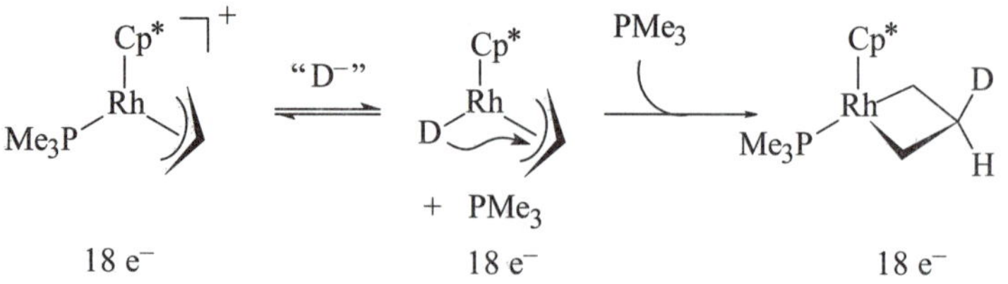

8-9 Since **26** is achiral, the alkyl halides will exist as a pair of enantiomers—a racemic modification.

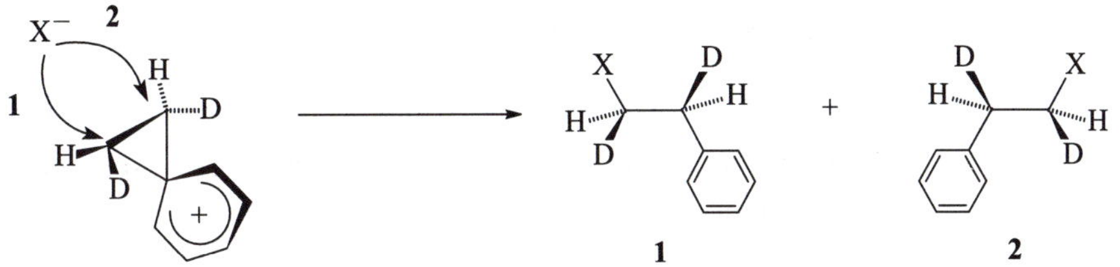

26 **1** **2**

Enantiomers

8-10 Starting with the 1,1-dideuterio Fe complex:

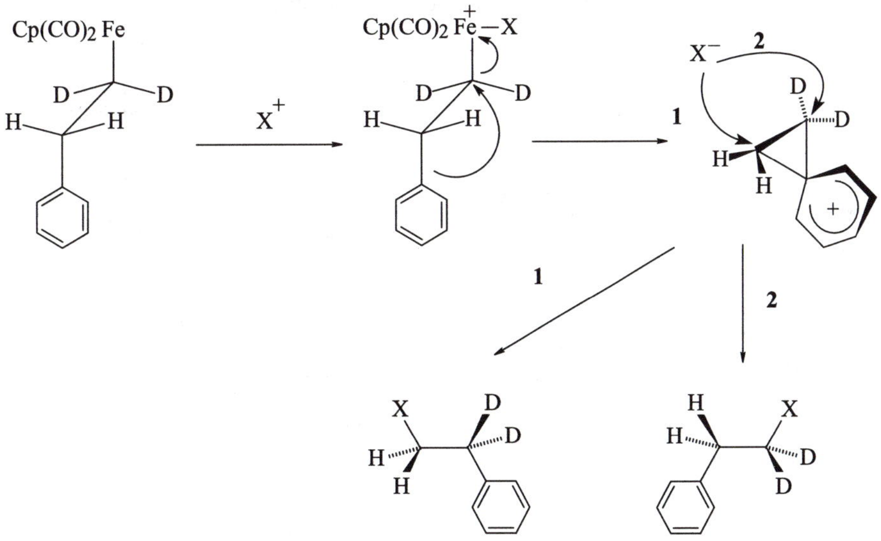

1 **2**

Constitutional isomers

Chapter 9

9-1   Increasing the concentration of PR_3 slows the reaction that produces the active catalyst:

$$H\text{-}(CO)RhL_3 \;\rightleftharpoons\; H(CO)RhL_2 + L$$
$$L = PR_3$$

On the other hand, excess PR_3 probably causes displacement of CO to give the following:

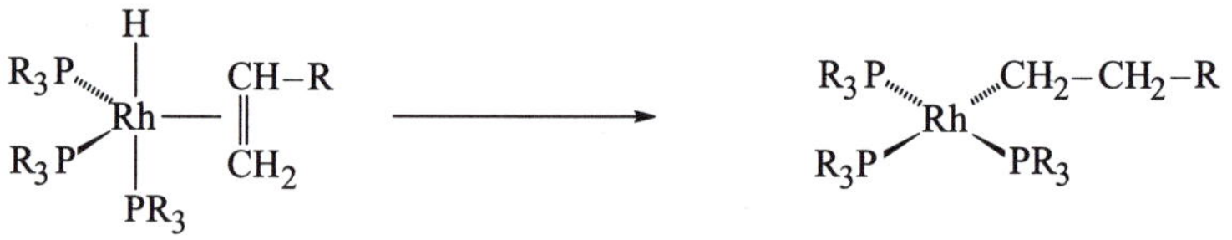

The increase in steric hindrance about the metal forces insertion to occur such that the metal is attached to the less substituted carbon.

9-2   The first step is as shown in Scheme **9.5**. Then the following occurs:

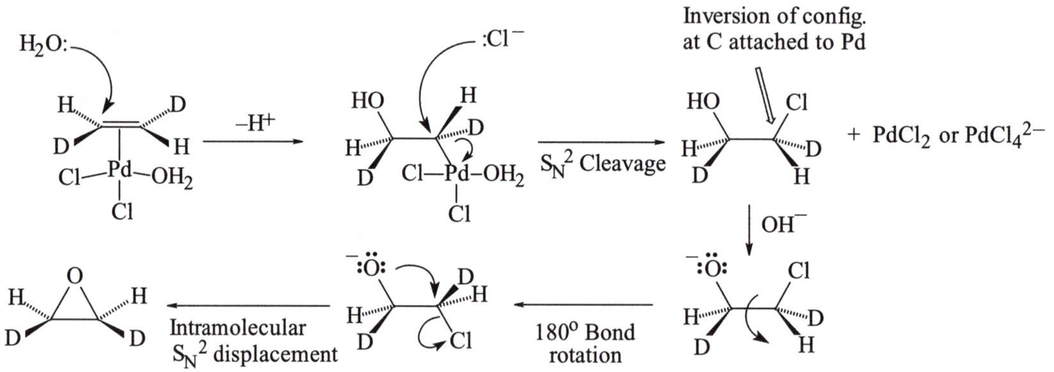

Inversion of configuration must occur at the carbon attached to Pd. If intramolecular 1,2-insertion occurred, the stereochemistry of **15** would be *trans*.

9-3

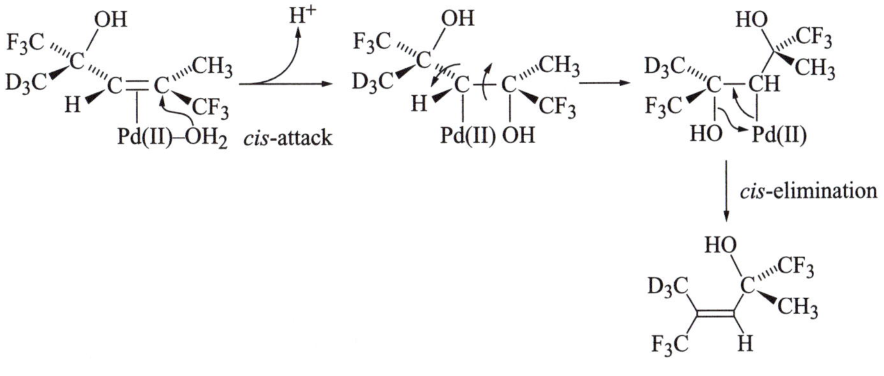

(S)-17

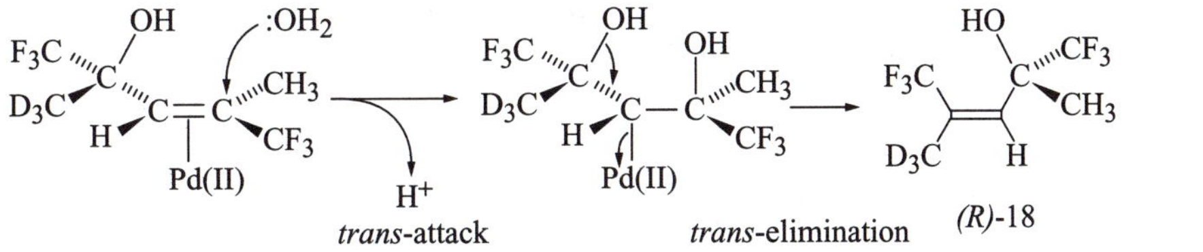

trans-attack   trans-elimination   (R)-18

9-4

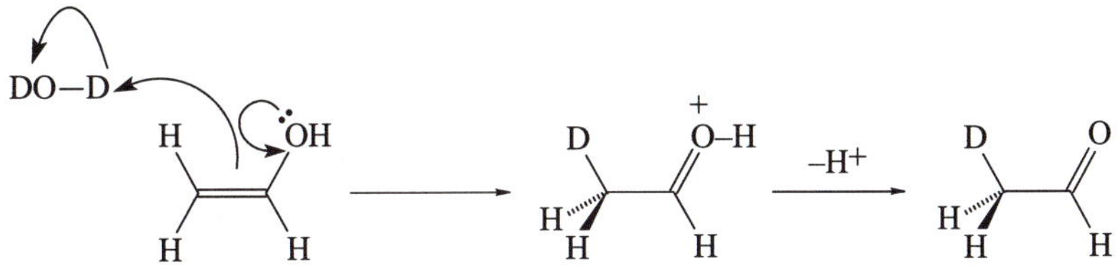

9-5   Complexation of the alkene is followed by attack of H_2O at the more substituted position.

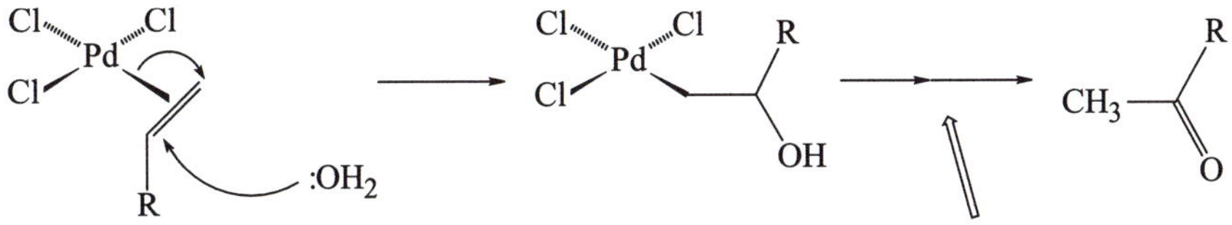

R = nonyl

As shown in Scheme **9.5**

9-6

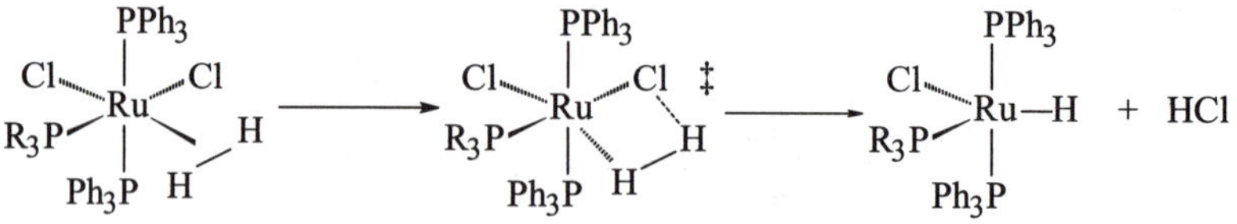

This is an example of σ bond metathesis. See Section **10-7**.

9-7

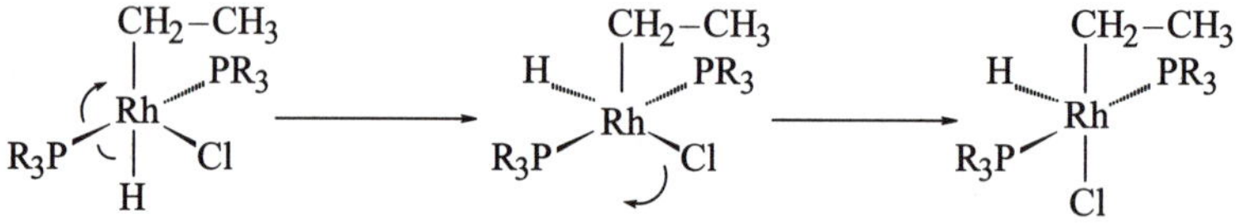

The reaction is energetically reasonable because it removes a H⁻ ligand from the position *trans* to Et (a ligand with a high *trans* influence) to the position *cis* to Et. Chloride ion, a ligand with weak *trans* influence, then moves *trans* to Et.

9-8 The rate-determining step involves OA of 1-iodopropane or 2-iodopropane to $[Rh(CO)_2I_2]^-$. The mechanism of this step is probably not S_N2, but rather one involving alkyl radicals. Since II° radicals are more stable than I°, 2-iodopropane (derived from 2-propanol) reacts faster than 1-iodopropane.

9-9 The bipyridyl ligand is rigid; loss of one binding interaction from bipyridyl to Ir is unlikely because of this rigidity and because of the chelate effect. Cp^* ligands also bind tightly to metals, and their loss is unlikely (slippage of Cp^* from η^5 to η^3 is possible, however).

9-10 In equation **9.47**, loss of CO_2 increases electron density on Ir. This process is facilitated by having electron withdrawing groups on bipyridyl. The greater the electron withdrawing ability of the X groups, the less the electron density available at the nitrogen atoms. Less electron density at nitrogen means poorer σ donating capability and therefore less electron density at Ir. Loss of CO_2 remediates the lack of electron density at Ir when X is electron withdrawing. Thus, the greater the electron withdrawing ability of X, the faster the loss of CO_2.

9-11

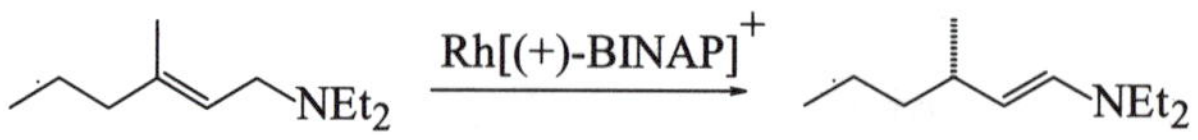

Chapter 10

10-1

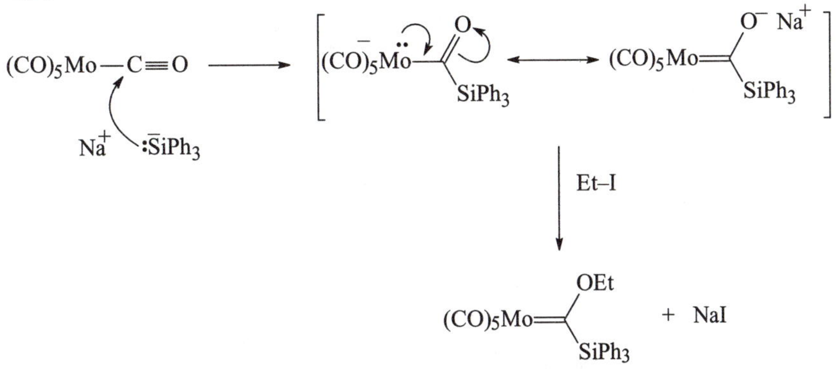

$(CO)_5Mo-C\equiv O$ → $\left[(CO)_5\overset{-}{Mo}-\overset{O}{\underset{SiPh_3}{C}} \longleftrightarrow (CO)_5Mo=\overset{O^- Na^+}{\underset{SiPh_3}{C}} \right]$

$Na^+ \quad :\bar{S}iPh_3$

$\Big\downarrow$ Et–I

$(CO)_5Mo=\overset{OEt}{\underset{SiPh_3}{C}}$ +   NaI

10-2

The first step of equation **10.6** involves first alkyl migration to CO followed by coordination of I⁻ to the metal.

10-3

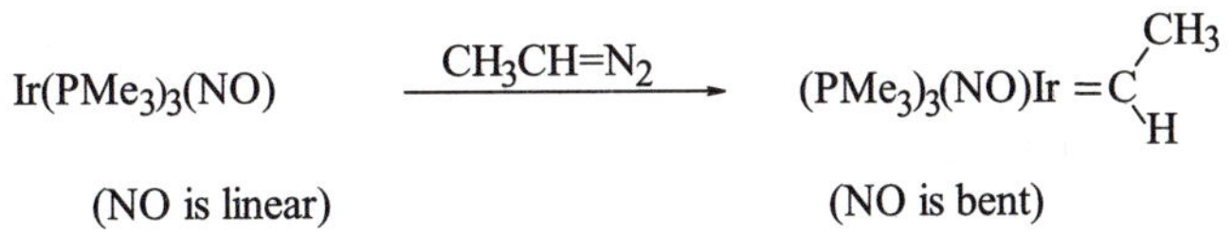

$Ir(PMe_3)_3(NO)$ $\xrightarrow{\ CH_3CH=N_2\ }$ $(PMe_3)_3(NO)Ir = C\overset{CH_3}{\underset{H}{\diagup}}$

(NO is linear)   (NO is bent)

10-4

$(CO)_5W = \overset{O}{\diagdown}$ $\xrightarrow{\ Bu–Li\ }$ $(CO)_5W = \overset{O}{\underset{Li^+}{\diagdown}}$ $\xrightarrow{\ Br\diagup\diagdown\diagup Br\ }$ $(CO)_5W = \overset{O}{\underset{Br}{\diagdown}}\;H$ + LiBr

A

$\Big\downarrow$ Bu–Li

$(CO)_5W = \overset{O}{\diagdown}$ ← $(CO)_5W = \overset{O}{\underset{Br}{\diagdown}}\; :^- Li^+$

B

+ LiBr

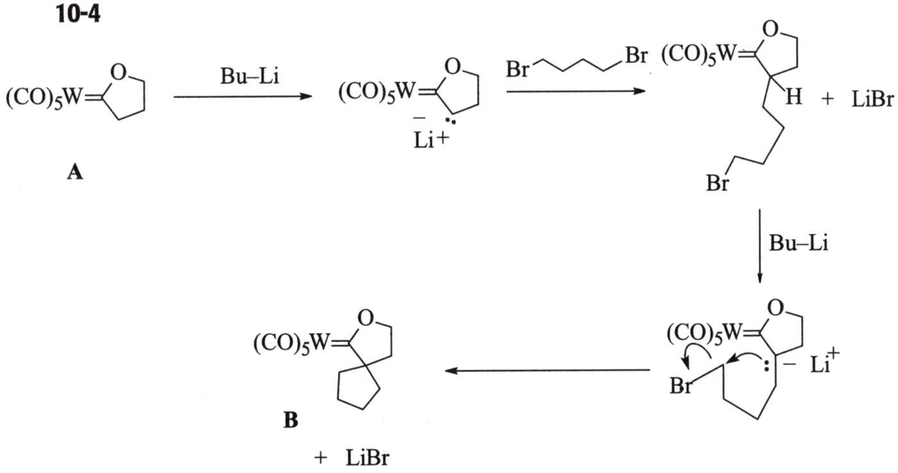

10-5

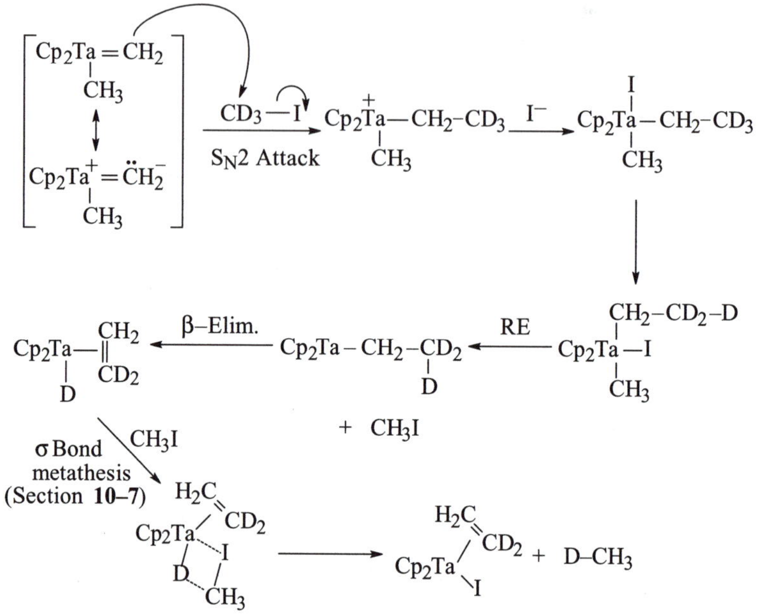
Cp2Ta=CH2
CH3
Cp2Ta+=CH2−
CH3
CD3—I
SN2 Attack
Cp2Ta—CH2-CD3
CH3
I−
I
Cp2Ta—CH2-CD3
CH3
CH2-CD2-D
Cp2Ta—I
CH3
RE
Cp2Ta-CH2-CD2
D
+ CH3I
β-Elim.
CH2
Cp2Ta—
CD2
D
σ Bond
metathesis
(Section 10–7)
CH3I
H2C
CD2
Cp2Ta
I
D
CH3
H2C
CD2
Cp2Ta
I
+ D-CH3

10-6

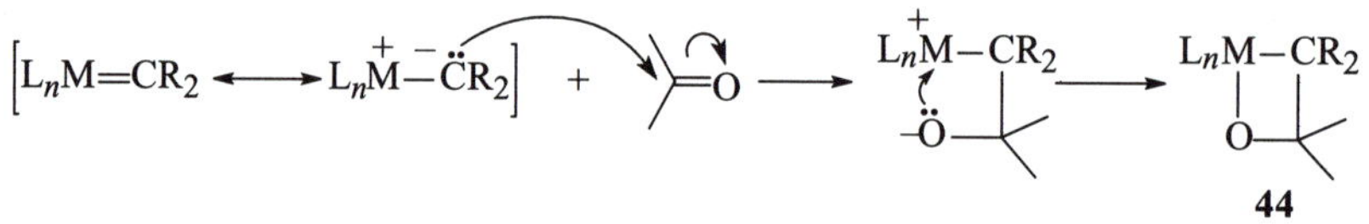
LnM=CR2
LnM—CR2
+
O
LnM—CR2
O
LnM—CR2
O
44

10-7

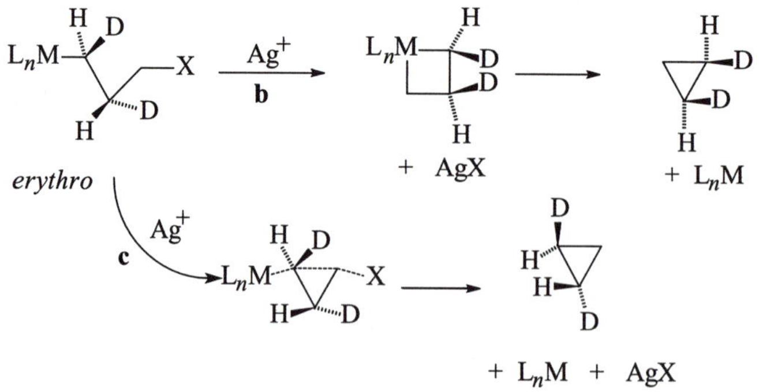
H D
LnM
X
D
H
erythro
Ag+
b
LnM
H
D
D
H
+ AgX
H
D
D
H
+ LnM
Ag+
c
H D
LnM
X
H
D
D
H
H
D
+ LnM + AgX

10-8

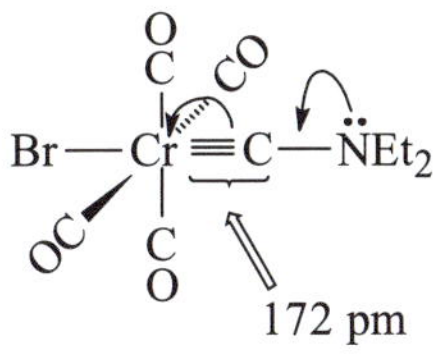

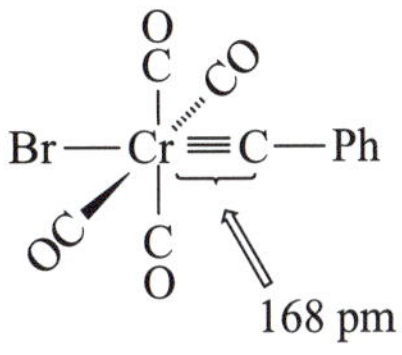

172 pm

168 pm

Resonance structures:

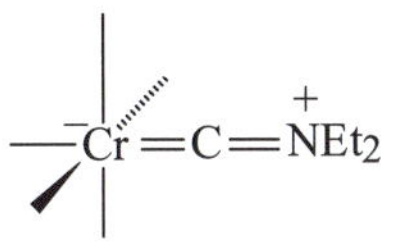

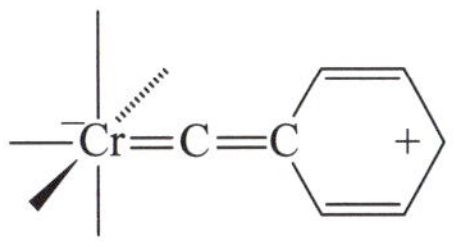

This structure is more stable than the analogous resonance structure of the phenyl compound. The availability of the lone pair on nitrogen yields a resonance structure in which the necessary octets are satisfied.

Structures such as this one are not as satisfactory, since one carbon is short of an octet.

10-9

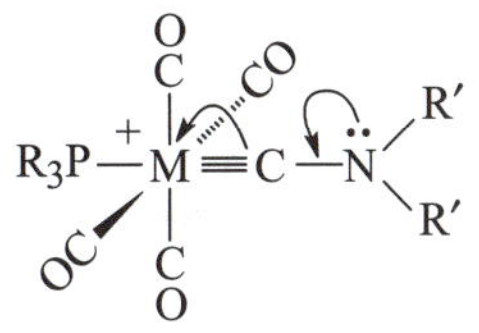

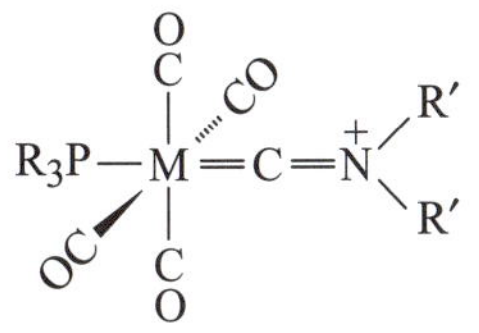

The cationic carbyne is resonance stabilized.

10-10 Metathesis involving 2-pentene and 2-hexene:

C–C=C–C–C   C–C=C–C–C–C
2-pentene   2-hexene

Possible alkenes:

C–C=C–C   2-butene
C–C=C–C–C   2-pentene
C–C=C–C–C–C   2-hexene
C–C–C=C–C–C   3-hexene
C–C–C=C–C–C–C   3-heptene
C–C–C–C=C–C–C–C   4-octene

10-11

Equation **10.46**:

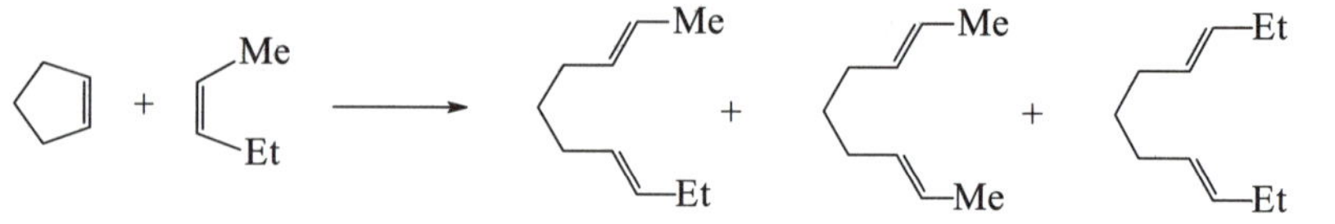

64   **65**   **66**

Pairwise mechanism:

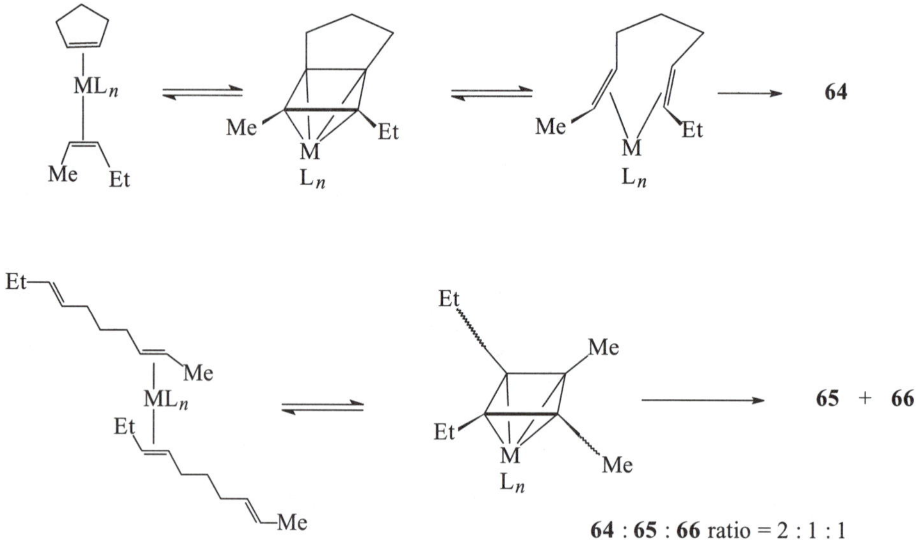

64 : **65** : **66** ratio = 2 : 1 : 1

10-12

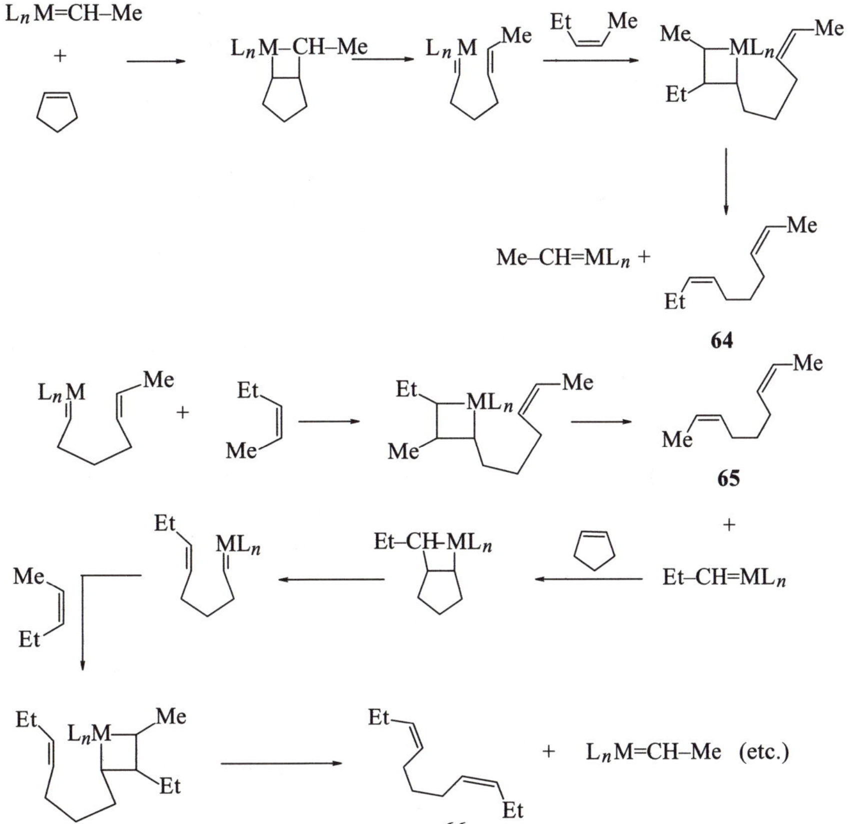
L_nM=CH–Me
+
L_nM–CH–Me
L_nM
Me
Et
Me
Me
ML_n
Me
Et
Me–CH=ML_n +
Me
Et
64
L_nM
Me
+
Et
Me
Et
ML_n
Me
Me
Me
Me
65
+
Et–CH=ML_n
Et–CH-ML_n
Et
ML_n
Me
Et
Et
L_nM
Me
Et
Et
Et
66
+ L_nM=CH–Me (etc.)

10-13

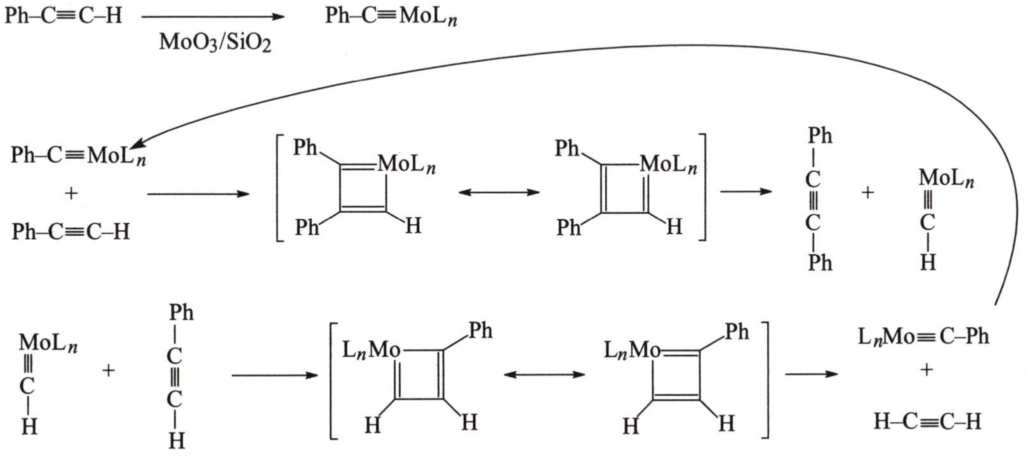
Ph–C≡C–H
MoO_3/SiO_2
Ph–C≡MoL_n
Ph–C≡MoL_n
+
Ph–C≡C–H
Ph
MoL_n
Ph
H
Ph
MoL_n
Ph
H
Ph
C
C
Ph
+
MoL_n
C
H
MoL_n
C
H
+
Ph
C
C
H
L_nMo
Ph
H
H
L_nMo
Ph
H
H
L_nMo≡C–Ph
+
H–C≡C–H

10-14 The Green-Rooney mechanism requires an α-elimination. Were it to occur, there should be a preference for polymerization of $CH_2=CH_2$ over $CD_2=CD_2$ because the rate of reaction for α-elimination is slower for:

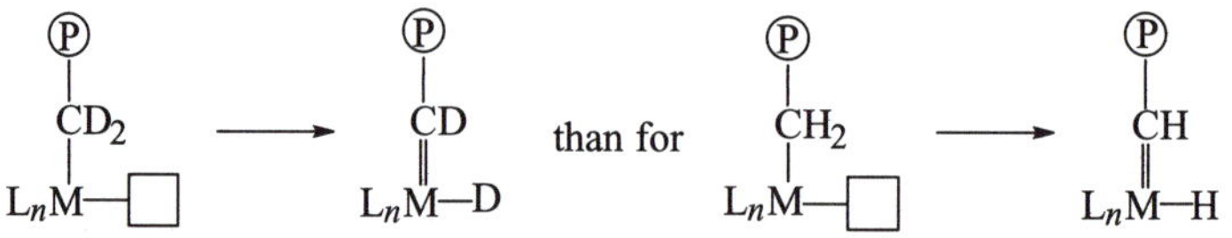

$$
\begin{array}{ccc}
\underset{L_nM-\square}{\overset{(P)}{\underset{|}{CD_2}}} \longrightarrow \underset{L_nM-D}{\overset{(P)}{\underset{\|}{CD}}} & \text{than for} & \underset{L_nM-\square}{\overset{(P)}{\underset{|}{CH_2}}} \longrightarrow \underset{L_nM-H}{\overset{(P)}{\underset{\|}{CH}}}
\end{array}
$$

The Cossee mechanism involves no C-H (D) bond breakage. Therefore, it doesn't matter whether $CH_2=CH_2$ or $CD_2=CD_2$ attaches to the growing chain.

10-15 Polymerization of propylene would give either syndiotactic or atactic polypropylene (PP) in the presence of **83**. Achirality of catalyst could not lead to isotactic PP. See Scheme **10.17**.

10-16

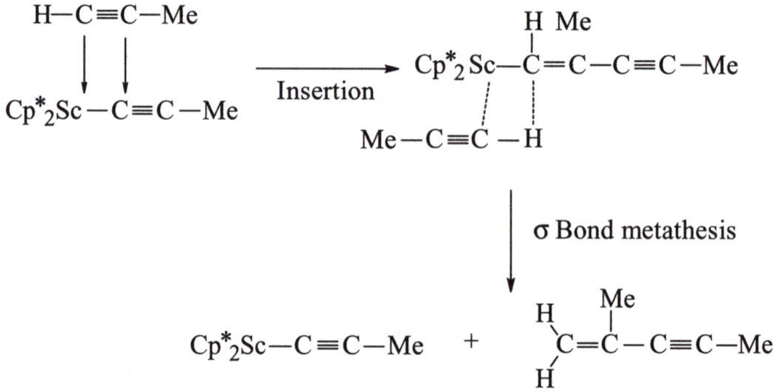

$$ H-C\equiv C-Me $$

$$ Cp^*_2Sc-C\equiv C-Me \xrightarrow{\text{Insertion}} Cp^*_2Sc-\overset{H}{\underset{}{C}}=\overset{Me}{\underset{}{C}}-C\equiv C-Me $$

$$ Me-C\equiv C-H $$

σ Bond metathesis

$$ Cp^*_2Sc-C\equiv C-Me \quad + \quad \overset{H}{\underset{H}{C}}=\overset{Me}{\underset{}{C}}-C\equiv C-Me $$

10-17

$$ H-CH=CH_2-CH_3 $$

$$ (P)-CH_2\{CH-CH_2\}_n ML_n \longrightarrow $$

with R substituent on CH

$$ L_nM-CH=CH_2-CH_3 $$

$$ + $$

$$ (P)-CH_2\{CH-CH_2\}_{n-1}CH-CH_2-H $$

with R substituents

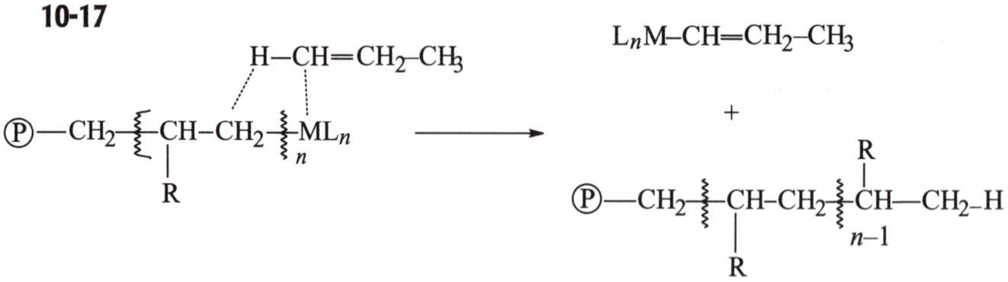

Yes, σ bond metathesis could be involved in chain transfer.

Chapter 11

11-1   1,2-Insertion for **8′** gives

8″ should then give the opposite stereochemistry.

11-2   If conversion of **7** to **8** is the rate-determining step, then the pathway leading to the minor product should proceed faster; enantioselectivity decreases.

11-3

11-4

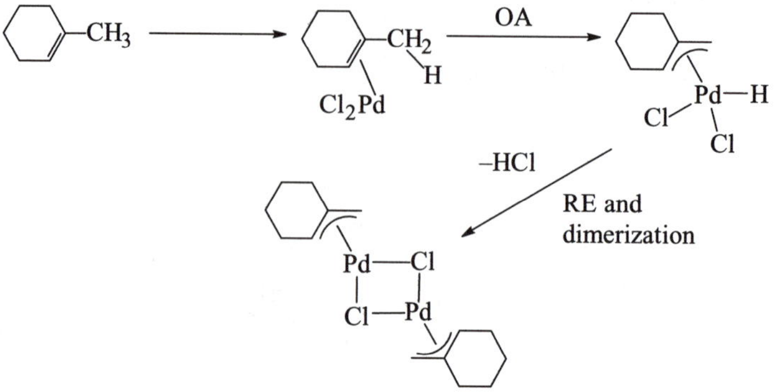

11-5

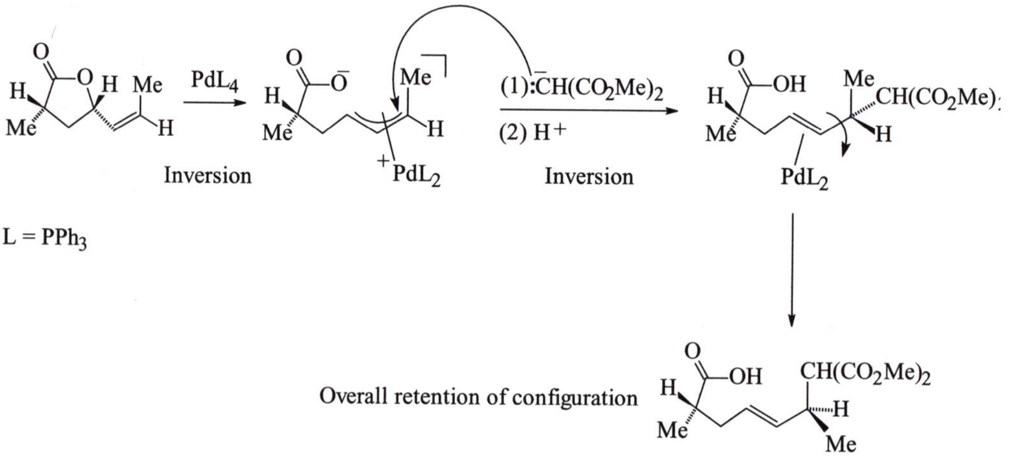

L = PPh$_3$

Overall retention of configuration

11-6

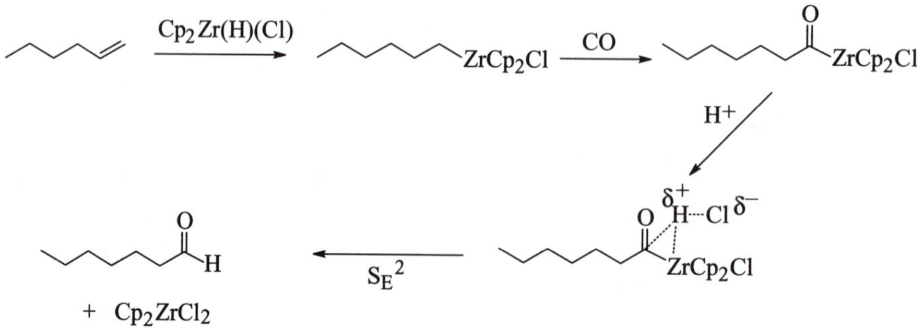

11-7

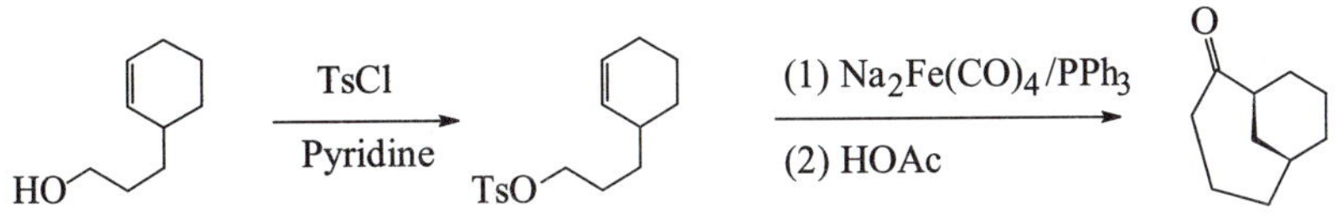

11-8

R = CO₂H

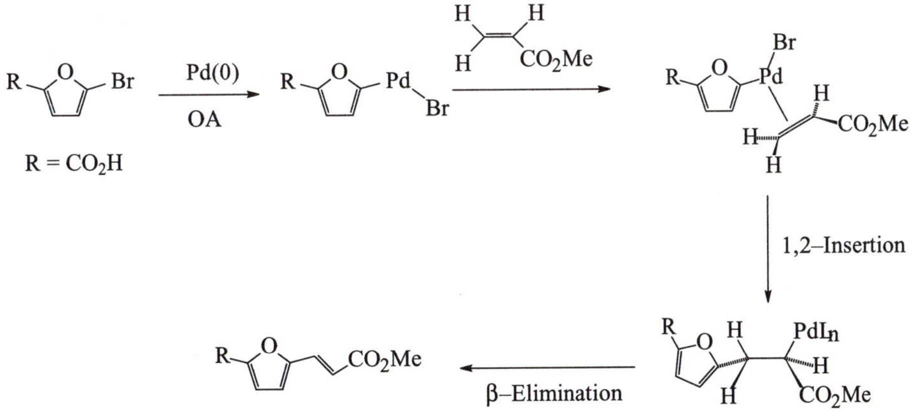

11-9

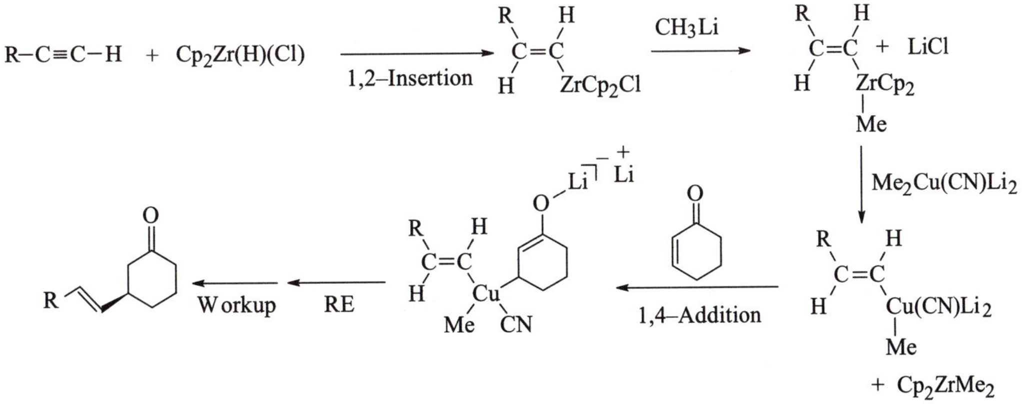

11-10

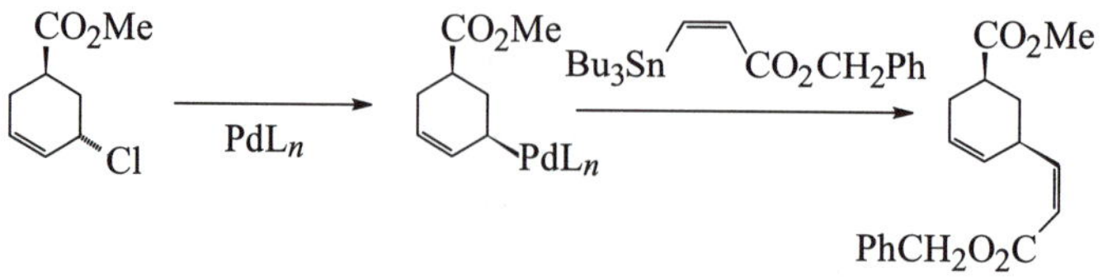

11-11

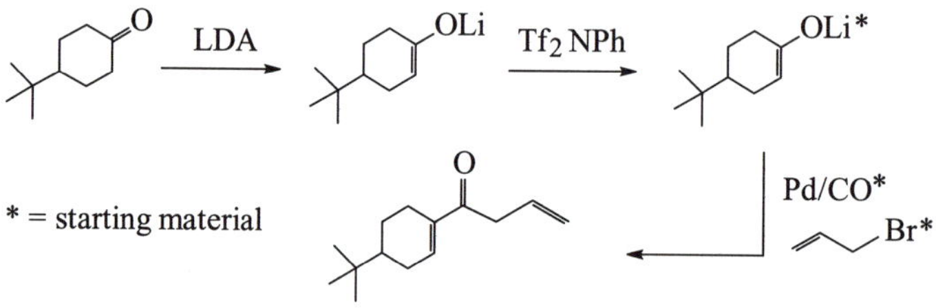

* = starting material

11-12

+ CO

Reductive extrusion

Electrocyclic
ring closure

Oxidation

Major product   Minor product

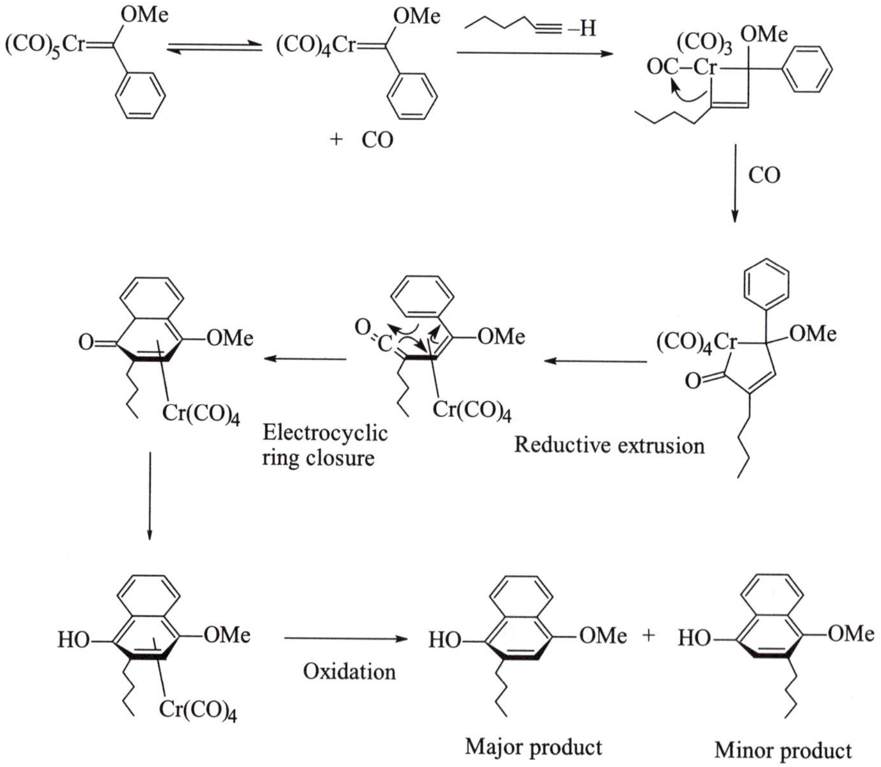

11-13

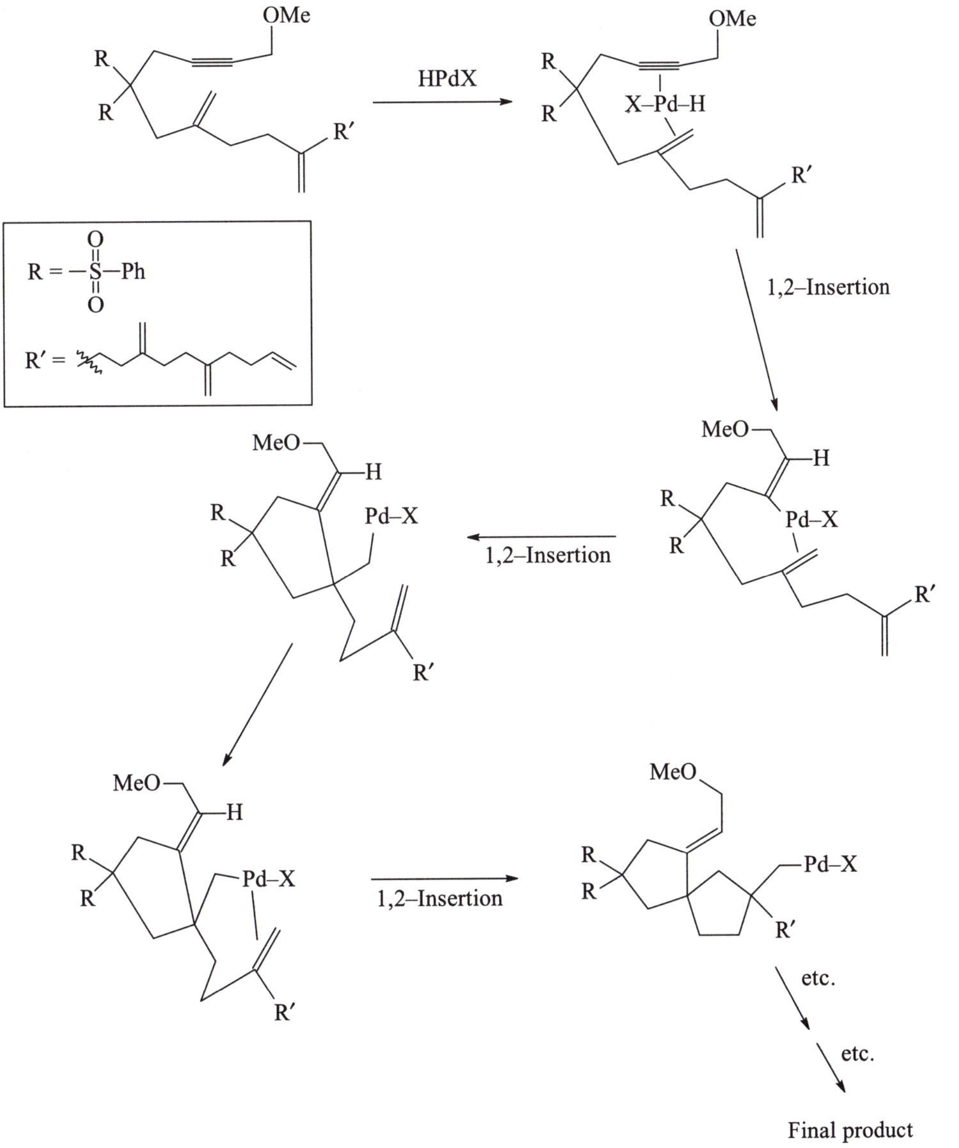

MeO

R = —S—Ph (with O double bonds)

R' = (structure shown)

1,2–Insertion

1,2–Insertion

1,2–Insertion

etc.

etc.

Final product

Chapter 12

12-1 Possible answers:

a. CH_2^+: $Re(CO)_4$ $(\eta^5\text{-Cp})Fe(CO)$

b. CH^-: $Pt(CO)_3$ $(\eta^5\text{-Cp})Co^{2-}$

c. CH_3: $Re(CO)_5$ $(\eta^5\text{-Cp})Mn(CO)_2^-$

12-2 Possible answers:

a. $(\eta^5\text{-Cp})Ni$ (15 e^-, 3 short of 18): CH

b. $(\eta^6\text{-}C_6H_6)Cr(CO)_2$ (16 e^-, 2 short of 18): CH_2

c. $[Fe(CO)_2(PPh_3)]^-$ (15 e^-, 3 short of 18): CH

12-3

a. $B_5H_8^- - 3\ H^+ = B_5H_5^{4-}$ Classification: *nido*

b. $B_{11}H_{11}^{2-}$ matches the formula $B_nH_n^{2-}$ Classification: *closo*

c. $B_{10}H_{18} - 8\ H^+ = B_{10}H_{10}^{8-}$ Classification: *hypho*

12-4

a. *closo*-$C_2B_3H_5 \rightarrow (BH)_2B_3H_5 = B_5H_7$

b. *nido*-$CB_5H_9 \rightarrow (BH)B_5H_9 = B_6H_{10}$

c. $SB_9H_9 \rightarrow (BH_3)B_9H_9 = B_{10}H_{12}$
or $SB_9H_9 \rightarrow (BH^{2-})B_9H_9 = B_{10}H_{10}^{2-}$ Classification: *closo*

d. $CPB_{10}H_{11} \rightarrow (BH)(BH_2)B_{10}H_{11} = B_{12}H_{14}$
or $CPB_{10}H_{11} \rightarrow (BH)(BH^-)B_{10}H_{11} = B_{12}H_{13}^-$
$B_{12}H_{13}^- - H^+ = B_{12}H_{12}^{2-}$ Classification: *closo*

12-5

a. $C_2B_7H_9(CoCp)_3 \rightarrow (BH)_2B_7H_9(BH)_3 = B_{12}H_{14}$
$B_{12}H_{14} - 2\ H^+ = B_{12}H_{12}^{2-}$ Classification: *closo*

b. $C_2B_4H_6Ni(PPh_3)_2 \rightarrow (BH)_2B_4H_6(BH) = B_7H_9$
$B_7H_9 - 2\ H^+ = B_7H_7^{2-}$ Classification: *closo*

12-6

a. $[Rh_7(CO)_{16}]^{3-}$

Valence electron count: $7(9) + 16(2) + 3 = 98$. For seven vertices this corresponds to two electrons short of the count (100) needed for a *closo* classification. Therefore, the classification is *capped closo* (see Table **12-11**), and the structure is a capped octahedron as shown.

b. $Rh_6(CO)_{16}$

Valence electron count: $6(9) + 16(2) = 86$. Classification: *closo*

c. $H_2Os_6(CO)_{18}$

Valence electron count: $2 + 6(8) + 18(2) = 86$. Classification: *closo*. (As footnote **a** to Table **12–11** mentions, the actual structure is capped *nido*, which has the same electron count as *closo*.)

d. $[Fe_4(CO)_{13}H]^-$

Valence electron count: $4(8) + 13(2) + 1 + 1 = 60$. Classification: *nido*. (Again, the actual structure differs from that predicted.)

Chapter 13

13-1 Nucleophilic abstraction.

13-2 Electrophilic abstraction by a metal ion.

13-3

Step 1: Sigma bond metathesis:

$$Cp^*_2Ln-CH(SiMe_3)_2 \;+\; H-BR_2 \longrightarrow Cp^*_2Ln\cdots CH(SiMe_3)_2 \,/\, H\cdots BR_2 \longrightarrow Cp^*_2Ln-H \;+\; CH(SiMe_3)_2-BR_2$$

Step 2: Ligand association followed by 1,2-insertion into an M-H bond:

$$Cp^*_2Ln-H \;+\; \overset{}{\underset{R}{=\!=}} \longrightarrow Cp^*_2Ln\cdots(\text{-R, H}) \longrightarrow Cp^*_2Ln-CH_2-CHR-H$$

Step 3: Sigma bond metathesis as in step 1.

13-4 The answer is shown in Figure **5-1**. The C=C bond would weaken due to back donation of electron density to the π^* bond of the alkene.

13-5

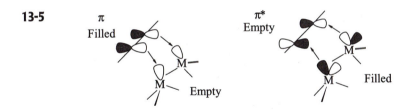

Index